Manufacturing Technologies and Production Systems

The book, which is part of a two-volume handbook set, presents a collection of recent advances in the field of industrial engineering, design, and related technologies. It includes state-of-the-art research conducted in the fields of Industry 4.0/5.0, smart systems/industries, robotics and automation, automobile engineering, thermal and fluid engineering, and its implementation.

Manufacturing Technologies and Production Systems: Principles and Practices offers a comprehensive description of the developments in industrial engineering, primarily focusing on industrial design, automotive engineering, construction and structural engineering, thermo-fluid mechanics, and interdisciplinary domains.

The book captures emerging areas of materials science and advanced manufacturing engineering and presents the most recent trends in research for emerging researchers, field engineers, and academic professionals.

Sustainable Manufacturing Technologies: Additive, Subtractive, and Hybrid

Series Editors: Chander Prakash, Sunpreet Singh, Seeram Ramakrishna, and Linda Yongling Wu

This book series offers the reader comprehensive insights of recent research break-throughs in additive, subtractive, and hybrid technologies while emphasizing their sustainability aspects. Sustainability has become an integral part of all manufacturing enterprises to provide various techno-social pathways toward developing environmental friendly manufacturing practices. It has also been found that numerous manufacturing firms are still reluctant to upgrade their conventional practices to sophisticated sustainable approaches. Therefore this new book series is aimed to provide a globalized platform to share innovative manufacturing mythologies and technologies. The books will encourage the eminent issues of the conventional and non-conventual manufacturing technologies and cover recent innovations.

3D Printing of Sensors, Actuators, and Antennas for Low-Cost Product Manufacturing
Edited by Rupinder Singh, Balwinder Singh Dhaliwal, and Shyam Sundar Pattnaik

Lean Six Sigma 4.0 for Operational Excellence Under the Industry 4.0 Transformation
Edited by Rajeev Rathi, Jose Arturo Garza-Reyes, Mahender Singh Kaswan, and Mahipal Singh

Handbook of Post-Processing in Additive Manufacturing
Requirements, Theories, and Methods
Edited by Gurminder Singh, Ranvijay Kumar, Kamalpreet Sandhu, Eujin Pei, and Sunpreet Singh

Manufacturing Engineering and Materials Science
Tools and Applications
Edited by Abhineet Saini, B. S. Pabla, Chander Prakash, Gurmohan Singh, Alokesh Pramanik

Manufacturing Technologies and Production Systems
Principles and Practices
Edited by Abhineet Saini, B. S. Pabla, Chander Prakash, Gurmohan Singh, Alokesh Pramanik

For more information on this series, please visit: www.routledge.com/Sustainable-Manufacturing-Technologies-Additive-Subtractive-and-Hybrid/book-series/CRCSMTASH

Manufacturing Technologies and Production Systems

Principles and Practices

Edited by Abhineet Saini, B. S. Pabla,
Chander Prakash, Gurmohan Singh, and
Alokesh Pramanik

CRC Press
Taylor & Francis Group
Boca Raton London New York

CRC Press is an imprint of the
Taylor & Francis Group, an **informa** business

Designed cover image: Unsplash

MATLAB® is a trademark of The MathWorks, Inc. and is used with permission. The MathWorks does not warrant the accuracy of the text or exercises in this book. This book's use or discussion of MATLAB® software or related products does not constitute endorsement or sponsorship by The MathWorks of a particular pedagogical approach or particular use of the MATLAB® software.

First edition published 2024
by CRC Press
2385 Executive Center Drive, Suite 320, Boca Raton FL 33431

and by CRC Press
4 Park Square, Milton Park, Abingdon, Oxon, OX14 4RN

CRC Press is an imprint of Taylor & Francis Group, LLC

© 2024 selection and editorial matter, Abhineet Saini, B. S. Pabla, Chander Prakash, Gurmohan Singh, Alokesh Pramanik; individual chapters, the contributors

ISBN: 978-1-032-42973-1 (hbk)
ISBN: 978-1-032-43412-4 (pbk)
ISBN: 978-1-003-36716-1 (ebk)

DOI: 10.1201/9781003367161

Typeset in Times
by Apex CoVantage, LLC

Contents

Editor Biographies

Dr Abhineet Saini is working as a Professor in the Department of Mechanical Engineering, Chitkara University, Punjab. He received his PhD degree in mechanical engineering from Panjab University, Chandigarh, India, and has more than 11 years of teaching and research experience in the field of manufacturing technology, which includes conventional machining, biomaterials, composite materials, additive manufacturing, CAD/CAM, and engineering optimization. He has been associated with a number of international and national conferences in the role of session chair for technical sessions. He has been the reviewer of many peer-reviewed journals and has authored/co-authored 20+ articles in peer-reviewed international journals of repute, conferences, and book chapters.

Dr B. S. Pabla is an experienced academician and researcher with 36 years of experience in academics and five years in industry. He earned his PhD in mechanical engineering from Panjab University, Chandigarh, India, and has served at various administrative positions and is presently working as a professor in the Mechanical Engineering Department at NITTTR Chandigarh. He has authored/co-authored books and research articles in various national and international journals of repute. He has also filed four patents, of which one has been granted and three are under process. Dr Pabla was awarded the Eminent Engineering Personality of the Year in 2014 by the Institution of Engineers (India) Haryana State Centre, India, and has visited a number of national and international universities/organizations for academic and research purposes.

Dr Chander Prakash is a Professor at the School of Mechanical Engineering, Lovely Professional University, Jalandhar, India. He has received a PhD in mechanical engineering from Panjab University, Chandigarh, India. His area of research is biomaterials, rapid prototyping, 3D printing, advanced manufacturing, modelling, simulation, and optimization. He has more than 15 years of teaching experience and six years of research experience. He has authored 100+ research papers and 20+ book chapters. He is also a guest editor of two journals, as well as a book series editor for the *Sustainable Manufacturing Technologies: Additive, Subtractive, and Hybrid* series for CRC Press/Taylor and Francis.

Dr Gurmohan Singh has over 12 years of teaching, research, and industry experience. His areas of interest include additive manufacturing, biomedical engineering, material science, and integrative research areas. He received his PhD degree in mechanical engineering from Chitkara University, Punjab, India, and has been associated with a number of international and national conferences as a session chair for technical sessions. He has also been a reviewer for many peer-reviewed journals and has authored/co-authored more than ten research articles/book chapters. He has also been part of organizing committees for three international conferences.

Dr Alokesh Pramanik is currently a senior lecturer in the Mechanical Engineering Department, School of Civil and Mechanical Engineering, Curtin University. He earned his PhD degree in mechanical engineering from the University of Sydney and has more than 15 years of research experience in the fields of manufacturing and composite materials at different universities. He has published more than 110 research articles, which include several books, many book chapters, and many reputed journal articles. His area of research is synthesis/development, surface modification, and advanced/precision machining of metallic and non-metallic biomaterials.

Contributors

Sangeetha Annam
Chitkara University Institute of
   Engineering and Technology
Chitkara University
Punjab, India

Jatin Arora
Chitkara University Institute of
   Engineering and Technology
Chitkara University
Punjab, India

Rashmi Arora
Chitkara College of Pharmacy
Chitkara University
Punjab, India

Ritchu Babbar
Chitkara College of Pharmacy
Chitkara University
Punjab, India

Renu Bala
Chitkara University Institute of
   Engineering and Technology
Chitkara University
Punjab, India

Puneet Bansal
I. K. G. Punjab Technical University
Kapurthala, Punjab, India and
Kurukshetra University
Kurukshetra, Haryana, India

Payal Bassi
Chitkara Business School
Chitkara University
Punjab, India

Jasdev Bhatti
Chitkara University Institute of
   Engineering and Technology
Chitkara University
Punjab, India

Tania Bose
Chitkara University Institute of
   Engineering and Technology
Chitkara University
Punjab, India

Bhawna Chetan
Chitkara University Institute of
   Engineering and Technology
Chitkara University
Punjab, India

Rishu Chhabra
Chitkara University Institute of
   Engineering and Technology
Chitkara University
Punjab, India

Himani Chugh
Chandigarh Group of Colleges
Landran, Mohali, Punjab, India

Brahma Deo
IIT Bhubaneswar
Jatni, Odisha, India

Atul Dutta
Chitkara School of Planning
   and Architecture
Chitkara University
Punjab, India

Rubina Dutta
Chitkara University Institute of
   Engineering and Technology
Chitkara University
Punjab, India

Harry Garg
CSIR-CSIO
Chandigarh, India

Meenu Garg
Chitkara University Institute of
  Engineering and Technology
Chitkara University
Punjab, India

Shubham Gargrish
Chitkara University Institute of
  Engineering and Technology
Chitkara University
Punjab, India

Geetanjali
Chitkara University Institute of
  Engineering and Technology
Chitkara University
Punjab, India

Rupali Gill
Chitkara University Institute of
  Engineering and Technology
Chitkara University
Punjab, India

Sandeep Singh Gill
National Institute of Technical Teachers
  Training & Research
Chandigarh, India

Cheenu Goel
Chitkara Business School
Chitkara University
Punjab, India

Nitika Goyal
Desh Bhagat University
Mandi Gobindgarh, India

Ganesh Gule
Assistant Professor
Lovely Professional University
Punjab, India

Abhishek Gupta
Shri Mata Vaishno Devi University
Kakryal, Katra, Jammu &
  Kashmir, India

Gourav Gupta
Lovely Professional University
Punjab, India

Isha Gupta
Chitkara University Institute of
  Engineering and Technology
Chitkara University
Punjab, India

Sheifali Gupta
Chitkara University Institute of
  Engineering and Technology
Chitkara University
Punjab, India

Gaurav Jain
Chitkara University Institute of
  Engineering and Technology
Chitkara University
Punjab, India

Manish Jain
Tata Steel Long Products Limited
Bileipada, Joda, India

Himanshu Jindal
Chitkara University Institute of
  Engineering and Technology
Chitkara University
Punjab, India

Poonam Jindal
Chitkara University Institute of
  Engineering and Technology
Chitkara University
Punjab, India

Mohit Kumar Kakkar
Chitkara University Institute of
  Engineering and Technology
Chitkara University
Punjab, India

Isha Kansal
Chitkara University Institute of
  Engineering and Technology
Chitkara University
Punjab, India

Arashmeet Kaur
University of Windsor
Windsor, Ontario, Canada

Gursleen Kaur
Punjabi University
Patiala, India

Harleen Kaur
Chitkara School of Planning
  and Architecture
Chitkara University
Punjab, India

Inderpreet Kaur
Chitkara University Institute of
  Engineering and Technology
Chitkara University
Punjab, India

Jaspreet Kaur
Desh Bhagat University
Mandi Gobindgarh, Punjab, India

Manjot Kaur
Desh Bhagat University
Mandi Gobindgarh, Punjab, India

Navnoor Kaur
Chitkara University Institute of
  Engineering and Technology
Chitkara University
Punjab, India

Puneet Kaur
Panjab University
Chandigarh, India

Ramanpreet Kaur
Chitkara College of Pharmacy
Chitkara University
Punjab, India

Ravneet Kaur
Chitkara University Institute of
  Engineering and Technology
Chitkara University
Punjab, India

Shaminder Kaur
Chitkara University Institute of
  Engineering and Technology
Chitkara University
Punjab, India

Pravin Kaushal
IIT Bhubaneswar
Jatni, Odisha, India

Ankush Khera
Chitkara University Institute of
  Engineering and Technology
Chitkara University
Punjab, India

Rajesh Kumar
Chitkara University Institute of
  Engineering and Technology
Chitkara University
Punjab, India

Sunil Kumar
Chitkara University Institute of
  Engineering and Technology
Chitkara University
Punjab, India

T. Mohit Kumar
IIT Bhubaneswar
Jatni, Odisha, India

Neelam Kumari
Chitkara University Institute of
  Engineering and Technology
Chitkara University
Punjab, India

Rajan Kumar Manchanda
Tata Steel Long Products Limited
Bileipada, Joda, India

Manpreet Singh Manna
Sant Longowal Institute of engineering
  & Technology
Longowal, Punjab, India

Archana Mantri
Chitkara University Institute of
    Engineering and Technology
Chitkara University
Punjab, India

Jaswinder Singh Mehta
Panjab University
Chandigarh, India

Tabasum Mirza
Chitkara University Institute of
    Engineering and Technology
Chitkara University
Punjab, India

Dimple Nagpal
Lovely Professional University
Phagwara, Punjab, India

Mithillesh Kumar P.
Amrita School of Engineering,
    Bengaluru
Amrita Vishwa Vidyapeetham, India

Kathuria Pankaj
Chandigarh University
Punjab, India

Mohit Pandey
Shri Mata Vaishno Devi University
Kakryal, Katra, Jammu & Kashmir,
    India

Atam Parkash
GNA University
Phagwara, India

Pravendra Patel
IIT Bhubaneswar
Jatni, Odisha, India

Yuvraj Singh Pathania
Chitkara University Institute of
    Engineering and Technology
Chitkara University
Punjab, India

Sandeep Phogat
Amity University Haryana
Gurugram, Haryana, India

Parteek Rana
Chitkara College of Pharmacy
Chitkara University
Punjab, India

Deepika Rani
Dr B. R. Ambedkar National Institute
    of Technology
Jalandhar, Punjab, India

Dev Sayal
Chitkara University Institute of
    Engineering and Technology
Chitkara University
Punjab, India

Anuranjan Sharda
GNA University
Phagwara, India

Asmita Sharma
Chitkara School of Planning
    and Architecture
Chitkara University
Punjab, India

Neha Sharma
Chitkara University Institute of
    Engineering and Technology
Chitkara University
Punjab, India

Nishant Sharma
Chitkara University Institute of
    Engineering and Technology
Chitkara University
Punjab, India

Preeti Sharma
Chitkara University Institute of
    Engineering and Technology
Chitkara University
Punjab, India

Ashish Sheje
Assistant Professor
Ajeenkya D Y Patil University
Pune, India

Abinash Singh
Chandigarh University
Mohali, Punjab, India

Amandeep Singh
Chitkara University Institute of
    Engineering and Technology
Chitkara University
Punjab, India

Aniran Singh
Chitkara University Institute of
    Engineering and Technology
Chitkara University
Punjab, India

Dhawan Singh
Chandigarh University
Mohali, Punjab, India

Luv Kumar Singh
Amity University Haryana
Gurugram, Haryana, India

Saravjeet Singh
Chitkara University Institute of
    Engineering and Technology
Chitkara University
Punjab, India

Bhavika Singla
Chitkara University Institute of
    Engineering and Technology
Chitkara University
Punjab, India

Neeraj Singla
Chitkara University Institute of
    Engineering and Technology
Chitkara University
Punjab, India

Kamal Srivastava
Research Scholar
Lovely Professional University
Punjab, India

Aditi Thakur
Eternal University
Himachal Pradesh, India

Neha Tuli
Chitkara University Institute of
    Engineering and Technology
Chitkara University
Punjab, India

Mukheshwar Yadav
Chitkara University Institute of
    Engineering and Technology
Chitkara University
Punjab, India

1 Online Failure Detection during Coal Injection by Wavelet Analysis of Pressure Signal

Pravin Kaushal, Pravendra Patel,
T. Mohit Kumar, Brahma Deo, Manish
Jain, Rajan Kumar Manchanda

1.1 INTRODUCTION

The coal-fired sponge iron rotary kiln is a continuous-time, huge, counter-current, high-temperature chemical reactor (green cylindrical box in Figure 1.1). The current sponge iron production capacity in India through the coal injection route is 30+ million tonnes per year. A 350 TPD (tonnes per day) kiln is approximately 80 metres long, cylindrical vessel, inclined (2–3 degrees), and 4+ metres in diameter. Input raw materials from the feed side are iron ore, coal (less than 50% of coal requirement) and dolomite. The remaining 40–50% of coal is injected from the discharge end (outlet end) through a stainless-steel pipe at a certain pressure using air as carrier gas. The main reason behind injecting coal from the outlet side is to spread the availability of the coal along the entire length of the kiln so that the reduction process can continue throughout the length. Another reason is to free up space by reducing the amount of feed coal on the feed side and increasing the amount of iron ore feed instead. This is done to increase the overall productivity of the kiln. The failure or malfunction of the injection pipe due to wear and/or deformation (due to high temperature inside the kiln) is a cause of concern as it directly impacts product quality and productivity of the kiln.

Several operational improvements have been done in the past to improve kiln productivity and consistency in product quality.[1–9] In the present work, the focus is on developing an automatic system of failure prediction of injection pipe based on monitoring of "pressure variation pattern" and converting it to wavelet transform. Wavelet analysis of a time series signal is a standard method[10] but its actual application to the analysis of results and its interpretation for control purposes, as explained in the present work, requires a special procedure. It has been successfully achieved in this work for the first time in the sponge iron industry in the world. Detailed trials at the plant site, first with analysis of historical data and then by online analysis

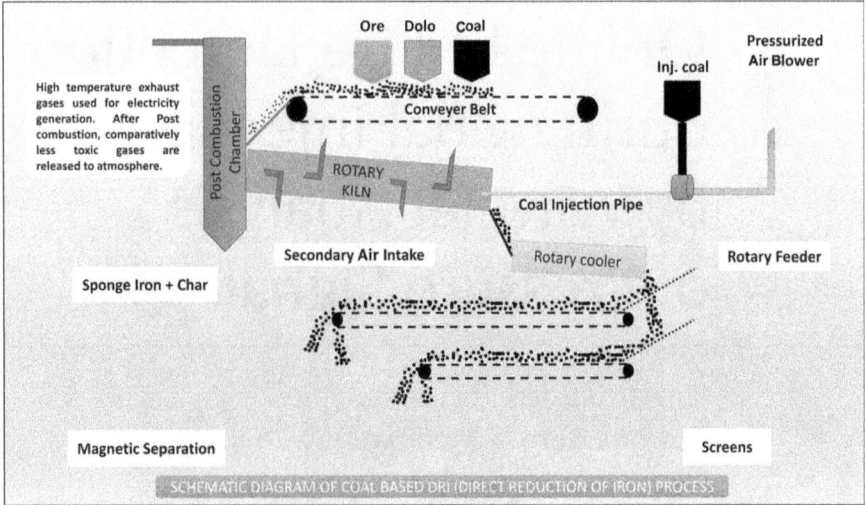

FIGURE 1.1 Schematic diagram of coal-based DRI process (created by the author).

of actual online data on 350 tonnes rotary kilns, were carried out. This procedure adopted in the present work can now be extended, in a similar way, to predict the failure of nozzles during solid fuel injection in rockets, as well as the gas injection through subsonic and supersonic nozzles in which the tip of the nozzle wears off due to friction or oxidation at elevated temperatures.

1.2 COAL INJECTION SYSTEM

Injection of coal from the outlet end of the kiln is done through a pipe in which pressurized air (as a career medium) is blown (shown by the arrow in Figure 1.2). Coal is brought into this pipe by using rotary feeders, and this coal (with the help of pressurized air) then travels through "mild steel pipe" followed, preferably, by stainless steel pipe into the kiln. The portion of the pipe protruding inside the kiln is 2+ metres long (depending upon kiln capacity). Besides the injection pipe, the cylindrical body of the rotary kiln has many secondary air blowers, placed at equal distances along the diameter and length. The airflow rate in each of these blowers, as well coal and ore feed rate, kiln rotation, primary air feed rate, and so on, are guided by control models developed jointly by IIT Bhubaneswar and Tata Steel Long Products Limited in the past.[1] The present investigation is about monitoring the coal injection lance itself to warn that the failure of the lance is approaching or the failure has just taken place.

A part of the coal burns at the tip of the lance and there is the generation of gases like CO and CO_2. The CO gas is required for the reduction process. The highest flame temperatures are produced in the vicinity of the tip of the coal injection lance, and this also greatly affects the wear of the tip. There is no physical and direct way to assess the wear of the tip inside the furnace except by taking it out. In the present work, a procedure has been developed to assess the same through measurement and

FIGURE 1.2 Coal injection side of the DRI kiln: two coal injection pipes and both rotary feeders can be seen (two long pipelines connecting the kiln with the rotary feeder are marked by arrow). Photograph by the author.

characterization of online pressure measurements followed by its analysis through wavelet transform.

1.3 FAILURE PREDICTION APPROACH AND ACTIONS TAKEN

In the initial part of the study only fast Fourier transform (FFT) was studied in great detail, but it did not help in clearly identifying either the failure or the case of approaching failure. Subsequently, continuous wavelet transformation (CWT) was used to prepare a scalogram, and by carefully observing and interpreting the changes in the scalogram, it was possible to set up a methodology of prediction both for the approaching failure and the failure itself. The computer program developed especially for this purpose gives the scalogram as output with the horizontal axis as "Time and date" and the severity index of the failure (also called coefficient value) on the frequency axis, which is represented by the colour spectrum. As an example, in the scalograms shown in Figure 1.3, the CWT value from 0.05 to 0.15 represents the normal functioning of the pipe, whereas the progressive change from value 0.15 and beyond till 0.40 implies the deviation of the pressure from its normal functioning and hence damage taking place inside the lance. The dates in Figure 1.3 on the x-axis refer to the

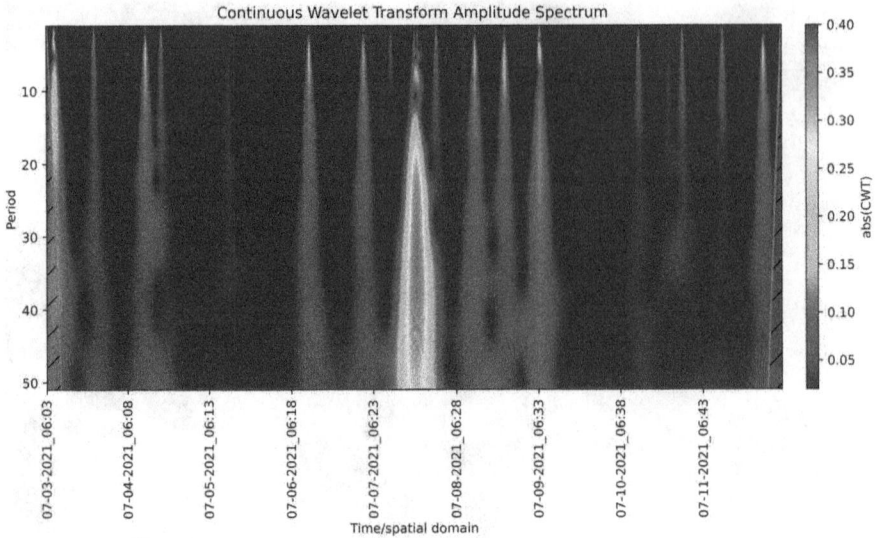

FIGURE 1.3   Continuous wavelet transformation of the pressure data for the period shown in the graph: day of failure 7 July 2021; the peak for CWT indicates lance failure.

particular day of plant operation. The dates are from 3 July to 11 July 2021. Various additional bright colour peaks (embedded in dark background) can also be seen in the scalogram. When the peak/patterns in the scalogram were matched with the record books of the plant they were found to be directly matching with the date of failure. A hole (puncture) occurred in the lance on 7 July in the reducer pipe (a part of the coal injection pipe).

It may be noted that the y-axis (abs (CWT)) axis has been restricted within the window of 0.02 to 0.40 to exclude (keep out) minor peaks and deviations. The choice of range (window) will depend on the production capacity of the kiln and has to be decided and set for each plant individually. Besides the light colour peaks in the centre appearing in Figure 1.3 are caused by manual cleaning actions taken on the lance on different days for lance maintenance purposes. The cleaning action of maintenance causes pressure fluctuations and thus has to be marked out accordingly and not confused with lance wear.

Once the computer code for analysis purposes was tested and ready, it was verified on other past (stored) pressure data taken from data archives of the DRI plant. The plant has well-established practices in data archiving and retrieval. As an example, a few investigative (post-mortem) results of documented failures (stored in the record) are given in Figure 1.4 and Figure 1.5.

In the scalogram (Figure 1.4), major peaks were observed on 13 May, 19 May, and 2 June. On all these dates there was some serious abnormality observed. On 13 May the stainless-steel portion of the coal injection pipe was damaged; on 19 May the coal bin was completely jammed restricting the coal flow and hence affecting the pressure, and on 2, June a puncture in the reducer pipe was noticed. Similarly

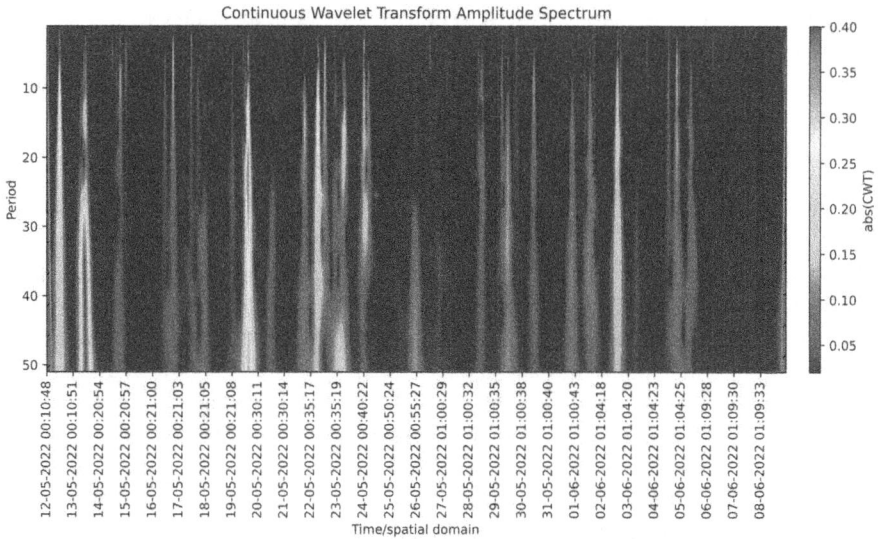

FIGURE 1.4 Continuous wavelet transformation of the pressure data for the period (12 May to 8 June 2022) shown in the graph; peaks having yellow or red colour indicate lance failure.

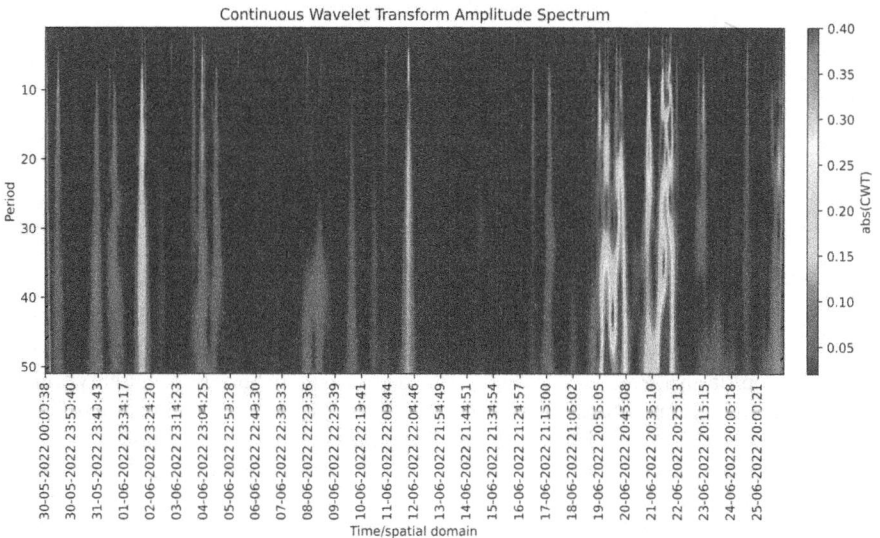

FIGURE 1.5 Continuous wavelet transformation of the pressure data for the period shown in the graph (30 May to 25 June 2022); peaks having light (white and off-white) indicate lance failure.

in Figure 1.5, major peaks can be seen on 2 June and 20 June. On 2 June there was a hole formed in the reducer pipe of the coal injection system. On 19 June, there was a big hole formed just below the rotary feeder of the coal injection system. Although the rotary feeder was replaced on the same day, it took a few more days of adjustment. According to the predictions of the scalogram, the rotary feeder was not functioning normally even after 3–4 days of the initial replacement done on 19 June. Thus, the scalograms can also indicate if the corrective or maintenance action done was adequate or not. After the success of these trials for the sufficiently large number of actual failure cases, the method developed is now being applied in practice.

1.4   CONCLUSION

The procedure, the prediction model (computer code), and the graphic user interface (GUI) developed in the present work can give an early warning about the malfunction/failure due to cracks and other aspects of coal injection lance. Early or timely recognition of lance failure helps to reduce resource wastage (both coal and ore). It also helps in improving plant productivity and maintaining the quality of the output by allowing efficient coal utilization, hence improving the carbon footprint.

In the present work, the focus was on the analysis of the coal injection pressure data only. It is planned to also study the effect of variation of surface moisture content of coal and the mean particle size variation of coal on injection pressure to link them to kiln process control.

1.5   ACKNOWLEDGEMENT

We are very grateful to the management of Tata Steel Long Products Limited (TSLPL), Joda, Odisha, India, now merged with Tata Steel, for providing on-site training and internship to our researchers. The constant guidance, help, and interaction between the plant engineers to our researchers at IIT Bhubaneswar is very much appreciated. We (at IIT Bhubaneswar) are also thankful to Tata Steel Long Products Limited (TSLPL) for promoting industry-academia collaboration.

REFERENCES

1. Puneet Choudhary, Gaurav Vishal, Shivam Shukla, Shah Chaitanya, Brahma Deo, Susil Kumar Sahoo, Parimal Malakar, Sourav Saran Bose, Gyanarajan Pothal, Partha Chattopadhyaya, Pressure Measurement and its Control in Sponge Iron Rotary Kiln in ASIA STEEL International February 6–9, 2018 at Bhubaneswar, Odisha, India.
2. Shivam Shukla, Ankur Kothari, Puneet Choudhary, Brahma Deo, Susil Kumar Sahoo, Parimal Malakar, Gyanrajan Pothal, Partha Chattopadhyaya, Dynamic Relationship between Surface Temperature and Thermocouple Temperature in Coal Fired Sponge Iron Rotary Kiln to Predict Accretion Profile, submitted in ASIA STEEL International Conference on February 6–9, 2018 at Bhubaneswar, Odisha, India.
3. Brahma Deo, Kinetics of Dissolution of Lime in Steelmaking Slags, The 3rd International Conference on Science and Technology of Iron and Steelmaking, December 11–13, 2017 IIT Kanpur India, pp. 123–126, 104; Brahma Deo, Optimal Operation of BOF Under Chaotic Conditions, ASIA STEEL International Conference on February 6–9, 2018 at Bhubaneswar, Odisha, India
4. Sampurna Borah, Shubhajit Mondal, Puneet Choudhary, Brahma Deo, Susil Kumar Sahoo, Parimal Malakar, Gyanranjan Pothal, Partha Chattopadhyay Operation of Coal Based Sponge Iron Rotary Kiln to Reduce Accretion Formation and Optimize

Quality and Power Generation, submitted in AISTech Conference on 6–9 May 2019 at Huntington Convention Centre of Cleveland, Cleveland, Ohio, USA.

5. Shubhajit Mondal, Sampurna Borah, Puneet Choudhary, Brahma Deo, Susil Kumar Sahoo, Parimal Malakar, Gyanranjan Pothal, Partha Chattopadhyay, Quality Prediction and Control in Coal-Fired Rotary Kilns at TATA Sponge Iron Ltd, in AISTech Conference on 6–9 May 2019 at Huntington Convention Centre of Cleveland, Cleveland, Ohio, USA, pp. 719–726.

6. Piyush Khatri, Puneet Choudhary, Brahma Deo, Parimal Malakar, Sourav Saran Bose, Gyanaranjan Pothal, Partha Chattopadhyaya, Determination of Surface Moisture and Particle Size Distribution of Coal Using On-Line Image Processing, International Mineral Processing Conference, Belgrade, Serbia, May 8–10, 2019.

7. Shibu Meher, Puneet Chowdhary, Himanshu Parida, Vijay Surya Vempati, Brahma Deo, Partha Chattopadhyay, Dynamic Quality Prediction and Control in Rotary Sponge Iron Kilns, June 2020, IOP Conference Series Materials Science and Engineering. http://doi.org/10.1088/1757-899X/872/1/012077

8. Himanshu Parida, Shibu Meher, Puneet Chowdhary, Brahma Deo, Parimal Malakar, Susil K. Sahoo, Partha Chattopadhyay, IOP Conference Series Materials Science and Engineering, June 2020. http://doi.org/10.1088/1757-899X/872/1/012078

9. Brahma Deo, Puneet Choudhary, Sourava Saran Bose, Control of Post Combustion to Stabilize Power Generation in a Sponge Iron Plant, Energetics, 2021. ISBN: 978-86-80593-73-9 195

10. Wavelets: Principles, Analysis and Applications, 2018, Mathematics and Statistics, Nova, Science and Technology, Theoretical and Applied Mathematics, Joseph Burgess (Editor), ISBN: 978-1-53613-374-5

2 The Growing Influence of the Digital Medium in Contemporary Indian Art

Kamal Srivastava, Dr Ganesh Gule, Ashish Sheje

2.1 INTRODUCTION

Everywhere we look, we notice the growing influence of computer digitization, which has restricted our social lives to computers, laptops, and mobile devices solely. In the same way that digital has supplanted the tempera colour medium in sports, literature, music, and other forms of communication, paintings and sculptures are not immune to the digital influences that have impacted other mediums.

Contemporary art has been dominated by printing techniques, such as lithographs, etchings, and even mixed media, but artists are now employing computer software to communicate their thoughts.

Throughout the history of India's visual arts, aesthetic principles have taken precedence over the medium. Artists have always been inspired by new media technologies to express their emotions, and this has never changed. Digital painting, graphic design, installation, and animation are just a few examples of the many ways in which it has influenced other types of art. In order to fully grasp the complexities of this advanced art form, one must encounter a variety of diverse procedures and approaches (Annum, 2014).

In the 1950s, artists began experimenting with computer-aided design for their own artistic purposes. The Howard Wise Gallery in New York hosted the first exhibition of computer-generated art, *Computer Generated Pictures*. In 1969, the Institute of Contemporary Art in London hosted a large-scale exhibition of computer art called Cybernetic Serendipity. Using geometric shapes in various random configurations, the artists developed what they called "digital or cybernetic art" for the majority of their works.

An artist like Harold Cohen (b. 1928) creates his own computer programs that randomly generates various forms of abstract drawings, which he subsequently enlarges and paints by hand. Computer art symbolizes not only photographs by pixels on the display but also creative expression of their expressiveness. Kinetic sculptures can also be controlled by computer programs (Gupta, 2018).

It is possible to describe computer fine art as art in which some or all of the artistic process is carried out using digital technology, such as a computer. This encompasses a wide range of possibilities.

DOI: 10.1201/9781003367161-2

2.2   OBJECTIVES

Using today's technology, the research aims to shed light on the beginning of a new period of painting in India by bringing together creativity and aesthetics in modern paintings. Many printed images can be seen all around us, yet they all demonstrate a predilection for a particular profession. However, the artist does not necessarily need a brush or a digital pen when utilizing this technique to combine those expressions through his imagination and art theory, which enhances the composition of the picture.

Aside from Photoshop and CorelDraw, forthcoming digital painting software must be included in this study. While digital painting is a complicated medium, it also has some restrictions, such as oil paint, acrylic paint, or any other type of media.

With an in-depth technical understanding of digital painting and an understanding of art critics' descriptions of current painting, we can begin to grasp how digital art has become increasingly dominant in contemporary painting in the 21st century.

2.3   NEED OF RESEARCH

We are aware of the powerful painting software Photoshop, which allows us to manipulate images with layers and merge them with others, and we can mix layers in a robust way, but the most important thing is to be creative, and without observation and art knowledge, we cannot work with any kind of tools. To change the medium and its visuals, tools can only be used.

Similarly, Corel Draw is a well-known illustration software that allows us to effortlessly draw a variety of complex forms, but without an understanding of composition, alignment, rhythm, and perspective, it is impossible to produce a quality logo or concept character. To reiterate, digital software is a medium for expressing our creativity, similar to etching, lithography, and lino cut, and it is difficult to develop if it is solely via the use of a computer. As a result of the many new applications for art-making software, it is sometimes difficult for viewers to discern whether or not a piece of art is a genuine oil painting or acrylic on canvas (or any other medium for that matter). The medium is digital. We do not know what other software artists are utilizing to create modern paintings in the digital medium, so we need to conduct research to learn more about the technological influence of digitalization in contemporary painting.

2.4   BACKGROUND OF RESEARCH

As a researcher, I am interested in gaining a better understanding of the history of traditional art. My fascination with art galleries has piqued my interest in researching digital transformation and identifying new job opportunities. The system is where art is created. The art world began to grow rapidly as a result of globalization.

Visual arts include drawing, painting, sculpture, crafts, photography, video, and architecture. The visual arts can be found in a variety of disciplines. It was once thought that the visual arts were limited to static compositions like painting and sculpture that the artist produced, and the observer immediately interpreted. A later iteration of this artist's work included computer-based technology into his work to allow for greater artistic freedom within the medium of the visual arts (Abrar, 2015).

Digital media art design has as one of its features the ability to create a one-of-a-kind creative space, complete with original art and a diverse range of imaginative thinking abilities. It is a combination of art and technology that is always evolving. Modern digital technology has introduced a new spirit to aesthetic design that affects its content, whether it is used to create language or express oneself. Compared to other forms of art and technology, it is innovative, practical, and cross-generational. Digital media design has a role to play in the advancement of science and technology in our country as a new form of art.

2.5   REVIEW OF LITERATURE

In the past few years, a lot of research has been done on the digital effects of art, but most of it has focused on graphic design, computer-made art, or printing. I believe that this art has to come in a modern form, and it should include modern abstract art, contemporary art, and art aesthetics, not just 2D illustration or card design. It is not enough to look at a painting and say that it is just a graphic, digital influence on art; you also have to write about the complexity of the painting and the medium it was created in.

A one pixel is the tiniest part of any picture. This device is used to create an image on a computer monitor. However, if you're seeking for a unique technique to express oneself when making an object by hand, the artist links numerous dots together to create a single picture. It is hard to make a picture without a lot of dots, but we can draw a line, for instance. As a result, we may then form a shape by joining several dots together.

Imagination, artistic fervour, and self-declaration all come together in digital art. Compared to conventional art, digital art is considerably more defined and allowed to develop. There will be no shortage of new, exciting, and one-of-a-kind developments in this sector of art.

I would like to speak of Sarah Gamboa in her research paper "The Impact of Technology Appreciation in the Arts and the Sustainability of Sales as a Factor in the Retail Art Industry," in which today, the word "technology" has taken on a very different meaning. It conjures up images of simple machines and hardworking carpenters instead of bundles of tornado lights and wires. So it is with "art." It is now digital technology enabled and activated, and its function has been completed. Where is better was the best example art technology communication, today computers are the new face of this association.

According to, Dr Archana Rani on digital technology its role in art creativity, mass production or digital media are both used in digital art, which is contemporary art in the 21st century. Traditional media outlets use digital art techniques to create visual effects for commercials and feature films. Despite the fact that desktop publishing is more closely associated with visual design, its influence on the publishing industry has been profound. Digital and conventional artists alike rely on a wide range of electronic resources and technologies to create their work.

According to Farha Abrar from Aligarh Muslim University on the use of digital art in contemporary Indian era, Photoshop, Corel Draw, and other computer programmes were used to create digital painting. In particular, digital painting techniques have made the art process more systematized and streamlined, giving artists a greater number of options for re-imagining their original concepts.

According to Dr Pothiti Kanellidi on the contemporary art market and how digital globalization is changing the art world, artists began experimenting with and inventing new mediums and ways of communicating with their audiences as a result of the technology revolution's impact on the arts. They grew interested in the areas governed by engineers and technicians, focusing on the "temporality" that distinguishes time art from other forms of art, such as sculpture. Artists can now use modern tools like the internet to spread the word about their work. Engage with new art audiences in an entirely new way, exploring a previously unknown world with them. Direct contact can be made over the internet. They were able to outperform the intermediaries because of the wide distribution of their artworks and the share of their material.

2.6 RESEARCH GAP

Computers and smartphones have become indispensable in our daily lives, as we have come to expect in an age marked by rapid advancement and the proliferation of new technology in virtually every aspect of our lives, including the arts. We all know that painting is about conveying one's thoughts as a matter of mind, and this picture does precisely that. Additionally, we use a variety of software programmes to create digital paintings, with Adobe Photoshop, Adobe Illustrator, and Corel Painter being among the most commonly used. 3D software like Maya, 3DS Max, and Blender is also used by artists.

Artistic creations will soon be shaped by the strength and scope of digital technology. The artist's fantasy world is a fantastic example of how technology is advancing and making art more sensible than ever before.

However long the artists take in realizing that computers are the best medium for their work is now up to them. Digital art inspires artists by providing enormous chances in digital media for them to express their artistic talent and use a combination of art and technology to do so.

2.7 INNOVATION AND IMPACT OF THE DIGITAL MEDIUM IN THE CONTEMPORARY ERA

Everywhere we look, we notice the growing influence of computer digitization, which has restricted our social lives to computers, laptops, and mobile devices solely.

In the same way that digital has supplanted the tempera colour medium in sports, literature, music, and other forms of communication, paintings and sculptures are not immune to the digital influences that have impacted other mediums (Belk, 2022).

Like modern painting, computer software and computer graphics are now being used by artists to convey their thoughts and feelings.

2.8   THE RISE OF DIGITIZATION IN PAINTING

In the recent decade, people have begun to appreciate the sophisticated and spectacular artworks generated with digital technology, which has led to an increase in attention and appreciation for digital art. More and more people are appreciating and appreciating digital art forms such illustrations, animations, 2D and 3D art, and other digital art forms because of its distinctive quality, depth, and impact.

In the 1950s, artists began using computers for the first time. The Howard Wise Gallery in New York hosted the first exhibition of computer-generated art, *Computer Generated Pictures*. In 1969, the Institute of Contemporary Art in London hosted a large-scale exhibition of computer art called *Cybernetic Serendipity* (Bhagwat, 2015).

At Bell Telephone Laboratories in Murray Hill, New Jersey, A. Michael Knoll designed a digital computer to output solely artistic patterns in the summer of 1962.

It is only fair to infer that digital art is on the increase because of the increasing popularity of this style of art today. There is a direct correlation between technological advancements and the growth of digital art. In the near future, new digital art forms and new digital artists will be born and recognized as some of the most important and valued artists of the new generation.

2.9   USE OF DIGITAL TECHNIQUES IN CONTEMPORARY PAINTING IN INDIA

The term "digital painting" usually refers to a painting made with computer technologies for expressing human emotions or feelings. To describe the new modern art form, "digital painting" is used.

Fractals and algorithmic art are examples of digital Indian art done only using computers. Images can also be altered with a scanner or created entirely from scratch with software and a computer mouse or tablet. The word "applicable" might be used in this context. Traditional media or hand techniques that have been digitally manipulated to create new works of art (Bhattacharjee, 2019).

Mixing non-digital photos with software on a computer to create digital paintings is called a mixed media artwork. They can either be printed on canvas or displayed digitally on a computer monitor. How common is photo editing? Highly customized images can now exercise their creative freedom with this software, which has been widely used, tested, and accepted (Budge, 2013).

Our lives revolve on 2D and 3D visuals and graphics in our technological era. Creativity has now surpassed all boundaries, and it can no longer be contained inside the confines of a brush and a canvas. Designing is on the rise, and art has become faultless, thanks to the introduction of numerous revolutionary tools. The combination of digital art and a working understanding of design tools allows us to enter the digital media sector. Today, as the world around us becomes increasingly digitalized, art is following suit. In modern Indian art, digital art is a widely acknowledged means of expression. This ground-breaking kind of art has given art, painting, sculpture, and communication a new dimension. Digital art is made by artists using a computer and software.

2.10   DEVELOPMENT AND CURRENT STATUS OF THE ONLINE CONTEMPORARY PAINTING MARKET

Online art sales and purchases have changed dramatically in the 21st century. In this way, India is not an exception. Selling paintings online has become a new trend in India in the last several years. Increasing numbers of young artists believe that selling their work online is the best option. The state's support of different organizations, such as an artist's studio, open art exhibits, and online galleries, has offered a broad platform for the contemporary era's development of the art profession.

When the National Gallery of Modern Art in New Delhi, India, opened in 1972, Michael Knoll displayed a series of "computer art" works. At the *State of Art* exhibition in Mumbai in 1991, Akbar Padamsee presented a group show called *A Group Show*. In 1996, Dhruv Mistry became the first Indian artist to post his work online (Colson, 2007).

Sharad Kumar Kavre coordinated a recent printmaking-based group show that investigates representation in the digital medium called Contemporary Art Exhibition at the New Delhi, India.

Internet art sales platforms are highly sought-after in the contemporary art market for two reasons:

They improve accessibility to and enjoyment from the art world. The relative ease with which desired artworks may be located is one advantage of online art sales platforms. Expert art collectors may find the necessary pieces of art online much more quickly than they would "by visiting various physical venues at random." Online art sales channels' convenience demands rapid access to popular pieces of art without placing customers on a waiting list. This second aspect is particularly important for younger art buyers who are intimidated by the "legendarily frigid greeting or complex processes" of high-profile galleries and actual galleries (Delaplaine, 2021).

Despite its accessibility and ease of use, online art sales channels have a hard time gaining confidence from older generations' significant clients. According to Hiscox 2019, the most challenging aspect for all online art businesses is gaining consumer trust. "Difficulties establishing the seller's reputation" was a big stumbling block for 60% of them in 2018 (Dhage, 2018). Online art sales platforms are currently grappling with two significant issues: authenticity and physical scrutiny of artworks. In 2018, 62% of online art buyers said they were worried about buying "a fake or an object that is not what it appears to be" online, and 74% said they could not inspect the artwork before buying (Drucker, 2013).

As a result, several contemporary artists have been quick to embrace the idea of digitization. According to a number of people, they use Microsoft's software, including Paint, to boost their art and design.

2.11   PROCESS AND INTRODUCTION OF NEW GROWING TECHNOLOGY FOR DIGITAL PAINTING

When it comes to traditional and digital painting, there are a lot of distinctions. The first distinction is that traditional painters use a brush and paint on a canvas or paper, whereas digital artists use a stylus and paint on a screen. This is known as parallax. A reduction in parallax in newer bullets has resulted in very small movement between the nib and the nib's output.

To get the desired shape and effect, traditional artists must carefully arrange their surfaces and media, but digital artists can make changes to their works as they go.

Software for digital painters includes a wide range of options. Programs like Corel Painter and Adobe Photoshop can be used to create images that resemble traditional paintings. It is up to the artist to decide how they want to work and what tools are most important to them when choosing a digital painting application. There are a number of factors to consider when it comes to the pricing of these programmes, such as the subscription-based approach that many of them have adopted (Frank, 2016).

As a result, there has been significant discussion in the community regarding what exactly defines art. Much of the issue rests on your perception of how physically active the artist has to be in the process, similar to the discussion in digital music creation (Gamboa, 2018). However, because art is wholly subjective, that reasoning may be refuted. Then there is the requirement for some loose framework to characterize each art form, which might be debated. In sum, the dispute is unlikely to be resolved very soon.

Digital art, regardless of its classification, is here to stay. Digital art is used in the following professions:

Design/animation in 2D
Design/animation in 3D animation for video games
Animation of characters
Animation of the face
Animation video
Drawing on a computer
Restorative art
Design for the eye
T-shirt and apparel design

From painters to engineers, tablets are crucial in practically every digital creative media.

Tablets can be used as a stand-alone device that substitutes the mouse, or they can be used in conjunction with other devices. The trackpad and stylus are the two essential components. In addition to basic note-taking, the trackpad enables for quick sketching and tracing. This is accomplished through the tablet's surface and the trackpad's pressure-sensitive technology.

Different settings and functionality are available for each trackpad and stylus. Individual settings may be assigned to specific trackpads via their interface, which might optimize your workflow. Some styluses have comparable characteristics (Jain, 2019).

Photographs, scans, satellite imagery, and other conceivable recordings of what existing are all used in digital photography to create pictures taken from reality. This portion frequently blurs the lines between what is and what is not, affecting our perception of reality. Artists like Nancy Burson, Daniel Canogar, Thomas Ruff, and Andreas Gorsky use traditional collage and element assembly techniques, as well as morphing technologies to overlay and merge visions (Hans, 2019).

Sculpture is created using computer-aided design tools and can be shown as actual objects/models or as virtual representations on displays. Computing enables the manipulation and management of complicated geometry, as well as their 3D representation, greatly increasing traditional design capabilities and fostering larger creative concepts (Joy, 2022). Artists including Tony Cragg, Wim Delvoye, Birch Cooper, Jon Rafman, and Anish Kapoor employ IT technology for the construction and assemblage of sophisticated and detailed pieces, as well as organic forms, as seen by Robert Lazzarini's usage of anamorphisms (Kanellidi, 2019).

The digital medium requires mixed media. In contrast to conventional production, computing suggests that components from many natural environments may be linked and coordinated to create a complete experience for the spectator (Zhong, 2021). As a result, artworks can incorporate still and moving images, augmented reality, sound, pictures, and other elements. One medium of the artwork can also be singled out, implying that a same digital invention might emerge in a variety of physical outputs, depending on the shared desires and goals of the commissioners, artists, and curators.

2.12 CONTRIBUTION OF DIGITAL ARTISTS IN CONTEMPORARY INDIAN PAINTING

Almost all young people nowadays consider themselves artistic types. While many people like traditional art, the digital media has gained in popularity in recent years. It is no surprise that so much art is done on computers, even professionally, because technology is where it is and so many young people are already proficient in it. As a young art collector, this person is becoming more active in obtaining art. It used to be an art gallery, but now it is a restaurant. Today, the online gallery is taking centre stage (Kefalidou, 2019).

Technology advancements have opened up new avenues in art design, and the new millennium has been given the gift of digital art. Digital artists, like traditional artists, observe, feel, and create art based on their observations of the world around them. Both in the digital and non-digital worlds, the same rules apply (Vishvakarma, 2020). Digital technology is the only difference between them and their analogue counterparts. Thus, the artist's personal perspective is mirrored in the artwork, as are cultural values. Digital artists are making great strides in discovering and developing a new culture of art as our society becomes increasingly digital (Kumar, 2018).

The most essential characteristic of digital art is that a large proportion of artists who use these new tools operate outside of the fine art realm. These painters admit to designing their works on a computer first, then copying them to canvas and painting them. As a result, artists might create a new invention by using more efficient and precise manufacturing procedures (Wurzer, 2022). The multimedia art of Paramjeet Singh, Ranbir Kaleka, Gogi Saroj Pal, Ananda Moy Benerji, Shovin Bhattachariya, Ved Nayar, Pushpamala, Sonia Khurana, Mandeep Singh Manu, Rakesh Chaudhary, Ravinder Singh, and many more reflects the important advances in this art form—reflecting new inventions. These artists provide an incredible range of digital artwork in a well-designed artist's style. They are astonishingly innovative and reasonably sensitive in their ways to highlighting the complicated concerns of current society (Lambert, 2021).

There are a number of well-known artists who are using digital technology to create contemporary paintings, such as Ashit Singh, Ankur Patar, Archana Nair, Jatin Roda, Medha Srivastava, Nikhil Shinde, Pranav Rajkumar, and Pratima Unde.

2.13   THE SCOPE AND IMPORTANCE OF DIGITAL ART

One of the advantages of digital media art design is its capacity to produce a singular creative environment, complete with original artwork and a wide range of imaginative thinking skills. It is a constantly changing fusion of art and technology. Whether it is utilized to produce language or express oneself, modern digital technology has given aesthetic design a new spirit that influences its content. It is novel, useful, and cross-generational in comparison to other types of art and technology. As a new genre of art, digital media design may contribute to the development of science and technology in our nation (Müller, 2019).

Interdisciplinary study in the social and scientific sciences is now the norm in most colleges that offer digital art courses. This trend demonstrates the impact of digital media on human practices and the widening range of art after the introduction of modern technology (Usha, 2019). Computer-aided design and illustration styles and methods are both used in digital art. Creativity is no longer limited to the brush and canvas of the olden days when life revolved around 3D images. Design is booming these days, and thanks to cutting-edge software, art is now flawlessly digital (Kanellidi, 2019).

Digital art is frequently divided into two categories by art historians: object-oriented artworks and process-oriented images. Digital technologies are a means to an end in the first scenario, serving as a tool for the creation of conventional things like as paintings, photos, prints, and sculptures. In the second situation, technology is the goal in and of itself, as artists explore the possibilities inherent in this new medium's own core (Rani, 2018). This last category, which is sometimes referred to as "new media," encompasses any computer-based art that is generated, saved, and delivered digitally. To put it another way, while some works employ digital tools to amplify an already-existing medium, others use digital technology as an integral and inseparable part of the artifact's creation (Sahu, 2011).

Digital media and the internet have made it easier and more effective than ever to preserve art, proving that the medium is not just for creating new images. The creative production process is strengthened, and the spectrum of colours and hues in an artwork are enlivened, through digitization (Singh, 2020). The number of new media and their uses is so large that it is impossible to list them all in a satisfactory amount of time. Some instances of new media types and how they were used by a few artists.

2.14   CONCLUSION

In contemporary Indian art, digital art is quickly becoming widely appreciated. Art, sculptures, and paintings have been given a new perspective thanks to this groundbreaking idea. Artists use digital technology and computer software to produce and modify this art form. As a result, the lines between design and high art are now quite obvious. It is the outcome of the collaboration of computer technology and human inventiveness. Young artists are increasingly employing technology to make their works of art. This development shows how digital media has affected everyday activities and how contemporary technology has broadened the field of art. As a result of the digital revolution, contemporary art has gained fresh perspectives and a broader scope. A vast range of artistic undertakings are covered and displayed by the phrase "digital art," and it has become an art trend that is nearly transparent and exceedingly wide. Some commonalities between a set of works of art can be discovered by using a digital art filter. Printmaking, photography, film, and video have all been embraced by artists in the creation of their work.

Twenty years ago, digital culture as we know it now was unimaginable; technology was employed exclusively in science and multimedia as a novelty, with CDs making their debut and mobile phones considered a luxury. As a contrast to the traditional manner of old movie posters and graphic scenes used as the backdrop for characters in films, today's matte painting is generated using computer software and serves as an illustration of how digital effects can be utilized on traditional painting. Matte painting, for example, is where digital painting thrives. Concept design for film, television, and video games is its most common use. An artist working with digital painting software like Corel Painter or Adobe Photoshop has access to the same resources as a real-world artist: a painting canvas, painting implements, colour palettes to combine, and an infinite number of colour options. Impressionism, realism, and watercolour are all forms of digital painting. Digital painting has both advantages and disadvantages. In spite of the fact that digital painting makes it easier for artists to work in a neat, clutter-free environment, some believe that artists who use physical brushes will always have better control over their work.

Digital painting does not, in the opinion of some artists, have the same character as traditional painting because of this. Many artists write blogs about the differences between digitally made work and traditional art. There is no denying that a marketing logic is at play here, particularly among younger collectors who invest money in art because it represents their status. Many of them have little knowledge of art or art history; thus, they rely on intermediaries to act as consultants and art advisers during the purchase process while choosing art items and artists.

REFERENCES

Abrar, F. (2015) Use of digital art in contemporary Indian era. *ShodhGanga*. https://core.ac.uk/download/pdf/144527512.pdf

Annum, G. Y. (2014) Digital painting evolution: A multimedia technological platform for expressivity in fine art painting. *Journal of Fine and Studio Art*, 4(1), 1–8.

Belk, R. (2022) Money, possessions, and ownership in the Metaverse: NFTs, cryptocurrencies, Web3 and Wild Markets. *Journal of Business Research*, 153, 198–205.

Bhagwat, N. (2015) *Development of contemporary art in western India Visuals*. Retrieved from SodhGanga: http://hdl.handle.net/10603/59331

Bhattacharjee, A. (2019) *Study on visual preferences for effective spotlight design: Reference to painting exhibition in Indian art galleries*. Retrieved from Lakshminath Bezbaroa Central Library Digital Repository: http://gyan.iitg.ernet.in/handle/123456789/1418

Budge, K. (2013) Virtual studio practices: Visual artists, social media and creativity. *Journal of Science and Technology of the Arts*, 5(1), 15–23.

Colson, R. (2007) *The fundamentals of digital art*. Bloomsbury Publishing.

Delaplaine, S. (2021) *The brave new virtual art world the evolution of digital art: NFTs and their effects on the art market in 2021*. Retrieved from Digital Commons at SIA: https://digitalcommons.sia.edu/stu_theses/93

Dhage, V. (2018) *Globalisation and its impact on contemporary Indian art*. Retrieved from Shodhganga: http://hdl.handle.net/10603/241565

Drucker, J. (2013) Is there a "digital" art history? *Visual Resources*, 29(1–2), 5–13.

Frank, F., Unver, E., & Benincasa-Sharman, C. (2017). Digital sculpting for historical representation: Neville tomb case study. *Digital Creativity*, 28(2), 123–140.

Gamboa, S. (2018) *The influence of technology in art appreciation and sales as a factor in the sustainability of the retail art industry*. Retrieved from University of South Florida: www.usf.edu/business/documents/undergraduate/honors/thesis-gamboa-sarah.pdf

Gupta, S. (2018) *Utility of Indian traditional art and mythology in digital art*. Retrieved from Shodhganga: http://hdl.handle.net/10603/292720

Hans, V. (2019) Digitalization in the 21st century. *Journal of Global Economy*, 15(1), 12–23.

Jain, S. G. (2019) The rise of digital art. *International Journal of Research – Granthaalayah*, 7(11), 161–164.

Joy, A. (2022) Digital future of luxury brands: Metaverse, digital fashion, and non-fungible tokens. *Wiley Online Library*, 31(3), 337–343.

Kanellidi, P. (2019) *The contemporary artmarket: How digital globalization is changing the art world*. Retrieved from International Hellenic University: https://core.ac.uk/reader/236205560

Kefalidou, A. (2019) *The importance of digital art technology for supporting people with vision*. Retrieved from Practices for a Museum Aiming at Social Inclusion: https://repository.ihu.edu.gr/xmlui/bitstream/handle/11544/29800/paperfinal%20%281%29%20%281%29.pdf?sequence=1

Kumar, R. (2018) *Evolution of design process and the impact of digital media on video art*. Retrieved from Sodhganga: http://hdl.handle.net/10603/279407

Lambert, N. (2021) Beyond NFTs: A possible future for digital art. *ITNOW*, 63, 8–10.

Müller, K. (2019) *Indian post-digital aesthetics*. Edizioni Museo Pasqualino.

Rani, A. (2018) Digital technology: It's role in art creativity. *Journal of Commerce & Trade*, 13(2), 61–65.

Sahu, S. (2011) *Chitrakala ke antargat vigyapan kala ke vaicharik saundayartamak samajik pahlu*. Retrieved from SodhGanga: http://hdl.handle.net/10603/264781

Singh, R. (2020) *Narrative approaches in Indian contemporary painting an analytical study*. Retrieved from Shodhganga: http://hdl.handle.net/10603/309399

Usha. (2019) *Role of patronage and art market in the development of contemporary Indian art A critical study*. Retrieved from Shodhganga: http://hdl.handle.net/10603/309401

Vishvakarma, R. R. (2020) An Indian digital medium artist's experience, exploration and exposition of digital art. *2020 joint international conference on digital arts, media and technology with ECTI northern section conference on electrical, electronics, computer and telecommunications engineering (ECTI DAMT & NCON)* (pp. 395–398). https://doi.org/10.1109/ECTIDAMTNCON48261.2020.9090759.

Wurzer, P. B. (2022) Work of art in the age of metaverse – Exploring digital art through augmented reality. *The 40th conference on education and research in computer aided architectural design in Europe (eCAADe 2022)* (pp. 447–456). CuminCad.

Zhong, M. (2021) Study of digital painting media art based on wireless network. *Wireless Communications and Mobile Computing*, 2021 (Article ID 4412294), 11p. https://doi.org/10.1155/2021/4412294

3 Automated Embedded IoT Framework Development for Electric Scooters

Dr Puneet Kaur

3.1 INTRODUCTION

An essential criterion for rapid growth of market for electric vehicles these days is the availability of environment-friendly and low-cost mobility solutions based on technological development in high-capacity batteries, efficient motors, and effective electronic controls. Many local and global scooter manufacturers have come up with electric scooter models of varying battery capacities. Most of the e-scooters follow a similar architecture and electrical blocks: electric motor with its associated power electronics drive, battery with its charger, speed control throttle, scooter lights, and battery level and speed indicators, as shown in Figure 3.1. All these blocks are controlled by a single computing unit of the scooter.

With easy availability of GPS trackers for collecting location of the vehicle and fetching some basic information like speed and time [1–3], the commercial applications of electric scooters has been enhanced. E-scooters hold a specific relevance for rental business across the country with dedicated applications for food delivery and leasing, which also requires effective traffic management and vehicle insurance. To accomplish these specific requirements, e-scooters need to be equipped with electronic hardware with IoT connectivity, which enables this information to be viewed from PC or mobile applications. There are few solutions available in the market which achieve the mentioned functionalities by integrating multiple electronic devices [4,5], but a single unified device achieving all the required features is not available off-the-shelf so far.

This chapter presents an integrated universal design of an embedded IoT platform for smart management of electric scooters which can be interfaced to the vehicle's computer to obtain the complete functional controls of the scooter. Design of this hardware-software platform integrates GPS, GPRS, Bluetooth Low Energy (BLE), battery management, and various input/output signal interfaces in a single hardware. It is not only compact, low-power, and in-vehicle device but also universal and flexible, which is scalable per requirement and portable to any electrical architecture of e-scooter. With acquisition and analysis of the various parameters, the battery's

DOI: 10.1201/9781003367161-3

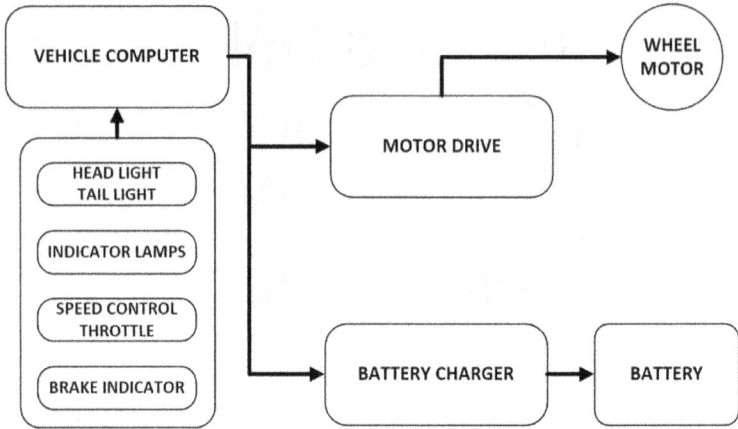

FIGURE 3.1   E-scooter electrical architecture.

health and driver's behaviour can also be accounted. Results from case studies have been added in this chapter to demonstrate the real-world implementation of this hardware-software platform.

3.2   ARCHITECTURE OF THE DEVELOPED IOT FRAMEWORK

A product-level block diagram, depicting the components of hardware layer, firmware layer (device application software), and cloud is shown in Figure 3.2. The various functions of the framework are handled collectively by the three layers. The developed firmware and the application software can be abstracted to any hardware and can be ported and reused on different microcontroller architectures [6]. This will support any future product revision if required in same segment.

3.3   DESIGN OF THE FRAMEWORK

Figure 3.3 shows the hardware layer designed around the PIC microcontroller PIC24FJ128GA106 [7] to achieve the desired smartness in the platform. It is ideal for battery-sensitive connectivity applications which demand low power consumption. It features multiple communication channels, good amount of SRAM, and flash memory. It collects e-scooter status and health data and provides information exchange over the Bluetooth Low Energy interface, as well as cloud server communication via GPRS. It also acquires location data via GPS and trigger other hardware-level actions based on the state of e-scooter. Further, integrated 16 channels, 10-bit ADC, and availability of +5 V tolerant digital input pins makes it a suitable choice for this application.

3.3.1   CONNECTIVITY INTERFACE

SIM868E module provides a 2G connectivity interface and additionally has integrated BLE 4.0 and GPS functionality which can be controlled by AT command

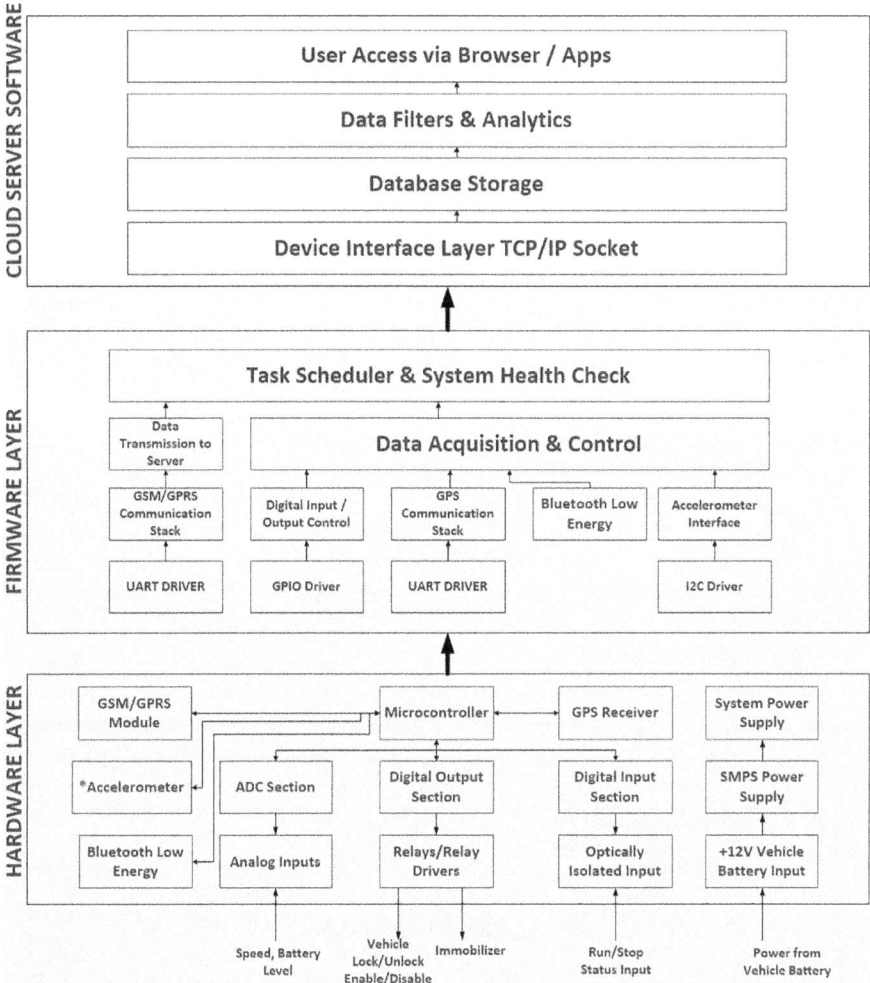

FIGURE 3.2 IoT platform block diagram.

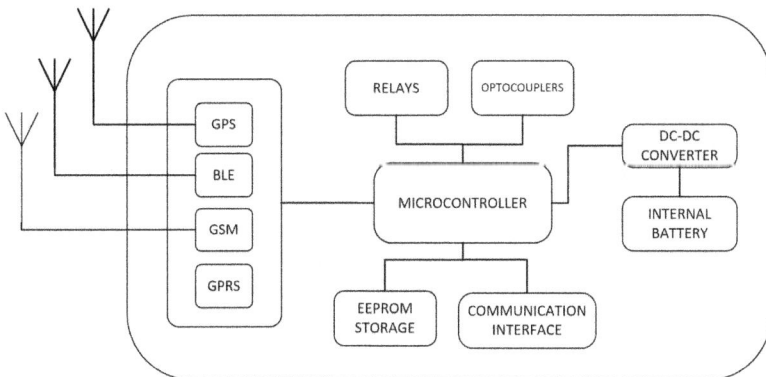

FIGURE 3.3 IoT platform hardware blocks.

set [8]. SIM868E modules is a Quad-band GSM/GPRS modem which supports 850/900/1,800/1,900 MHz frequencies [9]. Microcontroller is responsible for controlling and utilizing features of SIM868E module via UART interface.

3.3.2 E-SCOOTER ELECTRONICS INTERFACE

Interface between the IoT device and e-scooter has three major divisions [10–12]. The digital inputs interface involves reading the e-scooter parameters which are of digital type and provide information as on/off or active/inactive states. These signals are optically isolated to ensure minimum electrical interference. This circuit is used for all 12 V activation signals which include the following functions: start-stop indication, boot cover open-close indication, headlamp and tail lamp status, battery charging active-inactive status, scooter handle and scooter-stand position, and battery terminal removal indication status.

There are digital outputs, the on-off-type digital signals transmitted by the microcontroller to e-scooter electrical circuits to control its functions. Electromechanical relays are used to achieve this functionality. This interface controls the actions like start/stop, main power cut-off, e-scooter immobilize or wheel jam, trigger actuator action for boot cover opening, horn activation/deactivation, and headlamp and tail-lamp control.

Input signals like battery voltage, battery current, and speed signal are read via the analogue-to-digital converter section. These signals are scaled down to match the input voltage range of the analogue-to-digital converter section. A resolution of 10-bit is utilized for signal conversion. The battery interface via analogue inputs is essential for battery management. The IoT platform utilizes the battery information to allow cloud analytics over the battery usage data. Figure 3.4 shows the various interfaces of e-scooter with the IOT platform.

3.4 OVERVIEW OF SOFTWARE STATE DIAGRAM

- Application software for the IoT platform is loaded in Microcontroller PIC24FJ128GA106. It is designed keeping in view the requirements of the e-scooter rental business. A data connectivity diagram between the

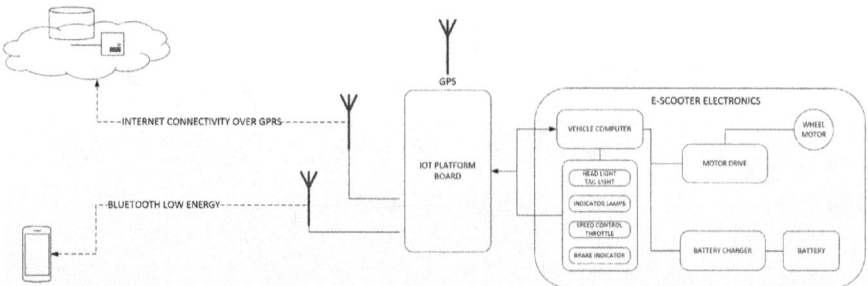

FIGURE 3.4 E-scooter interface with IoT platform board.

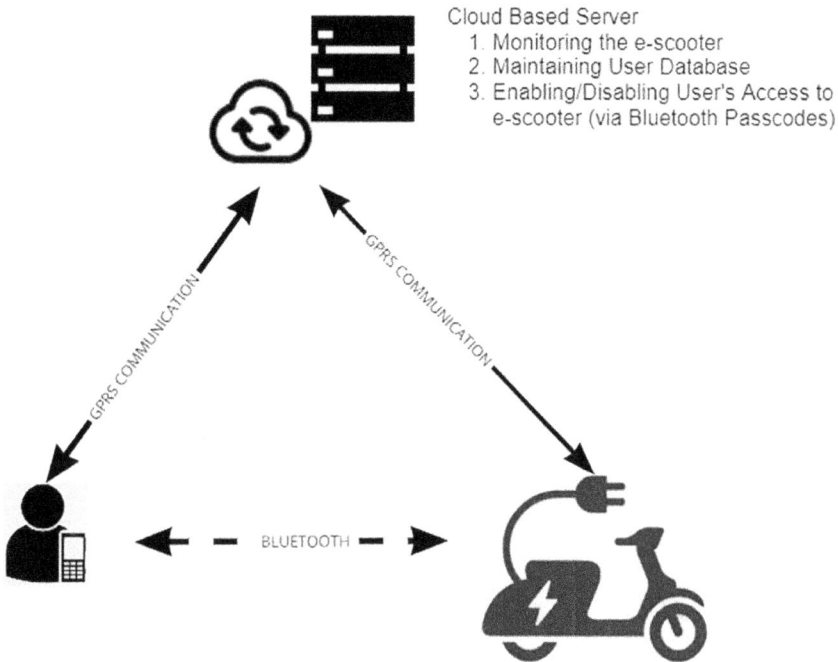

Cloud Based Server
1. Monitoring the e-scooter
2. Maintaining User Database
3. Enabling/Disabling User's Access to
   e-scooter (via Bluetooth Passcodes)

GPRS COMMUNICATION

GPRS COMMUNICATION

BLUETOOTH

FIGURE 3.5   Information exchange view.

e-scooter, the e-scooter rental business, and the end-user is shown in Figure 3.5. As seen in the Figure 3.5, the user's mobile phone can connect to server over GPRS and can also connect directly to e-scooter via the Bluetooth Low Energy interface.

The software diagram shown in Figure 3.6. represents the various states of the application software flow and includes following functionalities:

- E-scooter status is monitored at regular intervals, if status of any parameter changes, the information is conveyed via GPRS transmission.
- When the scooter is in a running state, GPS location is reported at a higher frequency of 3–5 seconds. If the scooter is in a stopped state, GPS location is reported once every hour.
- The IoT device continuously scans for incoming BLE connections. If a con nection is attempted with the e-scooter from an external mobile phone, the encrypted passcode is checked and communication is allowed, or else it will disconnect the BLE connection. For every attempted BLE connection, it will transmit the information to the business cloud.
- When the e-scooter is in a stopped state, GSM/GPRS/GPS is kept in low power mode. Also, the microcontroller runs at a lower clock frequency to reduce power consumption.

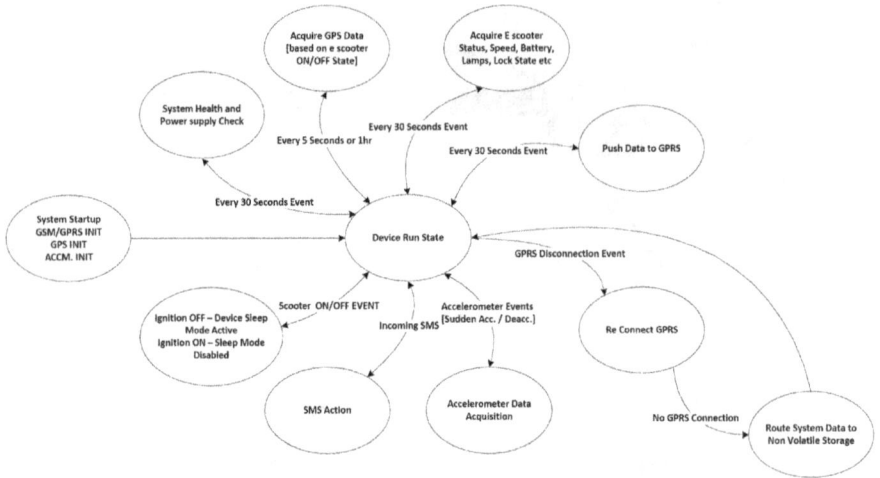

FIGURE 3.6 State flow diagram for the device (microcontroller) software.

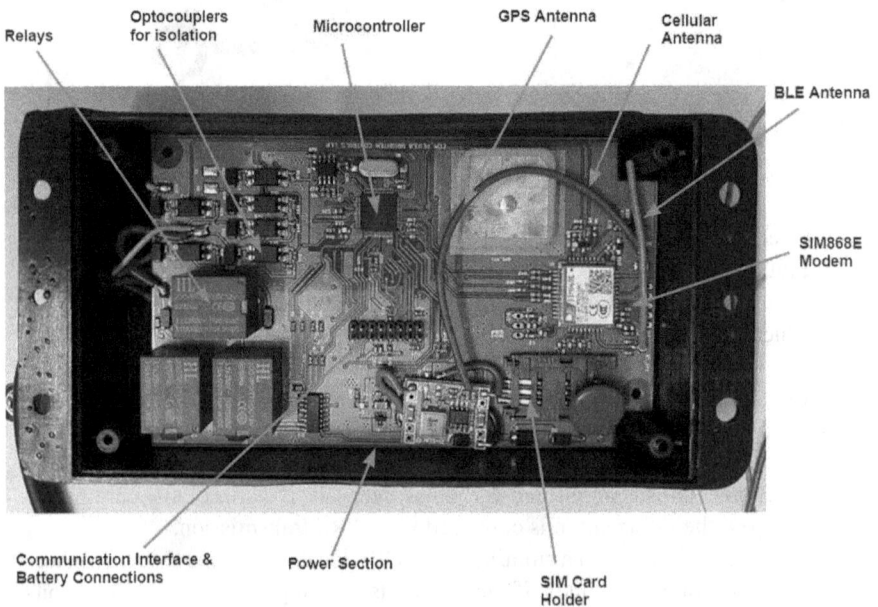

FIGURE 3.7 Picture of the designed and developed IoT platform for scooters.

The software keeps a persistent TCP connection with the business server and communicates via JSON formatted packets. The payload in JSON packets consists of commands, encrypted passcodes, and software update data.

The hardware of the discussed IoT platform was fabricated with multilayer PCB so that it occupies very less space in the vehicle. The captured picture of the developed platform is shown in Figure 3.7.

3.5 RESULTS AND DISCUSSION

IoT platform discussed here was installed in the market-available e-scooters, and data was collected over a period of 15 days to validate the product design.

3.5.1 DEVICE DEPLOYMENT

Once powered on, the IoT device establishes a connection with the cloud server over GPRS. This connection is established using a unique domain name, which is assigned to the cloud platform running server for the given IOT platform. Device-to-server communication is established using TCP socket, and data packets are JSON-formatted with key-value pairs of information. Data sent from device to server mostly comprise of location information, status information of e-scooter parameters and responses to the commands sent by server as shown in Figure 3.8. Server data packets sent to device comprise of commands to trigger various actions on e-scooter and to retrieve information. Information is exchanged between device and server allows following features to be exercised:

- Location information transmitted by devices allows the tracking of the e-scooter.
- Notifications sent from device to server, which includes alerts for events when the device is forcibly disconnected from the e-scooter, alerts for battery disconnection or removal from the e-scooter and boot cover open.
- The server is the only source responsible for setting Bluetooth access passcodes in the device. This ensures that only those mobile phones can access the e-scooter which have received IoT platform's Bluetooth passcode information from server.

In the Figure 3.8 JSON sample data packet sent by IoT device to server is shown. It provides the information about the latitude/longitude of the device. The "DI" field is the unique IMEI number, and the "STS" field has status data, which has digital bits listing the state of various e-scooter events represented by 0/1 (or active/inactive).

Following is the screenshot of actual data collected at server with GPS information of device:

FIGURE 3.8 Data sent by the IoT device to the server.

3.5.2 MOBILE APP DETAILS

Mobile application was developed keeping in view the usage requirements of a e-scooter driver or user. It provides all the controls and notifications which may be required by the user from the time e-scooter is rented/leased till the time user returns the e-sooter after completing the leasing period. Mobile application can get connected to both the cloud server (via GPRS) and the IoT platform via the Bluetooth interface.

Figure 3.9 and Figure 3.10 show the views of mobile application pages. The screenshots in the figures are the controls available in the mobile application and live notification panel. The control panel includes unlock, lock, immobilize, boot cover opening, battery cover opening, horn, and locate features. The live notification panel highlights the instantaneous status of the e-scooter, which includes lock state, battery charge state, boot cover, and battery cover status. Rows of notification panel change colours based on the parameter state.

3.5.3 SERVER-SIDE ADMIN DASHBOARD

The admin panel is a web-based tool and can be accessed using web browser. For the testing application it was customized for e-scooter rental business model. Figure 3.11 shows the admin panel; this allows control on the e-scooter from business or asset owner standpoint. It can be used to monitor state of the e-sooter and trigger control commands, which can override the commands of e-scooter user. Control commands are specifically helpful in case of e-sooter theft or tampering. In addition, admin panel also triggers the command for changing of Bluetooth passcodes which control the access permissions for ethe -scooter user. Figure 3.12 highlights the command panel and notification panel of the admin dashboard, which are used to control and monitor the vehicle.

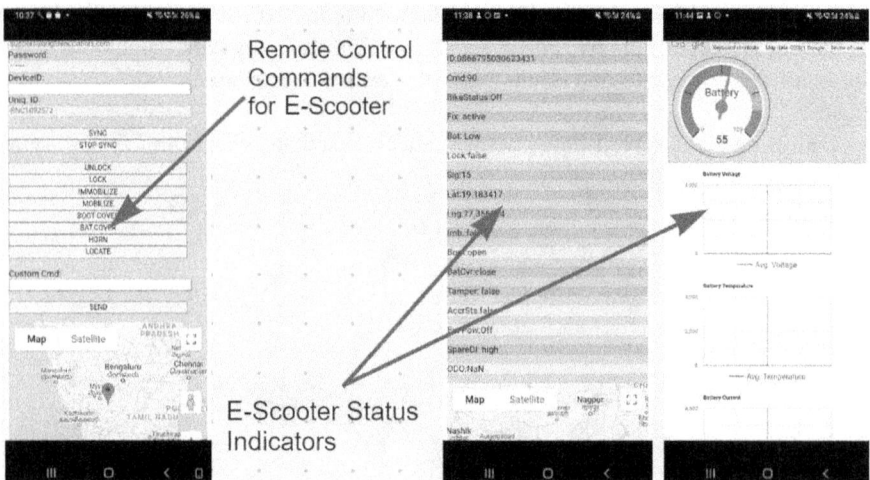

FIGURE 3.9 Mobile app controls and notification panels.

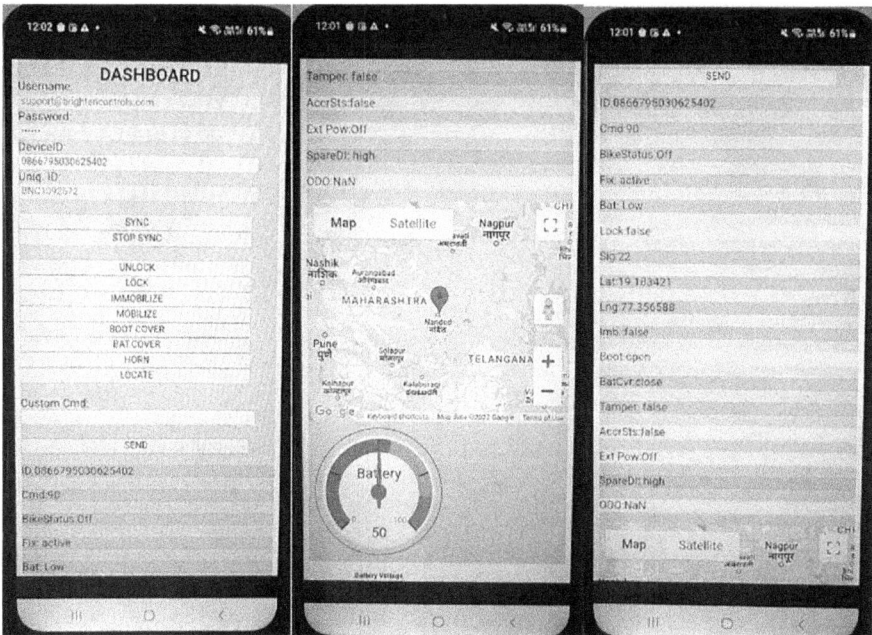

FIGURE 3.10 Android mobile application view.

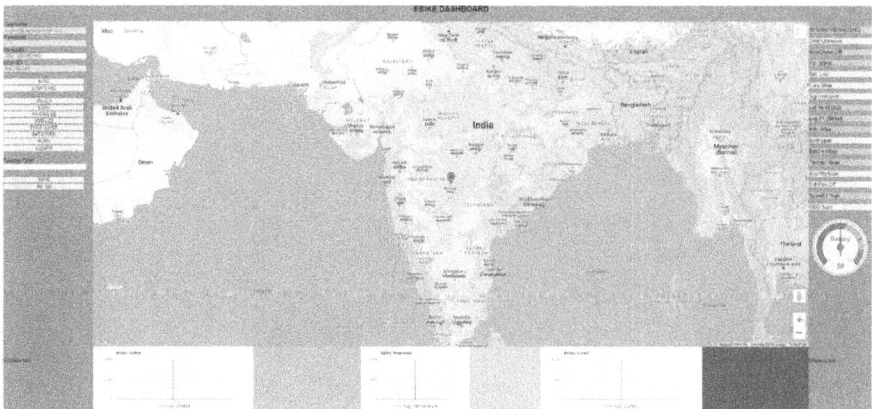

FIGURE 3.11 Web-based admin panel.

3.5.4 BATTERY MONITORING AND ANALYSIS

The IoT device monitors the battery voltage, current and temperature continuously. Battery charging and discharging cycle information is sent to the cloud. Figure 3.13 shows the web dashboard, where live battery status can be monitored. Information can also be downloaded as CSV file for offline analysis of data. This information

Username:
support@brightencontrols.com
Password:
••••••
DeviceID:
00____5030625402
Uniq. ID:
BNC1093366

SYNC
STOP SYNC

UNLOCK
LOCK
IMMOBILIZE
MOBILIZE
BOOT COVER
BAT.COVER
HORN
LOCATE

BLUETOOTH PASSCODE
?43?353689

SEND
Bat. Cal

Device/E-Scooter ID

Remote Control
Commands
for E-Scooter

Change Bluetooth
Passcode in Device

E-Scooter Status
Indicators

ID:0866795030625402
Cmd:90
BikeStatus:Off
Fix: active
Bat: Low
Lock:false
Sig:15
Lat:19.183421
Lng:77.356586
Imb :false
Boot:op.
BatCy close
Tamper: false
AccrSts:false
Ext Pow:Off
SpareDI: high
ODO:NaN

Battery

0   100
50

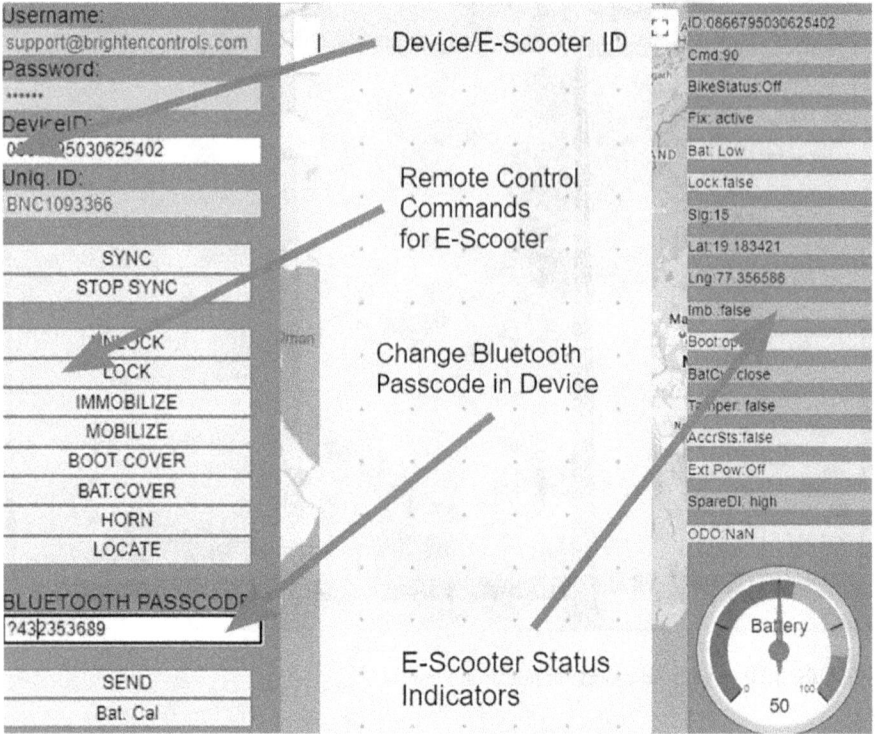

FIGURE 3.12 Control commands panel and notification panel.

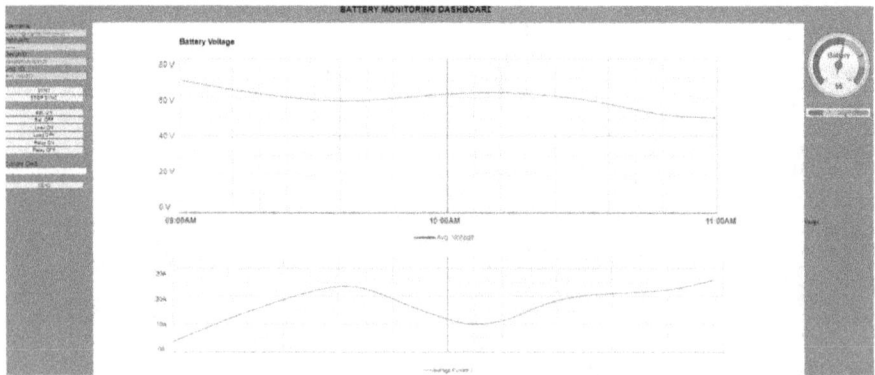

BATTERY MONITORING DASHBOARD

Battery Voltage

80 V
60 V
40 V
20 V
0 V
09:00AM   10:00AM   11:00AM

FIGURE 3.13 Battery monitoring dashboard.

includes instantaneous battery voltage and current, which is used to build software models in the cloud to predict battery health and end-of-life notifications [13–17]. Additional parameters are calculated in cloud server by corelating the historical trends of the battery charge and discharge curves.

By comparing the instantaneous "run time" rate-of-change of battery voltage to the historical trend of the same battery, the health of the battery was derived and categorized as good, average, or faulty.

For the "run time" battery analysis and prediction, the rate of change was used to continuously calculate the "deviation" and arrive at the conclusion about battery health by corelating it to the historical data.

3.6 SCOPE AND DISCUSSION

The present study in the chapter highlights the enhanced utility of the e-scooter with help various controls and monitoring functions provided by the telematics platform. The universal design of the platform can be very helpful for different applications of e-scooter in business of leasing, renting, goods/food delivery, and last-mile connectivity business segments. The author has highlighted in the chapter the effectiveness of the platform to give an indication about the health of the batteries used in e-scooters.

The scope of the presented design can be further enhanced by upgrading the connectivity options by using 4G-LTE or NB-IOT instead of 2G GPRS interface used in this study. The local connectivity option can also be upgraded to BLE 5.0 for a better Bluetooth performance and reduced power consumption.

The battery charge/discharge data accumulated in the cloud from multiple devices can be used to effectively model the battery behaviour even further, which can provide early indicators to the business about possible replacements of batteries in near future and its financial impact on the business model.

REFERENCES

[1] Iman M. Almomani, Nour Y. Alkhalil, Enas M. Ahmad, & Rania M. Jodeh, "Ubiquitous GPS Vehicle Tracking and Management System", *Jordan Conference on Applied Electrical Engineering and Computing Technologies (AEECT)*, Vol. 12, 1–6, 2011.

[2] Dat Pham Hoang, Drieberg Micheal, & Cuong Nyuven Chi, "Development of Vehicle Tracking System Using GPS and GSM Modem", *IEEE International Conference on Open Systems*, Vol. 12, 89–94, 2013.

[3] Lee Seok Ju, Girma Tewolde, & Kwon Jaerock, "Design and Implementation of Vehicle Tracking System Using GPS/GSM/GPRS Technology and Smartphone Application", *IEEE World Forum on Internet of Things*, Vol. 3, 353–358, 2014.

[4] Dong Yun Jung, Hyun Gyu Jang, & Minki Kim, "48-to-5/12 V Dual Output DC/DC Converter for High Efficiency and Small form Factor in Electric Bike Applications", *IEEE Electrical Design of Advanced Packaging and System Symposium (EDAPS)*, Vol. 1–3, 14–16, 2017.

[5] Manuel Fogue, "Automatic Accident Detection: Assistance through Communication Technologies and Vehicles", *IEEE Vehicular Technology Magazine*, Vol. 7, 90–100, 2012.

[6] P. Kaur, "State of Art of Smart Vehicle Management System Based on PIC Microcontroller with Accelerometer", *Journal of Electrical Engineering*, Vol. 2, 224–233, 2016.

[7] Microchip Corp, Datasheet of PIC24FJ256GA106 Microcontroller. Retrieved from www.microchip.com

[8] GPS Antenna, Datasheet of NEX2540 GPS Antenna. Retrieved from www.nexcomm-asia.com

[9] SIMCom, Datasheet of SIM 800H GPRS/GSM Module. Retrieved from www.simcom.com

[10] Chyi-Ren Dow, Van-Tung Bui, & Chao-ying, "An Energy Management System for e-Bikes", *8th Annual IEEE Annual Information Technology, Electronics and Mobile Communication Conference (IEMCON)*, IEEE. Retrieved from https://ieeexplore.ieee.org/, 2017.

[11] Florin Dumitrache & Marius Catalin, "E-Bike Electronic Control Unit", *IEEE 22nd International Symposium for Design and Technology in Electronic Packaging (SIITME)*, IEEE. Retrieved from https://ieeexplore.ieee.org/, 2016.

[12] Tehseen Iilhi, Tausif Zahid, & Muhammad Zahid, "Design Parameter and Simulation Analysis of Electric Bike Using Bi-Directional Power Converter", *IEEE International Conference on Electrical Communication and Computer Engineering (ICECCE)*. Retrieved from https://ieeexplore.ieee.org/, 2020.

[13] Abhijeet Chandratre, Himanshi Saini, & Sai Hunuma, "Battery Management System for E-Bike: A Novel Approach to Measure Crucial Parameters for a VRLA Battery", *IEEE India International Conference on Power Electronics 2010 (IICPE 2020)*, IEEE. Retrieved from https://ieeexplore.ieee.org/, 2011.

[14] Paris Ali Topan, M. Nisvo, & Ghufron Fathoni, "State of Charge (SOC) and State of Health (SOH) Estimation on Lithium Polymer Battery via Kalman Filter", *2nd IEEE International Conference on Science and Technology-Computer (ICST)*, IEEE. Retrieved from https://ieeexplore.ieee.org/, 2016.

[15] Rui Xiong, Youbzhi Zhang, & Ju Wang, "Lithium-Ion Battery Health Prognosis Based on a Real Battery Management System Used in Electric Vehicles", *IEEE Transactions on Vehicular Technology*, Vol. 68, 4110–4121, 2019.

[16] Fawad Ali Shah & Shehzar Shahzad Sheikh, "Battery Health Monitoring for Commercialized Electric Vehicles Batteries: Lithium-Ion", *IEEE International Conference on Power Generation Systems and Renewable Energy Technologies (PGSRET)*, IEEE. Retrieved from https://ieeexplore.ieee.org/, 2019.

[17] Matteo Corno & Gabriele Pozatto, "Active Adaptive Battery Aging Management for Electric Vehicles", *IEEE Transactions on Vehicular Technology*, Vol. 69, 2020.

4 Modelling the Barriers for the Implementation of Total Quality Management in Aircraft Maintenance Using ISM

Sandeep Phogat, Mukheshwar Yadav, Luv Kumar Singh

4.1 INTRODUCTION

Aircraft maintenance is a field that is highly controlled by different aviation regulation authorities around the world, like the European Aviation Safety Agency, France Aviation Agency, and Directorate of Civil Aviation. This field requires highly skilled personnel, and not everyone can maintain an aircraft. Only a licensed engineer, a technician from an airline, or maintenance repair organization (MRO) personnel having the required license can perform various maintenance activities by the book or maintenance manual, like inspection, repair, fault rectification, and applying MOD given by the manufacturer. As we all know, quality is governed by the customer. As such, TQM does not have a specific definition, but it can be defined as the technique to systematically analyse and improve an existing process in an organization. Quality improvement is a never-ending process. To improve a particular process, TQM (total quality management) involves everyone from the top management to the lowest level who are involved in that process by assuring proper training, funding, and spreading knowledge about quality and its benefits for an organization. TQM can be applied in an organization by various tools statistical process control, lean six sigma, PDCA cycle, or by following some internationally set standards.

4.2 LITERATURE REVIEW

A study on an aero-engine maintenance firm done by Vassilakis and Besseris (2010), the study is carried out over time of six months to gather crucial data for their research work and analysis to be done for the faults identified in the process. They presented a very simple method to implement statistical process control in the organization and concluded that the maintenance program of aero-engines could be benefited from progressive maintenance. An evaluated study of an aero-engine maintenance firm's assembly process of the exhaust nozzle had been done by Vassilakis and Besseris (2009). They analysed the gathered data for their research using some of the TQM tools like control charts, and fishbone diagrams. Finally, the authors concluded that certain parameters of the aero-engine maintenance process showed unsatisfactory results. Rhoades and Waguespack Jr (2005) compared the safety quality gap between the low-cost carrier (LCC) and full-cost carrier (FCC) by gathering the data from the Department of Transportation records. To calculate the safety and service rate of each carrier, the authors applied a simple mathematical formula between accidents, customer complaints, and yearly departures and calculated the mean and variance. The authors found out that the LCC has minimized the safety quality gap with FCC but failed to give basic service to its customers.

A mathematical model developed by Massoud Bazargan (2016) to reduce the cost of heavy-duty maintenance for the airlines suggested that more labour-intensive and costlier checks can be outsourced to have a low economic impact on the airline. Lockheed Martin and Taaffe and Allen (2014) together carried out a study on the performance indicators through which the company calculates its success. They used the matrix scorecard survey technique to analyse the gathered data. The inclusion of root cause analysis was suggested by the author, and a reduction in scorecard score was observed. Gališanskis (2004) highlighted the importance of the usage of quality assurance tools such as audits or implementation of some internationally accepted standards in the aviation industry to increase the safety of an aircraft in his study. The safety management system (SMS) between military aviation and civil aviation firms was compared by Chatzi (2018). The differences observed by the author in his research work between military and civil aviation firms were effective communication, just culture, and reporting culture; both the firms were compared based on philosophy, structure, and components. From his comparison between the military and civil aviation firms, the author concluded that military firms should adopt a SMS culture similar to that used by civil aviation firms. Mukwakungu et al. (2019) performed quantitative research on the effectiveness of quality assurance on rolling stock maintenance (RSM) at the largest South African rail company. In order to gather data for his quantitative research, the author interviewed 30 employees from the engineering division of the company. Effective training of the employees in the training division was suggested by the author in the conclusion of his final study. To calculate the quality assurance staffing effectiveness, a quantitative study on 16 air force fighter aircraft squadrons was performed by Moore et al. (2007). A few factors were considered by the authors to analyse and gather information. The information was gathered by the authors through interviews and Delphi surveys. After thoroughly analysing the factors, the authors concluded that the maintenance managers had no other option than to either leave the quality assurance personnel slots empty or to assign individuals with credentials that are not permitted. Goh and Lim (1996) undertook a long research work to improve the repair turn time and redo work on the engine blades at a Singapore-based aircraft maintenance firm that repairs the plane engine. The authors collected data from May 1994 to January 1995 to identify the faults that caused the delay in repairing blades. After identifying the faults and analysing them, the authors concluded to implement TQM and observed a reduction in the redo and repair turn time of engine blades.

4.3 IDENTIFICATION OF BARRIERS TO TOTAL QUALITY MANAGEMENT

Many factors are capable to hinder the application of TQM in aircraft maintenance. From the literature survey, these factors named barriers are found and discussed here:

1. Lack of initial support—Per the study by Goh and Lim (1996), lack of support from the top management at the starting days of TQM implementation has an adverse effect on an organization's quality movement.
2. Absence from meetings—Per the study by Goh and Lim (1996), absence in meetings related to quality improvement by employees and top management has a bad effect on quality as the top management's goal for quality improvement cannot be passed on to the employees.
3. Reluctance to disclose information—Per the study by Goh and Lim (1996), if the employee of an organization does not disclose the information related to areas causing a problem to respective departments, then the process cannot be improved, and hence, quality is affected.
4. Employee reluctance to complete work on time—Per the study by Vassilakis and Besseris (2010) and Taaffe and Allen (2014), the reluctance of an employee toward his work can lead to some defects in the services and products provided by the organization. It can also affect the closure of quality documents, hence affecting the quality information availability.
5. Consumer complaints—Per the study by Rhoades and Waguespack Jr's (2005), consumer/customer complaints are a big factor that determines the decreasing quality of the products or services provided by the organization.
6. Non-conformance—Per the study by Taaffe and Allen (2014), non-conformance to the quality procedures or steps by an employee leads to a decrease in the quality of the products and services provided by the organization.
7. Incompetence of employees—Per the study by Vassilakis and Besseris (2010) and Moore et al. (2007), the incompetence of employees leads to major accidents or major lapses in providing quality to the customers of the organization.
8. Improper planning—Per the study by Bazargan (2016), improper planning of the process layout or raw materials can cause an increase in cost to the organization and also affect the quality of the services provided by the organization.
9. Absence of root cause analysis—Per the study by Taaffe and Allen (2014), if root cause analysis is not done for an area showing problems, then the cause of the problem is not identified, which will affect the quality of the product

10. Improper paperwork—Per the study by Taaffe and Allen (2014), improper paperwork affects the quality of the product as the quality record is not made available to the respective team/department, which can lead to rejection of the product while inspection.
11. Absence of statistical process control/analysis—Per the study by Vassilakis and Besseris (2009) and Gališanskis (2004), the absence of statistical process control can lead to an increase in the cost of manufacturing the product, and the shortcomings are not identified, which affects the quality of the products.
12. Improper examination of audit results—Per the study by Gališanskis (2004), if the audit results are not worked on properly, it may affect the quality of the products or services.
13. Lack of communication—Per the study by Chatzi (2018), in a military organization, communication is governed by some rules which affect the development of reporting culture, ultimately affecting the safety culture in the organization.
14. Lack of training—Per the study by Mukwakungu et al. (2019), if the employee is not properly trained, then the employee will not be able to do the work assigned to him properly, which affects the quality of the service provided by the organization.

The major 14 barriers to TQM implementation are identified by doing the vast literature review and tabulated in Table 4.1.

TABLE 4.1
Barriers to Total Quality Management

S. No.	Barriers	Reference
1.	Lack of initial support	Goh and Lim (1996)
2.	Absence from meetings	Goh and Lim (1996)
3.	Reluctance to disclose information	Goh and Lim (1996)
4.	Employee reluctance to complete work on time	Vassilakis and Besseris (2010), Taaffe and Allen (2014)
5.	Consumer complaints	Rhoades and Waguespack Jr's (2005)
6.	Non-conformance	Taaffe and Allen (2014)
7.	Incompetence of employees	Vassilakis and Besseris (2010), Moore et al. (2007)
8.	Improper planning	Bazargan (2016)
9.	Absence of root cause analysis	Taaffe and Allen (2014)
10.	Improper paperwork	Taaffe and Allen (2014)
11.	Absence of statistical process control/ analysis	Vassilakis and Besseris (2009), Gališanskis (2004)
12.	Improper examination of audit results	Gališanskis (2004)
13.	Lack of communication	Chatzi (2018)
14.	Lack of training	Mukwakungu et al. (2019)

4.4 ISM RESEARCH METHODOLOGY

ISM stands for interpretive structural modelling (ISM). Undefined and badly articulated mental models of the system are modified into a visual, clear-cut model with the help of this approach. The structured model was prepared and analysed with the help of a computer in this approach. A judgement made after discussion with the expert explains the linkage of factors with each other. Based on interconnection a complete layout is made from the factors, hence making this approach a structural approach. The complete layout and exact interconnection of factors and systems being reviewed are portrayed in the digraph model in this approach. J. N. Warfield 1976 came up with a qualitative tool known as ISM. It is a popular approach for pinpointing and outlining the interconnection between specific factors which explain an obstacle and gives a mean that can create order in the complexity of such elements. It uses expert hands-on experience and understanding to convert a confusing system into various subsystems and develop a layered structural model, as shown in Table 4.2.

TABLE 4.2
Implementation of ISM

S. No.	Author and Year	Implementation of ISM
1.	Jain and Raj (2021)	Constraints of FMS
2.	Mittal et al. (2021)	TB barriers
3.	Priya et al. (2021a)	Factors of the global economy
4.	Priya et al. (2021b)	Assessment of government measures
5.	Jain and Ajmera (2020)	Enabler of Industry 4.0
6.	Ajmera and Jain (2019a)	Lean factors in the Indian healthcare industry
7.	Ajmera and Jain (2019b)	QOL suffering from diabetes
8.	Ajmera and Jain (2019c)	Barriers of Health 4.0
9.	Jain and Soni (2019)	Performance variables of the FMS
10.	Patri and Suresh (2018)	Lean implementation in healthcare organizations
11.	Jain and Ajmera (2018)	Used ISM for modelling medical tourism factors
12.	Chauhan et al. (2018)	Analysed the waste-recycling barriers
13.	Malviya and Kant (2017)	Modelling of green SCM
14.	Dube and Gawande (2016)	Analysed the barriers of green supply chain
15.	Jain and Raj (2016)	FMS performance factors
16.	Gupta and Ramesh (2015)	Used ISM for detecting the factors influencing HCSC in India
17.	Jain and Raj (2015a)	Used ISM for analysing flexibility in a flexible manufacturing system
18.	Jain and Raj (2015b)	Used TISM and fuzzy MICMAC for analysing flexibility in a flexible manufacturing system
19.	Jain and Raj (2014)	FMS productivity factors
20.	Sharma et al. (2013)	Variables of assembly line balancing
21.	Balasubramanian (2012)	Used ISM for evaluating the barriers of GSCM
22.	Colin et al. (2011)	Various stages of the supply chain
23.	Ramesh et al. (2010)	Variables affecting the Indian textile industries
24.	Raj et al. (2008)	Enablers of FMS
25.	Faisal et al. (2007)	Barriers in the supply chains of SMEs

The various steps of SIM are as follows:

1. Listed factors influencing the system are being reviewed like objectives, actions, individuals, etc.
2. After step 1, a circumstantial interrelationship is developed between factors due to which pairs of factors are scrutinized.
3. A structural self-interaction matrix (SSIM) is established for factors, that points out pair-wise interrelation between factors of the system that is being reviewed.
4. The reachability matrix (RM) is established from the SSIM, and the resultant matrix is looked over for transitivity. Transitivity in ISM is defined as if factor X is linked to Y and Y is linked to Z, then X is automatically linked to Z.
5. After completion of step 4, segmentation of RM is done into various levels.
6. After segmentation of RM, a graph is made and transitivity is deleted from the RM considering the relationship obtained between the variables.
7. After replacing the variables symbols with the actual factors ISM model is developed by making the graph
8. In the final step, the ISM model is thoroughly checked for any mistakes and recommended changes are made.

4.5   MODELLING THE ENABLERS FOR THE IMPLEMENTATION OF TOTAL QUALITY MANAGEMENT IN AIRCRAFT MAINTENANCE

The systematic approach to implementation of ISM model is done by following the steps:

4.5.1   VARIOUS FACTORS AND RELATIONSHIP BETWEEN THEM IS ESTABLISHED

Various disablers for the implementation of TQM in aircraft maintenance are identified in various authors' literature. A relationship based on the reviews of the experts in the field of aircraft maintenance, this relationship shown in Table 4.3 describes how the variables approach each other.

TABLE 4.3
Contextual Relationship

Symbol	Meaning
V	Variable "a" approaches to variable "b."
A	Variable "b" approaches to variable "c."
X	Variables "a" and "b" approach to each other
O	Variables "a" and "b" are not related to each other.

4.5.2 Evolution of SSIM

The matrix shown in Table 4.4 is the SSIM matrix after a discussion with the experts.

4.5.3 Evolution of Initial RM

After the SSIM is developed, RM is further developed from the SSIM by replacing the V, A, X, and O with binary digits 1 and 0. The initial RM shown developed is mentioned in Table 4.5 along with the formula for how the symbols are converted into binary digits. After that, the initial reachability matrix is derived as shown in Table 4.6.

4.5.4 Evolution of Final Reachability Matrix (RM)

The final RM is developed from the initial RM by establishing transitivity in it, shown in Table 4.7.

TABLE 4.4
SSIM

Variables	1	2	3	4	5	6	7	8	9	10	11	12	13	14
1	1	O	X	V	O	O	V	V	O	O	V	O	X	X
2		1	O	O	O	V	X	O	V	O	O	O	X	O
3			1	A	X	V	O	X	V	O	V	A	A	V
4				1	X	X	X	A	V	A	A	A	X	X
5					1	X	A	A	X	X	V	A	X	A
6						1	X	X	O	X	A	A	X	X
7							1	V	O	V	V	V	A	A
8								1	X	X	A	X	X	A
9									1	A	V	A	A	A
10										1	V	V	A	A
11											1	A	A	A
12												1	O	X
13													1	X
14														1

TABLE 4.5
RM Entries

Symbol	Cell (a, b) Entry	Cell (b, a) Etry
V	1	0
A	0	1
X	1	1
O	0	0

TABLE 4.6
Initial Reachability Matrix

Variables	1	2	3	4	5	6	7	8	9	10	11	12	13	14
1	1	0	1	1	0	0	1	1	0	0	1	0	1	1
2	0	1	0	0	0	1	1	0	1	0	0	0	1	0
3	1	0	1	0	1	1	0	1	1	0	1	0	0	1
4	0	0	1	1	1	1	1	0	1	0	0	0	1	1
5	0	0	1	1	1	1	0	0	1	1	1	0	1	0
6	0	0	0	1	1	1	1	1	0	1	0	0	1	1
7	0	1	0	1	1	1	1	1	0	1	1	1	0	0
8	0	0	1	1	1	1	0	1	1	1	0	1	1	0
9	0	0	0	0	1	0	0	1	1	0	1	0	0	0
10	0	0	0	1	1	1	0	1	1	1	1	1	0	0
11	0	0	0	1	0	1	0	1	0	0	1	1	0	0
12	0	0	1	1	1	1	0	1	1	0	1	1	0	1
13	1	1	1	1	1	1	1	1	1	1	1	0	1	1
14	1	0	0	1	1	1	1	1	1	1	1	1	1	1

TABLE 4.7
Final Reachability Matrix

Variables	1	2	3	4	5	6	7	8	9	10	11	12	13	14
1	1	1*	1	1	1*	1*	1	1	1*	1*	1	1*	1	1
2	1*	1	1*	1*	1*	1	1	1*	1	1*	1*	1*	1	1*
3	1	0	1	1*	1	1	1*	1	1	1*	1	1*	1*	1
4	1*	1*	1	1	1	1	1	1*	1	1*	1*	1*	1	1
5	1*	1*	1	1	1	1	1*	1*	1	1	1	0	1	1*
6	1*	1*	1*	1	1	1	1	1	1*	1	1*	1*	1	1
7	0	1	1*	1	1	1	1	1	1*	1	1	1	1*	1*
8	1*	1*	1	1	1	1	1*	1	1	1	1*	1	1	1*
9	0	0	0	1*	1	1*	0	1	1	0	1	0	0	0
10	0	0	1*	1	1	1	0	1	1	1	1	1	0	1*
11	0	0	0	1	0	1	0	1	0	0	1	0	0	0
12	1*	0	1	1	1	1	1*	1	1	1*	1	1	1*	1
13	1	1	1	1	1	1	1	1	1	1	1	1*	1	1
14	1	1*	1*	1	1	1	1	1	1	1	1	1	1	1

4.5.5 SEGMENTATION OF REACHABILITY MATRIX

mentioned steps are followed for segmenting the matrix, and Table 4.8 shows the final iterations and their levels:

TABLE 4.8
Iterations (Levels)

Variables	Reachability Set	Antecedent Set	Intersection Set	Level
1	1	1	1	Level 5
2	1,2,7,13	1,2,7,13	1,2,7,13	Level 4
3	1,3,7,10,12,13,14	1,2,3,7,10,12,13,14	1,3,7,10,12,13,14	Level 3
4	1,2,3,4,5,6,7,8,9,10,11, 12,13,14	1,2,3,4,5,6,7,8,9,10,11,12, 13,14	1,2,3,4,5,6,7,8,9,10,11,12, 13,14	Level 1
5	1,2,3,5,7,9,10,13,14	1,2,3,5,7,9,10,12,13,14	1,2,3,5,7,9,10,13,14	Level 2
6	1,2,3,4,5,6,7,8,9,10,11, 12,13,14	1,2,3,4,5,6,7,8,9,10,11,12, 13,14	1,2,3,4,5,6,7,8,9,10,11,12, 13,14	Level 1
7	2,7,13	1,2,7,13	2,7,13	Level 4
8	1,2,3,4,5,6,7,8,9,10,11, 12,13,14	1,2,3,4,5,6,7,8,9,10,11,12, 13,14	1,2,3,4,5,6,7,8,9,10,11,12, 13,14	Level 1
9	5,9	1,2,3,5,7,9,10,12,13,14	5,9	Level 2
10	3,10,12,14	1,2,3,7,10,12,13,14	3,10,12,14	Level 3
11	4,6,8,11	1,2,3,4,5,6,7,8,9,10,11,12, 13,14	4,6,8,11	Level 1
12	1,3,7,10,12,13,14	1,2,3,7,10,12,13,14	1,3,7,10,12,13,14	Level 3
13	1,2,7,13	1,2,7,13	1,2,7,13	Level 4
14	1,2,3,7,10,12,13,14	1,2,3,7,10,12,13,14	1,2,3,7,10,12,13,14	Level 3

1. Reachability and antecedent set are derived for each variable from the final RM.
2. Then the intersection of both sets is identified for each variable.
3. The top level is given to the variable whose reachability set and intersection set are the same.
4. In the next step, the variable, which has been given a level, is eliminated from both reachabilities and the antecedent set.
5. The previous four steps are repeated till every single variable is defined level.

4.5.6 FINDING DRIVING AND DEPENDENCE POWER

Driving power and dependence power are calculated for each factor and shown in Table 4.9. The driving power for each factor is the total number of factors (including itself) that it may help to achieve. On the other hand, dependence power is the total number of factors (including itself) that may help in achieving it.

4.5.7 MAKING OF ISM MODEL

The ISM model is made based on the relationship given in the RM. A directed graph, also known as a diagram, is made, and the transitive links are deleted, as shown

TABLE 4.9
Driving and Dependence Power

Variables	1	2	3	4	5	6	7	8	9	10	11	12	13	14	Driving Power
1	1	1*	1	1	1*	1*	1	1	1*	1*	1	1*	1	1	14
2	1*	1	1*	1*	1*	1	1	1*	1	1*	1*	1*	1	1*	14
3	1	0	1	1*	1	1	1*	1	1	1*	1	1*	1*	1	13
4	1*	1*	1	1	1	1	1	1*	1	1*	1*	1*	1	1	14
5	1*	1*	1	1	1	1	1*	1*	1	1	1	0	1	1*	13
6	1*	1*	1*	1	1	1	1	1	1*	1	1*	1*	1	1	14
7	0	1	1*	1	1	1	1	1	1*	1	1	1	1*	1*	13
8	1*	1*	1	1	1	1	1*	1	1	1	1*	1	1	1*	14
9	0	0	0	1*	1	1*	0	1	1	0	1	0	0	0	6
10	0	0	1*	1	1	1	0	1	1	1	1	1	0	1*	10
11	0	0	0	1	0	1	0	1	0	0	1	0	0	0	4
12	1*	0	1	1	1	1	1*	1	1	1*	1	1	1*	1	13
13	1	1	1	1	1	1	1	1	1	1	1	1*	1	1	14
14	1	1*	1*	1	1	1	1	1	1	1	1	1	1	1	14
Dependence Power	10	9	12	14	13	14	11	14	13	12	14	11	11	12	

in Figure 4.1, and the flowchart for the same is shown in Figure 4.2. This diagram represents the interrelation between the barriers to the implementation of TQM in aircraft maintenance. The arrows will be made between barriers if they are related to each other.

4.6 MICMAC ANALYSIS

The main aim of MICMAC analysis is to review the driving power and the dependence power of the factors. Multiplication abilities of matrices are the main philosophy for MICMAC analysis. It is carried out to pinpoint the key factors that guide the system in different varieties as shown in Figure 4.3. Based on their driving and dependence on power, the factors are categorized into four varieties:

1. **Autonomous factors**—factors having weak driving power and weak dependence power are known as autonomous factors. They might be having few linkages with the system that might be strong; otherwise, they are not connected to the system.
2. **Linkage factors**—factors whose driving and dependence power is strong are called as linkage factors. Any disturbance to these factors will affect the whole system, including the factors themselves.
3. **Dependent power**—factors having weak driving power but strong dependence power is known as dependent power.

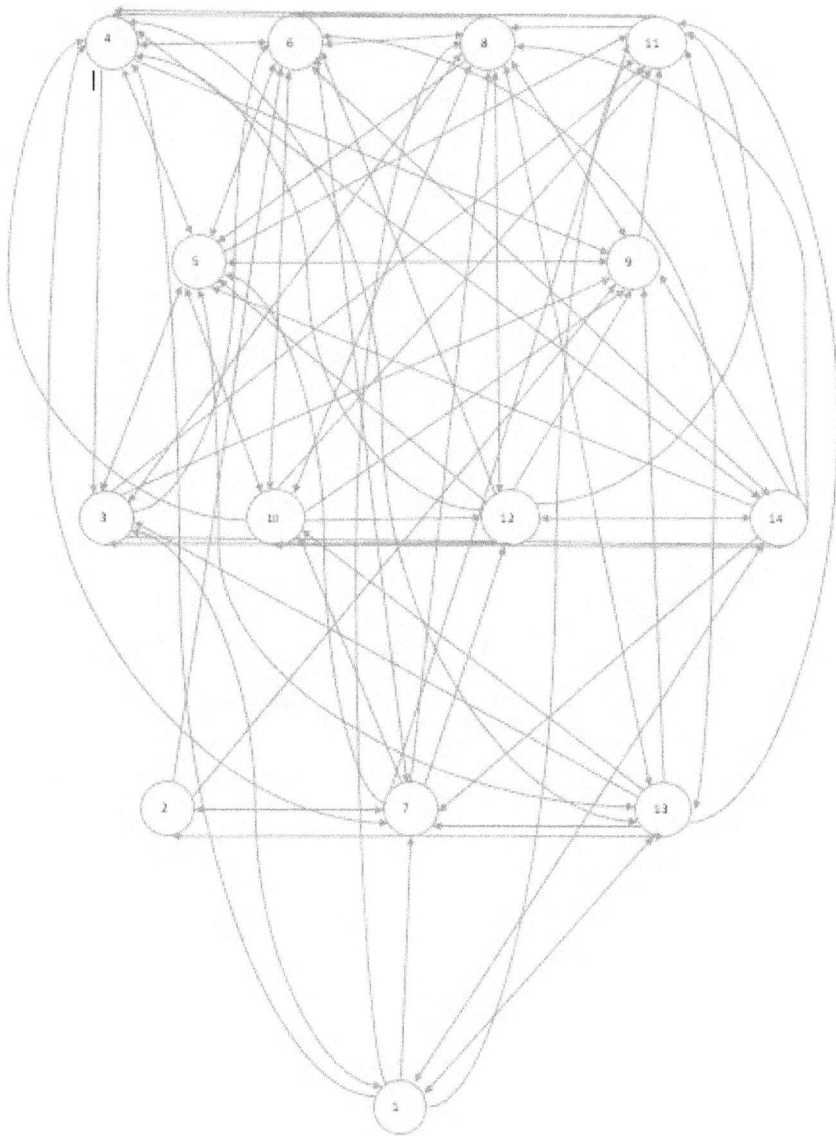

FIGURE 4.1 Diagram of barriers affecting the implementation of TQM in aircraft maintenance.

4. **Independent factors**—factors having strong driving power and weak dependence power are known as independent factors. A factor having a very strong driving power is known as a key factor and lies in the group of independent or linkage factors.

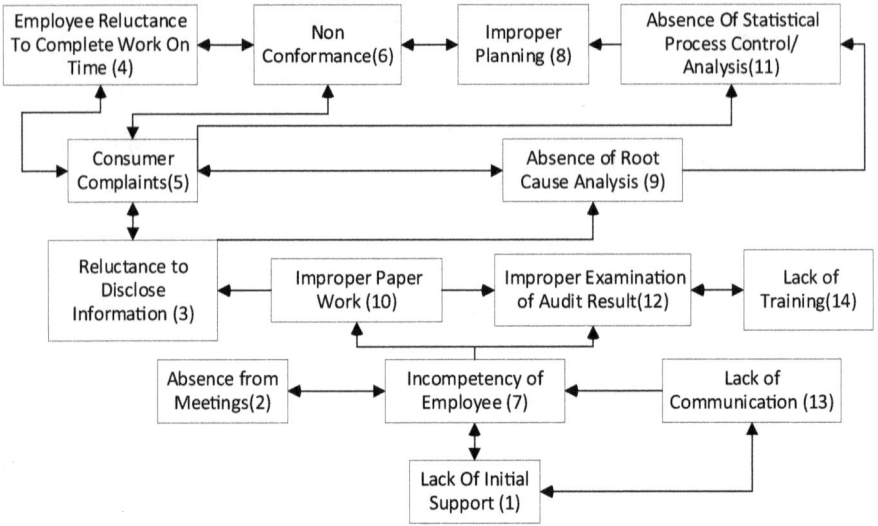

FIGURE 4.2 Flowchart of barriers affecting the implementation of TQM in aircraft maintenance.

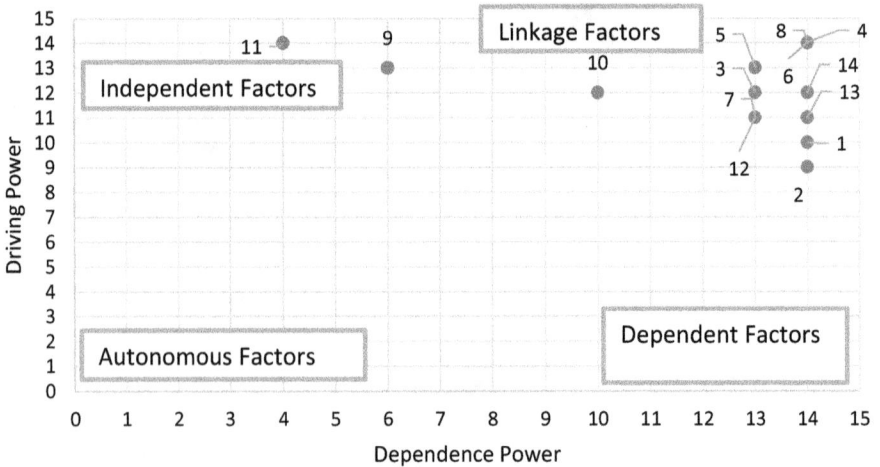

FIGURE 4.3   MICMAC analysis.

4.7   DISCUSSION

The objective of this study was to identify the various barriers affecting the implementation of TQM in aircraft maintenance and make a structured model using the interpretive structural methodology (ISM) technique. ISM technique put in place a ladder of various activities that may be used in the various departments and organizations within an airline or a maintenance repair organization facility. All the staff such as engineers, technicians, ground staff, and other supporting staff of aircraft maintenance should get knowledge about the tricky difficulties and their interrelationship from the driving and dependency power. The MICMAC analysis done in our study depicts that the barriers factors—(1) lack of initial support, (2) absence from meeting, (3) reluctance to disclose information, (4) employee reluctance to complete work on time, (5) consumer complaints, (6) non-conformance, (7) incompetency of employee, (8) improper planning, (9) absence of root cause analysis, (10) improper paperwork, (11) absence of statistical process control/analysis, (12) improper examination of audit result, (13) lack of communication, and (14) lack of training—have a crucial impact on the implementation of TQM in aircraft maintenance.

1. **Autonomous factors**—factors having weak driving power and weak dependence power are known as autonomous factors. They might be having few linkages with the system that might be strong; otherwise, they are not connected to the system. No factor lies in the autonomous factor region in MICMAC analysis. The factor that falls in this region does not affect the system so much.
2. **Linkage factors**—factors whose driving and dependence power is strong are called linkage factors. Any disturbance to these factors will affect the whole system, including the factors themselves. Most of the factors—(1) lack of initial support, (2) absence from the meeting, (3) reluctance to disclose information, (4) employee reluctance to complete work on time, (5) consumer complaints, (6) non-conformance, (7) incompetency of employee, (8) improper planning, (10) improper paperwork, (12) improper examination of audit result, (13) lack of communication, and (14) lack of training—fall in this region.
3. **Dependent power**—factors having weak driving power but strong dependence power is known as dependent power. In our research work, no factor falls in this region.
4. **Independent factors**—factors having strong driving power and weak dependence power are known as independent factors. A factor having a very strong driving power is known as a key factor. Factors (9) absence of root cause analysis and (11) absence of statistical process control/analysis fall in this region. These factors are considered as the root cause for all the difficulties being faced in the implementation of TQM in aircraft maintenance; expeditious and very firm actions should be taken for removing these factors from the system.

4.8 CONCLUSION

After going through various previous research work, the barriers to TQM in aircraft maintenance have been identified in this chapter. Consumer complaints, employee reluctance to complete work on time, non-conformance, incompetence of employees, initial lack of support, inability to attend meetings, information withheld due to apprehension, improper planning, lack of root cause analysis, improper paperwork, absence of statistical process control/analysis, improper examination of audit results, lack of communication, and lack of training are the barriers of TQM that can prevent an organization or a company from providing quality products to its customers. These factors, if not worked upon by an organization, can also lead to the loss of market share, customer base, and finally less profit. The ISM model developed showcases how all the barriers affect the implementation of TQM in aircraft maintenance. This study showcases how the implementation of TQM in aircraft maintenance affects the Indian aviation industry. There is a positive increase in the Indian aviation industry and to maintain that positive increase in Indian aviation the personnel associated with aircraft maintenance may refer to the crucial barriers and their relationships with each other to develop a high-quality aircraft maintenance environment. The same can be used for future work by taking the list to a field-/practical-based project or a case-study-based research work.

REFERENCES

Ajmera, P. and Jain, V. (2019a) A fuzzy interpretive structural modeling approach for evaluating the factors affecting lean implementation in Indian healthcare industry. *International Journal of Lean Six Sigma*, Vol. 11, pp. 376–397.

Ajmera, P. and Jain, V. (2019b) Modeling the factors affecting the quality of life in diabetic patients in India using total interpretive structural modeling. *Benchmarking: An International Journal*, Vol. 26, pp. 951–970.

Ajmera, P. and Jain, V. (2019c) Modelling the barriers of Health 4.0–the fourth healthcare industrial revolution in India by TISM. *Operations Management Research*, Vol. 12, pp. 129–145.

Balasubramanian, S. (2012) A hierarchical framework of barriers to green supply chain management in the construction sector. *Journal of Sustainable Development*, Vol. 5, pp. 15–27.

Bazargan, Massoud (2016) Airline maintenance strategies—in-house vs. outsourced—an optimization approach. *Journal of Quality in Maintenance Engineering*, Vol. 22(2), pp. 114–129.

Chatzi, Anna V. (2018) Safety management systems: An opportunity and a challenge for military aviation organisations. *Aircraft Engineering and Aerospace Technology*, Vol. 91(1), pp. 190–196.

Chauhan, A., Singh, A. and Jharkharia, S. (2018) An interpretive structural modeling (ISM) and decision-making trail and evaluation laboratory (DEMATEL) method approach for the analysis of barriers of waste recycling in India. *Journal of the Air & Waste Management Association*, Vol. 68, pp. 100–110.

Colin, J., Estampe, D., Pfohl, H.C., et al. (2011) Interpretive structural modeling of supply chain risks. *International Journal of Physical Distribution & Logistics Management*, Vol. 41 pp. 839–859.

Dube, A.S. and Gawande, R.S. (2016) Analysis of green supply chain barriers using integrated ISM-fuzzy MICMAC approach. *Benchmarking: An International Journal*, Vol. 23, pp. 1558–1578.

Faisal, M.N., Banwet, D. and Shankar, R. (2007) Supply chain risk management in SMEs: Analysing the barriers. *International Journal of Management and Enterprise Development*, Vol. 4, pp. 588–607.

Gališanskis, A. (2004) Aspects of quality evaluation in aviation maintenance. *Aviation*, Vol. 8(3), pp. 18–26.

Goh, Mark and Lim, Fang-Seng (1996) Implementing TQM in an aerospace maintenance company. *Journal of Quality in Maintenance Engineering*, Vol. 2(2), pp. 3–20.

Gupta, U. and Ramesh, A. (2015) Analyzing the barriers of health care supply chain in India: The contribution and interaction of factors. *Procedia-Social and Behavioral Sciences*, Vol. 189, pp. 217–228.

Jain, V. and Ajmera, P. (2018) Modeling the factors affecting Indian medical tourism sector using interpretive structural modeling. *Benchmarking an International Journal*, Vol. 70, pp. 1461–1479.

Jain, V. and Ajmera, P. (2020) Modeling the enablers of industry 4.0 in the Indian manufacturing industry. *International Journal of Productivity & Performance Management*, Vol. 70, pp. 1233–1262.

Jain, V. and Raj, T. (2014) Modeling and analysis of FMS productivity variables by ISM, SEM & GTMA approach. *Frontiers of Mechanical Engineering*, Vol. 9, pp. 218–232.

Jain, V. and Raj, T. (2015a) A hybrid approach using ISM & modified TOPSIS for the evaluation of flexibility in FMS. *International Journal of Industrial & System Engineering*, Vol. 19, pp. 389–406.

Jain, V. and Raj, T. (2015b) Modeling and analysis of FMS flexibility factors by TISM and fuzzy MICMAC. *International Journal of System Assurance Engineering and Management*, Vol. 6, pp. 350–371.

Jain, V. and Raj, T. (2016) Modeling and analysis of FMS performance variables by ISM, SEM and GTMA approach. *International Journal of Production Economics*, Vol. 171, pp. 84–96.

Jain, V. and Raj, T. (2021) Study of issues related to constraints in FMS by ISM, fuzzy ISM and TISM. *International Journal of Industrial & System Engineering*, Vol. 37, pp. 197–221.

Jain, V. and Soni, V.K. (2019) Modeling and analysis of FMS performance variables by fuzzy TISM. *Journal of Modeling in Management*, Vol. 14, pp. 2–30.

Malviya, R.K. and Kant, R. (2017) Modeling the enablers of green supply chain management: An integrated ISM–fuzzy MICMAC approach. *Benchmarking: An International Journal*, Vol. 24, pp. 536–568.

Mittal, P., Ajmera, P., Jain, V., et al. (2021) Modeling and analysis of barriers in controlling TB: Developing countries' perspective. *International Journal of Health Governance*, ahead-of-print.

Moore, Terry D., Johnson, Alan W., Rehg, Michael T. and Hicks, Michael J. (2007) Quality assurance staffing impacts in military aircraft maintenance units. *Journal of Quality in Maintenance Engineering*, Vol. 13(1) pp. 33–48.

Mukwakungu, S.C., Sibeko, Z. and Mbohwa, C. (2019, December). The effectiveness of rolling stock maintenance on quality assurance at the largest South African rail company. In *2019 IEEE International Conference on Industrial Engineering and Engineering Management (IEEM)* (pp. 416–420). IEEE.

Patri, R. and Suresh, M. (2018) Factors influencing lean implementation in healthcare organizations: An ISM approach. *International Journal of Healthcare Management*, Vol. 11, pp. 25–37.

Priya, M.S., Jain, V., Kabiraj, S., et al. (2021a) Modelling the factors affecting global economy during COVID-19 using ISM approach. *International Journal of Services, Economics and Management*, Vol. 12, pp. 294–316.

Priya, S.S., Priya, M.S., Jain, V., et al. (2021b) An assessment of government measures in combatting COVID-19 using ISM and DEMATEL modelling. *Benchmarking: An International Journal*, ahead-of-print.

Raj, T., Shankar, R. and Suhaib, M. (2008) An ISM approach for modelling the enablers of flexible manufacturing system: The case for India. *International Journal of Production Research*, Vol. 46, pp. 6883–6912.

Ramesh, A., Banwet, D. and Shankar, R. (2010) Modeling the barriers of supply chain collaboration. *Journal of Modelling in Management*, Vol. 5, pp. 176–193.

Rhoades, Dawna L. and Waguespack, Blaise Jr. (2005) Strategic imperatives and the pursuit of quality in the US airline industry. *Managing Service Quality: An International Journal*, Vol. 15(4) pp. 344–356.

Sharma, P., Thakar, G. and Gupta, R.C. (2013). Interpretive structural modeling of functional objectives (Criteria's) of assembly line balancing problem. *International Journal of Computer Applications*, Vol. 83, pp. 14–22.

Taaffe, Kevin M. and Allen, Robert William (2014) Performance metrics analysis for aircraft maintenance process control. *Journal of Quality in Maintenance Engineering*, Vol. 20(2), pp. 122–134.

Vassilakis, E. and Besseris, G. (2009) An application of TQM tools at a maintenance division of a large aerospace company. *Journal of Quality in Maintenance Engineering*, Vol. 15(1), pp. 31–46.

Vassilakis, E. and Besseris, G. (2010) The use of SPC tools for a preliminary assessment of an aero engines' maintenance process and prioritization of aero engines' faults. *Journal of Quality in Maintenance Engineering*, Vol. 16(1), pp. 5–22.

5 Investigation of the Effect of Throat Diameter on Thrust Force Developed in CD Rocket Nozzle Using CFD

Atam Parkash, Anuranjan Sharda

5.1 INTRODUCTION

The rocket nozzle is a mechanical device consisting of a pipe of varying cross-sectional area that is used to direct or modify the flow of a fluid (liquid and/or gas) leaving the combustion chamber. It is thus a simple tube that is used to convert the thermochemical energy generated in the combustion chamber into kinetic energy. Here a low-velocity, high-pressure, and high-temperature gas is converted into a high-velocity, lower-pressure, and low-temperature gas. For a convergent-divergent (CD) nozzle, the area decreases from the inlet section to the throat and then increases from the throat to the outer section. In this chapter, CFD analysis is carried out using two software: CATIA V5, which is used for designing the nozzle, and ANSYS Fluent 21.0, which is used for analysing the flows in the nozzle and for generating various contours such as pressure, density, velocity contour, and so on.

A CD nozzle is designed for achieving speeds that are greater than the speed of sound. The ratio of attaining speed and speed of sound is known as the Mach number. The design of this nozzle is obtained from the area-velocity relation:

$$\frac{dA}{dV} = -\frac{A}{V} \times \left(1 - M^2\right) \tag{5.1}$$

where, M = Mach number (which means the ratio of the speed of flow to the speed of sound), A = area, V = velocity, dA = change in area, and dV = change in velocity.

The major function of the CD nozzle is to accelerate the flue gases produced inside the combustion chamber. The maximum thrust can be developed only when the combustion of fuel takes place adequately, with high combustion efficiency, and simultaneously when the CD nozzle is capable of increasing velocity to a sufficient level. The thrust produced by a CD nozzle in a rocket engine depends on the exit velocity of the flue gases, the density of flue gases at the exit, and the pressure at the exit of the rocket nozzle. These three flow variables (the exit velocity, density, and pressure at the exit) are all determined by the rocket nozzle design. For a firmly operating rocket propulsion system moving in a homogeneous atmosphere, the total thrust can be calculated using equation (5.2):

$$F = \dot{m} \cdot v_e + \left(p_e - p_0 \right) \cdot A_e \qquad (5.2)$$

where $\dot{m}$ = mass, v_e = exit velocity, p_e = exit pressure, p_0 = pressure outlet, and A_e = exit area.

5.2   LITERATURE REVIEW AND RESEARCH OBJECTIVES

The thrust developed by the rocket engine depends on the mass flow rate, the exit velocity of the exhaust gases, and the pressure at the outlet of the engine. The values of these three flow variables determine the amount of thrust produced by the nozzle. Researchers have carried out CFD simulations to obtain optimal results for various parameters: convergence and divergence angle, throat diameter, Mach number, and so on. These research and discussion of results of these researches are elaborated in the following paragraphs.

Narayana et al. (2016) conducted a numerical study of element sizing in the mesh generation. It was observed by the authors that dynamic pressure reduces from 7,786.104 Pa to 1,572.61 Pa by reducing the element size. The Mach number, however, increased from 2.784976 to 3.105635 on increasing the number of mesh divisions. Thus, the study concluded that if the element size of the mesh is reduced, then an optimal result can be achieved and elements with large mesh sizes disturb the results [1]

Roy et al. (2016) carried out computational fluid dynamics (CFD) studies by varying the convergence angle, divergence angle, and throat radius of the nozzle to optimize the static pressure and Mach number of CD nozzles for rocket engines. For carrying out optimization of the performance parameters, the 3 × 3 Taguchi design was used with the following input parameters: (1) convergence angle varying at 30° at first level, 45° at the second level, and 60° at the third level; (2) divergence angle varying at 7.5° at first level, 15° at second level, and 30° at third level; (3) throat radius varying as 0 mm at first level, 130 mm at the second level, and 225 mm at third level. Inlet diameter, outlet diameter, and throat diameter were fixed by the authors as 1.00 m, 0.75 m, and 0.40 m, respectively. While carrying out the simulations for analysis, the authors considered the inlet temperature to be 3,600 K and the combustion chamber temperature was determined at two different values of pressures (bar) that is 42.2 and 51.5 bars. After optimization, the authors predicted the Mach number to be 3.73 at a convergent angle of 60° and divergence angle of 15° for 42 bar pressure and 3.92 at a convergent angle of 30° and divergence angle of 7.5° for 52 bar pressure [2].

Singh et al. (2017) studied the normal shock wave in the divergence section and concluded that the shock wave moves toward the exit or divergent section of an increasing pressure ratio. If the NPR (0.88) is low, the shock wave occurs near the throat, and if the NPR is high, the shock wave moves toward the exit section in the divergent section [3].

Sushma et al. (2017) carried out numerical simulations, using the software ANSYS Fluent so as to study and understand the performance of the conical nozzle and the scarfed nozzle. The turbulence models—the SA (Spalart-Allmaras) model and the SST model—were used to calculate the nozzle jet field. From the analysis, it was concluded that the Mach number of the scarf nozzle was relatively low compared to that of the conical nozzle. Scarf nozzles were preferred due to missile configuration restrictions. The k-epsilon turbulence model can accurately predict the flow characteristics in the nozzle. Due to the result of the external flow from the outlet of the scarfed nozzle, the Mach number was lower than the conical nozzle [4].

Meena et al. (2021) carried out numerical studies to examine the flow through the contour nozzles and conical nozzles. In this study, the authors varied divergence angles at five different levels, 10°, 12°, 14°, 16°, and 18°; the throat diameters were also varied at two levels, 0.304 m and 0.404 m; whereas the inlet diameter and the outlet diameter were fixed as 1 m and 0.861 m. SolidWorks version 2020 was used to design the nozzle and ANSYS was used for the simulation to represent the Mach number, velocity, and pressure data. The authors found that the simulated k-epsilon turbulence model was the closest turbulence model with results and inferences close to the experimental results. The authors observed that higher turbulence degrees are detected for contour nozzles compared to respective conically shaped nozzles having the same end geometry. It was also concluded that higher divergence angles create impingement for CD nozzles and conical nozzles [5].

Joshi et al. (2020) carried out research on design optimization of Mach speed, temperature, and pressure for conical CD nozzle and bell nozzle. It was concluded that the profile of the bell-shaped nozzle had better pressure and velocity profiles than the profile of the conical nozzle [6].

Nayeem et al. (2020) carried out a numerical analysis of the CD rocket nozzle to determine the optimal convergence and divergence angle using the software ANSYS 19.0 (R3). For carrying out the study the authors considered different values of the angle of convergence (35°, 40°, 45°) and the angle of divergence (15°, 20°, 25°) and plotted various contours such as velocity contours, temperature contours, pressure contours, and Mach number contours. These contours were used as steady-state outputs and respective parameters were evaluated using flexible software. The authors concluded that optimal values were achieved for the convergence angle of 40° and the divergence angle of 25°. For these angles, the optimal results obtained for different parameters were pressure = 1.1518 MPa, temperature = 1428 K, velocity = 1925 m/s, and Mach number = 2.3. [7].

Khan et al. (2021) studied the compressible flow in a converging-diverging nozzle using CFD and simulated various parameters such as inlet diameter of 25.9 mm, throat diameter of 7.7 mm, outlet diameter of 10 mm, convergence length of 25 mm, divergence length of 13.2 mm, extension length 60 mm, convergence angle 20°, divergence angle 5°, NPR 6, 7, 7.82 & 8.2, inlet pressure = 691036.5 Pa, inlet temperature = 300 K, outlet pressure = 0 Pa, and outlet temperature = 300 K. The output parameters which were evaluated through these studies were pressure flow, density flow, and temperature flow. The authors concluded that the flow in the nozzle must have a supersonic speed with a Mach number greater than 1. It was observed that shock occurs for nozzles with a Mach number of 2 and a nozzle pressure ratio (NPR) of 7 or less. It was observed that for Mach, M = 2, the NPR required for the correct measurement was 7.82. The authors concluded that when the NPR was greater than 7.82, the flow from the nozzle does not expand sufficiently whereas when the NPR was less than 7.72, the flow from the nozzle was excessively expanded [8].

Tolentino & Mirez (2022) carried out a flow field study for off-design conical nozzles with non-circular cylindrical throat sections. The authors simulated the flow field with the RANS model in ANSYS-Fluent R16.2 code for 2D domains. All the four conservation equations viz. conservation of mass, momentum, energy, and state were used as governing equations. For simulating the over-expanded flow turbulence in the nozzle, Sutherland's viscosity equation and the Spalart-Allmaras turbulence model was applied. The authors observed fluctuations in pressure and Mach number with the increase in throat length. The authors concluded that the flow accelerates without any internal shock in the throat section for the length ranging from 5 to 15% of the throat diameter [9].

In the light of the detailed literature review, the gaps were identified and objectives of current research were defined. It was observed from the literature review conducted that most of the research was carried out on CD nozzle by varying the convergent angle, divergent angle of nozzle, throat diameter, length-to-diameter ratio, expansion ratio, micro-jets, nozzle pressure ratio (NPR), and shockwave with an aim to determine the Mach number, pressure contour, density contour, and velocity contour of CD nozzle with numerical results. It was also concluded from the literature that the smaller size of the mesh was important for optimal results. It was also observed that very little work has been carried out to analyse the effect of the throat diameter on the thrust force for the rocket nozzle through simulations followed by optimization. Thus, the main focus of this research is cantered on analysing the performance of a CD rocket nozzle by determining the effect of throat diameter on the thrust developed.

5.2.1 MODELLING OF CD NOZZLE IN CATIA V5

After defining the objective of the research, the next step is modelling the rocket nozzle and selection of the parameters for this model. Since the conical CD nozzle is symmetric, the decision was obvious to carry out simulations for 2D domains with axial symmetry on the x-axis. The 2D nozzle model is designed in CATIA V5 software in the part modelling module and GSD (generative-shaped design) tool. The dimensions of the nozzle model for the analysis are deduced by employing method of characteristics [5]. The dimensions so obtained are listed in Table 5.1. For the current

TABLE 5.1

Nozzle Dimensions

Inlet Diameter (mm)	**1,000 mm**
Convergent Angle (degree)	30°
Throat Diameter (mm)	300 mm to 410 mm (with 10 mm steps)
Divergent Angle (degree)	18°
Exit Diameter or Divergent Exit Diameter (mm)	4 times the throat diameter

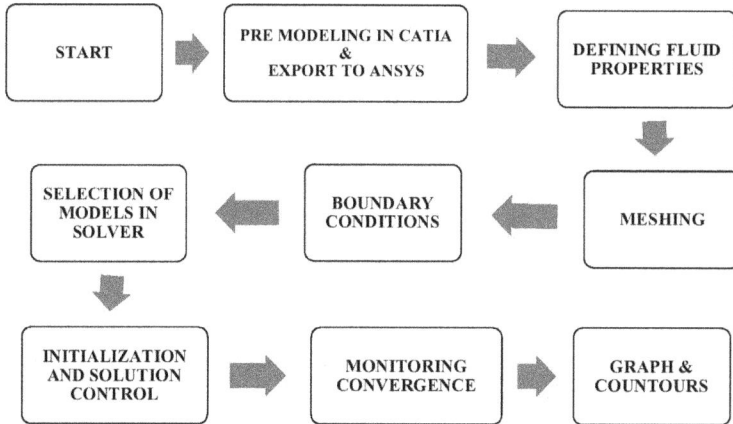

FIGURE 5.1   Flowchart outlining the simulation methodology.

study, the divergent angle is fixed at 18°, and the exit diameter is determined considering expansion ratio (diameter based) to be 4 as defined for Merlin 1D engine used
by Space X on their Falcon rocket [10]. In this analysis, the CD nozzle considered
variation of Mach number from 4.20 to 4.25.

5.2.2   Modelling Procedure on ANSYS 21.0

One area of fluid mechanics called computational fluid dynamics (CFD) makes use
of numerical techniques and algorithms to solve and examine issues involving fluid
flows. The countless computations needed to simulate how fluids and gases interact with the intricate surfaces utilized in engineering are carried out by computers.
Writing down the CFD codes is how commercial software like ANSYS works with
CFD. The numerical algorithms that can be used to solve fluid issues form the basis
of CFD codes. Figure 5.1 shows the flowchart presenting full detail of the methodology to be followed while carrying out flow simulation in ANSYS software.

In order to provide easy access to its solving capabilities, CFD codes in ANSYS
cover three main elements as summarized here.

5.2.2.1   Pre-processing

A flow problem is input into the pre-processor using an interface that is user-friendly,
and this input is then transformed into a format that the solver can use. The following
activities are carried out in the pre-processing stage:

1. Pre-modelling in CATIA and Export to ANSYS
   For the simulation in ANSYS, a .stp extension file is required. Extension
   with .stp made in the CATIA. The Flow Fluent software is dragged into the
   ANSYS workspace, and the Nozzle Design is exported to ANSYS Fluent
   for analysis.

TABLE 5.2

Fluid Properties and Boundary Conditions

Material	Fluid: Air
	Density: Ideal gas
	Viscosity: Sutherland
Boundary condition	Inlet pressure: 5,150,000 Pa
	Inlet temperature: 3,600 K
	Outlet pressure: 23,842.30 Pa
	Outlet temperature: 300 K

2. Defining Fluid Properties, Boundary Conditions, and Meshing

The fluid material and boundary conditions (Prosun Roy et al.) are then defined in Table 5.2.

In the meshing section, the nozzle is defined according to a condition like where is input and where is output means which part of the nozzle attaches to the combustion chamber and which part of the nozzle attaches to the atmosphere. The size of the convergent section for all the designs is fixed due to the same inlet diameter and equal convergent angle, but the divergent section is different for each design due to the change in the throat diameter that is mentioned in Table 5.1. Meshing elements are to be sized by the "sizing method," and each element size is taken as 5 mm for all designs. The face meshing tool is also used for structuring the meshing. Total centre axis length of the rocket nozzle is taken between 1,100 mm and 1,700 mm, and the meshing size is taken as 5 mm for all simulations.

5.2.2.2   Solver

1. Selection of Models in Solver

For each analysis, the solver is selected as the general solver. The type of general solver can be density-based or pressure-based. For current analysis density-based solver is used because flue gases coming out of nozzle behave like a compressible fluid.

The standard kinetic energy (k) and its dissipation rate (ε); that is, the k-epsilon model is a semi-empirical model based on model transport equations for the turbulence. This model is best suited for flow away from the wall like that of a free surface region. On the other hand, the k-omega model is best suited when the flow occurs near the wall where adverse pressure gradient is developed. SST is combination of k-omega and k-epsilon turbulence model. For the best result, SST (shear stress transport model is selected as it contains combined benefits of both k-omega and k-epsilon turbulence model.

2. CFD Analysis Initialization and Solution Control

For the current study, hybrid initialization and solution monitoring control is chosen in which the solution after CFD calculations is constantly monitored and checked

TABLE 5.3
Simulation Setup in ANSYS

General Solver	Type: Density-based
	Velocity: Absolute
	Time: Steady 2D
	Space: planar
Model	Energy: On
	Viscous: SST k-ω turbulence model
Reference Values	Compute from: Inlet
	Reference zone: Compute from the inlet
Initialization	Hybrid initialization
Solution	Run calculations

for convergence. In case of non-convergence, the mesh or solution parameters are changed till converged solution is attained. CFD setups were initiated with 2,000 numbers of iterations and iterations were increased till the solution was converged and the target convergence level was observed to be 10^{-6}.

All the important selections to be defined in the solver are listed in Table 5.3.

3. Post-processing
a. Examine the Flow Contours

During the post-processing, the data which is required as output after solution, like pressure, Mach number, velocity, density, thermal conductivity, and so on are defined. Per the meshing/solver selections and defined boundary conditions, tests were carried out on the ANSYS Fluent, and the contour profiles were generated for each analysis. For the current study, a total of 12 tests were performed and four contours for each of the tests were generated. These four contour profiles include pressure, velocity, density, and Mach number contour profiles.

The throat diameters for the current study are 300, 310, 320, 330, 340, 350, 360, 370, 380, 390, 400, and 410 mm. The ranges of these parameters are selected by pilot run performed during the thesis work. These throat diameter values are for Mach number 4.25, which is usually encountered in the CD rocket nozzle.

5.3 RESULT AND ANALYSIS

Figure 5.2 and Figure 5.3 depict the pressure and velocity contours respectively of the CD nozzle for a divergence angle of 18° and a throat diameter of 300 mm. It is observed from the contours that pressure starts decreasing from the inlet towards the outlet of the nozzle from 5.15×10^6 Pa to 1.23×10^4 Pa. It is also observed that velocity increases from the inlet towards the outlet of the nozzle from 0 m/s to 2,440 m/s continuously. If the flow is supersonic or the Mach number is greater than 1, the diverging section work like a convergent nozzle means

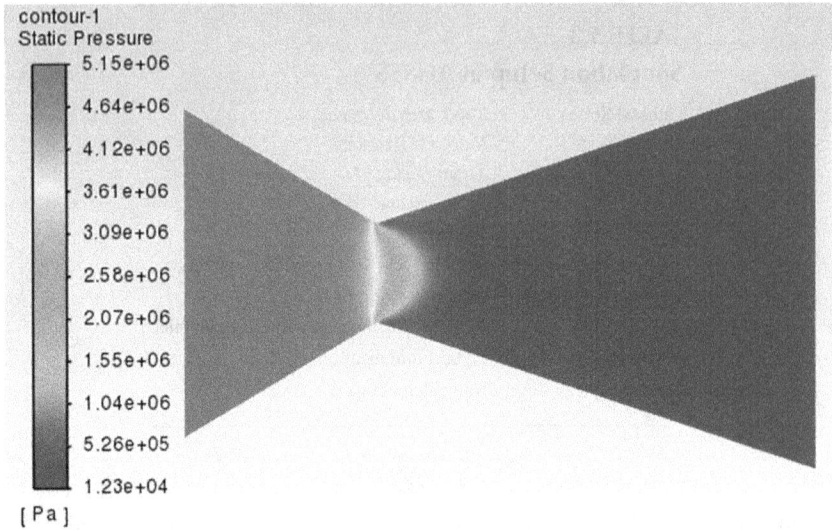

FIGURE 5.2 Pressure contour.

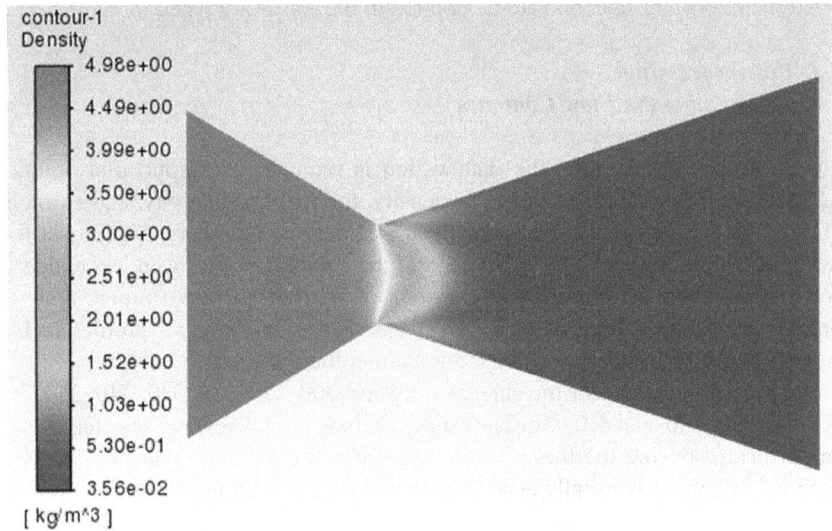

FIGURE 5.3 Density contour.

that velocity increases with increasing the diameter and the pressure reduces with increasing the diameter. In another case, if the flow is subsonic or the Mach number is less than 1 then the diverging section works like a diffuser means that velocity decreases with increasing the diameter of the diffuser. Figures 5.4 and Figure 5.5 show the density and Mach number contours of the CD nozzle for a divergence angle of 18° and length of 300 mm. It is observed that density starts

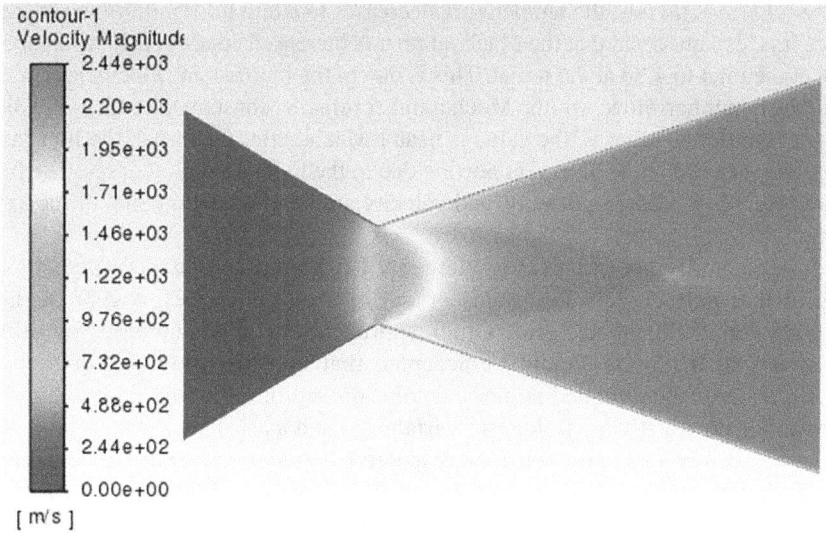

FIGURE 5.4   Velocity contour.

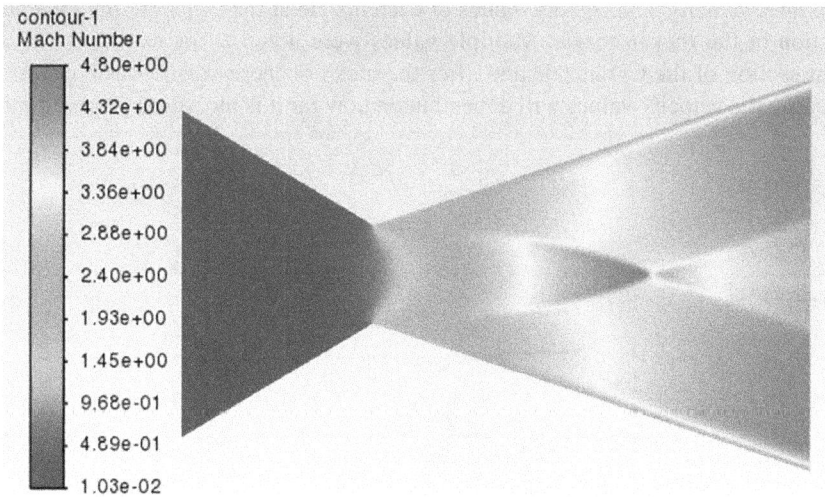

FIGURE 5.5   Mach number contour.

decreasing from 4.98 kg/m³ at the inlet to 3.56×10^{-2} kg/m³ at the outlet of the nozzle.

The temperature at the convergent section is more as compared to the outlet temperature because at the inlet the temperature is higher due to the presence of hot gases in the combustion chamber or reservoir and at the outlet, the temperature is lower (171.5 K). Thus, the temperature decreases to attain the equilibrium. In addition, it is also observed that the Mach number is increased continuously from 1.03×10^1 at the inlet to 4.80 at the outlet. This is due to the increase in velocity and hence the Mach number. However, the Mach number remains constant at a value of 9.89×10^{-1} at the throat—that is, the value is near 1 Mach. After the throat, the flue gases suddenly expand in the divergent portion due to the high outflow of gases. The flow parameters like pressure, density, and velocity are shown as a function of the axial length.

It is generally not practical to view raw CFD simulation data, especially on a grid that may contain thousands or millions of grid nodes, except perhaps a fairly small "mapped" grid. Other approaches to alphanumeric reporting; however, such reports should be accepted that may be useful for qualitative verification of the obtained numerical solution and/or extraction of quantitative results for post-analysis. Important variables, such as surface flows, forces, and integrals, can be calculated on each respective boundary spanning the computational domain [11].

Pressure, velocity, and density values were collected at the extreme edge of the divergent section. Figure 5.6 to Figure 5.9 show the distribution of pressure, density, velocity, and Mach number, respectively, along the x-axis. After carrying out all simulations, data was collected from ANSYS results in the form of the pressure, velocity, and density values of each nozzle at the exit from the divergent section of the rocket nozzle. Multiple values were noted at the exit of the divergent section of the CD nozzle and after the mean of these values was considered because the velocity values will depend upon how far it is measured from the wall.

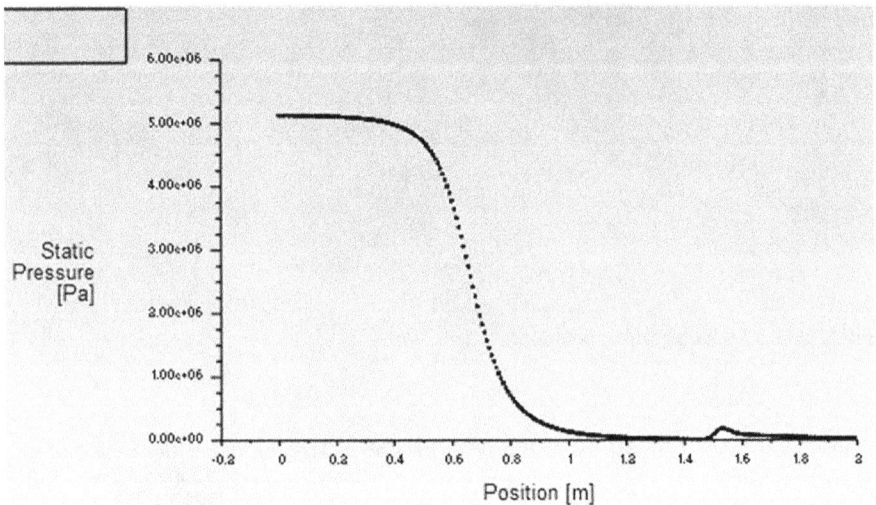

FIGURE 5.6 Pressure distribution along the x-axis.

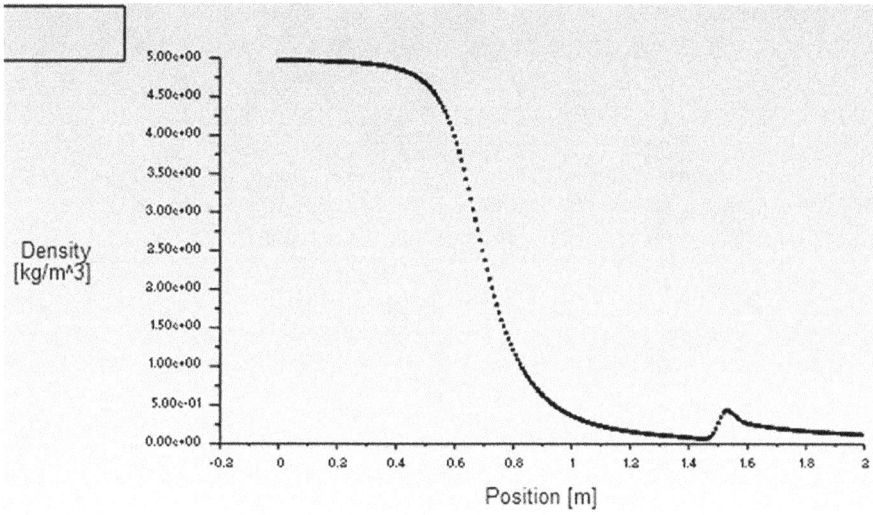

FIGURE 5.7 Density distribution along the x-axis

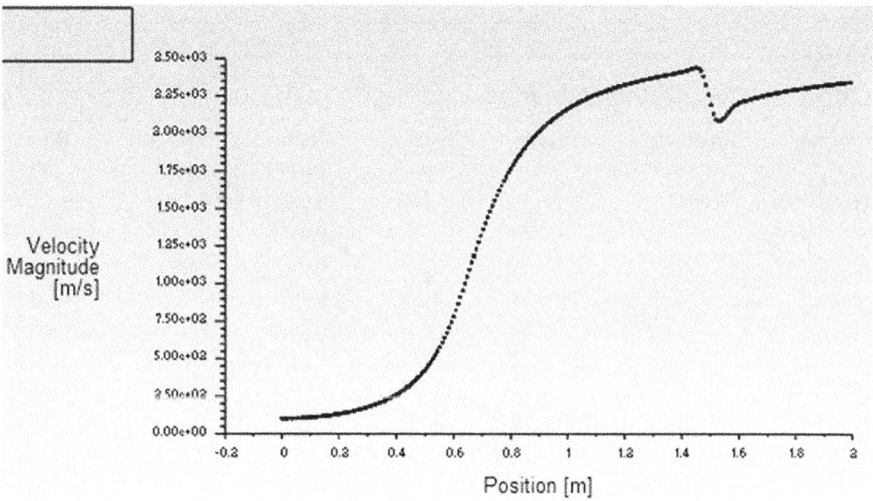

FIGURE 5.8 Velocity distribution along the x-axis.

Velocity near the axis was expected to be maximum, and near the wall, it must be minimum; hence, the mean of these values was considered in the calculation of thrust. Table 5.4 shows the calculations and observations thus obtained from ANSYS. From Table 5.4, it is observed that the thrust force increases with the increase in the throat diameter.

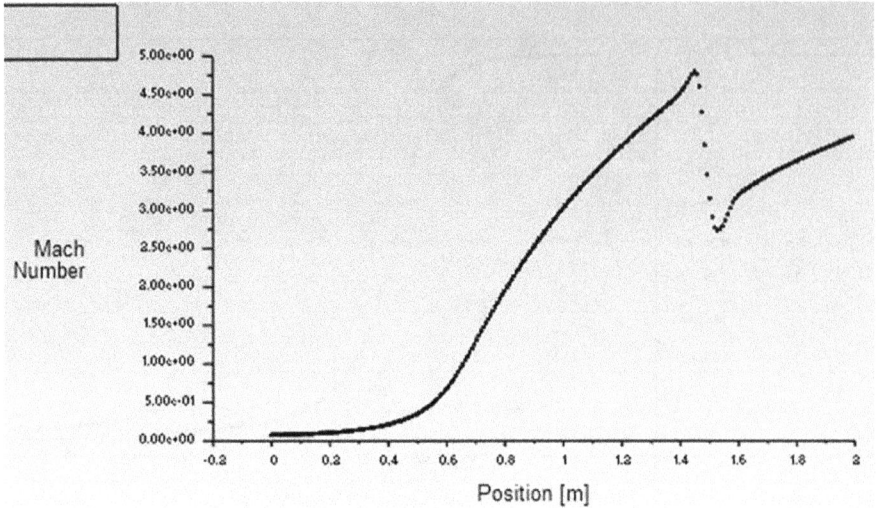

FIGURE 5.9   Mach number distribution along the x-axis.

TABLE 5.4
Calculation and Observation from ANSYS

Exp. No.	Throat Dia. (mm)	Pressure (Pa)	Velocity (m/s)	Density (kg/m³)	Exit Area (m²)	Thrust (N)
1	300	22,496.29	2,346	0.098	0.07068	38,027.12
2	310	22,482.64	2,347.25	0.0985	0.07547	40,854.49
3	320	22,485.45	2,347.99	0.098	0.08042	43,340.17
4	330	22,448.87	2,349.21	0.0987	0.08552	46,463.95
5	340	22,421.68	2,350.01	0.0986	0.09079	49,308.28
6	350	22,413.29	2,350.84	0.0985	0.09621	52,234.93
7	360	22,385.29	2,351.7	0.0984	0.10178	55,240.43
8	370	22,398.39	2,352.4	0.0985	0.10752	58,451.53
9	380	22,389.52	2,352.91	0.098	0.11341	61,365.4
10	390	23,390.92	2,353.68	0.0985	0.11945	65,126.51
11	400	22,395.26	2,354.13	0.0985	0.12566	68,413.43
12	410	23,362.6	2,355.03	0.0984	0.132	71,974.72

5.4   CONCLUSION AND FUTURE SCOPE

A nozzle is a device that controls the flow of fluid; for example, in the case of a rocket engine, the flow of a fluid-like flue gas is controlled. The CD nozzle is most commonly used to increase the velocity of the flue gases in rocket nozzles. In the current study, the design and analysis of a CD nozzle are carried out using CATIA V5 and ANSYS Fluent to conduct the flow field analysis. From this analysis, various flow properties like pressure, velocity, density, and Mach number are determined. Thus, it is concluded from the simulation result that the thrust force increases with an increase in the throat diameter because of the increase in mass flow rate. This increase in thrust force varies from 38,027.12 N to 71,974.72 N when the throat diameter is increased from 300 mm to 410 mm.

Based on the current CFD study, the following future studies may be carried out:

- Optimization techniques can be used for optimizing single- or multiple-response parameters such as Mach number, thrust force, drag force, lift force, and so on.
- Design optimization may be selected for further study by changing the various design parameters like minimum and maximum length for the converging section so as to ensure that flue gases have sufficient time to generate the flow. Similarly, the minimum and maximum lengths for the diverging section may be varied, and optimal results are obtained for proper expansion.
- Different pressure ranges may also be selected for future work. CFD studies may be carried out by changing the internal or chamber pressure and temperature so as to maximize the thrust force.
- Practical or experimental work may be also used to validate the numerical simulations carried in current study.

REFERENCES

[1] Narayana, K. P. S. S., & Reddy, K. S. (2016). Simulation of Convergent-Divergent Rocket-Nozzle Using CFD-Analysis. *IOSR Journal of Mechanical and Civil Engineering*, 13(4), 58–65. https://doi.org/10.9790/1684-1304015865

[2] Roy, P., Mondal, A., & Barai, B. (2016). CFD Analysis of Rocket Engine Nozzle. *International Journal of Advanced Engineering Research and Science (IJAERS)*, 3(1). www.ijaers.com

[3] Singh, P. K., & Tripathi, A. (2017). CFD Analysis of De-Laval Nozzle. *Ijariie*, 3(2), 4390–4411. www.ijariie.com

[4] Sushma, L., Udaya Deepika, A., Kumar Sunnam, S., & Madhavi, M. (2017). CFD Investigation for Different Nozzle Jets. *Materials Today: Proceedings*, 4(8), 9087–9094. https://doi.org/10.1016/j.matpr.2017.07.263

[5] Meena, L., Niranjan, M. S., Aman, Gautam, Gagandeep, Kumar, G., & Zunaid, M. (2021). Numerical Study of Convergent-Divergent Nozzle at Different Throat Diameters and Divergence Angles. *Materials Today: Proceedings*, 46, 10676–10680. https://doi.org/10.1016/j.matpr.2021.01.432

[6] Joshi, P., Gandhi, T., & Parveen, S. (2020). Critical Designing and Flow Analysis of Various Nozzles Using CFD Analysis. *International Journal of Engineering and Technical Research*, 9(2), 421–424. https://doi.org/10.17577/ijertv9is020208

[7] Nayeem, S., Chaitanya, K. L. V. B. S. S. S., Murali, G., Dileep, C., & Ramana, D. V. (2020). Optimization of Convergent–Divergent Taper Angle with Combustion Chamber of Rocket Engine through Numerical Analysis. *International Journal of Innovative Technology and Exploring Engineering*, 9(6), 76–81. https://doi.org/10.35940/ijitee.f3509.049620

[8] Khan, S. A., Ibrahim, O. M., & Aabid, A. (2021). CFD Analysis of Compressible Flows in a Convergent-Divergent Nozzle. *Materials Today: Proceedings*, 46, 2835–2842. https://doi.org/10.1016/j.matpr.2021.03.074

[9] Tolentino, S. L., & Mirez, J. (2022). Throat Length Effect on the Flow Patterns in Off-Design Conical Nozzles. *FME Transactions*, 50(2), 271–282. https://doi.org/10.5937/fme2201271T

[10] *Little Bit More Detailed Analysis of the Space X's Merlin 1D Engine*. Retrieved November 13, 2022, from Space X's Merlin 1D engine analysis: www.studocu.com/en-us/document/berkeley-college/introduction-to-regression-and-analysis-of-variance/space-xs-merlin-1d-engine-analysis/5225116

[11] Birtcher, K., & Dynamics, C. F. (2018). *Computational Fluid Dynamics*. US [Online]. https://doi.org/10.1016/B978-0-08-101127-0.00001-5

6 A Study on the Feasibility of a Hybrid District Heating System in the Indian North Region

Yuvraj Singh Pathania, Rajesh Kumar

6.1 INTRODUCTION

The northern Himalayas have an extremely frigid winter environment. As a result, to counteract the frigid temperatures, we would need consistent energy sources that can utilized both during the day and night. This hybrid district heating (HDH) system is advantageous, particularly in the northern Himalayas. District heating (also known as heat or thermal networks), or DH, is a method that uses insulated pipes to deliver heat supplied to a centrally located heating system. It is used for various commercial and residential purposes due to heating and cooling demands. Therefore, heat is generated for DH using multiple sources, such as biomass, heat boilers, etc. This district heating system (DHS) will operate more efficiently and contain emissions than conventional boilers [1]. The DHS has been trialed in the late 19th century, approximately 150 years ago. The first financially successful gadget in the world was created in Lockport, New York. The Lockport project, created by Birdsill Holly in 1877, provided a small number of nearby residences and other users with steam from a central boiler facility. By 1890, DHSs were swiftly erected in several upstate New York's small towns and communities [2].

The DHS was expanded to certain other smaller cities around the US, particularly mostly in country's industrialized northern states. The DHS is already a reality in significant cities, including Chicago, Pittsburgh, and Baltimore. Most early systems used steam delivered by reciprocating engines, primarily employed to produce electricity. In the early designs, steam was created using the heat lost during the producing process. Long into the 20th century, electrical turbines remained the main source of power for DH. A few enquiries revealed that DH with combined heat and power (CHPDH) has one of the lowest carbon footprints of all fossil-era facilities and is the most affordable way to reduce carbon outflows [3]. Its goal was to attain a solar fraction of more than 90%. Working papers include essential operational information for the machine and implementation and a description and computation of success measures over ten years of operation [4].

DOI: 10.1201/9781003367161-6

As global warming accelerates, environmental destruction is getting worse day by day. This is due to increasing pollution worldwide caused by deforestation, fossil fuel use, and other factors. You can reduce global warming by using renewable energy.

As a result, this work demonstrates the renewable energy potential of DHSs. In this DHS, steam or water is used to move heat through insulated pipes. This study demonstrates that northern India has a large number of hybrid resources for DHSs, which is crucial during the winter.

6.2 NEED AND AVAILABILITY

Climate change, pollution, and energy insecurity are the most pressing issues of our day, and we must employ renewable energy sources to address them. This affects how we design buildings and infrastructure. Because these energies are abundant in our solar and environmental systems and non-polluting to our atmosphere, we may employ them in DHSs.

6.2.1 SOLAR-BASED ENERGY TECHNOLOGIES

When Garcia Saez et al. [5] looked at the viability of integrating solar panels with a small-scale organic Rankine cycle system, they came to the conclusion that it was. The findings justify the deployment of these systems on a small scale for residential applications, especially in cold climates. With a payback period of 3.1 years, the internal rate of return is better than 15%.

In Lagos, Nigeria, Ugulu et al. [6] look at household willingness to pay for and take part in off-grid solar PV adoption as well as the long-term expansion of new communities. The data indicate that families, regardless of tenancy type, are interested in PV-generated energy. They also determined that there was a general desire to pay, that it was more robust in the presence of government backing, and that many families would participate in the feed-in-tariff system if it were made available and given the opportunity.

6.2.2 BIOMASS-BASED ENERGY TECHNOLOGIES

Waste management systems' key aims are Energy and material recovery and residue disposal. On the other hand, the most acceptable waste management methods are tied to resource recovery, economic needs, or rubbish destruction capabilities and the quest for an environmental regulatory framework in the region of concern. Consequently, the best waste management system was chosen, which satisfies all of the requirements for effective operation [7]. Various waste transformation methods make use of the three most widely available technologies. Figure 6.1 tells us how municipal solid waste is converted into valuable items using various waste-to-energy systems. Thermal conversion, biological conversion, and landfilling with gas recovery are examples of these [8].

6.2.3 WIND-BASED ENERGY TECHNOLOGIES

Siura et al. [9] begins with the observation that wind turbine power curves under laboratory conditions perform unreasonably well, leading to overly optimistic profitability

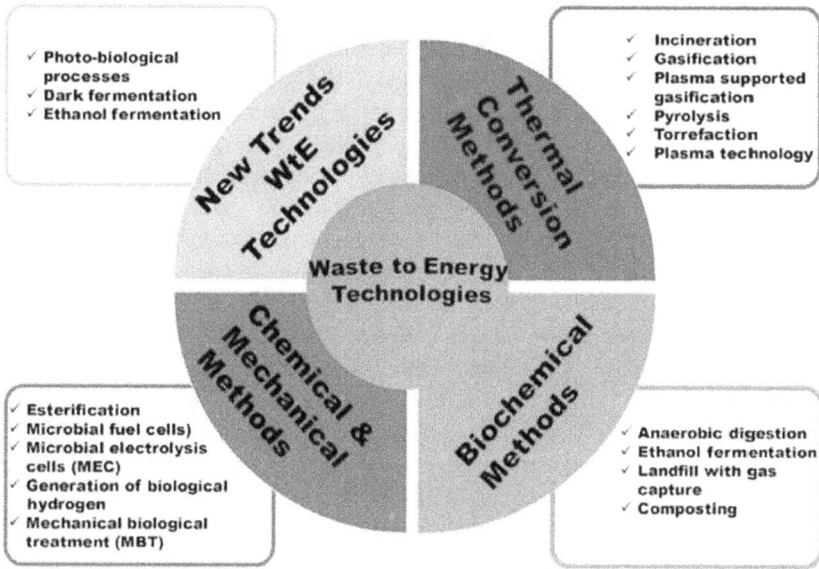

New Trends WtE Technologies
- ✓ Photo-biological processes
- ✓ Dark fermentation
- ✓ Ethanol fermentation

Thermal Conversion Methods
- ✓ Incineration
- ✓ Gasification
- ✓ Plasma supported gasification
- ✓ Pyrolysis
- ✓ Torrefaction
- ✓ Plasma technology

Chemical & Mechanical Methods
- ✓ Esterification
- ✓ Microbial fuel cells)
- ✓ Microbial electrolysis cells (MEC)
- ✓ Generation of biological hydrogen
- ✓ Mechanical biological treatment (MBT)

Biochemical Methods
- ✓ Anaerobic digestion
- ✓ Ethanol fermentation
- ✓ Landfill with gas capture
- ✓ Composting

Waste to Energy Technologies

FIGURE 6.1 Municipal solid waste is converted into valuable items using various waste-to-energy systems [8].

forecasts for installations at specific sites. Instead, he is creating a very accurate wind turbine mode based on artificial neural network (ANN) that, when compared to his SCADA readings of the actual wind turbine, perfectly forecasts the power curve of a wind turbine.

A low-carbon transition to a sustainable energy system was the focus of Miguel and colleagues' [10] investigation of wind energy planning. According to them, the majority of European nations are going through a sizable transition away from carbonized electric energy sources and toward renewable and sustainable options to fulfil rising demand. For example, renewable energy will provide 27% of Spain's electricity needs by 2030. Due to its availability, efficiency, and availability, wind energy is predicted to be one of the sources of power generation in this sustainable transition both high performance and affordable. The study focused on the value of existing wind farms and the efficiency with which existing wind sources were used. Additionally, they compared total wind farm depletion to existing wind farm recharge using the most recent technology using a novel judgement method that relics on cost-benefit analysis. Re-energizing is a highly alluring substitute as a result. The long-term sustainability of renewable sources of energy may benefit both the environment and the economy when properly implemented.

6.2.4 Technology-Based on Geothermal Energy

Yildirim et al. [11] investigation of the potential for geothermal wells to produce heat. Although temperatures and reservoirs may be favourable for removing heat for the purpose of generating electricity, local factors may make it impossible by allowing for insufficient flow rates. Utilizing a heat exchanger and another loop system to remove the heat is one approach to get around this. The review highlights the potential of this strategy and the $46/MWh prospective generating cost.

In order to determine whether a city could be totally powered by renewable energy, In Frederikshavn, North Denmark, Ostergaard and Lund [12] look into a low-temperature geothermal renewable energy system for heating. Numerous renewable energy sources are exploitable in the region, including waste, low-temperature geothermal energy potential, and offshore wind power. An energy system is constructed and studied. The findings show that the absorption heat pump and geothermal energy combination is advantageous for this city's sustainable energy system.

Saeidi and Noorolahi [13] studied mathematical models for earth-heating systems for cooling and heating in an original piece of work. This kind was selected because of its large diameter, shallow depth, and prominent metal spiral fins. The results revealed that its blades securely grasped the pipe from inside of the ground, enhancing rate of heat transfer by up to 31% by maximizing area of contact and heat capacity. This might be a crucial and efficient way to cut the price of installing a geothermal heat pump.

The cascading application of geothermal resources for long-term power generation in northwest Iran is taken into consideration by Noorolahi et al. [14]. Four geothermal power consumption scenarios were created and modelled for this study's analysis and comparison of the effects on energy sustainability. The double flash cycle can supply 27.5 MW of power and 76.1 MW of heat straight from wells drilled for geothermal applications, according to efficiency studies. Additionally, compared to employing fossil fuel-based power units with heating systems, the ecological consequences of producing both heat and electricity in the development of the proposed results in a decrease of 696,200 tonnes of CO_2 emissions.

6.3 APPLICATION AND ADVANCEMENTS IN ENERGY

Knowing that resources for energy are decreasing every day, hybrid systems could be used in DHSs such as large cities and shopping malls. The water or steam heated by this combination of renewable system is used to warm the building. In 2020, COVID-19 will spread rapidly around the world, major effects on the energy industry, as well as the global economic and health sectors [15]. In nations without fossil fuels, the utilization of renewable energy sources is becoming more and more crucial [16]. Environmental safety and emission reduction are now on the rise around the world [17]. Most countries are promoting renewable energy day by day by using from fossil fuels.

Different types of furnaces and boilers have been documented by Vicente and Alves [18] for different types of biomass sources. Agalihani et al. [19] first proposed a plant-assisted biomass gasification bioremediation method to assess post-pollution impacts of poplar biomass. Soltero et al. [20] highlighted core issues such as heat loss, heat storage, and convergence of energy sources, which are very difficult and need to be focused on analysis. Electricity usage in the construction industry is limited and there are great opportunities to save energy. For a wide range of construction uses,

Mazzarella [21] researched conventional energy-saving practises. Internal vapour barriers were examined and evaluated by Ferrari and Riva [22] in order to help with envelope heating. In their study of various ventilation system settings, Blasquez et al. [23] established energy-saving benchmarks for thermal comfort, energy efficiency, sales, and so on [24] recommended changing the double and triple glass windows in British hotels with ones that may greatly lessen heat loss.

Various temporary sources of renewable energy, such as the sun and wind, have a significant effect on the supply fluctuations of power plants and grids, as they usually prevent their efficient use. As a result, energy storage devices are thought to be the most viable approach to closing the supply-demand imbalance. The application of direct heating via thermal energy storage was looked into by Haste et al. [25]. This allows us to generate significantly more green power and reduce fluctuation effects. Xu and Wang [26] created a thermal mass circuit including a charging subsystem as well as a conditioning subsystem. The machine's intensity glide may be increased at a constant flow rate degree.

6.3.1 Wind-Biomass

The massive implementation of renewable energy supplies is crucial in reducing CO_2 emissions associated with power production. A practical alternative to traditional generations of electricity primarily from fossil fuels can be provided by wind energy. However, the restrictive operating rules of energy markets are influenced by wind turbines since wind generators are inherently variable. Thus, they have significant problems producing exact production schedules on a day beforehand and meeting planned requirements of real-time service [27]. Remote regions heavily depend on whether the alternative is selected for importing fossil fuels. Given fossil fuel shortages, air emissions, and high transport prices, local renewable resources have drawn considerable interest worldwide for the availability of energy in rural areas [28]. Studies on green hybrid energy systems are analysed, covering different configurations, optimizing technology, and planning requirements in both grid-connected and independent modes [29]. Figure 6.2 suggests a hybrid biogas generator system and wind turbine array for district heating system.

6.3.2 Wind-Solar

This device is planned for power generation using solar panels and small wind turbines. Wind and solar power are compatible, making the electricity-generating operation almost yearly. Wind-solar hybrid system's key elements are the wind and turrets generators, photovoltaic solar panels, batteries, wires, charging controllers, and converters. In addition, the wind-solar hybrid system provides electricity for battery charging, and AC systems can be operated using inverters. The wind aero generator is mounted on a tower with a minimum height of 18 metres from the ground floor [30]. This converted energy from wind and solar is further used in DHSs for the heating purpose of the centralized system. Figure 6.3 shows the block diagram of the solar flares hybrid system, which is used to calculate the power production of the wind turbines and solar panels.

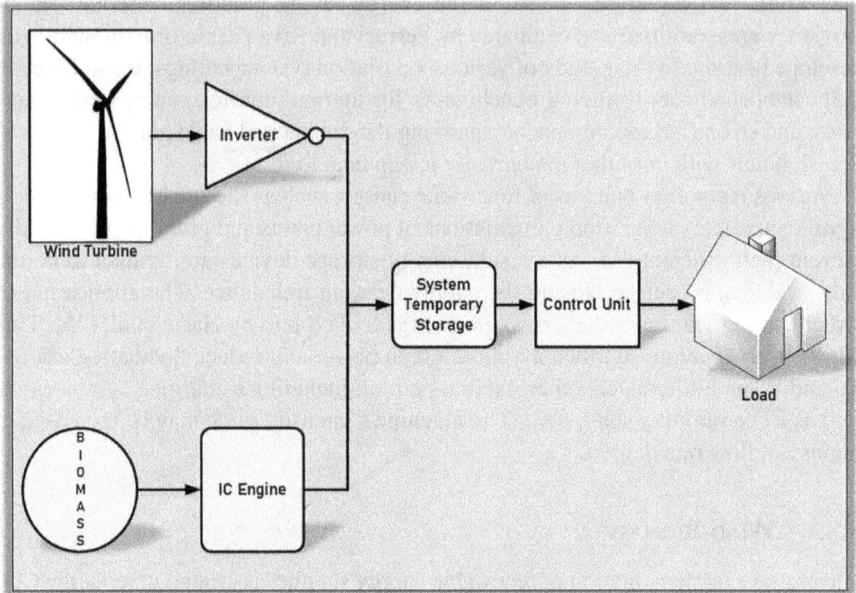

FIGURE 6.2　Suggested Hybrid Biogas Generator System and Wind Turbine Array.

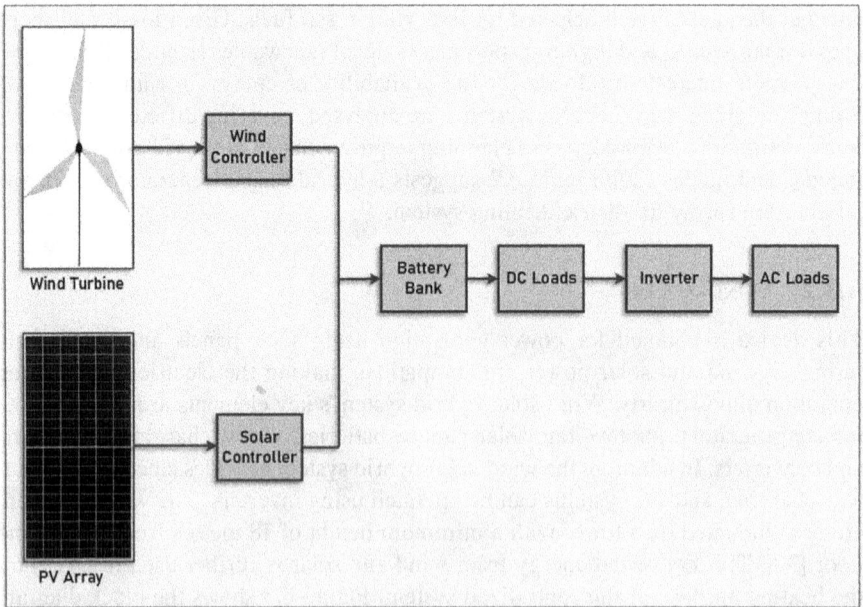

FIGURE 6.3　Solar-Wind Hybrid System For Power Output.

According to many authorities on renewables, there are many advantages over a single system from a limited "hybrid" electricity system that incorporates household wind and solar electricity (photovoltaic). In most of the United States, wind rates are vital in the summer as the sun shines best and longest. When less sunshine is available, the wind is high in winter. Because wind and solar systems have peak operation times at various times of the day, hybrid systems are more likely to generate electricity as needed. Many hybrid systems are standalone, "off-grid" systems unrelated to a power delivery system. Most hybrid systems provide power by batteries and/or a generator powered by traditional fuels such as diesel when wind and solar systems are not generated. The motor generator will control and recharge the battery when the batteries are down. The addition of a motor generator complicates the operation, but these devices can be operated automated by modern electronic controls. An engine-generator can also reduce the scale of the other machine elements. Be aware that the storage space must be sufficiently high during non-charge to provide electricity. Battery banks usually supply the electric load for one to three days [31].

6.4 VARIOUS BIOMASS KINDS FOR ENERGY PRODUCTION

More practical activities and technology will be unable to stop the rise in world energy use. For this reason developing new, more affordable energy sources, as well as alternative fuels, energy recovery, and ultimately sources of renewable electricity like biofuels, were crucial. Sunlight is used in the synthesis of the plant-based substance known as biomass. It comprises organic waste from human and animal sources, animal and plant garbage (like timber from forest areas and garbage from forestry and farming activities), and industrial effluents. As a result, there are more new installations using regional resources and other types of wastage. Manure is also increasingly being used to produce heat and electricity [32]. Bioenergy is a type of renewable power produced through biological conversion of bioenergy, or heat. Biomass may be produced from a variety of materials, such as wood from vegetation, forestry byproducts, straw, stover, sugarcane waste, and green agricultural waste, to name a few; Figure 6.4 [33] depicts the production waste from food production together with wastewater, MSW (urban solid waste), dark liquor from the paper sector, molasses, and rice husks as samples of farming, livestock, and industrial waste. Additionally, the biofuels utilized in the DH system to generate heat are where this energy comes from.

The possibility for using biomass as a green raw material to produce many types of energy is immense. However, in order to deal with fossil fuels, trustworthy conversion procedures must be applied. Not all of the solutions presented for a certain device part and power generation are effective or practical from an economic standpoint [34]. As a result, choosing suitable process combinations that result in optimum plant design is a crucial part of this research. Aside from deciding the optimal design and machinery sizing, determining the optimal production capability, duration, and the plant's bulk is crucial. An exchange between grouped large plants and dispersed tiny plants must be considered before installation selections are made. Under this

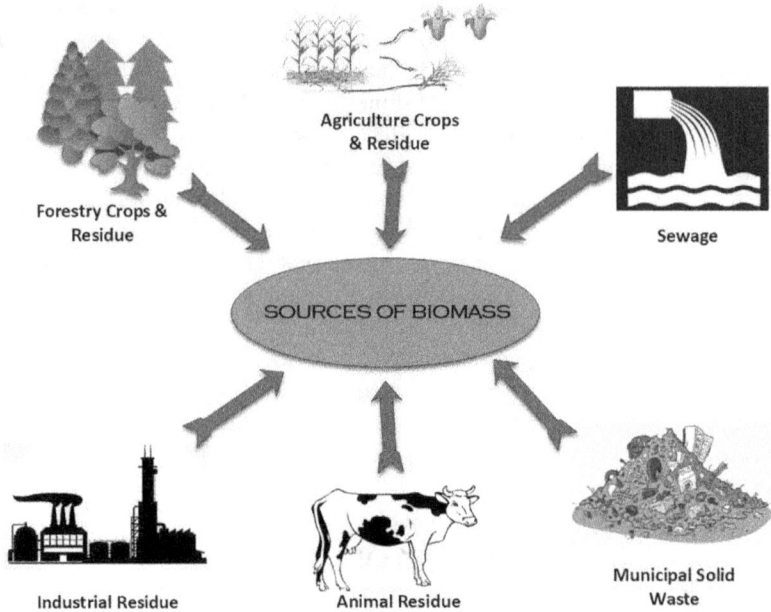

FIGURE 6.4   A Brief Look At Some Standard Biomass Tools [33].

regard, many research studies are now being conducted, taking technical, monetary, and environmental factors into consideration [35].

6.4.1 BIOCHEMICAL CONVERSION

Anaerobic digestion and fermentation are also biochemical (or biological) conversion mechanisms. AD uses bacteria to convert organic matter into gaseous materials, providing fair economies and a wide range of applications worldwide. In contrast, yeast turns the contained sugar into ethanol during feedstock fermentation. AD is the direct conversion to a biogas gas of raw material and is mostly methane and carbon dioxide, where other gases, such as hydrogen sulphide, are traced [36]. The material (also known as the substrate) is a lignocellulose-free organic substance (other than wood) that is metabolized by microorganisms when there is no air.

Its transformation method, similar to oxygen annealing in organic matter decomposition, provides economically interesting and stable compounds. Bacteria in anaerobic environments transform the biomass, releasing gases for a fuel energy of around 20–40% of the basic material's lower calorific value (NCV) [37]. An AD plant's two primary products are as follows:

• Biogas is a renewable energy source; 40% to 50% carbon dioxide, 50% to 60% methane, and quantities of all other polluting chemicals are present.

- The term "bio-fertilizer" refers to fertiliser manufactured from plants (dig estate). This pure, inert moisture contains valuable phytonutrients and organic humus. For usage on land or secondary refinement, it is definable split either liquid or solid components [38]. The methods for creating hybrid renewable energy systems, including biomass conversion, are summarized in Table 6.1.

6.4.2 THERMO-CHEMICAL CONVERSION

Bioenergy is generated from biomass and is a sustainable and clean energy source. As fossil fuel supplies are depleting and greenhouse gas emissions from fossil fuel consumption continue to rise, this issue is gaining much attention. As a result of photosynthesis, biomass refers to all organic components generated from green plants [50]. Bioenergy may be produced from biomass using one of two methods: thermochemical or biochemical/biological processes [51]. Pyrolysis, gasification, liquefaction, and combustion are also methods of thermo-chemically converting biomass. Since pyrolysis involves chemical reactions that generate solid, liquid, and gaseous molecules without oxygen, it is the first stage in all thermo-chemical processes. The superior ability to destroy the majority of organic compounds and the shorter latency make thermochemical operations generally more efficient than biochemical/

TABLE 6.1

Design Techniques for Biofuel Production in a Hybrid Power System

References	Methodology	Objectives of the Paper
Kahraman and Kaya [39]	Fuzzy AHP	To make an accurate prediction of future energy demand
Zhou et al. [40]	Modelling of superstructures based on MILP	To design an optimum distributed energy system for China
Nakata et al. [41]	NLP	To create a sustainable energy system with applications in rural Japan
Gupta et al. [42], Saini and Sharma [43]	LP	To design a renewable energy system for remote areas
Gupta et al. [44]	MILP	Mainly, to calculate a hybrid energy system's ideal cost
Gupta et al. [45]	Simulation	Mainly to figure out how much a hybrid energy system should cost with the help of calculation
Gupta et al. [46]	MILP simulation	To calculate the cost of a hybrid energy system using MILP simulation
San Cristóbal [47]	Combination of MCDM, VIKOR, and AHP	To select the best renewable sources for the energy project in Spain
Hakimi and Moghaddas-Tafreshi [48]	Simulation	To utilize a hybrid power system for district heating and then calculate the cost of the system in Iran's southeast
Rubio-Maya et al. [49]	LP	To create a two-step optimization technique for a hybrid energy poly generation unit

biological processes. For instance, it is typically believed that lignin molecules are non-fermentable and hence unable to be completely destroyed by biological processes, yet they are decomposable by thermo-chemical techniques [52].

6.4.3 PHYSICO-CHEMICAL CONVERSION

The physico-chemical waste conversion includes many techniques for improving the physical and chemical characteristics of solid waste. The fuel component of the trash is transformed into high-energy fuel pellets that may be utilized to generate steam. First, the garbage is dried up to reduce the excessive humidity. Before the waste is compacted and turned into fuel pellets, sand, grains, and other incombustible materials should be removed mechanically. There are numerous unique advantages to fuel pellets compared to coal and wood since they are cleaner and non-combustible, have reduced ash and humidity, are consistent in dimensions, and are economical and eco-friendly [53]. Lignocellulosic biomass refers primarily to dry plants' cellulose, hemicellulose, and lignin composition [54]. The available energy-driven lignocellulosic biomass feedstocks are mainly in agriculture, forestry, and industrial areas and its types and examples are listed in Table 6.2.

6.4.4 FLUIDIZED BED COMBUSTOR FOR DHS

Fluidization converts solid fuel particles into a fluid-like condition for combustion, which provides the benefits of high heat transfer rate, compact boiler design, fuel flexibility, low-grade fuel combustion, and decreased pollutants, such as SOx and NOx [55]. A fraction of biomass, which was previously defined as organic non-fossil material of biological origin, may be used to produce energy. Examples of biomass include

TABLE 6.2
Lignocellulosic Biomass Feedstocks Available for Energy Applications

Supply Sector	Type	Example
Agriculture	Plants that produce energy from lignocellulosic biomass	Plants that are herbaceous (e.g., switchgrass, miscanthus, reed)
	Crop residue	Crop of straw (e.g., rice straw, wheat straw, corn stalk, cotton stalk)
	Energy-producing plants include those for sugar, starch, and oil	Rapeseed, sugarcane, corn
Forest	Specific forestry	Plantations with rapid rotation
	Forestry byproducts	Logs from thinning, woodblocks, woodchips from heads and limbs, and woodchips from thinning
Industry	Agro-industrial lignocellulosic residues	Bagasse from sugarcane, corn cob, and rice husks
	Wood waste from industry	Wood waste from industry and sawdust from sawmills
Other	Lignocellulosic waste	Park and garden lignocellulosic residues (e.g., pruning, grass)

wood waste, agricultural waste, specialised energy crops, and industrial and municipal trash with a plant origin. Burning is the most common method of converting biomass into heat and power [56]. Recent years have seen the construction of numerous new biofuel energy plants, and the modelling process is still going on. Fluidized bed combustion (FBC), one of these methods, is necessary for utility-scale power plants using biomass. This is because fluidization technology has excellent fuel versatility, low pollutant levels, and high combustion efficiency [57]. Fluidized boilers are the most popular boiler for biomass fuel blazed within a heated bed of inert, usually sandy particles. The combustion air in the bed suspends the particle-fuel mix. As a result, the gas/solid blend exhibits liquid properties as speeds rise. In addition, the scrubbing effect of the bed material on the fuel often improves the combustion process by "removing the CO_2 and the solid residue (char) that usually forms around the fuel particles to allow oxygen to enter the fuel material more easily and improve combustion rates and performance." This method also improves heat transfer and allows low temperatures to work: bed temperatures vary between 1,400 and 1,600°F, much lower than 2,200°F for the boiler. The lower boiler temperatures also emit less nitrogen oxide, which is of environmental and regulatory benefit because wood and biomass fuels are high in nitrogen. Sulphur dioxide is usually negligible from wood waste and biomass, but if sulphur is a contaminant, the added calcareous material in the fluid base can neutralize it [58].

Fluidized bed combustion (FBC) is a well-established fire, electricity, and combined technology. This chapter's primary focus is on two fluidized bed boilers: bubble fluidized boilers (BFBs) and fluidized boilers (CFBs), which work under atmospheric conditions. The most important market share in installed capacity (MW) of CFBs is that large CFBs are large CFBs. In contrast, BFBs are found primarily in smaller CHP boilers in regional or industrial heating systems [59].

A fluidized bed gasifier and boiler can better process as typical for biomass materials with higher ash content. This conversion device is generally more suitable for large-scale operations, as shown in Figure 6.5. Gases at more incredible speed are

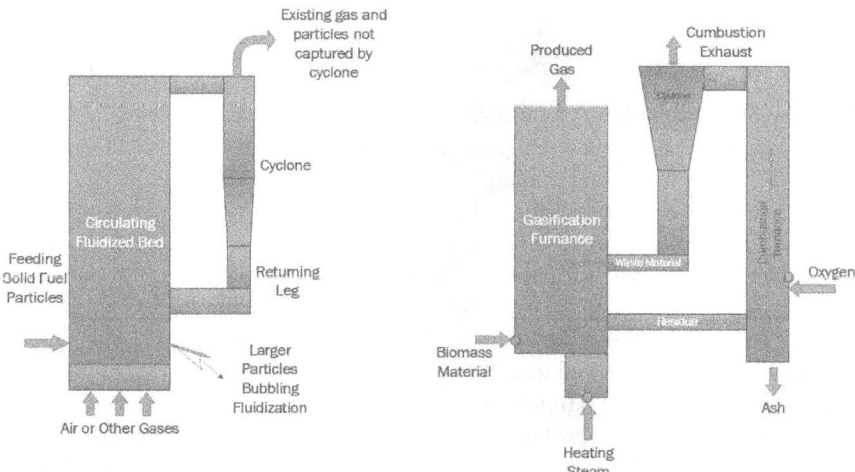

FIGURE 6.5 Fluidized Bed Conversion of Biomass and Waste [60].

necessary to suspend and elevate the particle bed in rotating fluidized beds (CFBs) due to the enormous kinetic energy of the fluid. As such, the bed surface is less smooth, and more contaminants can be drawn from the bed by nature than from stationary beds. Qualified particles are returned to the reactor bed by an external loop. The particles in a cyclone separator can be labelled and removed from or returned to the bed, depending on the phase, according the size of the particles cut [60].

6.5   UPCOMING DIFFICULTIES FOR HDH SYSTEM

The new problems that arise during modelling, operation, and design are the major emphasis in relation to modern DHS. We will look at the first two forms of modelling today, active consumer and consumer modelling, because it is nearly difficult to have every specific model for everyone in today's culture. Consider, for instance, a collection of apartments, each of which has a distinct consumption pattern. It is possible that some elements of a DHS are unknowable or poorly understood. The identification of parameters, modelling, and estimation are the next steps. The model and calculation used to create the DHS for the consumer will be impacted. The heating system is then enhanced for ultimate control in the future. Future optimization techniques will focus on creating stable operating circumstances that guarantee consumers receive heat while abiding by all operational restrictions already in place [61].

It has been discovered that the business model's client side has changed due to external issues. The most challenging problem, however, is the transition of vital resources. The current company model lives longer when external factors are considered, but this does not make it more competitive. Businesses have a window of time to convert their essential resources after prolonged life. DH will be a part of the new power infrastructure if the shift is effective. The destiny becomes less likely if the adjustment is not made [62]. Using biomass heaters with supplemental power sources or linkages to the pipelines through integrated heat and energy sources plants, several new DH projects are being created (CHP). Recent studies have concentrated on better load prediction and low-carbon technologies.

6.6   CONCLUSION

The adaptability of different hybrid systems for generating heat in winter is the focus of this chapter's investigation in regards to the possibilities of DHS in northern India. The development of renewable energy sources for heat in northern India will greatly benefit from the introduction of this hybrid system. In addition, HDH systems contribute to the development of circular economy systems, reduce carbon emissions, and increase energy self-sufficiency. This chapter's major goal is to develop solar and biomass energy sources for producing heat. By combining solar and biomass energy, the research shows how various DHS sources operate. A pleasant interior atmosphere may be produced thanks to this technique. This feasibility study of several energy sources in the Himalayas of northern India demonstrates how they may be used in combination to heat the interior space. This report examines heating during winter in a hybrid system in the upper regions of northern India. This research focuses on a solar-biomass hybrid system and how it can be used in the harsh winters of northern India. The deployment of this hybrid system would substantially aid the growth of renewable energy resources for heat in northern India.

REFERENCES

[1] D. A. Orchard William, "Carbon footprints of various sources of heat—biomass combustion and CHPDH comes out lowest," 2009. https://claverton-energy.com/carbon-footprints-of-various-sources-of-heat-chpdh-comes-out-lowest.html.

[2] A. R. Mazhar, S. Liu, and A. Shukla, "A state of art review on the district heating systems," *Renew. Sustain. Energy Rev.*, vol. 96, pp. 420–439, 2018, doi:10.1016/j. rser.2018.08.005.

[3] H. Averfalk and S. Werner, "Economic benefits of fourth generation district heating," *Energy*, vol. 193, p. 116727, 2020, doi:10.1016/j.energy.2019.116727.

[4] L. Mesquita, D. McClenahan, J. Thornton, J. Carriere, and B. Wong, "Drake Landing solar community: 10 years of operation," *ISES Sol. World Congr. 2017—IEA SHC Int. Conf. Sol. Heat. Cool. Build. Ind. 2017, Proc.*, pp. 333–344, 2017, doi:10.18086/swc.2017.06.09.

[5] I. Garcia-Saez, J. Méndez, C. Ortiz, D. Loncar, J. A. Becerra, and R. Chacartegui, "Energy and economic assessment of solar Organic Rankine Cycle for combined heat and power generation in residential applications," *Renew. Energy*, vol. 140, pp. 461–476, 2019, doi:10.1016/j.renene.2019.03.033.

[6] A. I. Ugulu and C. Aigbavboa, "Assessing urban households' willingness to pay for stand-alone solar photovoltaic systems: A case study of lagos, nigeria," *J. Sustain. Dev. Energy, Water Environ. Syst.*, vol. 7, no. 3, pp. 553–566, 2019, doi:10.13044/j.sdewes.d7.0274.

[7] G. Ali, S. Abbas, and F. Mueen Qamer, "How effectively low carbon society development models contribute to climate change mitigation and adaptation action plans in Asia," *Renew. Sustain. Energy Rev.*, vol. 26, pp. 632–638, 2013, doi:10.1016/j.rser.2013.05.042.

[8] K. A. Kalyani and K. K. Pandey, "Waste to energy status in India: A short review," *Renew. Sustain. Energy Rev.*, vol. 31, pp. 113–120, 2014, doi:10.1016/j.rser.2013.11.020.

[9] T. Rasheed et al., "Valorisation and emerging perspective of biomass based waste-to-energy technologies and their socio-environmental impact: A review," *J. Environ. Manage.*, vol. 287, no. March, p. 112257, 2021, doi:10.1016/j.jenvman.2021.112257.

[10] G. Ciulla, A. D'Amico, V. Di Dio, and V. Lo Brano, "Modelling and analysis of real-world wind turbine power curves: Assessing deviations from nominal curve by neural networks," *Renew. Energy*, vol. 140, pp. 477–492, 2019, doi:10.1016/j.renene.2019.03.075.

[11] M. De Simón-Martín, Á. De La Puente-Gil, D. Borge-Diez, T. Ciria-Garcés, and A. González-Martínez, "Wind energy planning for a sustainable transition to a decarbonized generation scenario based on the opportunity cost of the wind energy: Spanish Iberian Peninsula as case study," *Energy Procedia*, vol. 157, pp. 1144–1163, 2019, doi:10.1016/j.egypro.2018.11.282.

[12] N. Yildirim, S. Parmanto, and G. G. Akkurt, "Thermodynamic assessment of downhole heat exchangers for geothermal power generation," *Renew. Energy*, vol. 141, pp. 1080–1091, 2019, doi:10.1016/j.renene.2019.04.049.

[13] P. A. Østergaard and H. Lund, "A renewable energy system in Frederikshavn using low-temperature geothermal energy for district heating," *Appl. Energy*, vol. 88, no. 2, pp. 479–487, 2011, doi:10.1016/j.apenergy.2010.03.018.

[14] R. Saeidi, Y. Noorollahi, and V. Esfahanian, "Numerical simulation of a novel spiral type ground heat exchanger for enhancing heat transfer performance of geothermal heat pump," *Energy Convers. Manag.*, vol. 168, no. May, pp. 296–307, 2018, doi:10.1016/j. enconman.2018.05.015.

[15] Y. Noorollahi, H. Gholami Arjenaki, and R. Ghasempour, "Thermo-economic modeling and GIS-based spatial data analysis of ground source heat pump systems for regional shallow geothermal mapping," *Renew. Sustain. Energy Rev.*, vol. 72, no. December 2016, pp. 648–660, 2017, doi:10.1016/j.rser.2017.01.099.

[16] E. Ghiani, M. Galici, M. Mureddu, and F. Pilo, "Impact on electricity consumption and market pricing of energy and ancillary services during pandemic of COVID-19 in Italy," *Energies*, vol. 13, no. 13, 2020, doi:10.3390/en13133357.

[17] H. Eroğlu, "Effects of Covid-19 outbreak on environment and renewable energy sector," *Environ. Dev. Sustain.*, vol. 23, no. 4, pp. 4782–4790, 2021, doi:10.1007/s10668-020-00837-4.

[18] S. Jin, "COVID-19, climate change, and renewable energy research: We are all in this together, and the time to act is now," *ACS Energy Lett.*, vol. 5, no. 5, pp. 1709–1711, 2020, doi:10.1021/acsenergylett.0c00910.

[19] E. D. Vicente and C. A. Alves, "An overview of particulate emissions from residential biomass combustion," *Atmos. Res.*, vol. 199, pp. 159–185, 2018, doi:10.1016/j.atmosres.2017.08.027.

[20] A. Aghaalikhani, E. Savuto, A. Di Carlo, and D. Borello, "Poplar from phytoremediation as a renewable energy source: Gasification properties and pollution analysis," *Energy Procedia*, vol. 142, pp. 924–931, 2017, doi:10.1016/j.egypro.2017.12.148.

[21] V. M. Soltero, R. Chacartegui, C. Ortiz, J. Lizana, and G. Quirosa, "Biomass District heating systems based on agriculture residues," *Appl. Sci.*, vol. 8, no. 4, 2018, doi:10.3390/app8040476.

[22] L. Mazzarella, "Energy retrofit of historic and existing buildings the legislative and regulatory point of view," *Energy Build.*, vol. 95, pp. 23–31, 2015, doi:10.1016/j.enbuild.2014.10.073.

[23] S. Ferrari and A. Riva, "Insulating a solid brick wall from inside: Heat and moisture transfer analysis of different options," *J. Archit. Eng.*, vol. 25, no. 1, p. 04018032, 2019, doi:10.1061/(asce)ae.1943-5568.0000334.

[24] T. Blázquez, S. Ferrari, R. Suárez, and J. J. Sendra, "Adaptive approach-based assessment of a heritage residential complex in southern Spain for improving comfort and energy efficiency through passive strategies: A study based on a monitored flat," *Energy*, vol. 181, pp. 504–520, 2019, doi:10.1016/j.energy.2019.05.160.

[25] R. Salem, A. Bahadori-Jahromi, A. Mylona, P. Godfrey, and D. Cook, "Investigating the potential impact of energy-efficient measures for retrofitting existing UK hotels to reach the nearly zero energy building (nZEB) standard," *Energy Effic.*, vol. 12, no. 6, pp. 1577–1594, 2019, doi:10.1007/s12053-019-09801-2.

[26] A. Hast, S. Rinne, S. Syri, and J. Kiviluoma, "The role of heat storages in facilitating the adaptation of district heating systems to large amount of variable renewable electricity," *Energy*, vol. 137, pp. 775–788, 2017, doi:10.1016/j.energy.2017.05.113.

[27] Z. Y. Xu and R. Z. Wang, "A sorption thermal storage system with large concentration glide," *Energy*, vol. 141, pp. 380–388, 2017, doi:10.1016/j.energy.2017.09.088.

[28] A. Pérez-Navarro, D. Alfonso, C. Álvarez, F. Ibáñez, C. Sánchez, and I. Segura, "Hybrid biomass-wind power plant for reliable energy generation," *Renew. Energy*, vol. 35, no. 7, pp. 1436–1443, 2010, doi:10.1016/j.renene.2009.12.018.

[29] J. Li, P. Liu, and Z. Li, "Optimal design and techno-economic analysis of a solar-wind-biomass off-grid hybrid power system for remote rural electrification: A case study of west China," *Energy*, vol. 208, p. 118387, 2020, doi:10.1016/j.energy.2020.118387.

[30] K. Shivarama Krishna and K. Sathish Kumar, "A review on hybrid renewable energy systems," *Renew. Sustain. Energy Rev.*, vol. 52, pp. 907–916, 2015, doi:10.1016/j.rser.2015.07.187.

[31] "Small Wind Energy and Hybrid Systems Programme." 2017. www.mahaurja.com/meda/en/off_grid_power/small_wind_solar_hybrid.

[32] "Hybrid Wind and Solar Electric Systems," *Energy Saver.* www.energy.gov/energysaver/buying-and-making-electricity/hybrid-wind-and-solar-electric-systems.

[33] M. Mikeska, J. Najser, V. Peer, J. Frantík, and J. Kielar, "Quality assessment of gas produced from different types of biomass pellets in gasification process," *Energy Explor. Exploit.*, vol. 38, no. 2, pp. 406–416, 2020, doi:10.1177/0144598719875272.

[34] S. Zafar, "Renewable energy in Algeria," *Africa, Renew. Energy*, 2017. http://www.ecomena. org/renewables-algeria.

[35] A. A. Bazmi and G. Zahedi, "Sustainable energy systems: Role of optimization modeling techniques in power generation and supply—A review," *Renew. Sustain. Energy Rev.*, vol. 15, no. 8, pp. 3480–3500, 2011, doi:10.1016/j.rser.2011.05.003.

[36] S. Yilmaz and H. Selim, "A review on the methods for biomass to energy conversion systems design," *Renew. Sustain. Energy Rev.*, vol. 25, pp. 420–430, 2013, doi:10.1016/j.rser.2013.05.015.

[37] D. Deublein and A. Steinhauser, *Biogas from Waste and Renewable Resources: An Introduction*, 2nd ed., pp. 2–5, 2010, doi:10.1002/9783527632794.

[38] P. Adams, T. Bridgwater, A. Ross, and I. Watson, *Biomass Conversion Technologies.* Elsevier Inc., 2018.

[39] C. Kahraman, "A fuzzy multicriteria methodology for selection among energy alternatives," *Expert Syst. Appl.*, vol. 37, pp. 6270–6281, 2010, doi:10.1016/j.eswa.2010.02.095.

[40] Z. Zhou, P. Liu, Z. Li, and W. Ni, "An engineering approach to the optimal design of distributed energy systems in China," *Appl. Therm. Eng.*, vol. 53, no. 2, pp. 387–396, 2013, doi:10.1016/j.applthermaleng.2012.01.067.

[41] T. Nakata, K. Kubo, and A. Lamont, "Design for renewable energy systems with application to rural areas in Japan," *Energy Policy*, vol. 33, pp. 209–219, 2005, doi:10.1016/S0301-4215(03)00218-0.

[42] A. Gupta, R. P. Saini, and M. P. Sharma, "Steady-state modelling of hybrid energy system for off grid electrification of cluster of villages," *Renew. Energy*, vol. 35, no. 2, pp. 520–535, 2010.

[43] R. P. Saini and M. P. Sharma, "Integrated renewable energy systems for off grid rural electrification of remote area," *Renew. Energy*, vol. 35, no. 6, pp. 1342–1349, 2010, doi:10.1016/j.renene.2009.10.005.

[44] A. Gupta, R. P. Saini, and M. P. Sharma, "Modelling of hybrid energy system d Part I: Problem formulation and model development," *Renew. Energy*, vol. 36, no. 2, pp. 459–465, 2011, doi:10.1016/j.rcncnc.2010.06.035.

[45] A. Gupta, R. P. Saini, and M. P. Sharma, "Modelling of hybrid energy system d Part II : Combined dispatch strategies and solution algorithm," *Renew. Energy*, vol. 36, no. 2, pp. 466–473, 2011, doi:10.1016/j.renene.2009.04.035.

[46] A. Gupta, R. P. Saini, and M. P. Sharma, "Modelling of hybrid energy system d Part III: Case study with simulation results," *Renew. Energy*, vol. 36, no. 2, pp. 474–481, 2011, doi:10.1016/j.renene.2009.04.036.

[47] J. R. S. Cristóbal, "Multi-criteria decision-making in the selection of a renewable energy project in Spain: The Vikor method," *Renew. Energy*, vol. 36, no. 2, pp. 498–502, 2011, doi:10.1016/j.renene.2010.07.031.

[48] S. M. Hakimi, "Optimal sizing of a standalone hybrid power system via particle swarm optimization for Kahnouj area in southeast of Iran," *Renew. Energy*, vol. 34, no. 7, pp. 1855–1862, 2009, doi:10.1016/j.renene.2008.11.022.

[49] C. Rubio-maya, J. Uche, and A. Martínez, "Sequential optimization of a polygeneration plant," *Energy Convers. Manag.*, vol. 52, no. 8–9, pp. 2861–2869, 2011, doi:10.1016/j.enconman.2011.01.023.

[50] G. Redman, *A Detailed Economic Assessment of Anaerobic Digestion Technology and Its Suitability to UK Farming and Waste Systems.* The Andersons Centre.

[51] P. McKendry, "Energy production from biomass (part 1): Overview of biomass," *Bioresour. Technol.*, vol. 83, no. 1, pp. 37–46, 2002, doi:10.1016/S0960-8524(01)00118-3.

[52] M. Ni, D. Y. C. Leung, M. K. H. Leung, and K. Sumathy, "An overview of hydrogen production from biomass," *Fuel Process. Technol.*, vol. 87, no. 5, pp. 461–472, 2006, doi:10.1016/j.fuproc.2005.11.003.

[53] B. Jenkins, L. Bexter, T. Miles Jr., and T. Miles, "Combustion properties of biomass flash," *Fuel Process. Technol.*, vol. 54, pp. 17–46, 1998.

[54] F. X. Collard and J. Blin, "A review on pyrolysis of biomass constituents: Mechanisms and composition of the products obtained from the conversion of cellulose, hemi-celluloses and lignin," *Renew. Sustain. Energy Rev.*, vol. 38, pp. 594–608, 2014, doi:10.1016/j.rser.2014.06.013.

[55] R. Inder and R. Kumar, "Current status and experimental investigation of oxy-fired fluidized bed," *Renew. Sustain. Energy Rev.*, vol. 61, pp. 398–420, 2016, doi:10.1016/j.rser.2016.04.021.

[56] L. Gustavsson, P. Börjesson, B. Johansson, and P. Svenningsson, "Reducing CO2 emissions by substituting biomass for fossil fuels," *Energy*, vol. 20, no. 11, pp. 1097–1113, 1995, doi:10.1016/0360-5442(95)00065-O.

[57] J. L. Easterly and M. Burnham, "Overview of biomass and waste fuel resources for power production," *Biomass Bioenergy*, vol. 10, no. 2–3, pp. 79–92, 1996, doi:10.1016/0961-9534(95)00063-1.

[58] W. Lin, K. Dam-Johansen, and F. Frandsen, "Agglomeration in bio-fuel fired fluidized bed combustors," *Chem. Eng. J.*, vol. 96, no. 1–3, pp. 171–185, 2003, doi:10.1016/j.cej.2003.08.008.

[59] M. Crawford, "Fluidized-bed combustors for biomass boilers," *Am. Soc. Mech. Eng.* www.asme.org/topics-resources/content/fluidized-bed-combustors-for-biomass-boilers.

[60] J. Corella, J. M. Toledo, and G. Molina, "A review on dual fluidized-bed biomass gasifiers," *Ind. Eng. Chem. Res.*, vol. 46, no. 21, pp. 6831–6839, 2007.

[61] N. N. Novitsky, Z. I. Shalaginova, A. A. Alekseev, V. V. Tokarev, O. A. Grebneva, A. V. Lutsenko, . . . and M. Chertkov, "Smarter smart district heating," *Proc. IEEE*, vol. 108, no. 9, pp. 1596–1611, 2020.

[62] K. Lygnerud, "Challenges for business change in district heating," *Energy. Sustain. Soc.*, vol. 8, no. 1, 2018, doi:10.1186/s13705-018-0161-4.

7 Numerical Analysis of Thermomagnetic Convection Flow in a Closed Loop for Cooling Application

*Jaswinder Singh Mehta, Rajesh Kumar,
Harmesh Kumar, Harry Garg*

7.1 INTRODUCTION

The heat transfer phenomenon is of prime concern in many industrial applications and wide variety of convection heat transfer techniques have been developed over a period of years to tackle the problem of heat dissipation in engineering devices. Technological advances in the field of electronics led to multi-fold increase in the computational ability of these devices as it has resulted in faster processing time but simultaneously subjecting the system to higher thermal loads. Conventional techniques fail to provide a satisfactory solution to the problem of increased heat flux. So there is utmost need to provide a satisfactory solution of high-heat-flux-load problem for the safe working of such devices and to increase their working life. Ferrofluid-based cooling device can provide a potential solution to the problem of high heat flux [1–4].

S. Shyam et al. [5] investigated both experimentally and computationally convective heat transfer characteristics of ferrofluid in a tube subjected to constant and alternating magnetic field. Under the influence of magnetic field, augmentation in heat transfer was observed. Formation of spiked-hump-like structure led to flow disturbance in the tube, resulting in augmentation of energy transfer.

M. Goharkhah et al. [6] performed a computational study on a ferrofluid flowing through a miniature heat sink in order to compare the accuracy of magnetic force models in ferrohydrodynamics. Magnetic force models predicted different behaviours with regards to volume fraction and Reynolds number and the error was found to exaggerate with rise in field intensity.

N. Gan Jia Gui et al. [7] measured experimentally forced convective heat transfer of single phase ferrofluid flow in microchannels. A magnetic flux was applied perpendicular to the flow direction. In contrast to the findings of established work, decrease of heat transfer rate with increase in magnetic flux was observed.

DOI: 10.1201/9781003367161-7

H. Jafari et al. [8] performed numerical investigation using COMSOL MultiPhysics to study the effect of non-uniform magnetic field on convective heat transfer characteristics of ferrofluid in a parallel plate channel. Heat transfer was found to significantly improve with increase in volume fraction of nanoparticles, Re (Reynolds Number), and field intensity.

R. Zanella et al. [9] analysed the convection heat transfer of an electromagnetic system comprising of a copper coil immersed in an oil-based ferrofluid. It was found using numerical analysis that with increase in the volume fraction, drop in maximum temperature of the system was observed due to setting up of thermomagnetic convection current in the fluid.

S. Dalvi et al. [10] numerically examined the mutual effect of thermomagnetic convection and magnetocaloric effect on kerosene-based fluid flowing through a circular annulus. It was found that with rise in remnant flux density, increase in average Nusselt number was observed at fixed L/D ratio of the annulus.

Bozhko and Putin [11] analysed the setting of thermomagnetic convection in a kerosene-based ferrofluid layer, when subjected to uniform transverse magnetic field generated by Helmholtz coils. Nusselt number was found to have direct relation with magnetic field intensity.

M. Barzegar Gerdroodbary et al. [12] examined the effect of magnetic field on the flow and heat characteristics of a ferrofluid inside a micro T-junction numerically. A finite volume method was applied and when subjected to magnetic field, amplification of heat transfer and drop in system temperature was noticed compared to absence of field.

M. Sheikholeslami et al. [13] numerically analysed the effect of non-uniform magnetic field on flow characteristics and convective heat transfer of ferrofluid inside a 90° elbow channel. With increase in nanoparticles concentration and at higher Reynolds number, augmentation of heat transfer was observed in the elbow channel. Deceleration of mainstream flow and induction of circulations in the vicinity of the inner wall of the elbow channel, when subjected to non-uniform magnetic field, were another important finding of the study.

The present study aims to analyse the flow behaviour of ferrofluid flowing through a closed loop for heat transfer application owing to thermomagnetic convection principle. An FEA solver, COMSOL MultiPhysics was used for the computational study, and it was revealed that the cooling system effectively dissipate the heat flux.

7.2 MATHEMATICAL MODELLING AND NUMERICAL SIMULATION

7.2.1 GEOMETRY DESCRIPTION

A 3D model used for numerical study is shown in Figure 7.1. An oval-shaped closed loop has been considered in this study, the main parts of which have been highlighted through which ferrofluid flows.

The system was subjected to heat flux using heating substrate as shown in Figure 7.1 (labelled as heat source), and heat was dissipated to the surrounding fluid through annular fins attached at the heat sink portion of the tube. The surrounding fluid was considered as air with initial temperature of 303 K.

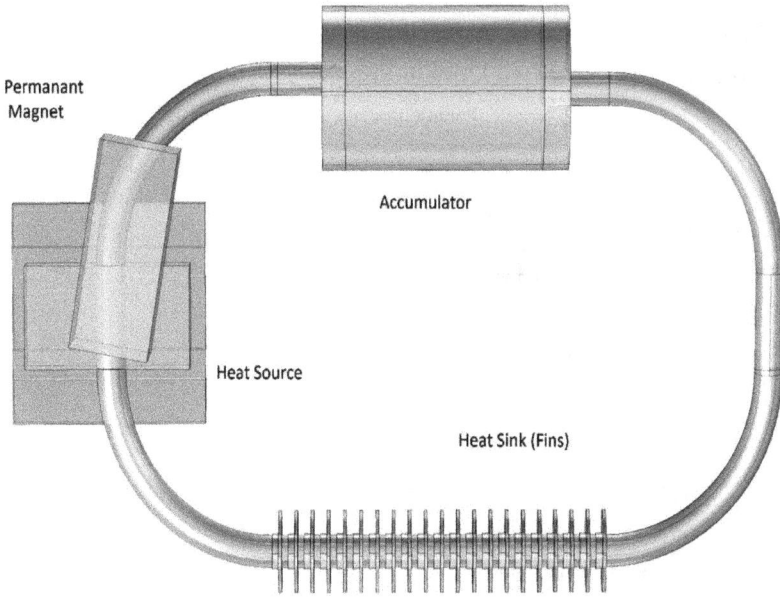

FIGURE 7.1 Three-dimensional geometrical model.

7.2.2 Governing Equations

Following equations [14] govern the flow of single phase ferrofluid through the closed loop under consideration:

Continuity equation:

$$\frac{\partial \rho}{\partial t} + \nabla . \left(\rho \vec{u} \right) = 0 \tag{2.1}$$

Momentum equation:

$$\rho \partial \vec{u} / \partial t + \rho \vec{u} . \nabla \vec{u} = -\nabla p + \nabla . (\mu (\nabla \vec{u} + \left(\nabla \vec{u} \right)^{\mathsf{T}}) + \vec{F} \tag{2.2}$$

Energy equation:

$$\rho c_p \left(\frac{\partial T}{\partial t} + \vec{u} . \nabla \mathrm{T} \right) = k \nabla^2 T \tag{2.3}$$

Magnetic induction:

$$\vec{B} = \mu_\circ \left(\vec{H} + \vec{M} \right) \tag{2.4}$$

Driving force (Kelvin body force):

$$\vec{F} = \left(\vec{M} . \nabla \right) \vec{B} \tag{2.5}$$

TABLE 7.1

Input Properties of Ferrofluid Used in the Analysis

Sr. No.	Parameters	Value
1	Viscosity, μ	2 cP
2	Density, ρ	910 kg/m^3
3	Thermal conductivity, k	0.174 W/(m-K)
4	Curie temperature of the fluid	45°C
5	Surrounding temperature	30°C
6	Magnetic susceptibility	0.386
7	Relative permeability of fluid	1.386

Table 7.1 represents the specifications of ferrofluid used in the study.

7.2.3   BOUNDARY CONDITIONS

Fluid was considered as incompressible and flow to be laminar in nature and was assumed to behave as single-phase Newtonian fluid. No-slip boundary conditions at the inner walls of the tube were assumed. Initial temperature of the ferrofluid has been considered same as surrounding temperature. The surrounding medium is air at an initial temperature of 303 K. Loop placement is in horizontal plane and effect of gravity thereby is neglected.

7.2.4   MESH DETAILS

Figure 7.2 shows the typical meshing in the computational domain. Grid was formed using tetrahedral mesh elements using predefined mesh settings available in COMSOL MultiPhysics. Finer meshing was employed in fluid flow domain and on the inner walls of the tube so that data may be accurately predicted and to minimize the error in measurement of data. However, fine grid elements were used for fins, magnet and heat source fixture and free tetrahedral meshing was applied for meshing the remaining geometry.

7.3   RESULT AND DISCUSSION

Figure 7.3(a–f) shows the temperature contour plots for different length of time. As the time progresses, the temperature differential in the fluid that initially existed near the heat source travelled to distant points in the loop. The presence of temperature gradient in the loop clearly demonstrates the movement of fluid under thermomagnetic convection principle in the presence of magnetic field generated by permanent magnet.

When heat input was applied, the fluid column near the heat source was subjected to differential heating. Under the influence of magnetic field, there exists difference

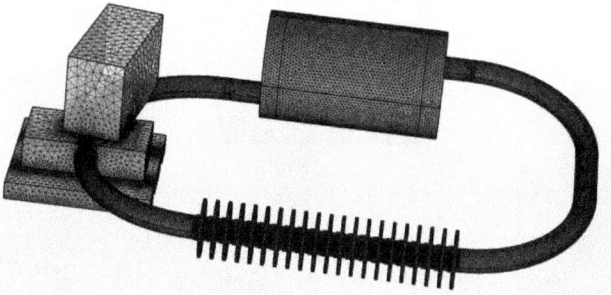

FIGURE 7.2 Mesh details of computational domain.

of magnetization due to which Kelvin body force is induced in the fluid and it acts in the direction of higher temperature. This thrust force is thus responsible for the movement of the fluid in the loop owing to the thermomagnetic convection principle. With time, the fluid as it moves travels to other part of the loop (labelled as heat sink), where it will dissipate its heat to the surrounding fluid. Fins have been incorporated in heat sink area to enhance the rate of heat dissipation. The lower-temperature fluid thus moves towards the accumulator section, where it get mixed with the cold reserved fluid and its temperature further drops down. The cold fluid is thus ready to be circulated in the loop for the next cycle.

Figure 7.4 shows the temperature curves at two different positions along the loop—namely, at heater inlet and other at heater outlet for varying lengths of time.

As heat flux was applied, fluid temperature begins to increase. During the initial transient period, the temperature difference between the two points begins to rise aptly under the effect of magnetic field. The rate of temperature rise at downstream of the heat source is quite large in comparison to temperature change at upstream of the heat source and this temperature gradient is responsible for the generation of thrust force, termed Kelvin body force. As the magnetization of the fluid decrease with temperature according to Curie's law, net thrust force thus acts on the fluid in the direction of high temperature section due to magnetization difference and the fluid velocity gradually increases initially and thereafter as the steady state is being attained, the velocity stabilizes as represented in Figure 7.5. This clearly shows the evidence of advective fluid transport from low temperature towards high temperature.

7.4 CONCLUSION

From the numerical analysis, it can be concluded that ferrofluid can be employed as a coolant as the fluid can flow on its own without any requirement of external device, such as a pump, and closed loop system was successful in dissipation of the heat flux. Such type of systems can be developed for cooling applications in future and reliability of such systems is high as they do not have any moving parts, such as pumps and valves. Thus, the operation would be noiseless and without any

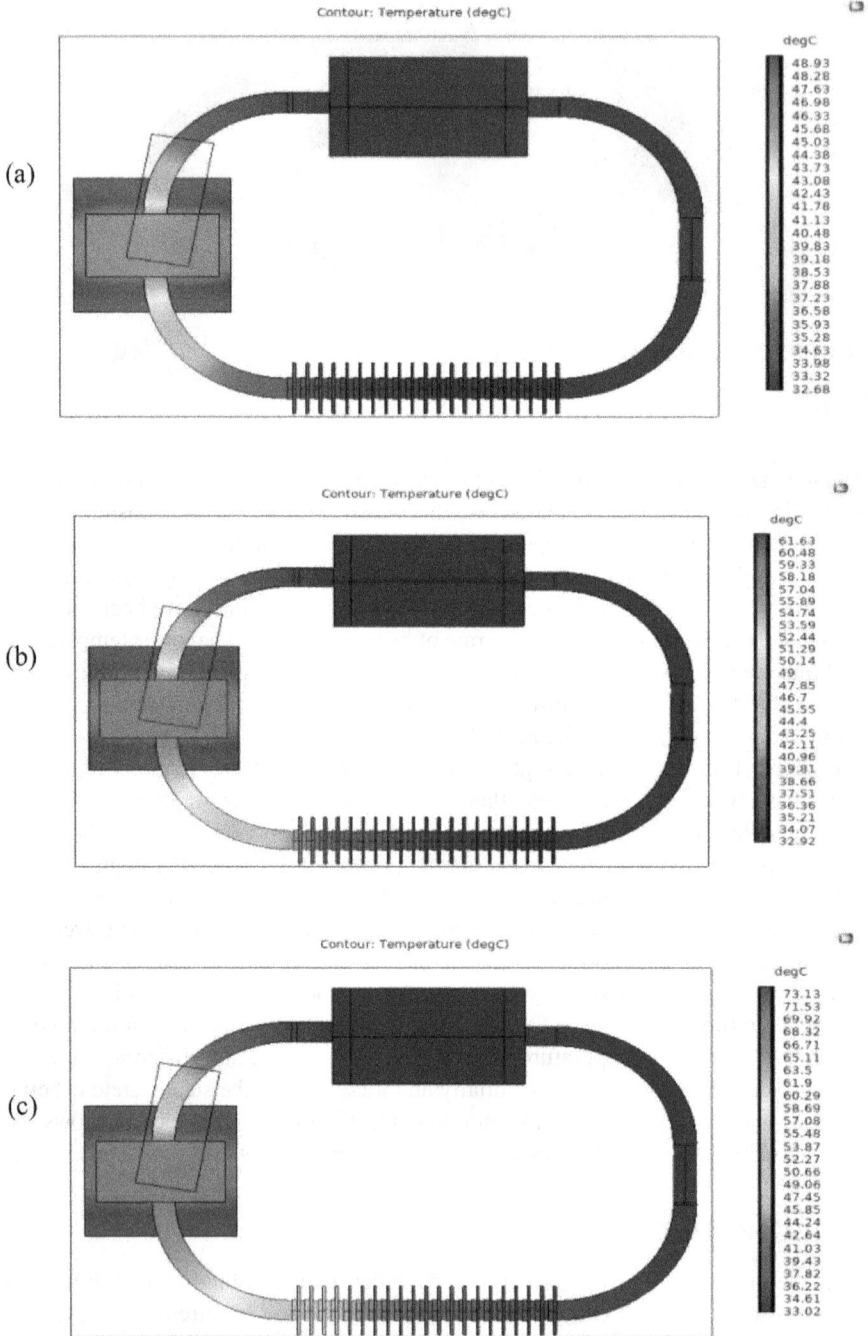

FIGURE 7.3 Temperature contour plots for time: (a) 30s, (b) 60s, (c) 90s, (d) 150s, (e) 180s, (f) 240s.

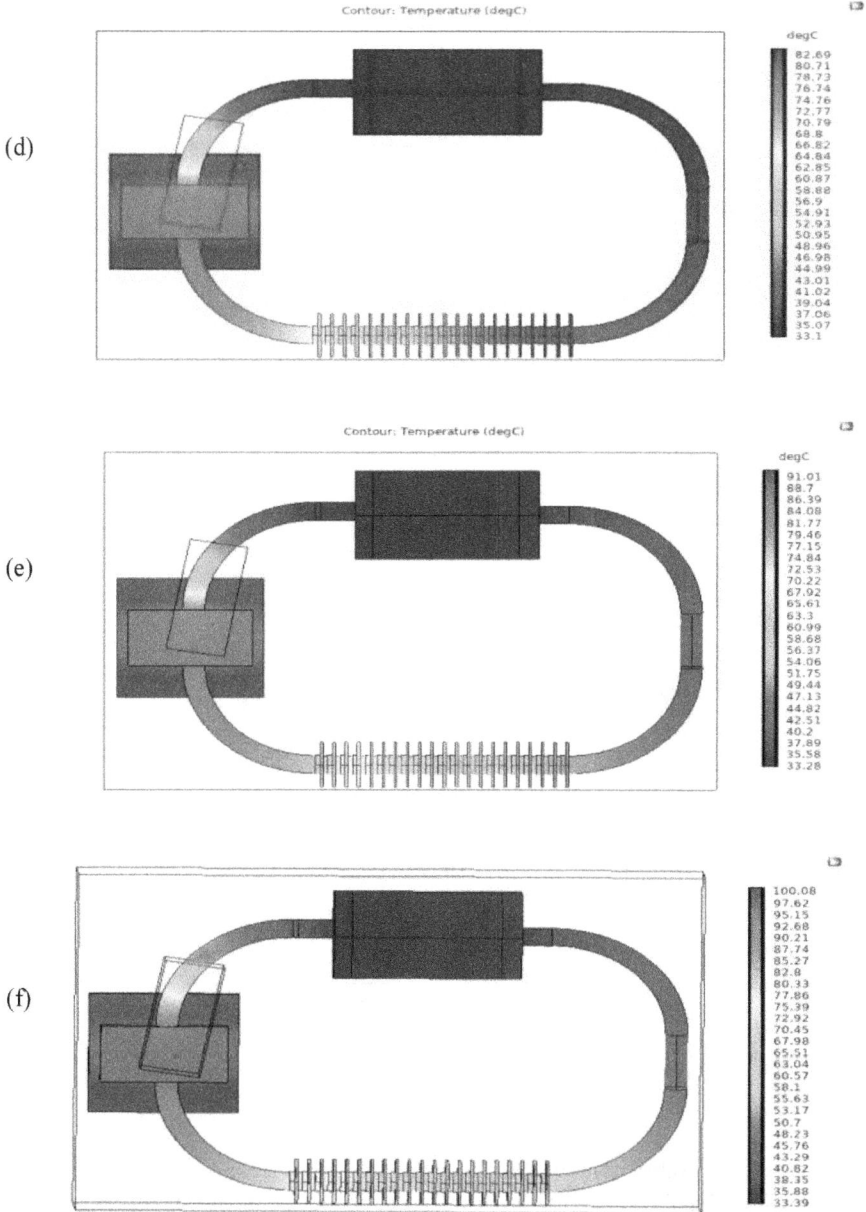

FIGURE 7.3 Continued

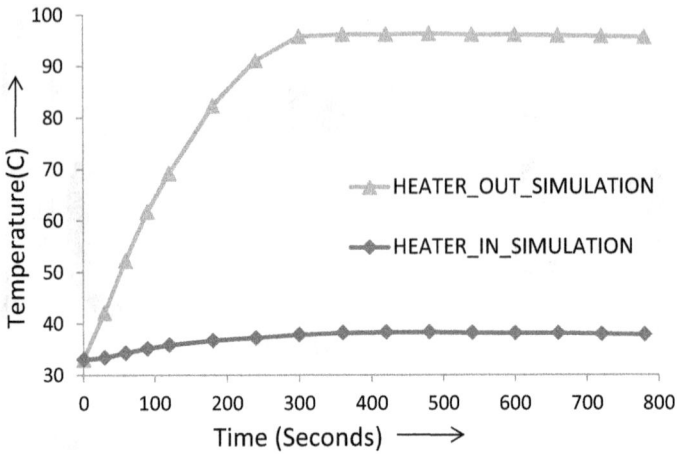

FIGURE 7.4   Temperature curve at varying times for two positions: heater inlet and heater outlet.

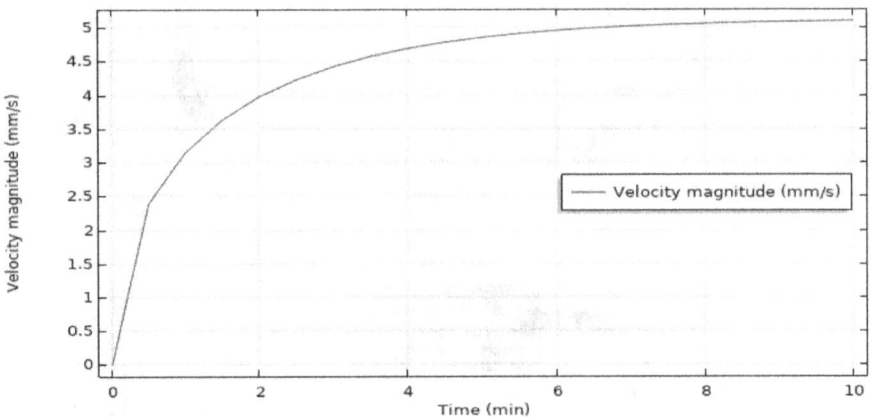

FIGURE 7.5   Variation of velocity as a function of time.

vibrations. Temperature contours and velocity plot provide the clear evidence of circulation of fluid in the loop without any mechanical device, thus lowering the system requirement as the system is passive in nature.

REFERENCES

[1] Yarahmadi, M., Moazami, H., and Goudarzi, S. M. B. (2015). Experimental investigation into laminar forced convective heat transfer of ferrofluids under constant and oscillating magnetic field with different magnetic field arrangements and oscillation modes. *Exp. Therm. Fluid Sci.*, Vol. 68, pp. 601–611.

[2] Aursand, E., Gjennestad, M. A., Yngve Lervag, K., and Lund, H. (2016). A multi-phase ferrofluid flow model with equation of state for thermomagnetic pumping and heat transfer. *J. Magn. Magn. Mater.*, Vol. 402, pp. 8–19.

[3] Sesen, M., Teksen, Y., Şendur, K., Pinar Mengüç, M., Öztürk, H., Yağc Acar, H. F., and Koşar, A. (2012). Heat transfer enhancement with actuation of magnetic nanoparticles suspended in a base fluid. *J. Appl. Phys.*, Vol. 112(6), p. 064320-1-6.

[4] Singh Mehta, J., Kumar, R., Kumar, H., and Garg, H. (2017). Convective heat transfer enhancement using ferrofluid: A review. *ASME. J. Thermal Sci. Eng. Appl.*, Vol. 10(2), p. 020801-020801-12.

[5] Shyam, S., Mehta, B., Mondal, P. K., and Wongwises, S. (2019). Investigation into the thermo-hydrodynamics of ferrofluid flow under the influence of constant and alternating magnetic field by InfraRed Thermography. *Int. J. Heat Mass Transf.*, Vol. 135, pp. 1233–1247.

[6] Goharkhah, M., Bezaatpour, M., and Javar, D. (2020). A comparative investigation on the accuracy of magnetic force models in ferrohydrodynamics. *Powder Technol.*, Vol. 360, pp. 1143–1156.

[7] Gui, N. Gan Jia, Stanley, C., Nguyen, N.-T., and Rosengarten, G. (2018). Ferrofluids for heat transfer enhancement under an external magnetic Field. *Int. J. Heat Mass Transf.*, Vol. 123, pp. 110–121.

[8] Jafari, H., and Goharkhah, M. (2020). Application of electromagnets for forced convective heat transfer enhancement of magnetic fluids. *Int. J. Therm. Sci.*, Vol. 157, p. 106495.

[9] Zanella, R., Nore, C., Bouillault, F., Guermond, J.-L., and Mininger, X. (2019). Influence of thermomagnetic convection and ferrofluid thermophysical properties on heat transfers in a cylindrical container heated by a solenoid. *J. Magn. Magn. Mater.*, Vol. 469, pp. 52–63.

[10] Dalvi, S., Karaliolios, E. C. J., van der Meer, T. H., and Shahi, M. (2020). Thermomagnetic convection in a circular annulus filled with magnetocaloric nanofluid. *Int. Commun. Heat Mass Transf.*, Vol. 116, p. 104654.

[11] Bozhko, A., and Putin, G. F. (2004). Magnetic action on convection and heat transfer in ferrofluid. *Indian J. Eng. Mater. Sci.*, Vol. 11, pp. 309–314.

[12] Gerdroodbary, M. B., Sheikholeslami, M., Mousavi, S. V., Anazadehsayed, A., and Moradi, R. (2018). The influence of non-uniform magnetic field on heat transfer intensification of ferrofluid inside a T-junction. *Chem. Eng. Process. Process Intensif.*, Vol. 123, pp. 58–66.

[13] Sheikholeslami, M., Gerdroodbary, M. B., Mousavi, S. V., Ganji, D. D., and Moradi, R. (2018). Heat transfer enhancement of ferrofluid inside an 90° elbow channel by non-uniform magnetic field. *J. Magn. Magn. Mater.*, Vol. 460, pp. 302–311.

[14] Rosensweig, R. E. (1985). *Ferrohydrodynamics*, Cambridge University Press, New York.

8 A Systematic Review on Solar Cells

Dhawan Singh, Abinash Singh,
Aditi Thakur, Himanshu Jindal

8.1 INTRODUCTION

With the rise in global population, the energy consumption rate also increased at rapid speed. Due to sudden decline in fossil fuels, the adaptation of renewable energy has been initiated keeping in view of some advantages like clean energy, abundance in availability, environment friendliness, and so on. Vast research has been done in this area for efficient harnessing of solar energy [1–4]. As fossil fuels become increasingly scarce, power is becoming a big worry. Currently, sunlight is the most accessible and dependable source of renewable energy. Moreover, compared to traditional energy production technologies, solar energy requires a lot fewer human costs. Cost and efficiency are the main obstacles to progress of any development. Despite the fact that solar energy is free and accessible everywhere, there is a one-time cost associated with the equipment needed to gather it.

In order to select the rightful solar cell for an exact terrestrial location, the user needs to recognize the basic mechanisms along with different widely studied solar technologies. Although this energy has many benefits, but it also has a few restrictions, like its intermittent nature in day due to changing weather conditions and unavailability of solar radiation at night. A tremendous amount of research has gone into creating solar cells and panels with high conversion efficiencies so as to efficiently capture the energy. A solar panel is a massive assembly of small solar cells that are organized in a certain geometric pattern to generate a specific quantity of power. Solar cells are also known as PV (photovoltaic) systems and can be broadly classified as follows: first-generation solar cells are wafer-based, second-generation solar cells are thin-film-based, and the third-generation solar cells use new emerging technologies [5–7].

Due to their high price, challenging manufacturing procedures, and detrimental environmental effect, the usage of first- and second-generation cells has now been restricted. The development of solar cells is mostly influenced by the price of the materials and the effectiveness of power conversion. The amount of solar energy that can be turned into electricity is known as the photovoltaic conversion efficiency, often known as the efficiency of solar PV modules. With high power conversion efficiency (PCE) of third-generation silicon-based solar cells have made them dominant these days. At present, solar panels can have a 22% radiation efficiency [8–9]. Research on storage of generated electric power is still not properly accomplished.

 DOI: 10.1201/9781003367161-8

Solar/PV systems are often used to generate electricity for larger applications, supplying electricity to homes and businesses, but are also used to give power to small electronic devices, such as calculators and watches. The need for solar energy can be attributed to the fact that it has been proven to be economical, environmentally friendly, and highly efficient.

This chapter aims to present a comprehensive and systematic review of recent developments in solar cells, as well as their uses, and discuss about their future trends and aspects. Section 8.2 discusses the fundamental operating principles of solar cells. The details about different generation of solar cells are covered in Sections 8.3, 8.4 and 8.5.

8.2 BASIC WORKING PRINCIPLES OF SOLAR CELLS

Solar cells are a type of semiconductor device that converts light into electrical current and is used to produce electricity from the sun's energy. They are composed of many layers of semiconductor. Figure 8.1 shows the evolution of solar cell materials, either in a single piece or made up of small devices, as in the case with the amorphous-based type. Research on solar cells began around 1939 [10–12]. A solar cell roughly has 0.5-to-0.6-volt energy output. Solar cells are one of the major applications of the photovoltaic effect. This phenomenon describes how voltage and current are produced when exposed to light.

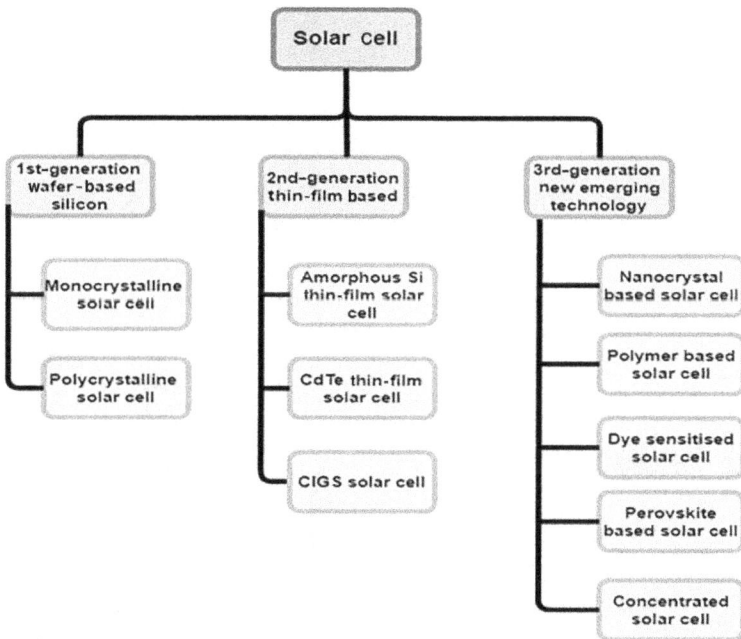

FIGURE 8.1   Classification of solar cells.

In a PV cell, it is the absorption of sunlight that triggers an electric current. Silicon is the most common type of solar cell is being widely used as its semiconductor properties allow it to absorb light without damaging the material or creating too much heat. A light-absorbing material called a layer of p-type silicon on the inside of the semiconductor is where electrons are created. A layer of n-type silica is located outside the semiconductor, then absorbs the charge generated by sunlight. These layers are called the epitaxial layers and are made up of either pure silicon or boron doped silicon, but both materials allow for electrons to be formed quickly and efficiently. There are a few different types of solar cells (given in Figure 8.2) [13–14].

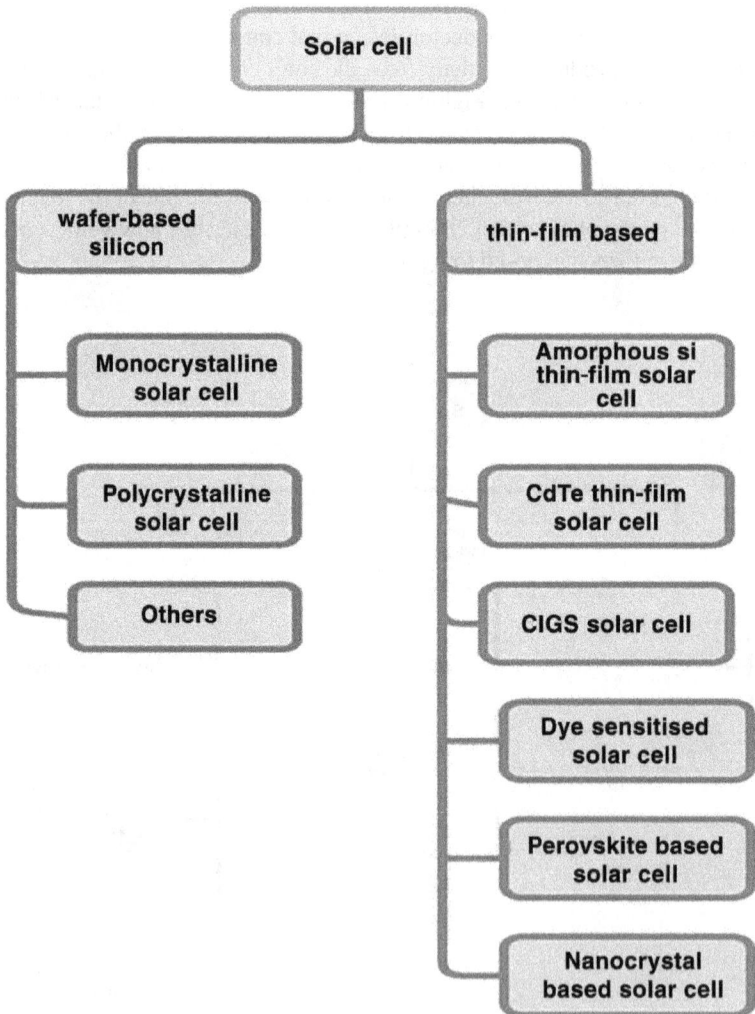

FIGURE 8.2   Wafer- and thin-film-based solar cells.

8.3 WAFER-BASED, OR FIRST-GENERATION, SOLAR CELLS

Wafer-based cells are widely used as solar cells because of the huge cost reduction. The most important characteristic is the cell area, which is around 200 mm^2 [15]. Wafers are sheets of semiconducting material that are thicker than a sheet of paper but thinner than comparable-sized silicon ingots or wafers. The first-generation device consists of 20 active layers on top of a back contact layer (BCL). Types of wafer-based solar cells are briefly compared in Figure 8.3.

Monocrystalline-type cell is made from high-quality (100% pure) crystals of semiconducting silicon. Single-junction type are one-layer-based solar cell. Monocrystalline solar cells have the highest efficiency, up to 24%. This means that they can convert 24% of the sunlight that hits them into electricity. If you are looking for an efficient and sustainable solar cell, look no further than the polycrystalline silicon solar cell. It is also known as a water-based silicon solar cell and is made from a material that is derived from sand—silicon. The polycrystalline silicon solar cell does not use any toxic chemicals in its production. This makes it a more environmentally friendly option. It can convert more sunlight into electricity, making it ideal for use in a variety of settings. Polycrystalline silicon is made from many small pieces of silicon or other materials that are then melted together to form a crystallized material. Layers of silicon that have been heated and melted produce amorphous silicon. The water-based silicon solar cell uses a thin film of silicon to absorb sunlight and create an electrical current. The water helps to keep the silicon film from drying out and provides a way to remove any impurities that may be present in the silicon.

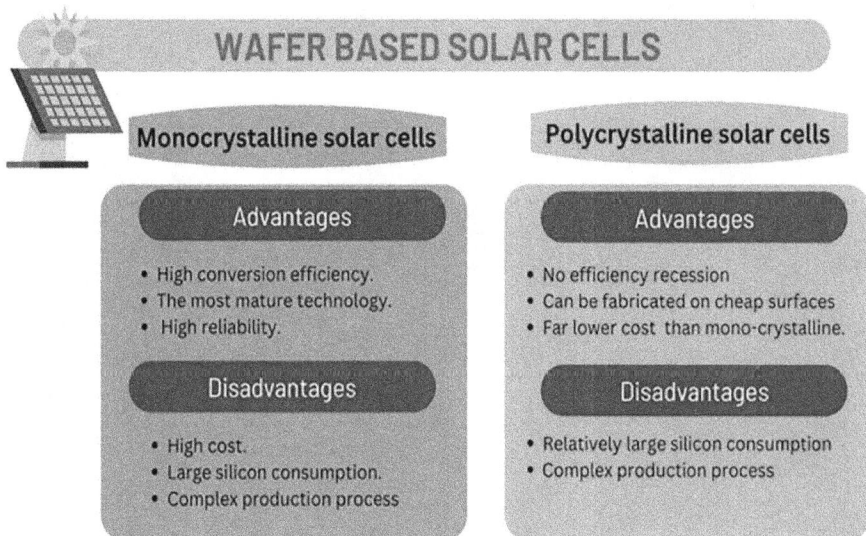

WAFER BASED SOLAR CELLS

Monocrystalline solar cells

Advantages
- High conversion efficiency.
- The most mature technology.
- High reliability.

Disadvantages
- High cost.
- Large silicon consumption.
- Complex production process

Polycrystalline solar cells

Advantages
- No efficiency recession
- Can be fabricated on cheap surfaces
- Far lower cost than mono-crystalline.

Disadvantages
- Relatively large silicon consumption
- Complex production process

FIGURE 8.3 Types of wafer-based solar cells.

8.4 THIN-FILM, OR SECOND-GENERATION, SOLAR CELLS

One or more thin layers of PV material are deposited on a substrate to create these types of solar cells [16]. Despite being less effective than silicon solar cells, this generation cells are more readily available. Solar cells can also be made up of copper indium gallium (di)selenide (CIGS) and cadmium telluride (CdTe) [17]. These materials are so thin—just a few micrometres thick. An amorphous silicon solar cell is made from a layer of silicon and layer of graphene. When acting as semiconductor, amorphous silicon cell can be used as photoelectrode. Due to lack of crystallinity, they are low in PV performance. Figure 8.4. presents types of second-generation solar cells [18]. They have a number of advantages over their generation competitors, like their high absorption coefficient, which allows them to absorb more sunlight than other materials. Additionally, CdTe films are less expensive to produce than other thin-film solar cell materials.

CIGS solar cells make use of readily accessible and affordable components. They are thin, flexible, and often transparent, which allows the light to pass through them without being absorbed. These are some of the reasons that CIGS solar cells have been shown to be efficient at converting sunlight into electricity. It is relatively easy for these solar cell materials to be mass-produced in thin wafers, and they can be applied to various surfaces, such as windows, cars, and window frames. The cost is still a large concern when it comes to replacing current energy sources with these renewable resources. Shellfish (the source of ciguatera toxin) are frequently consumed in tropical areas, especially along medium-warm latitudes where they abound. If people ingest these toxins without knowing it, they will experience a range of symptoms such as diarrhoea, vomiting, and neurological problems. Recently, these are used as a device for detection of ciguatera toxin.

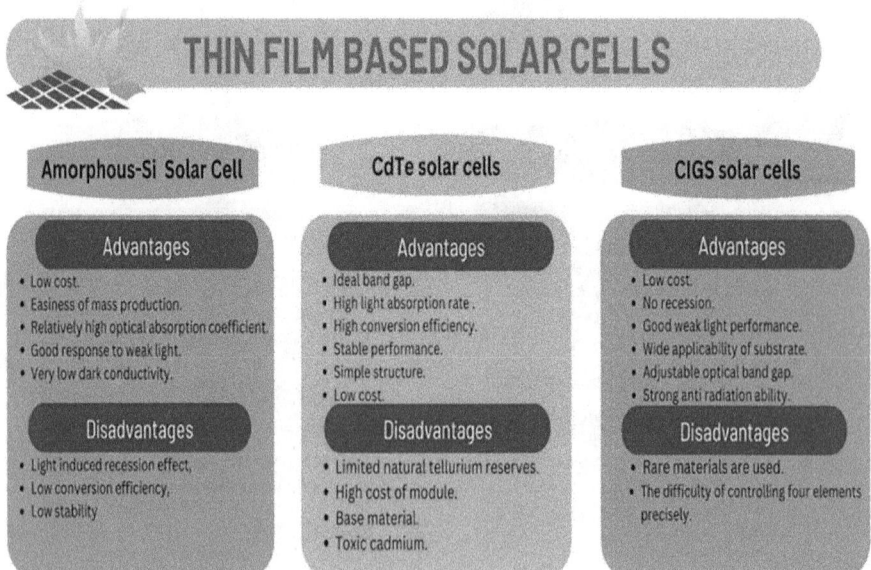

FIGURE 8.4 Thin-film-based solar cells.

8.5 EMERGING-TECHNOLOGY-BASED, OR THIRD-GENERATION, SOLAR CELLS

The new-and-emerging-technology-based solar cells are illustrated in Figure 8.5 and are explained in this section.

8.5.1 POLYMER-BASED SOLAR CELL

By depositing a very thin layer of active material onto a substrate, these cells are created as shown in Figure 8.6. The advantage of this approach is that it uses less material than traditional solar cells, and so it is cheaper to manufacture [19]. The

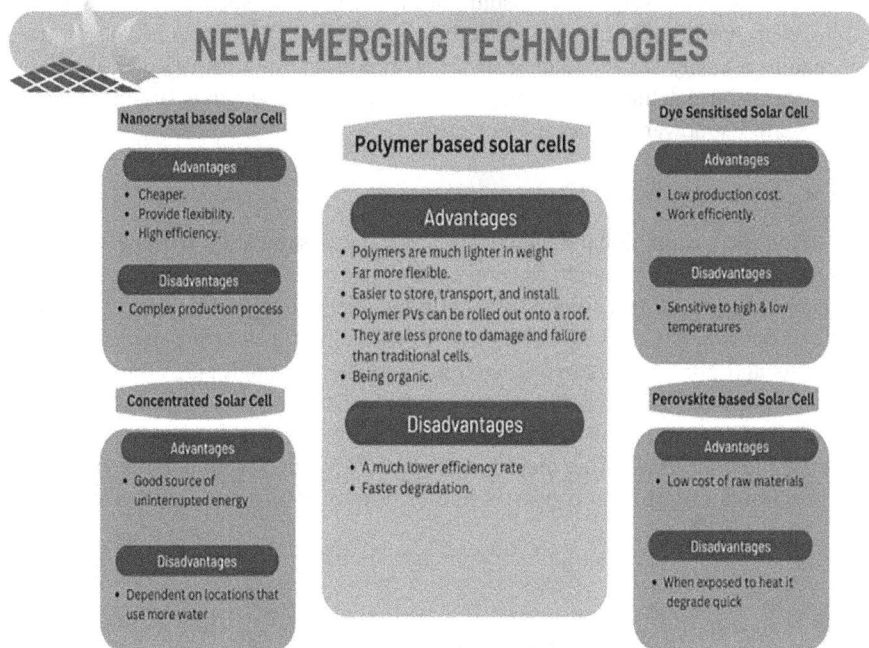

FIGURE 8.5 Emerging-technology-based solar cells.

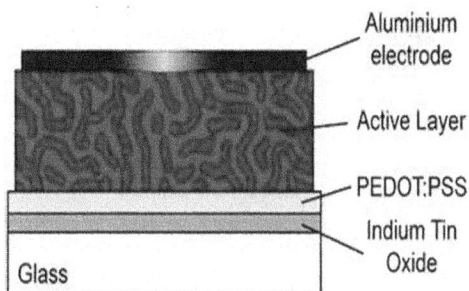

FIGURE 8.6 Polymer-based solar cell structure.

polymer-based solar cells can also be placed in thin-film-type solar cells, but it is advanced as compared to them. An active polymer layer is deposited onto a substrate to create these cells. This method has the benefit of being easy and inexpensive to create. Yet these cells' efficiency is still not as great as that of conventional solar cells.

8.5.2   CONCENTRATED SOLAR CELL

Concentrated solar cells—also known as concentrated photovoltaic (CPV)—use mirrors and lenses to concentrate a large area of sunlight onto a small area, increasing the amount of light that reaches the cell [20]. This approach, which is less widely used than PV systems based on large-area modules made from silicon, can produce higher efficiencies with a lower cost for some applications. A solar tower is the most common method of concentrating sunlight in concentrated solar power plants. The CSP system has higher efficiency than the conventional PV system because it does not require a costly semiconductor material to convert light into electrical energy. CSP systems have achieved an average efficiency of 30%, which is similar to that of utility-scale PV power stations in the world.

8.5.3   PEROVSKITE SOLAR CELL

Due to the increasing global demand for power, perovskite solar cells have become an important topic of research. Recently, they were reported to be more efficient as compared to conventional silicon solar cells and thus could provide an economically viable alternative [21–22]. Moreover, these will help in decreasing greenhouse gas emissions and provide sustainable energy solutions. Although perovskite solar cell materials are still being researched on a large-scale level, it is expected that their commercial production will bring about promising results for society at large.

Perovskites are crystalline materials (Figure 8.7) that have the ability to absorb a large amount of light and convert it into electricity, which makes them extremely

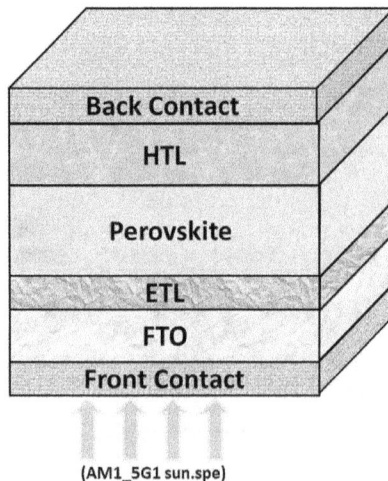

(AM1_5G1 sun.spe)

FIGURE 8.7   Perovskite solar cell structure.

attractive for solar energy conversion. Perovskites are semiconductors, which do not require an external application of voltage or current to initiate conduction; thus, they would be excellent alternatives for PV cells. Perovskites are composed of ions with a net-positive charge and undergo a reversible electronic transition.

8.5.4 NANOCRYSTAL-BASED SOLAR CELL

The different types of solar cells utilize different methods to catch and convert the energy of sunlight into electricity. As with all semiconductors, the solar cell layers are made up of atoms that have multiple positive and negative electrical charges. In this case, however, the electrons in these layers are shared among many atoms. Each solar cell has its own lattice structure, which becomes more complex as more layers are added. The semiconductor layer is composed of a gas called an electron donor that provides electrical charge carriers while also acting as a material with which to link the individual solar cell layers together and form a single unit [23]. This material is usually called an acceptor because it accepts electrons from the donor gas and gives them up to the active material. The electron donor and acceptor are usually gases. The layers are then sandwiched between two electrodes. These electrodes are usually composed of metal, such as aluminium, that can easily transmit electrons while allowing the cell to emit electrons onto one electrode and absorb them from the other.

8.5.5 DYE-SENSITIZED SOLAR CELL (DSSC)

A DSSC, uses a film of indium tin oxide as an electrolyte between two electrodes [24]. In this cell, the light-absorbing dye is anchored onto a titanium dioxide surface. In a thin-film dye-sensitized solar cell the substrate is made from flexible plastic. When light strikes the dye, electrons are set free, which then flow through an external wire to create electricity.

It has been used successfully in a small prototype device, but its cost makes it impractical for utility-scale use due to the need for large amounts of expensive rare earth elements. Dye-sensitized solar cells offer several advantages over traditional PV cells: they are more durable and they require less material to create—both of which are important considerations when constructing power plants.

8.5.6 TANDEM SOLAR CELL

When sunlight hits a solar cell, the power generated is so small that it is easily lost in the surrounding area. In an effort to increase efficiency, scientists have paired solar cells with each other using a technique known as tandem cells. "Tandem cells are two separate solar panels sitting close together on top of each other" [25–27]. They create more power and can be seen from further away because of their higher efficiency than single panels. Tandem cells feature a cell straddling the middle of each panel. When sunlight shines on one side, electrons are sent to the conductive material in one battery. The electrons then make their way to the opposite cell, where they take energy from a chemical and transmit it to another nearby cell, powering that panel as well. A second, separate set of batteries sit atop these sister panels and transfer energy back and forth between them using an electrochemical process called galvanic corrosion at their contacts. Tandem cells can be used in a variety of applications, from rooftop panels to larger scale manufacturing.

8.6   A COMPARISON OF VARIOUS TYPES OF SOLAR CELLS

Different solar cell types considered in this research paper have been compared on basis of efficiency and energy bandgap. The efficiency of a solar cell is stated as the amount of irradiation that falls on its surface and is converted into electricity. The efficiency of the solar cells in combination with the available irradiation has a significant impact on costs, but overall system efficiency is important. Bandgap of absorber layer can be made tunable by using different composition of perovskite layers. Theoretically, CdTe-type solar cells and nanocrystal-type solar cells are considered most efficient, but in reality, perovskite type solar cells have been proven most efficient, with up to 25% of efficiency (Figure 8.8(a)). The bandgap is highest for polycrystalline and lowest for monocrystalline solar cells as compared to rest of types in Figure 8.8(b).

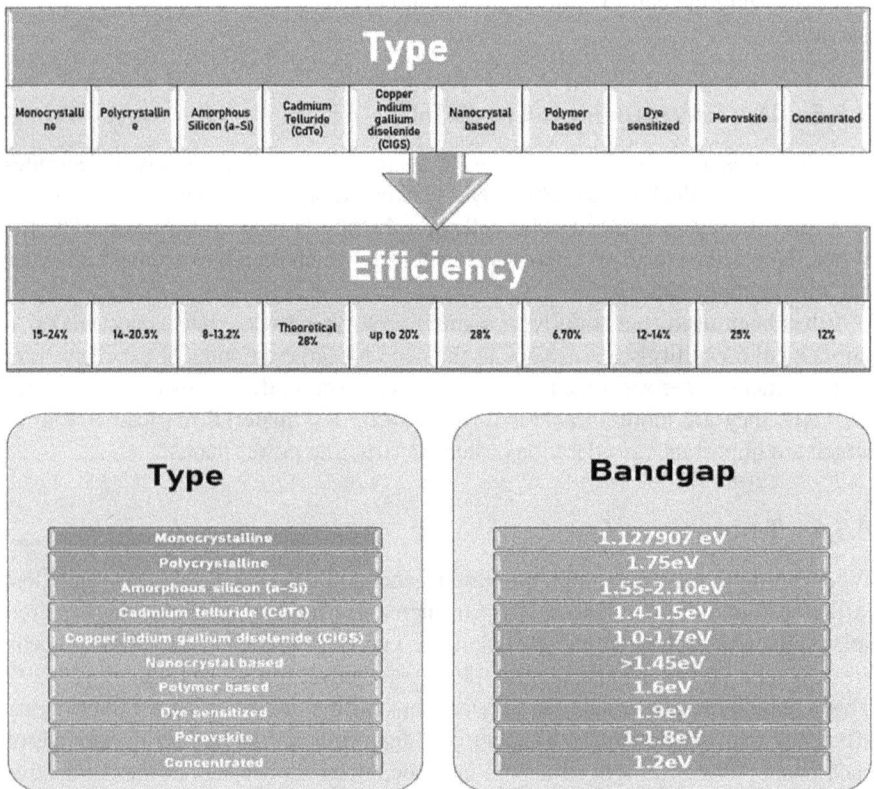

				Type					
Monocrystalline	Polycrystalline	Amorphous Silicon (a-Si)	Cadmium Telluride (CdTe)	Copper indium gallium diselenide (CIGS)	Nanocrystal based	Polymer based	Dye sensitized	Perovskite	Concentrated

				Efficiency					
15-24%	14-20.5%	8-13.2%	Theoretical 28%	up to 20%	28%	6.70%	12-14%	25%	12%

Type	Bandgap
Monocrystalline	1.127907 eV
Polycrystalline	1.75eV
Amorphous silicon (a-Si)	1.55-2.10eV
Cadmium telluride (CdTe)	1.4-1.5eV
Copper indium gallium diselenide (CIGS)	1.0-1.7eV
Nanocrystal based	>1.45eV
Polymer based	1.6eV
Dye sensitized	1.9eV
Perovskite	1-1.8eV
Concentrated	1.2eV

FIGURE 8.8   Comparison on the basis of (a) efficiency and (b) bandgap.

8.7 CONCLUSION AND FUTURE SCOPE

This work proposed a systematic and comprehensive review of early PV cells and recent advances in the field of solar cells in this review paper. Vast research has been done in this area for efficient harnessing of solar energy. Despite the fact that solar energy is free and accessible everywhere, there is a one-time cost associated with the equipment needed to gather it. In order to opt the rightful solar cell for an exact terrestrial location, the user needs to recognize the basic mechanisms along with widely studied different solar technologies. Although this energy has many benefits, it also has a few restrictions, like its intermittent nature by day due to changing weather conditions and unavailability of solar radiation at night. Solar cells are also known as PV system, and this chapter aims to present its classification. Finally, a comparison of different types of solar cells are performed on the basis of efficiency and bandgap. The future will involve emerging solar technology, which is still in its early stages. In the coming years, silicon alternatives are likely to appear on our solar farms and rooftops implementing renewable energy. These improvements will continue to be made possible by increased bulk solar cell production and new technologies that make the cells less expensive and more effective. Solar cells have potential applications in residential, commercial, industrial, tactical, and space applications.

REFERENCES

[1] Ahmad, F. (2020). *Optoelectronic Modeling and Optimization of Graded-Bandgap Thin-Film Solar Cells*. The Pennsylvania State University.
[2] Kannan, N., & Vakeesan, D. (2016). Solar energy for future world: A review. *Renewable and Sustainable Energy Reviews*, 62, 1092–1105.
[3] Wang, X., & Wang, Z. M. (2014). High-efficiency solar cells. *Physics, Materials, and Devices. Springer Series in Materials Science*, 190.
[4] Thakur, A., Singh, D., & Gill, S. K. (2022, August). Current scenario and perspective of wind-solar cogeneration in India. In *2022 Third International Conference on Intelligent Computing, Instrumentation and Control Technologies (ICICICT)*, IEEE, Accepted.
[5] Coskun, H. (2020). *Ion Diffusion and Interlayers in Perovskite Solar Cells* (Doctoral dissertation, National University of Singapore, Singapore).
[6] Bagdaş, A. C. (2017). *X-Ray Photoelectron Spectroscopy Analysis of Magnetron Sputtered Cu2ZnSnS4 Based Thin Film Solar Cells with CdS Buffer Layer* (Doctoral dissertation, Izmir Institute of Technology, Turkey).
[7] Ling, J., Wali, Q., Al-Douri, Y., & Jose, R. (2022). Fundamentals of solar cells. In *Renewable Energy: Analysis, Resources, Applications, Management, and Policy* (pp. 5–1). AIP Publishing LLC.
[8] Liu, Y., Akin, S., Pan, L., Uchida, R., Arora, N., Milić, J. V., & Grätzel, M. (2019). Ultrahydrophobic 3D/2D fluoroarene bilayer-based water-resistant perovskite solar cells with efficiencies exceeding 22%. *Science Advances*, 5(6), 2543.
[9] Fan, S., Wang, X., Cao, S., Wang, Y., Zhang, Y., & Liu, B. (2022). A novel model to determine the relationship between dust concentration and energy conversion efficiency of photovoltaic (PV) panels. *Energy*, 252, 123927.
[10] Chen, C. J. (2011). *Physics of Solar Energy*. John Wiley & Sons.
[11] Meinel, A. B., &Meinel, M. P. (1977). Applied solar energy: An introduction. *NASA STI/Recon Technical Report A*, 77, 33445.

[12] Kannan, N., & Vakeesan, D. (2016). Solar energy for future world: A review. *Renewable and Sustainable Energy Reviews*, 62, 1092–1105.

[13] Luft, W., Stafford, B., Von Roedern, B., & DeBlasio, R. (1992). Prospects for amorphous silicon photovoltaics. *Solar Energy Materials and Solar Cells*, 26(1–2), 17–26.

[14] Li, B. J. (2014). *The Study of Grain Boundaries in Polycrystalline Thin-Film Solar Cells*. Stanford University Press.

[15] Green, M. A. (2011). Ag requirements for silicon wafer-based solar cells. *Progress in Photovoltaics: Research and Applications*, 19(8), 911–916.

[16] Chopra, K. L., Paulson, P. D., & Dutta, V. (2004). Thin-film solar cells: An overview. *Progress in Photovoltaics: Research and applications*, 12(2–3), 69–92.

[17] Kessler, F., & Rudmann, D. (2004). Technological aspects of flexible CIGS solar cells and modules. *Solar Energy*, 77(6), 685–695.

[18] Ferekides, C. S., Balasubramanian, U., Mamazza, R., Viswanathan, V., Zhao, H., & Morel, D. L. (2004). CdTe thin film solar cells: Device and technology issues. *Solar Energy*, 77(6), 823–830.

[19] Mayer, A. C., Scully, S. R., Hardin, B. E., Rowell, M. W., & McGehee, M. D. (2007). Polymer-based solar cells. *Materials Today*, 10(11), 28–33.

[20] Yang, Y., Yang, W., Tang, W., & Sun, C. (2013). High-temperature solar cell for concentrated solar-power hybrid systems. *Applied Physics Letters*, 103(8), 083902.

[21] Thakur, A., Singh, D., & Gill, S. K. (2022, July). Comparative performance analysis and modelling of tin based planar perovskite solar cell. In *2022 International Conference on Intelligent Controller and Computing for Smart Power (ICICCSP)* (pp. 1–5). IEEE.

[22] Thakur, A., Singh, D., & Gill, S. K. (2022). Numerical simulations of 26.11% efficient planar CH3NH3PbI3 perovskite n-i-p solar cell. *Materials Today: Proceedings*, doi:10.1016/J.MATPR.2022.08.423.

[23] Kumar, S., & Scholes, G. D. (2008). Colloidal nanocrystal solar cells. *Microchimica Acta*, 160(3), 315–325.

[24] Alizadeh, A., Roudgar-Amoli, M., Bonyad-Shekalgourabi, S. M., Shariatinia, Z., Mahmoudi, M., & Saadat, F. (2022). Dye sensitized solar cells go beyond using perovskite and spinel inorganic materials: A review. *Renewable and Sustainable Energy Reviews*, 157, 112047.

[25] Chen, B., Ren, N., Li, Y., Yan, L., Mazumdar, S., Zhao, Y., & Zhang, X. (2022). Insights into the development of monolithic perovskite/silicon tandem solar cells. *Advanced Energy Materials*, 12(4), 2003628.

[26] Madan, J., Singh, K., & Pandey, R. (2021). Comprehensive device simulation of 23.36% efficient two-terminal perovskite-PbS CQD tandem solar cell for low-cost applications. *Scientific Reports*, 11(1), 1–13.

[27] Al-Mousoi, A. K., Mohammed, M. K., Pandey, R., Madan, J., Dastan, D., Ravi, G., & Sakthivel, P. (2022). Simulation and analysis of lead-free perovskite solar cells incorporating cerium oxide as electron transporting layer. *RSC Advances*, 12(50), 32365–32373.

9 A Brief Review on ANN Approach towards the Physicochemical Properties of Biodiesel

Gaurav Jain, Sunil Kumar, Rajesh Kumar

9.1 INTRODUCTION

The ever-growing population, automation, rapid growth in automobile sector, climate change, and continuous need to meet energy demands have created an opportunity for many to give viable energy solutions. Fossil fuels have been continuously expending and they pollute the environment too. This has created a need to find a long-term solution to this problem. Renewable energy is a feasible solution to the ever-increasing energy demands. One of the best alternatives for this problem is biodiesel, which can meet the ever-increasing energy demands. Biodiesel is the most suitable option for diesel engines, as it is environmentally friendly, inexhaustible, and renewable. Biodiesels are mono-alkyl esters of long-chain fatty acids derived from different generation feedstocks, according to ASTM. These oils react with oils react with an alcohol, producing methyl, ethyl, or propyl ester. Fatty-acid alkyl esters are produced from different oils like vegetable oils, animal fats, and waste cooking oils [1].

Biodiesel is divided into four generations. Biodiesel which are produced from edible oils, such as coconut oil, palm oil, and olive oil, are termed as first-generation biodiesels [2]. Second-generation biodiesels are produced from the non edible vegetable, oils such as jatropha, neem oil, karanja oil, and rubber seed oil [3,4]. Waste oils and microalgae which are found locally are used to produce third-generation biodiesels. The commonly used sources are waste cooking oils and fish oils [5]. Fourth-generation biodiesels are produced from synthetic biological technology, and still lot of work is required in this generation, although raw material is available in abundance and is cheap. Photobiological solar fuels and electro-fuels are considered in fourth generation of biodiesels [6].

9.2 PHYSICOCHEMICAL PROPERTIES OF BIODIESEL

Various characteristics of first- and second-generation feedstock have been reviewed in Table 9.1. Properties like density, pour point, acid value, cloud point, cetane

DOI: 10.1201/9781003367161-9

97

TABLE 9.1

Properties of First- and Second-generation Biodiesel Oils

Oil	Density at 15°C	Pour Point °C	Cloud Point °C	Cetane Number	Flash Point °C	Acid Value mg/g	Heating Value MJ/kg	Viscosity at 40°C (mm2/s)	Iodine Number	Ref. No.
Soya bean	882	−3.2	0.0	44.7	140.1	0.2	35.4	4.2	117.7	[9]
Coconut	867	−8.3	−1.6	64.7	113.8	0.2	35.2	3.2	—	[10]
Hazelnut	896	−6.0	−7.7	63.0	172.7	0.4	39.6	4.8	109.0	[11]
Canola	878	−8.0	−3.3	54.0	172.4	0.5	—	4.4	113.6	[11]
Palm	870	14.3	14.3	60.2	176.7	0.2	34.4	4.5	50.5	[12]
Mustard	879	−18.0	16.0	56.0	169.2	0.2	40.4	5.5	128.0	[13]
Rapeseed	879	−11.0	−3.5	48.3	169.5	0.3	35.8	4.4	112.0	[9,14]
Sunflower	869	−2.0	1.3	45.7	180.3	0.4	34.7	4.3	128.7	[15]
Cotton seed	887	−12.5	1.7	48.1	210.0	0.5	39.8	4.2	120.0	[16]
Ground nut	920	3.0	8.0	59.9	132.0	—	39.8	4.4	71.8	[17]
Jojoba	866	—	—	63.5	80.5	0.8	44.8	2.2	49.0	[18]
Sesame	867	−4.0	0.5	59.0	176.7	0.3	40.3	4.2	83.5	[19]
Peanut	878	11.5	12.6	58.2	176.0	—	35.3	4.7	67.5	[20]
Kusum	875	−2.0	—	—	152.0	0.4	—	5.3	37.6	[21]
Bitter almond	884	−6.0	4.5	45.2	169.0	0.3	—	4.6	117.3	[22]
Castor	922	−20.0	−11.2	37.6	178.6	0.1	38.1	17.1	85.5	[23]
Karanja	889	6.4	13.3	56.6	157.4	—	36.6	4.8	89.0	[24]
Babassu	872	—	4.0	63.3	117.0	0.4	31.8	4.2	—	[25]
Rice bran	889	−6.8	0.6	65.0	161.0	—	38.2	5.2	106.0	[26]
Camelina	885	−6.3	2.5	48.9	150.0	0.2	45.2	4.1	146.5	[27]
Linseed	852	−9.6	2.4	34.6	241.0	0.3	37.5	4.0	178.0	[28]
Jatropha	865	6.0	5.7	55.4	175.5	0.2	40.8	4.5	95.8	[29]
Mahua	895	4.3	4.3	55.0	129.5	0.4	36.9	4.8	74.2	[30]
Neem	886	7.0	14.5	51.3	144.8	—	39.8	6.1	46.8	[31]
Tobacco	865	−12.0	—	51.5	165.0	—	42.2	3.6	136.0	[32]
Rubber	875	−7.0	3.1	53	173.4	0.12	39.2	5.6	144.0	[33]

number (CN), flash point (FP), calorific value, and viscosity have been reviewed rigorously of 26 different oils.

Density being a crucial physical property is used to obtain quantity of fuel given by injection process for complete burning of oil. From Table 9.1, density of castor oil is found to be maximum (922 kg/m³), while linseed has the minimum value (852 kg/m³). Cloud point has negative and positive values. It is defined as the minimum possible temperature at which wax present in oil starts to be formed in crystals and look like clouds [7]. From Table 9.1, mustard oil has the maximum value (16°C), while castor oil has the minimum (11.2°C). High CN means the ability to self-ignite after fuel has been sent to burning chamber. The low value shows that the exhaust has high emission which leads to increase in knocking of engine. From Table 9.1, coconut oil has the highest value of CN (64.7), and linseed oil has lowest value (34.6).

Flash point (FP) is high in biodiesels as compared to conventional diesel. It is more than 150°C in biodiesel and 55–65°C in conventional diesel [8]. From Table 9.1, the value of FP of biodiesel oil derived from jojoba is less as compared to the rest of the oils. The minimum value of FP is for jojoba (80.5°C) and the maximum is for linseed (241°C). Pour point is found highest for palm oil (14.3°C) and lowest for castor oil (–20°C) per Table 9.1. Acid value is defined amount of free fatty acids in fuel sample. Its high value can lead to rusting in fuel delivery passage. According to Table 9.1, acid value is lowest for castor oil (0.1) and highest for jojoba (0.8).

Calorific value is defined as the amount of energy, which is given when a unit quantity of fuel is burned. Biodiesel has high oxygen level as compared to diesel, so it has less calorific value. From Table 9.1, camelina has highest value of CV (45.2 MJ/kg), and babassu has the lowest value (31.8 MJ/kg). Viscosity, the capability to flow, it is found to be more for biodiesel when compared with traditional diesel. High viscosity creates insufficient atomization, which leads to the deposit of impurities and other foreign particles. Per Table 9.1, it is maximum for castor oil (17.1 mm²/s) and minimum for jojoba (2.2 mm²/s). Iodine number (IN) is defined as the measure of unsaturation of an oil or fat. According to Table 9.1, it is maximum for linseed oil (178) and minimum for kusum (37.6).

9.3   ANN ARCHITECTURE

In recent years, ANN have been widely employed to anticipate the behaviour of complex systems. ANN can readily handle the complexity of the interaction of independent and dependent values. In many industrial problems, the forecasts made using ANN are highly accurate and reliable [34, 35]. Process optimization has gotten much easier as a result of the introduction of new ANN algorithms. The ANN technique has been employed by a number of researchers for prediction, optimization, and validation [36, 37]. The construction of an ANN is made up of layers: input, hidden, and output, each with a different number of neurons. Depending on the problem's complexity, the number of hidden layers can be one or many. In most cases, a multilayer neural network with a single hidden layer is used to make predictions, as shown in Figure 9.1.

ANN flowchart for predictions is shown in Figure 9.2. First of all, input data is collected from the experiments or literature. This data is normalized/standardized

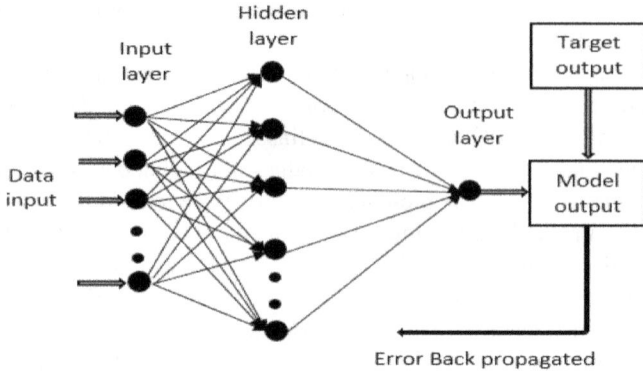

FIGURE 9.1   Schematic of ANN technique.

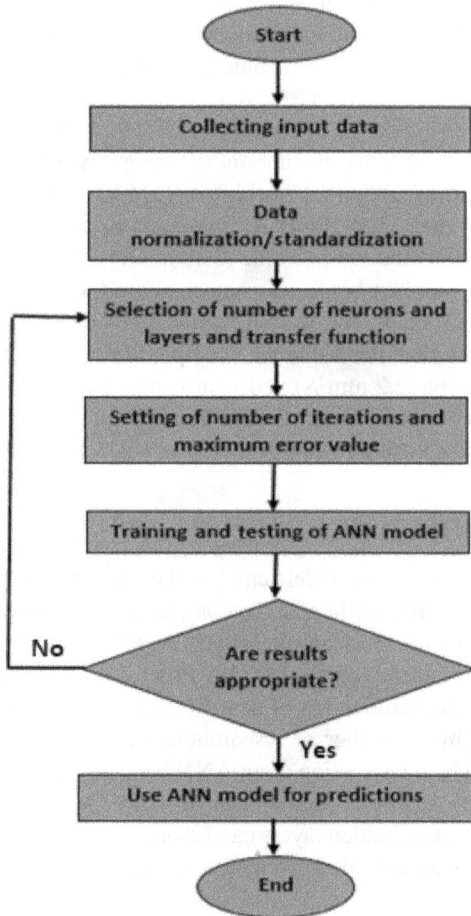

FIGURE 9.2   ANN flowchart for predictions.

before using the ANN model. For a particular ANN model, the numbers of neurons, hidden layers, and the type of transfer function are selected based upon the nature of the given input and required output. Then, the setting of the number of iterations and maximum error value has been done. Training and testing of the ANN model have been performed. The predictions are made only if the training and testing results are appropriate.

9.4 ANN APPLICATIONS TO BIODIESEL PHYSICOCHEMICAL PROPERTIES

Based on fatty-acid compositions, linear regression model and ANN were used to estimate viscosity and CN of biodiesels. In comparison, ANN model's predictions were more accurate. As previously assumed, the forecasted values are far closer to the experimental results [38]. A three-layered backpropagation ANN model has been designed by Najafi et al. [39] to predict the heating value of biodiesel. Biodiesel was made from vegetable oils (corn, sunflower, olive, canola, soybean, rice bran, and grapeseed oils) via the transesterification process. The content of fatty acid ethyl esters (FAEE) in biodiesels was determined using gas chromatography-mass spectrometry analysis. Biodiesel samples were made after the biodiesels were combined in various weight ratios. The heating values predicted by ANN were found to be substantially closer to experimental values. Percentage error, percentage accuracy, and R^2 were calculated and found to be 0.033, 99.8, and 0.9725, respectively. Hence, the ANN model was suggested for heating value predictions.

Piloto-Rodríguez [40] used the multiple linear regression (MLR) model to predict CN based on 10 FAMEs compositions in biodiesel and found that it could predict CN with an accuracy of 89%, whereas a single hidden layered backpropagation ANN could predict CN with an accuracy of 92%. In ANN predictions, the Levenberg–Marquardt algorithm was applied. ANN was chosen above MLR for the given set of data. Sánchez-Borroto et al. [41] suggested two multilayer Perceptron ANN models for the predictions of CN and ignition delay. A single hidden layered backpropagation network (11:4:1) employing the conjugate gradient descend algorithm was the best model for CN predictions. The 5:2:1 ANN was chosen to anticipate the ignition delay. The proposed networks were shown to be extremely useful in predicting biodiesel CN and ignition delay.

Rocabruno-Valdés et al. [42] used the ANN model to estimate the density, KV, and CN of biodiesel. The Levenberg-Marquardt method was employed in conjunction with the backpropagation network. The ANN model showed the potential for predicting biodiesel properties with low mean square error. Using logsig and purelin activation functions for hidden layers, Giwa et al. [43] created an ANN model to predict biodiesel characteristics. R^2 values of 0.967, 0.958, 0.991, and 0.994 for CN, KV, FP, and density, respectively, were very close to the experimental data. For CN, KV, FP, and density, the average absolute deviation percentages were 1.637, 1.689, 0.997, and 0.149, respectively.

ANN was used by Eryilmaz et al. [44] to forecast the KV of biodiesels. The seeds of wild mustard and safflower were employed in the transesterification process to produce biodiesels. Density, specific gravity, KV, calorific value, FP, water content, colour, cloud point, pour point, cold filter plugging point, copper strip corrosion, and pH were determined. Regression analyses were done and R^2, correlation constants, and RMSE were determined. ANN model with two hidden layers (1:7:7:3) with a backpropagation algorithm was developed. The performance of the ANN model was compared with the regression model. It was found that ANN comparatively gave better predictions with R^2 of 0.9999 and a maximum absolute percentage error (MAPE) of 0.34. Hence, ANN models were found to be best suitable for KV predictions. Barradas Filho et al. [45] used ANN to forecast biodiesel KV, iodine value, and induction period. Biodiesel samples containing 13 different types of FAMEs were used as ANN inputs. The outcomes of the ANN were compared using linear methods. In comparison to MLR, the ANN showed more predictability.

Cotton biodiesel density and viscosity were studied experimentally and predicted using ANN. The results revealed that increasing temperature lowered density and viscosity while increasing biodiesel concentration in the mixture increased density and viscosity. Best predictions for density were obtained using ANN with 0.02% MAPE, while the worst prediction results belong to linear regression of viscosity with 16.89% MAPE [46]. De Oliveira et al. [47] proposed the feedforward backpropagation ANN to predict FP, cetane index and sulphur content with MAPE of 4.6%, 0.4% and 3.3%, respectively. KV of waste cooking oil biodiesel blends were predicted using ANN by Gülüm et al. [48]. In this study, the rational model was found better than ANN model.

The ANN-predicted densities of various biodiesel-diesel-alcohol ternary blends were found to be quite close to the experimental density data, as were the exponential and linear models [49]. Yu and Zhao [50] proposed a backpropagation ANN model with generic technique for modelling the non-linear connections of biodiesel physicochemical parameters. The BPNN-GA model was able to map the attributes connected with FAMEs compositions. In comparison to other known regression models, the suggested BPNN-GA model predicts results with smaller errors. Arce et al. [51] employed an ANN model to forecast the biodiesel's physicochemical qualities. ANN produced highly accurate results for the given set of data. ANN and MLR were used by Ganeshmoorthy et al. [52] to predict the density and KV of fish oil biodiesel blends with diesel. R^2 and MAPE for MLR and ANN were calculated to check the superiority of the model. The ANN model has given the superior results. Eryilmaz et al. [53] investigated the prediction of KV of hazelnut oil methyl ester using regression analysis (RA) and ANN. Percentage error was found to be less for ANN compared to RA. R^2 for ANN was greater compared to RA.

Balabin et al. [54, 55] used MLR, principal component regression (PCR), and ANN for biodiesel properties predictions. The ANN approach was found to be superior to MLR and PCR. Baghban et al. [56] discussed CN predictions based on FAMEs employing PSO-ANN and TLBO-ANN algorithms. TLBO-ANN proved greater performance compared to PSO-ANN for a given set of data. Agarwal et al. [57] experimentally investigated various physicochemical properties of biodiesel produced from vegetable oils. The MLR and the ANN approach were used for predictions. It was found that ANN predictions were more accurate than MLR for a given set of data. Using four ANN models, Zheng et al. [58] estimated the viscosity of biodiesel blends. The models were trained and tested using 693 experimental data points. The CF-NN was shown to be the best model among four neural networks: GR-NN, RBF-NN, CF-NN, and MLP-NN. ANN was utilized by Ramadhas et al. [59] to predict the CN of biodiesel. The CN was predicted using a multi-layer feedforward, radial base, GR-NN, and recurrent network. Biodiesel has a predicted CN that was compared to the actual CN. This approach proved successful in predicting the CN of biodiesel.

Based on the composition of 12 FAMEs from 131 different biodiesel samples, Miraboutalebi et al. [60] developed RF and ANN machine learning algorithms to estimate the value of CN. Both models have great accuracy. For the prediction of the biodiesel KV, Meng et al. [61] built an ANN model. To compare the findings with ANN, the Knothe-Steidley and Ramrez-Verduzco approaches were applied. According to the results, the ANN technique has predicted the best results. With an R^2 of 0.9774, the ANN approach had the highest accuracy. Hosseini et al. [62] used a perceptron ANN model with a single hidden layer to predict the dynamic viscosity of biodiesels. For training, testing, and validating the findings, the proposed ANN model was found to be effective.

9.5 DISCUSSION AND FUTURE SCOPE

In the continuous biodiesel production business, determining biodiesel's physico-chemical properties is a time-consuming operation. Based on the fatty-acid content, ANN models are created to predict the attributes. The use of ANN to predict the properties of biodiesel is extremely useful in terms of reducing number of experiments. The accuracy of the model depends upon the proper selection of neurons and suitable transfer function. The predicted physicochemical characteristics of biodiesel are of the utmost importance in the diesel sector.

In recent years, exponential growth has been observed in the research field of biodiesel production, process optimization and predictions using ANN. Figure 9.3

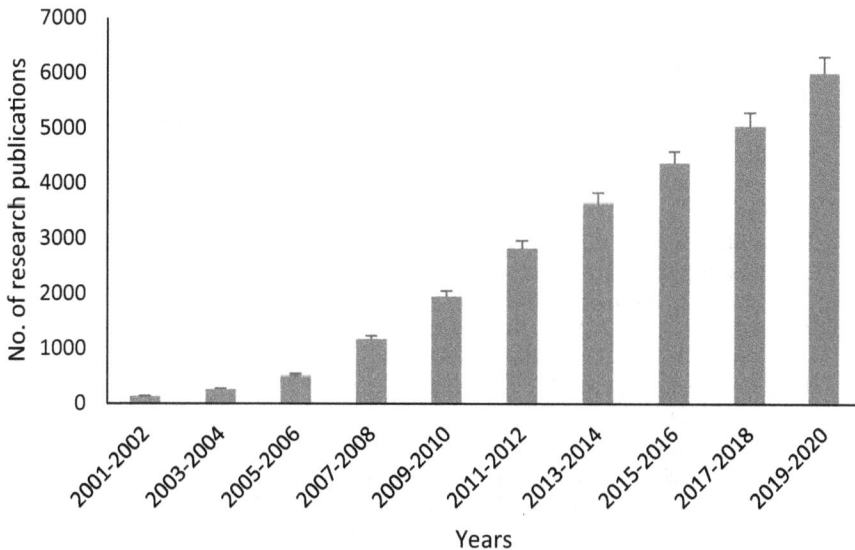

FIGURE 9.3   Research growth in biodiesels over the years.

shows the number of research published on ANN applications on biodiesels over the years per the Google Scholar database. Due to the advancement in computational technology, more ANN algorithms are in the developing stage, which would decrease the processing time and increase the accuracy of the results. The development of new ANN models will be highly beneficial to the industry and research community.

9.6 CONCLUSION

It has been concluded by critically reviewing the literature that predictions of biodiesel physicochemical properties are very important to reduce the number of experiments. ANN has been proven as a promising technique to predict the properties. These predictions are very accurate (low RMSE and high R^2). The feedforward backpropagation ANN model was widely used by the researchers. The most used ANN algorithms for predicting the biodiesel properties are the conjugate gradient descend algorithm, intelligent genetic algorithm, and Levenberg-Marquardt. Logsig, purelin, and tan hyperbolic transfer functions were used depending upon the nature of the required output.

REFERENCES

[1] Yusuf, N. N. A. N., Kamarudin, S. K., & Yaakub, Z. (2011). Overview on the current trends in biodiesel production. *Energy Conversion and Management*, 52(7), 2741–2751.
[2] Mahdavi, M., Abedini, E., & hosein Darabi, A. (2015). Biodiesel synthesis from oleic acid by nano-catalyst (ZrO2/Al2O3) under high voltage conditions. *RSC Advances*, 5(68), 55027–55032.
[3] Peer, M. S., Kasimani, R., Rajamohan, S., & Ramakrishnan, P. (2017). Experimental evaluation on oxidation stability of biodiesel/diesel blends with alcohol addition by rancimat instrument and FTIR spectroscopy. *Journal of Mechanical Science and Technology*, 31(1), 455–463.
[4] Shameer, P. M., & Ramesh, K. (2017). Green technology and performance consequences of an eco-friendly substance on a 4-stroke diesel engine at standard injection timing and compression ratio. *Journal of Mechanical science and Technology*, 31(3), 1497–1507.
[5] Tariq, M., Ali, S., & Khalid, N. (2012). Activity of homogeneous and heterogeneous catalysts, spectroscopic and chromatographic characterization of biodiesel: A review. *Renewable and Sustainable Energy Reviews*, 16(8), 6303–6316.
[6] Singh, D., Sharma, D., Soni, S. L., Sharma, S., Sharma, P. K., & Jhalani, A. (2020). A review on feedstocks, production processes, and yield for different generations of biodiesel. *Fuel*, 262, 116553.
[7] Sakthivel, R., Ramesh, K., Purnachandran, R., & Shameer, P. M. (2018). A review on the properties, performance and emission aspects of the third generation biodiesels. *Renewable and Sustainable Energy Reviews*, 82, 2970–2992.
[8] Bhuiya, M. M. K., Rasul, M. G., Khan, M. M. K., Ashwath, N., & Azad, A. K. (2016). Prospects of 2nd generation biodiesel as a sustainable fuel—Part: 1 selection of feedstocks, oil extraction techniques and conversion technologies. *Renewable and Sustainable Energy Reviews*, 55, 1109–1128.

[9] Mihaela, P., Josef, R., Monica, N., & Rudolf, Z. (2013). Perspectives of safflower oil as biodiesel source for South Eastern Europe (comparative study: Safflower, soybean and rapeseed). *Fuel*, 111, 114–119.

[10] Abollé, A., Loukou, K., & Henri, P. (2009). The density and cloud point of diesel oil mixtures with the straight vegetable oils (SVO): Palm, cabbage palm, cotton, groundnut, copra and sunflower. *Biomass and Bioenergy*, 33(12), 1653–1659.

[11] Öztürk, E. (2015). Performance, emissions, combustion and injection characteristics of a diesel engine fuelled with canola oil–hazelnut soapstock biodiesel mixture. *Fuel Processing Technology*, 129, 183–191.

[12] Moser, B. R., & Vaughn, S. F. (2010). Evaluation of alkyl esters from Camelina sativa oil as biodiesel and as blend components in ultra low-sulfur diesel fuel. *Bioresource Technology*, 101(2), 646–653.

[13] Sanjid, A., Masjuki, H. H., Kalam, M. A., Abedin, M. J., & Rahman, S. A. (2014). Experimental investigation of mustard biodiesel blend properties, performance, exhaust emission and noise in an unmodified diesel engine. *APCBEE Procedia*, 10, 149–153.

[14] Ramadhas, A. S., Jayaraj, S., & Muraleedharan, C. (2005). Biodiesel production from high FFA rubber seed oil. *Fuel*, 84(4), 335–340.

[15] Younis, K. A., Gardy, J. L., & Barzinji, K. S. (2014). Production and characterization of biodiesel from locally sourced sesame seed oil, used cooking oil and other commercial vegetable oils in Erbil-Iraqi Kurdistan. *American Journal of Applied Chemistry*, 2(6), 105–111.

[16] No, S. Y. (2011). Inedible vegetable oils and their derivatives for alternative diesel fuels in CI engines: A review. *Renewable and Sustainable Energy Reviews*, 15(1), 131–149.

[17] Bello, E. I., & Daniel, F. (2015). Optimization of groundnut oil biodiesel production and characterization. *Applied Science Reports*, 9(3), 172–180.

[18] Canoira, L., Alcantara, R., García-Martínez, M. J., & Carrasco, J. (2006). Biodiesel from Jojoba oil-wax: Transesterification with methanol and properties as a fuel. *Biomass and Bioenergy*, 30(1), 76–81.

[19] Saydut, A., Duz, M. Z., Kaya, C., Kafadar, A. B., & Hamamci, C. (2008). Transesterified sesame (Sesamum indicum L.) seed oil as a biodiesel fuel. *Bioresource Technology*, 99(14), 6656–6660.

[20] Singh, S. P., & Singh, D. (2010). Biodiesel production through the use of different sources and characterization of oils and their esters as the substitute of diesel: A review. *Renewable and Sustainable Energy Reviews*, 14(1), 200–216.

[21] Sharma, Y. C., & Singh, B. (2010). An ideal feedstock, kusum (Schleichera triguga) for preparation of biodiesel: Optimization of parameters. *Fuel*, 89(7), 1470–1474.

[22] Fadhil, A. B., Aziz, A. M., & Altamer, M. H. (2016). Potassium acetate supported on activated carbon for transesterification of new non-edible oil, bitter almond oil. *Fuel*, 170, 130–140.

[23] Sánchez, N., Sánchez, R., Encinar, J. M., González, J. F., & Martínez, G. (2015). Complete analysis of castor oil methanolysis to obtain biodiesel. *Fuel*, 147, 95–99.

[24] Sahoo, P. K., Das, L. M., Babu, M. K. G., Arora, P., Singh, V. P., Kumar, N. R., & Varyani, T. S. (2009). Comparative evaluation of performance and emission characteristics of jatropha, karanja and polanga based biodiesel as fuel in a tractor engine. *Fuel*, 88(9), 1698–1707.

[25] Abreu, F. R., Lima, D. G., Hamú, E. H., Wolf, C., & Suarez, P. A. (2004). Utilization of metal complexes as catalysts in the transesterification of Brazilian vegetable oils with different alcohols. *Journal of Molecular Catalysis A: Chemical*, 209(1–2), 29–33.

[26] Kong, W., Kang, Q., Feng, W., & Tan, T. (2015). Improving the solvent-extraction process of rice bran oil. *Chemical Engineering Research and Design*, 104, 1–10.

[27] Özçelik, A. E., Aydoğan, H., & Acaroğlu, M. (2015). Determining the performance, emission and combustion properties of camelina biodiesel blends. *Energy Conversion and Management*, 96, 47–57.

[28] Dixit, S., & Rehman, A. (2012). Linseed oil as a potential resource for bio-diesel: A review. *Renewable and Sustainable Energy Reviews*, 16(7), 4415–4421.

[29] Ganapathy, T., Gakkhar, R. P., & Murugesan, K. (2011). Influence of injection timing on performance, combustion and emission characteristics of Jatropha biodiesel engine. *Applied Energy*, 88(12), 4376–4386.

[30] Saravanan, N., Nagarajan, G., & Puhan, S. (2010). Experimental investigation on a DI diesel engine fuelled with Madhuca Indica ester and diesel blend. *Biomass and Bioenergy*, 34(6), 838–843.

[31] Agarwal, A. K., Khurana, D., & Dhar, A. (2015). Improving oxidation stability of bio-diesels derived from Karanja, Neem and Jatropha: Step forward in the direction of commercialisation. *Journal of Cleaner Production*, 107, 646–652.

[32] Usta, N., Aydoğan, B., Çon, A. H., Uğuzdoğan, E., & Özkal, S. G. (2011). Properties and quality verification of biodiesel produced from tobacco seed oil. *Energy Conversion and Management*, 52(5), 2031–2039.

[33] Onoji, S. E., Iyuke, S. E., Igbafe, A. I., & Nkazi, D. B. (2016). Rubber seed oil: A potential renewable source of biodiesel for sustainable development in sub-Saharan Africa. *Energy Conversion and Management*, 110, 125–134.

[34] Kumar, S., Kumar, V., & Singh, A. K. (2020). Prediction of maximum pressure of journal bearing using ANN with multiple input parameters. *Australian Journal of Mechanical Engineering*, 1–10.

[35] Kumar, S., Kumar, V., & Singh, A. K. (2020). Predictions of minimum fluid film thickness of journal bearing using feed-forward neural network. In *Proceedings of International Conference in Mechanical and Energy Technology* (pp. 229–237). Springer, Singapore.

[36] Liu, R., Yang, B., Zio, E., & Chen, X. (2018). Artificial intelligence for fault diagnosis of rotating machinery: A review. *Mechanical Systems and Signal Processing*, 108, 33–47.

[37] Kumar, S., Kumar, V., & Singh, A. K. (2021). Artificial neural network model development for the analysis of maximum pressure of hole entry journal bearing using SciLab. In *Emerging Trends in Mechanical Engineering* (pp. 19–29). Springer, Singapore.

[38] Cheenkachorn, K. (2004). Predicting properties of biodiesels using statistical models and artificial neural networks. In *Proceedings of the Joint International Conference on Sustainable Energy and Environment* (Vol. 13), Hua Hin, Thailand.

[39] Najafi, B., Fakhr, M. A., & Jamali, S. (2011). Prediction of heating value of vegetable oil-based ethyl esters biodiesel using artificial neural network. *Tarım Makinaları Bilimi Dergisi*, 7(4), 361–366.

[40] Piloto-Rodríguez, R., Sánchez-Borroto, Y., Lapuerta, M., Goyos-Pérez, L., & Verhelst, S. (2013). Prediction of the cetane number of biodiesel using artificial neural networks and multiple linear regression. *Energy Conversion and Management*, 65, 255–261.

[41] Sánchez-Borroto, Y., Piloto-Rodriguez, R., Errasti, M., Sierens, R., & Verhelst, S. (2014). Prediction of cetane number and ignition delay of biodiesel using artificial neural networks. *Energy Procedia*, 57, 877–885.

[42] Rocabruno-Valdés, C. I., Ramírez-Verduzco, L. F., & Hernández, J. A. (2015). Artificial neural network models to predict density, dynamic viscosity, and cetane number of biodiesel. *Fuel*, 147, 9–17.

[43] Giwa, S. O., Adekomaya, S. O., Adama, K. O., & Mukaila, M. O. (2015). Prediction of selected biodiesel fuel properties using artificial neural network. *Frontiers in Energy*, 9(4), 433–445.

[44] Eryilmaz, T., Yesilyurt, M. K., Taner, A., & Celik, S. A. (2015). Prediction of kinematic viscosities of biodiesels derived from edible and non-edible vegetable oils by using artificial neural networks. *Arabian Journal for Science and Engineering*, 40(12), 3745–3758.

[45] Barradas Filho, A. O., Barros, A. K. D., Labidi, S., Viegas, I. M. A., Marques, D. B., Romariz, A. R., . . . & Marques, E. P. (2015). Application of artificial neural networks to predict viscosity, iodine value and induction period of biodiesel focused on the study of oxidative stability. *Fuel*, 145, 127–135.

[46] Özgür, C., & Tosun, E. (2017). Prediction of density and kinematic viscosity of biodiesel by artificial neural networks. *Energy Sources, Part A: Recovery, Utilization, and Environmental Effects*, 39(10), 985–991.

[47] de Oliveira, F. M., de Carvalho, L. S., Teixeira, L. S., Fontes, C. H., Lima, K. M., Câmara, A. B., . . . & Sales, R. V. (2017). Predicting cetane index, flash point, and content sulfur of diesel–biodiesel blend using an artificial neural network model. *Energy & Fuels*, 31(4), 3913–3920.

[48] Gülüm, M., Onay, F. K., & Bilgin, A. (2018). Comparison of viscosity prediction capabilities of regression models and artificial neural networks. *Energy*, 161, 361–369.

[49] Gülüm, M., Onay, F. K., & Bilgin, A. (2019). Measurement and estimation of densities of different biodiesel–diesel–alcohol ternary blends. *Environmental Progress & Sustainable Energy*, 38(6), e13248.

[50] Yu, W., & Zhao, F. (2019). Prediction of critical properties of biodiesel fuels from FAMEs compositions using intelligent genetic algorithm-based back propagation neural network. *Energy Sources, Part A: Recovery, Utilization, and Environmental Effects*, 1–14.

[51] Arce, P. F., Guimarães, D. H., & de Aguirre, L. R. (2019). Experimental data and prediction of the physical and chemical properties of biodiesel. *Chemical Engineering Communications*, 206(10), 1273–1285.

[52] Ganeshmoorthy, V., Muthukannan, M., & Thirugnanasambandam, M. (2019, August). Comparison of linear regression and ANN of fish oil biodiesel properties prediction. In *Journal of Physics: Conference Series* (Vol. 1276, No. 1, p. 012074). IOP Publishing.

[53] Eryilmaz, T., Arslan, M., Yesilyurt, M. K., & Taner, A. (2016). Comparison of empirical equations and artificial neural network results in terms of kinematic viscosity prediction of fuels based on hazelnut oil methyl ester. *Environmental Progress & Sustainable Energy*, 35(6), 1827–1841.

[54] Balabin, R. M., Lomakina, E. I., & Safieva, R. Z. (2011). Neural network (ANN) approach to biodiesel analysis: Analysis of biodiesel density, kinematic viscosity, methanol and water contents using near infrared (NIR) spectroscopy. *Fuel*, 90(5), 2007–2015.

[55] Balabin, R. M., & Safieva, R. Z. (2011). Near-infrared (NIR) spectroscopy for biodiesel analysis: Fractional composition, iodine value, and cold filter plugging point from one vibrational spectrum. *Energy & Fuels*, 25(5), 2373–2382.

[56] Baghban, A., Kardani, M. N., & Mohammadi, A. H. (2018). Improved estimation of Cetane number of fatty acid methyl esters (FAMEs) based biodiesels using TLBO-NN and PSO-NN models. *Fuel*, 232, 620–631.

[57] Agarwal, M., Singh, K., & Chaurasia, S. P. (2010). Prediction of biodiesel properties from fatty acid composition using linear regression and ANN techniques. *Indian Chemical Engineer*, 52(4), 347–361.

[58] Zheng, Y., Shadloo, M. S., Nasiri, H., Maleki, A., Karimipour, A., & Tlili, I. (2020). Prediction of viscosity of biodiesel blends using various artificial model and comparison with empirical correlations. *Renewable Energy*, 153, 1296–1306.

[59] Ramadhas, A. S., Jayaraj, S., Muraleedharan, C., & Padmakumari, K. (2006). Artificial neural networks used for the prediction of the cetane number of biodiesel. *Renewable Energy*, 31(15), 2524–2533.

[60] Miraboutalebi, S. M., Kazemi, P., & Bahrami, P. (2016). Fatty acid methyl ester (FAME) composition used for estimation of biodiesel cetane number employing random forest and artificial neural networks: A new approach. *Fuel*, 166, 143–151.

[61] Meng, X., Jia, M., & Wang, T. (2014). Neural network prediction of biodiesel kinematic viscosity at 313 K. *Fuel*, 121, 133–140.

[62] Hosseini, S. M., Pierantozzi, M., & Moghadasi, J. (2019). Viscosities of some fatty acid esters and biodiesel fuels from a rough hard-sphere-chain model and artificial neural network. *Fuel*, 235, 1083–1091.

10 Effect of Rotation in a Maxwell Fluid in Porous Medium Heated Underside

Renu Bala, Tania Bose, Inderpreet Kaur

Table of Parameters

a	Overall horizontal wave number
d	Height of the porous layer
g	Gravitational acceleration
R	Rayleigh number
T_a	Taylor number
P_r	Prandtl number
η	Porous parameter
σ	Complex growth rate
R_s	Rayleigh number for stationary convection.
R_a^{osc}	Oscillatory Rayleigh number

10.1 INTRODUCTION

The thermal variability of a fluid layer heated from below, also known as Rayleigh-Benard convection. Many researchers like Benard [1], Rayleigh [2], Rajagopal et al. [3], and Sekhar and Jayalatha [4] have been investigated on this. Thermal instability plays a vital role in many branches like oceanography, geophysics, atmospheric physics, and so on. Chandrasekhar [5], Drazin and Reid [6], and Bejan [7] did a detailed theoretical and experimental studies of Rayleigh-Benard convection in Newtonian fluids.

Thermal convection in a fluid-saturated porous medium has attracted the interest of engineers and scientists for a long time due to its numerous applications. Understanding of thermal convective behaviour of fluids in porous medium is important for controlling many processes, such as geothermal energy utilization, oil reservoir modelling, catalytic packed beds, filtration, building thermal insulation, and nuclear waste disposal. Horton and Rogers [8] were the first who studied the thermal convection in a porous medium as an extension of the Rayleigh-Bénard instability phenomenon and after that based on Darcy's law it was studied by Lapwood [9].

Many other researchers Katto and Masuoka [10], Bejan [11], Mckibbin and O'Sullivan [12], Khaled and Vafai [13], Agra et al. [14], and Gupta et al. [15] studied different aspects of thermal convection of Newtonian fluids in porous media. For an overview of thermal convection in porous medium, one may be referred to Vafai [16] and Nield and Bejan [17]. Straughan [18] analysed the occurrence of overstabilities in porous layers. Maxwell fluid is a viscoelastic fluid having the properties both of elasticity and viscosity. The Maxwell fluid model was originally developed by Maxwell [19] to describe the elastic and viscous response of air. Maxwell considered a model of fluid flow that was in essence a spring in series with viscous drag. The applied stress was felt by both components, and the response was the sum of the responses of both components. Maxwell Jeffrey model is used by Rudraiah et al. [20] and Bertola and Cafaro [21] to investigate a viscoelastic fluid in a porous layer heated from the underside.

The origin of study on the rotation effect on convection instability was studied by Chandrasekhar [22], Vadasz [23], and Veronis [24] and gave theoretical and practical applications. Bhatia and Steiner [25] show that the rotation has destabilizing effects on the overstability mode convection. These results are contradictory with stabilizing influence of rotation in the case of an ordinary viscous fluid. The Coriolis effect on thermal convection, in which gravity buoyancy was neglected was investigated by Vadasz [26] and Govender [27]. Laroze et al. [28] studied thermal convection in a rotating binary viscoelastic liquid mixture.

The modified Darcy-Brinkman-Maxwell model based on local volume averaging technique was introduced by Tan and Masuoka [29]. Double-diffusive natural convection of Maxwell fluid with porous medium which is based upon the modified Darcy-Maxwell model having effect of solutal buoyancy forces was studied by Wang and Tan [30]. Wang and Tan [31] performed linear and nonlinear stability analyses on the double-diffusive convection in a Maxwell fluid-saturated porous medium with the Soret effect. Yin et al. [32] investigated the effects of hydrodynamic boundary and constant flux heating conditions on a Maxwell fluid in a horizontal porous layer heated from below.

In this chapter, the results of the linear stability analysis have been extended for thermal convection in a Maxwell fluid layer when heated from below, under the influence of uniform vertical rotation. The objective in this chapter is to study the behaviour of various parameters like Prandtl number, Taylor number, and porous parameter on the onset of both stationary and oscillatory convection. The results indicate that for the state of stationary convection the porous parameter and Taylor number has stabilizing effect on the system and the state of oscillatory convection the Taylor number and Prandtl number has destabilizing effect on the system.

10.2   MATHEMATICAL FORMULATION

We have consider an infinite horizontal layer of Maxwellian viscoelastic Boussinesq fluid of thickness 'd' bounded by plane $z = 0$ and $z = d$ in porous medium of constant porosity ε and medium permeability K. The layer is rotating with angular velocity

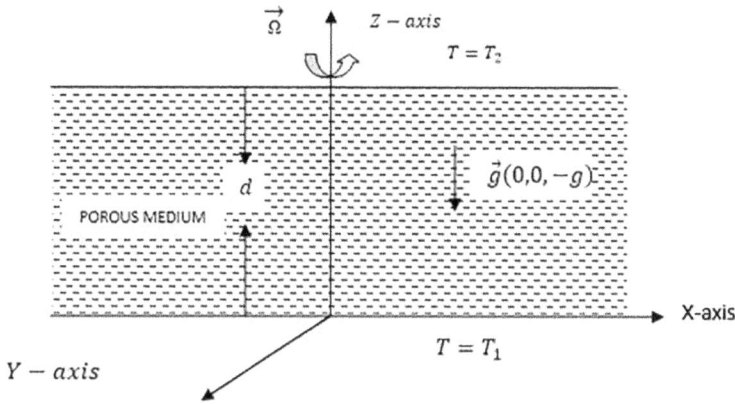

FIGURE 10.1 Physical configuration.

$\Omega(0,0,\Omega)$. The porous layer is heated from below. The modified Darcy-Brinkman-Maxwell model has been used to analyse the problem (Tan and Masuoka) [29].

The governing equations for the system are given by

$$\nabla.V = 0, \tag{10.1}$$

$$\rho_0 \left(1 + t_m \frac{\partial}{\partial t}\right) \frac{\partial V}{\partial t} = -\left(1 + t_m \frac{\partial}{\partial t}\right) \left[\nabla p - \vec{k} g \rho_0 \alpha T - 2\rho_0 \left(V \times \Omega\right)\right] + \mu \nabla^2 V - \frac{\mu}{K} V, \tag{10.2}$$

$$\varepsilon \frac{\partial T}{\partial t} + V.\nabla T = \kappa \nabla^2 T, \tag{10.3}$$

where $V = (u, v, w)$, $g, \kappa, \vec{k}, p, \rho_0, T, \mu, \alpha$ denote the volume average velocity, acceleration due to gravity, thermal diffusivity, unit vector along z direction which is vertically upward, pressure, density, temperature, viscosity, and thermal coefficient of expansion of fluid, respectively. The equation of state is

$$\rho = \rho_0 \left(1 - \alpha \left(T - T_0\right)\right), \tag{10.4}$$

where ρ_0, T_0 stands for density and temperature at lower boundary $z = 0$.

The steady state solution is

$$V_b = (0,0,0), \ \rho = \rho_b(z), \ p = p_{b(z)}, \ T = T_{b(z)}. \tag{10.5}$$

To study the stability of the system, the following perturbation equations are used

$$V = V_b + V', \ \rho = \rho_b + \rho', \ p = p_b + p', \ T = T_b + T', \tag{10.6}$$

where the primes denote the perturbation quantities related to the basic state indicated by the subscript 'b'. Now after substituting equation (10.6) in equations (10.1)–(10.4) and neglecting all higher order terms of the small quantities, we have

$$\nabla.V' = 0, \tag{10.7}$$

$$\rho_0\left(1+t_m\frac{\partial}{\partial t}\right)\frac{\partial V'}{\partial t} = -\left(1+t_m\frac{\partial}{\partial t}\right)\left[\nabla p' - \vec{k}g\rho_0\alpha T' - 2\rho_0\left(V\times\Omega\right)\right]$$

$$+\mu\nabla^2 V' - \frac{\mu}{K}V', \tag{10.8}$$

$$\varepsilon\frac{\partial T'}{\partial t} - \frac{\Delta T}{d}w' = \kappa\nabla^2 T'. \tag{10.9}$$

In Cartesian form, the equations (10.7)–(10.9) can be written as

$$\frac{\partial u'}{\partial x} + \frac{\partial v'}{\partial y} + \frac{\partial w'}{\partial z} = 0, \tag{10.10}$$

$$\rho_0\left(1+t_m\frac{\partial}{\partial t}\right)\frac{\partial u'}{\partial t} = -\left(1+t_m\frac{\partial}{\partial t}\right)\left[\frac{\partial p'}{\partial x} - 2\rho_0\Omega v\right] + \mu\nabla^2 u' - \frac{\mu}{K}u', \tag{10.11}$$

$$\rho_0\left(1+t_m\frac{\partial}{\partial t}\right)\frac{\partial v'}{\partial t} = -\left(1+t_m\frac{\partial}{\partial t}\right)\left[\frac{\partial p'}{\partial y} + 2\rho_0\Omega v\right] + \mu\nabla^2 v' - \frac{\mu}{K}v', \tag{10.12}$$

$$\rho_0\left(1+t_m\frac{\partial}{\partial t}\right)\frac{\partial w'}{\partial t} = -\left(1+t_m\frac{\partial}{\partial t}\right)\left[\frac{\partial p'}{\partial z} - g\rho_0\alpha T'\right] + \mu\nabla^2 w' - \frac{\mu}{K}w', \tag{10.13}$$

$$\varepsilon\frac{\partial T'}{\partial t} - \frac{\Delta T}{d}w' = \kappa\nabla^2 T'. \tag{10.14}$$

Operating equations (10.11) and (10.12) by $\dfrac{\partial}{\partial x}$ and $\dfrac{\partial}{\partial y}$, respectively, adding and using equation (10.10).

Then eliminating p' between this resulting equation and equation (10.13), we get

$$\left[\rho_0\left(1+t_m\frac{\partial}{\partial t}\right)\frac{\partial}{\partial t} - \mu\nabla^2 + \frac{\mu}{K}\right]\nabla^2 w' = \left(1+t_m\frac{\partial}{\partial t}\right)$$

$$\left[g\rho_0\nabla_1^2\alpha T' - 2\rho_0\Omega\frac{\partial\zeta}{\partial z}\right]. \tag{10.15}$$

Now operating equation (10.11) by $-\dfrac{\partial}{\partial y}$ and equation (10.12) by $\dfrac{\partial}{\partial x}$, and adding, we get an equation for vorticity as

$$\rho_0\left(1+t_m\frac{\partial}{\partial t}\right)\frac{\partial\zeta}{\partial t}\nabla^2 w' = \left(1+t_m\frac{\partial}{\partial t}\right)2\rho_0\Omega\frac{\partial w'}{\partial z} + \mu\nabla^2\zeta - \frac{\mu}{K}\zeta, \tag{10.16}$$

where $\zeta = \dfrac{\partial v'}{\partial x} - \dfrac{\partial u'}{\partial y}$ is z component of vorticity.

Now equations (10.14)–(10.16) are non-dimensionalized by using following non-dimensional quantities defined by

$$x = x^* d, \ y = y^* d, \ z = z^* d, \ w = \frac{w^* \kappa}{d}, \ t = \frac{t^* d^2}{\kappa}, \ T = T^* \Delta T, \ \zeta = \frac{\zeta^* \kappa}{d^2}, \ \nabla^2 = \frac{\nabla^{*2}}{d^2},$$

$$\nabla_1^2 = \frac{\nabla_1^{*2}}{d^2}, \ \eta = \frac{d^2}{K}, \ P_r = \frac{v}{\kappa}, \ \lambda = \frac{t_m \kappa}{d^2}, \ \text{to yield}$$

$$\left[\frac{1}{P_r} \left(1 + \lambda \frac{\partial}{\partial t} \right) \frac{\partial}{\partial t} - \nabla^2 + \eta \right] \nabla^2 w = \left(1 + \lambda \frac{\partial}{\partial t} \right) \left[R \nabla_1^2 T - T_a^{1/2} \frac{\partial \zeta}{\partial z} \right], \qquad (10.17)$$

$$\left[\frac{1}{P_r} \left(1 + \lambda \frac{\partial}{\partial t} \right) \frac{\partial}{\partial t} - \nabla^2 + \eta \right] \zeta = T_a^{1/2} \frac{\partial w}{\partial z}, \qquad (10.18)$$

$$\left[\varepsilon \frac{\partial}{\partial t} - \nabla^2 \right] T = w, \qquad (10.19)$$

where $R = \dfrac{g \alpha \Delta T d^3}{\kappa v}$ and $T_a = \dfrac{4 \Omega^2 d^4}{v^2}$ are the Rayleigh number and the Taylor number, $P_r = \dfrac{v}{\kappa}$ is Prandtl number and $\eta = \dfrac{d^2}{K}$ is porous parameter.

10.3   LINEAR STABILITY ANALYSIS

To study the linear stability analysis, we use the normal mode analysis procedure in which we consider the solution of the form

$$(w, T, \zeta) = (w, \theta, Z)(z) exp(ilx + imy + \sigma t), \qquad (10.20)$$

where the horizontal wave numbers in x and y direction are l and m, and $\sigma = \sigma_r + i\sigma_i$ is the complex growth rate.

Substituting equation (10.20) in equations (10.17)–(10.19), we obtain

$$\left[\frac{1}{P_r} (1 + \lambda \sigma) \sigma - (D^2 - \beta^2) + \eta \right] (D^2 - \beta^2) W =$$
$$(1 + \lambda \sigma)(-R\beta^4 \theta - T_a^{1/2} DZ), \qquad (10.21)$$

$$\left[\frac{1}{P_r} (1 + \lambda \sigma) \sigma - (D^2 - \beta^2) + \eta \right] Z = T_a^{1/2} DW, \qquad (10.22)$$

$$\left[\sigma - (D^2 - \beta^2) \right] \theta = W, \qquad (10.23)$$

where $D = \dfrac{d}{dz}$, is the differential operator in z direction and $\beta^2 = l^2 + m^2$ is the horizontal wave number.

Equations (10.21)–(10.23) are solved for eigenvalues by using the following free boundary conditions:

$$w = D^2 w = \theta = DZ = 0 \text{ at } z = 0 \text{ and } z = 1. \tag{10.24}$$

The solutions of equations (10.21)–(10.23) satisfying the boundary condition (10.24) should be a periodic wave of the form:

$$W = A \sin \pi z, \theta = B \sin \pi z, \; Z = -\dfrac{C}{\pi} \cos \pi z, DZ = C \sin \pi z. \tag{10.25}$$

Substituting equation (10.25) in equations (10.21)–(10.23), the condition for the existence of a non-trivial eigenvalue is found:

$$\begin{vmatrix} \pi^2 T_a^{1/2} & 0 & \Gamma^2 + \eta + \dfrac{1}{P_r}(1+\lambda\sigma)\sigma \\[2mm] \Gamma^2 \left[\Gamma^2 + \eta + \dfrac{1}{P_r}(1+\lambda\sigma)\sigma \right] & -(1+\lambda\sigma)R\beta^2 & -(1+\lambda\sigma)T_a^{1/2} \\[2mm] -1 & \left(\Gamma^2 + \sigma \right) & 0 \end{vmatrix} = 0,$$

where $\Gamma^2 = \pi^2 + \beta^2$. After expanding the determinant, we get

$$R = \dfrac{\pi^2 T_a \left[\left(\Gamma^2 + \sigma \right)(1+\lambda\sigma) \right] + \left[\Gamma^2 + \eta + \dfrac{1}{P_r}(1+\lambda\sigma)\sigma \right] \left\{ \Gamma^2 \left(\Gamma^2 + \sigma \right) \left[\Gamma^2 + \eta + \dfrac{1}{P_r}(1+\lambda\sigma)\sigma \right] \right\}}{\left[\Gamma^2 + \eta + \dfrac{1}{P_r}(1+\lambda\sigma)\sigma \right](1+\lambda\sigma)\beta^2}, \tag{10.26}$$

10.4 STATIONARY CONVECTION

Now after taking $\sigma = 0$ for the marginally stable steady convection, equation (10.26) yields equation (10.27) given as

$$R_s = \dfrac{\Gamma^2}{\beta^2} \left[\dfrac{\pi^2 T_a + \Gamma^2 \left(\Gamma^2 + \eta \right)^2}{\left(\Gamma^2 + \eta \right)} \right], \tag{10.27}$$

where R_s is the Rayleigh number for stationary convection.

10.5   OSCILLATORY CONVECTION

For oscillatory motion σ is represented in the form $\sigma = \sigma_r + i\sigma_i$. As at the marginal stability state $\sigma_r = 0$. Thus, by using $\sigma = i\sigma_i$ in equation number (10.26) and after removing the complex quantities from the denominator, we get

$$R_a = \Delta_1 + i\sigma_i \Delta_2 \qquad (10.28)$$

In appendix, the expressions for Δ_1 and Δ_2 are given. Here R_a is real since it is a physical quantity. Hence, from equation (10.28), it follows that either for principle of exchange of stabilities, $\sigma_i = 0$ and for oscillatory motion $\Delta_2 = 0, \sigma_i \neq 0$.

For oscillatory convection, the dispersion relation is of the form as given in equation (10.29).

$$b_0 \left(\sigma_i^2\right)^3 + b_1 \left(\sigma_i^2\right)^2 + b_2 \left(\sigma_i^2\right) + b_3 = 0, \qquad (10.29)$$

where the expressions of b_0, b_1, b_2 and b_3 are shown in the appendix.
Now equation (10.28) yields

$$R_a^{osc} = \frac{a_0 \left(\sigma_i^2\right)^4 + a_1 \left(\sigma_i^2\right)^3 + a_2 \left(\sigma_i^2\right)^2 + a_3 \left(\sigma_i^2\right) + a_4}{\beta^2 \left[\left(\Gamma^2 + \eta - \dfrac{2\lambda\sigma_i^2}{P_r}\right)^2 + \sigma_i^2 \left(\dfrac{1}{P_r} + \lambda\Gamma^2 + \eta\lambda - \dfrac{\lambda^2\sigma_i^2}{P_r}\right)^2\right]}, \qquad (10.30)$$

where R_a^{osc} is the oscillatory Rayleigh number, with $a_i's$ as in the appendix from equation (10.30).

10.6   RESULTS AND DISCUSSION

In former section, the instability analysis of a Maxwell fluid in a porous medium heated from the underside is studied in the presence of uniform vertical rotation via linear stability analysis. The expressions for both stationary and oscillatory critical Rayleigh number are obtained analytically which characterize the stability of the system.

Here, the numerical results in terms of Rayleigh number and wave number for both stationary and oscillatory mode for various values of Taylor number, Prandtl number and porous parameter are presented graphically. The connectedness allows the linear stability criteria to be expressed in terms of the critical Rayleigh number, below which the system is stable and above which the system is unstable. Further, dependence of frequency of overstability on the wave number and the comparison of critical Rayleigh numbers for stationary and oscillatory modes are also shown separately.

Figures 10.2–10.3 shows the neutral curves for stationary mode for different values of R_s for $\eta = 0$ and $\eta = 20$ for the fixed value of $T_a = 100$ and R_s for $T_a = 0, T_a = 100, T_a = 500, T_a = 1000$ for the fixed value of $\eta = 20$.

Figure 10.2 shows the effect of porous parameter (η) on neutral stability curve for stationary mode for fixed value of $T_a = 100$. From the figure, it is clear that an increase in the value of porous parameter (η) increases the minimum Rayleigh number of the stationary mode, indicating that the porosity has a stabilizing effect on the onset of convection in rotatory Maxwell fluid heated from below.

Figure 10.3 shows the effect of Taylor number (T_a) on neutral stability curve for stationary mode for fixed values of parameters. From the figure it is clear that an increase in the value of Taylor number (T_a) increases the minimum Rayleigh number of the stationary mode, indicating that the Taylor number (T_a) postpones the onset of convection in Maxwell fluid heated from below.

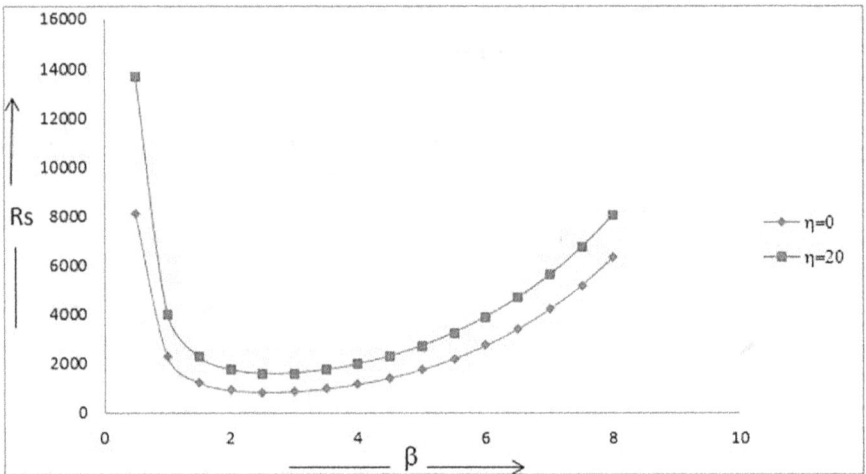

FIGURE 10.2 R_s as a function of wave number for $\eta = 0$ and $\eta = 20$ for the fixed value of $T_a = 100$.

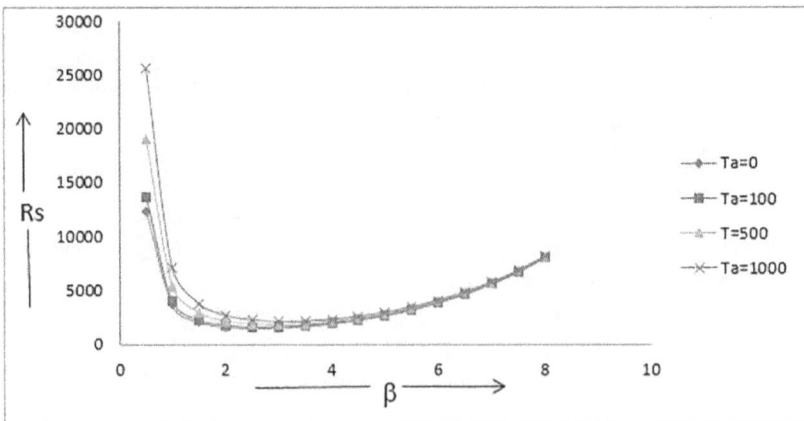

FIGURE 10.3 R_s as a function of wave number for $T_a = 0$, $T_a = 100$, $T_a = 500$, $T_a = 1000$ for the fixed value of $\eta = 20$.

Figure 10.4 shows the effect of Taylor number (T_a) on neutral stability curve for oscillatory mode for fixed values of other parameters $\eta = 20$, $\lambda = 0.5$, $P_r = 10$. From the figure, it can be seen that critical Rayleigh number for oscillatory mode decreases with the increase in the value of Taylor number (T_a), indicating that the effect of increasing T_a is to destabilize the system.

Figure 10.5 shows the effect of Prandtl number (P_r) on neutral stability curve for oscillatory mode for fixed values of other parameters $\eta = 20$, $\lambda = 0.5$, $T_a = 100$. From the figure it can be seen that critical Rayleigh number for oscillatory mode decreases

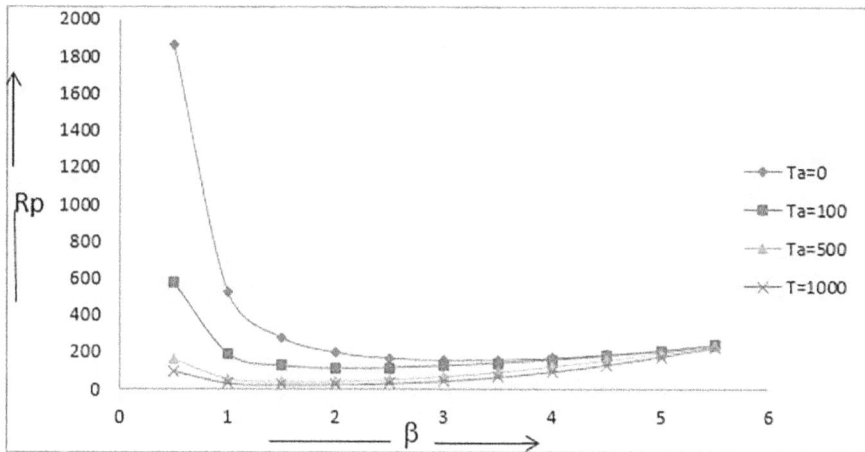

FIGURE 10.4 R_p as a function of wave number for $T_a = 0$, $T_a = 100$, $T_a = 500$, $T_a = 1000$ for the fixed value of $\eta = 20$, $\lambda = 0.5$, $P_r = 10$.

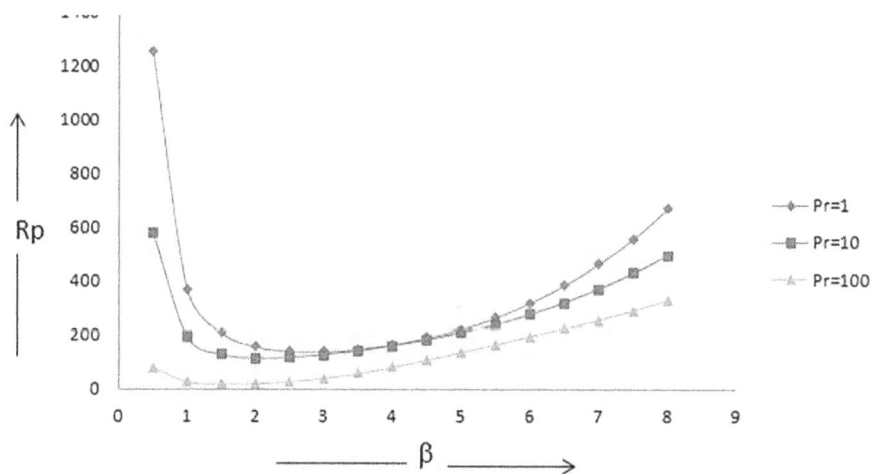

FIGURE 10.5 R_p as a function of wave number for $P_r = 1$, $P_r = 10$, $P_r = 100$, for the fixed value of $\eta = 20$, $\lambda = 0.5$, $T_a = 100$.

with the increase in the value of Prandtl number (P_r), indicating that the effect of increasing P_r is to destabilize the system.

10.7   CONCLUSIONS

The onset of convection in the Maxwell fluid layer heated from below under the influence of uniform vertical rotation has been investigated using the linear stability analysis. The onset criterion for stationary and oscillatory convection is derived analytically. Further, the behaviour of various parameters like Prandtl number, Taylor number, and porous parameter on the onset of both stationary and oscillatory convection has been analysed numerically using Scientific WorkPlace and Mathematica 7.0. The results indicate that a stabilizing effect is visible for the state of stationary convection, with respect to the porous parameter as well as Taylor number, and a destabilizing effect on the system is visible for the state of oscillatory convection with respect to the Taylor number and Prandtl number. In limiting cases, after removing the rotation, previous results for stability analysis of a Maxwell fluid in a porous medium heated from below are recovered from our results.

Appendix

$$\Delta_1 = \frac{a_0\left(\sigma_i^2\right)^4 + a_1\left(\sigma_i^2\right)^3 + a_2\left(\sigma_i^2\right)^2 + a_3\left(\sigma_i^2\right) + a_4}{\beta^2\left[\left(\Gamma^2 + \eta - \dfrac{2\lambda\sigma_i^2}{P_r}\right)^2 + \sigma_i^2\left(\dfrac{1}{P_r} + \lambda\Gamma^2 + \eta\lambda - \dfrac{\lambda^2\sigma_i^2}{P_r}\right)^2\right]},$$

$$\Delta_1 = \frac{b_0\left(\sigma_i^2\right)^3 + b_1\left(\sigma_i^2\right)^2 + b_2\left(\sigma_i^2\right) + b_3}{\beta^2\left[\left(\Gamma^2 + \eta - \dfrac{2\lambda\sigma_i^2}{P_r}\right)^2 + \sigma_i^2\left(\dfrac{1}{P_r} + \lambda\Gamma^2 + \eta\lambda - \dfrac{\lambda^2\sigma_i^2}{P_r}\right)^2\right]},$$

$$a_0 = \frac{-\Gamma^2\lambda^4}{P_r^3},$$

$$a_1 = \frac{-2\Gamma^2\lambda^2}{P_r^3} + \frac{3\Gamma^4\lambda^3}{P_r^2} + \frac{3\Gamma^2\lambda^3\eta}{P_r^2},$$

$$a_4 = \pi^2 T_a\Gamma^4 + \Gamma^{10} + 3\Gamma^8\eta + 3\Gamma^6\eta^2 + \pi^2 T_a\Gamma^2\eta + \Gamma^4\eta^3,$$

$$b_0 = \frac{\Gamma^4\lambda^4}{P_r^3},$$

$$b_1 = \frac{2\Gamma^4\lambda^2}{P_r^3} + \frac{\Gamma^4\lambda^2}{P_r^2} - \frac{3\Gamma^6\lambda^3}{P_r^2} + \frac{\Gamma^2\eta\lambda^2}{P_r^2} - \frac{3\Gamma^4\eta\lambda^3}{P_r^2} - \frac{\pi^2 T_a\lambda^3}{P_r},$$

$$b_3 = -\frac{\pi^2 T_a \Gamma^2}{P_r} + \frac{\Gamma^8}{P_r} + \frac{2\Gamma^6 \eta}{P_r} + \frac{\Gamma^4 \eta^2}{P_r} - \Gamma^{10}\lambda - 3\Gamma^8 \eta \lambda - 3\Gamma^6 \eta^2 \lambda -$$
$$\Gamma^4 \eta^3 \lambda + \pi^2 T_a \Gamma^2 + \Gamma^8 + 3\Gamma^6 \eta + 3\Gamma^4 \eta^2 + \pi^2 T_a \eta + \Gamma^2 \eta^3,$$

REFERENCES

1. Benard, H., 1900, Les tourbillions cellulaires dans une nappe liquid. *Revenue generale des Sciences, Pures et Appliqués* 11, 1261–1271, 1309–1328.
2. Rayleigh, L., 1916, On the convective currents in a horizontal layer of fluid when the higher temperature is on the upper side. *Phil. Mag.* 32, 529–546.
3. Rajagopal, K. R., Saccomandi, G., Vergori, L., 2009, Stability analysis of the Rayleigh–Bénard convection for a fluid with temperature and pressure dependent viscosity. *J. Appl. Math. Phys. (ZAMP)* 60(4).
4. Sekhar, G. N., Jayalatha, G., 2010, Elastic effects on Rayleigh-Bénard convection in liquids with temperature-dependent viscosity. *Int. J. Therm. Sci.* 49(1), 67–75.
5. Chandrasekhar, S., 1961, *Hydrodynamic and Hydromagnetic Stability*, Clarendon, Oxford.
6. Drazin, P., Reid, W., 1981, *Hydrodynamic Stability*, Cambridge University Press, Cambridge.
7. Bezan, A., 2004, *Convection Heat Transfer*, 3rd edn., John Wiley and Sons, Inc., Hoboken, NJ.
8. Horton, C., Rogers, F., 1945, Convection currents in a porous medium. *J. Appl. Phys.* 16(6), 367–370.
9. Lapwood, E. R., 1946, Convection of a fluid in a porous medium. *Proc. Camb. Philol. Soc.* 44, 508–521.
10. Katto, Y., Masuoka, T., 1967, Criterion for the onset of convective flow in a fluid in a porous medium. *Int. J. Heat Mass Transf.* 10(3), 297–309.
11. Bejan, A., 2003, Simple methods for convection in porous media: Scale analysis and the intersection of asymptotes. *Int. J. Energy Res.* 27(10), 859–874.
12. Mckibbin, R., O'Sullivan, M. J., 1980, Onset of convection in a layered porous medium from below. *J. Fluid Mech.* 96(2), 375–393.
13. Khaled, A. R. A., Vafai, K., 2003, The role of porous media in modeling flow and heat transfer in biological tissues. *Int. J. Heat Mass Transf.* 46(26), 4989–5003.
14. Angra, S., Sharma, B., Sharma, K. D., 2022, Amalgamation of virtual reality, augmented reality and machine learning: A review, *2nd International Conference on Advance Computing and Innovative Technologies in Engineering (ICACITE)*, pp. 2601–2604, doi:10.1109/ICACITE53722.2022.9823716.
15. Gupta, K., Bose, T., Rattan, M., Chamoli, N., 2022, Modeling creep on account of residual stress of thermally graded rotating disc. *ECS Trans.* 107, doi:10.1149/10701.9805ecst. https://iopscience.iop.org/issue/1938-5862/107/1
16. Vafai, K., 2006, *Handbook of Porous Medium*, CRC Press, Taylor and Francis Group, Boca Raton, FL.
17. Nield, D., Bejan, A., 2006, *Convection in Porous Media*, 3rd edn., Springer, Berlin.
18. Straughan, B., 2006, Global nonlinear stability in porous convection with a thermal non—equilibrium model. *Proc. Roy. Soc. A.* 462, 409–418.
19. Maxwell, J. C., 1867, On the dynamical theory of gases. *Philos. Trans. R. Soc.* 157, 49–88, doi:10.1098/rstl.1867.0004
20. Rudraiah, N., Siddhesshwar, P. G., Masuoka, T., 2003, Non-linear convection in porous media: A review. *J. Porous Media* 6.

21. Bertola, V., Cafaro, E., 2006, Thermal instability of viscoelastic fluids in horizontal porous layers as initial problem. *Int. J. Heat Mass Transfer.* 49, 4003–4012.

22. Chandrasekhar, S., 1953, The instability of a layer of fluid heated below and subject to Coriolis forces. *Proc. R. Soc. A* 217, 306–327.

23. Vadasz, P., 1998, Coriolis effect on gravity-driven convection in a rotating porous layer heated from below. *J. Fluid Mech.* 376, 351–375.

24. Veronis, G., 1959, Cellular convection with finite amplitude in a rotating fluid. *J. Fluid Mech.* 5, 401–435.

25. Bhatia, P. K., Steiner, J. M., 1972, Convective instability in a rotating viscoelastic fluid layer. *ZAMM* 52, 321–327.

26. Vadasz, P., 2001, Coriolis effect on free convection in a long rotating porous box subject to uniform heat generation. *Int. J. Heat Mass Transfer.* 38, 2011–2018.

27. Govender, S., 2003, Coriolis effect on the linear stability of convection in a porous layer placed far away from the axis of rotation. *Transp. Porous Media.* 51, 315–326.

28. Laroze, D., Martinez-Mardones, J., Bragard, J., 2007, Thermal convection in a rotating binary viscoelastic liquid mixture. *Eur. Phys. J. Spec. Top.* 146, 291–300.

29. Tan, W., Masuoka, T., 2007 Stability analysis of a Maxwell Fluid in a Porous medium heated from below. *Phys. Lett. A.* 360, 454–460.

30. Wang, S., Tan, W., 2008, Stability analysis of double-diffusive convection of Maxwell fluid in a porous medium heated from below. *Phys. Lett. A.* 372(17), 3046–3050.

31. Wang, S., Tan, W., 2011, Stability analysis of soret-driven double-diffusive convection of Maxwell fluid in a porous medium. *Int. J. Heat Fluid Flow.* 32(1), 88–94.

32. Yin, C., Fu, C., Tan, W., 2012, Onset of thermal convection in a Maxwell fluid saturated porous medium. The effects of hydrodynamic boundary and constant flux heating conditions Trans. *Porous. Med.* 91, 777–790.

11 Augmented-Reality-Based Mobile Learning Environments
A Review

Tabasum Mirza, Neha Tuli, Archana Mantri

11.1 INTRODUCTION

Augmented reality (AR) is the visualization technology which combines real life and virtual objects to create enhanced user experiences. It works by adding virtual objects to real environments and enable visualization and real-time interaction with the objects in hybrid environment [1].

11.2 AR IN EDUCATION

AR technology has wide applications in various areas of life and its popularity as an instructional tool is increasing in the field of education. AR tools can yield pedagogical benefits when designed and developed suitably in orientation with age, curriculum, and learning objectives of the target learners. AR-based environment can provide new learning opportunities for the learner. An ARLE (augmented-reality-based learning environment) can help the learner in visualizing abstract concepts and complex phenomena with the help of overlaid 2D or 3D virtual objects on real surroundings [2]. Based on social cognitive theory of learning by Albert Bandura, the attention, motivation, retention, observational learning, and production process are essential steps to ensure effective learning and learning occurs in the social context of dynamic interaction between learner and the environment [3].

AR can help develop engaging and motivating virtual learning environments that helps provide effective learning experience to students [4,5]. As a new generation learners being digital natives are more drawn to 3D modelling of educational content with the help of technology rather than reading books ARLE can provide suitable engaging and motivating learning environment to them. AR technology can be used in STEM (science, technology, engineering, and mathematics), language, social sciences, and medical education and trainings [6–8]. There is significant research supporting the positive impact of AR tools on learner motivation, engagement, and retention [9,10].

DOI: 10.1201/9781003367161-11

11.3   HOW TO USE AR IN EDUCATION

AR can be used in education in the form of AR books, games, training kits, object modelling, and discovery-based learning [11–13]. The use of AR has been explored at different stages of education—kindergarten, elementary, secondary, professional, and higher education. It can aid in learning of children with special needs [14,15]. AR has applications as a part of the Industry 4.0 movement in engineering and technology [16,17]. At early stages of education, AR can help in cognitive development of young children by providing interactive and visually stimulating environment [18]. Other applications include AR-based colouring and story books, handwriting tools, and discovery tools [19] that combine entertainment and education [20,21]. AR has been used in teacher education and astronomy education [22,23]. It is a useful tool for facilitating informal learning [24], language education, history education, geo education, and learning of mathematics and science [25–27]. These tools can be helpful in teaching of scientific and mathematical concepts at the elementary level to support inquiry based learning and develop scientific temper in students [28].

11.4   MOBILE AUGMENTED REALITY

With the progress in mobile hardware and technology, increasing the use of mobile devices, revolutionary processing speed, and power of mobile devices, mobile-based augmented reality (MAR) is gaining popularity. MAR makes use of smartphones, tablets, and PDAs (personal digital assistants), leading to its exponential growth as an easy and affordable learning tool [29]. Amid the COVID-19 crisis, the global AR market may reach the high of $230.6 billion by 2027 [30].

11.5   NEP-2020 AND AR IN EDUCATION

National Education Policy (NEP-2020), in alignment with the UNESCO Sustainable Development Goals of 2030, is aimed at the universalization of education in India and provides guidelines for accessible, equitable, affordable, and inclusive education. The NEP-2020 focuses on critical thinking and experiential learning and discourages rote learning to transform the Indian education system into a knowledge-based society. It lays emphasis on holistic development and equipping students with 21st-century skills. AR tools can help in achievement of NEP goals by fostering creativity, scientific temperament, and experiential learning in learners [31]

11.6   COVID-19 AND AR IN EDUCATION

The unplanned and instant migration to the online environment as an alternative mode of learning amidst COVID-19 crisis has changed the outlook on teaching/learning process. One major challenge in online mode of learning is lack of learner motivation, engagement, and real-time interaction. AR-based learning environments can help providing engaging and effective learning experience leading to knowledge retention and learner satisfaction [32]

11.7   RESEARCH QUESTIONS

This study is concerned with the following research questions:

1. What are the recent emerging trends of AR in field of education?
2. What are the main challenges involved with using mobile-augmented reality in education?
3. What pedagogical and design principles are required to develop effective AR tools?
4. What are the areas or subjects in which AR is being used in education?
5. How can AR be used for achievement of learning outcomes effectively?

11.8   METHODOLOGY

The study methodology was divided into two phases: research and review processes. During the first phase, a keyword search was carried out on the databases using keywords "AR in education," "mobile augmented reality," "MAR challenges and limitations," "use of AR in STEM," and "use of AR in higher/primary/secondary education." The articles were analysed manually according to their title and summary, then only items that met the inclusion criteria or were linked to the research questions were considered for review.

The PRISMA flow diagram shown in Figure 11.1 depicts the process of systematic review conducted in this study. The totals of 132 articles related to AR in education were identified from the database, out of which 88 items did not qualify for inclusion and had to be excluded from the study. Full texts of two articles were not available, so they were also excluded, and 42 articles were selected for synthesis.

A total of 42 publications were analysed in this review, which included articles from different databases—SCOPUS, IEEE, and SCI—published in the time period of 2011–2021. The articles were grouped on the basis of various parameters like authors, year of study, AR technology used, methodology used in the study (quantitative or qualitative), level of education (preschool, primary, secondary, or higher), and areas or subjects for which AR tools are used.

11.9   EXTRACTION OF INFORMATION

The useful information was extracted using content analysis and qualitative synthesis methods. The information extracted included research contributions to AR in education, problems or challenges identified in current AR technology, and limitations and scope for future research.

11.10   LITERATURE REVIEW

Ronald T. Azuma provides overview of AR systems, characteristics, requirements of AR, systems, registration and tracking in outdoor environments, suggestions to overcome challenges, and social and political issues in use of AR[1]. Krevelen

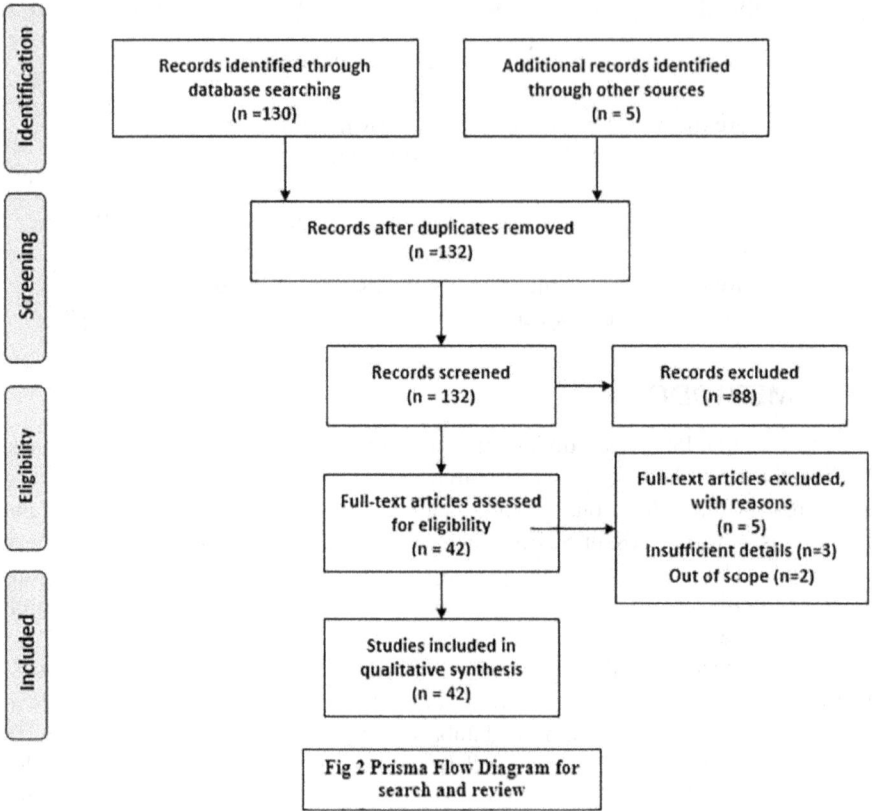

```
Identification    Records identified through        Additional records identified
                    database searching               through other sources
                       (n =130)                            (n = 5)


Screening                      Records after duplicates removed
                                       (n =132)


Eligibility               Records screened          →      Records excluded
                            (n = 132)                          (n =88)


                          Full-text articles assessed   →   Full-text articles excluded,
                              for eligibility                      with reasons
                                (n = 42)                             (n = 5)
                                                          Insufficient details (n=3)
                                                             Out of scope (n=2)

Included                    Studies included in
                          qualitative synthesis
                                (n = 42)


                          Fig 2 Prisma Flow Diagram for
                               search and review
```

FIGURE 11.1 PRISMA flow diagram for literature review on AR in education.

et al. discuss the brief background of AR and the technological framework, evolution, and major breakthroughs in the AR field. It also highlights the applications of AR in various areas like a military operation, advertising and marketing, entertainment, tourism, medical science and education [33]. Silvia Lee et al. discusses the importance of AR as an instructional tool and technological, pedagogical challenges in its use. It also highlights the need for innovative solutions to create effective AR learning tools [34]. Julie Carmigniani et al. provides an overview of MAR technology its applications, challenges, and requirements [35]. Misty Antonioli et al. analyse the theoretical foundations of using AR in education as a pedagogical tool that can help mainstream education inside and outside classroom, learners with special needs, teaching of STEM subjects etc [36,37]. Kangdon Lee et al. discuss application of AR in education and trainings, with reference to teaching of different subjects like mathematics and geometry, augmented biology, physics education and astronomy, its potential impact on the future of education to create effective learning environments [38]. Ammar H. Safar et al. in his study indicated significant improvement in effectiveness of learning using AR applications in comparison to traditional approaches

[39]. Mehdi Mekni et al. surveys the potential of MAR, its technical and ergonomic limitations, provides insights into the current trends and infrastructure of MAR, its application in different areas, usability issues, and technical, social acceptance challenges [40]. Tasneem Khan et al. studies the effect of AR on the students at the University of Cape Town with increased learning outcomes [41].

Dimitris Chatzopoulos et al., in the study on MAR, analyse the basics of MAR, core systems of MAR, importance of data management in MAR systems, categorization of application fields, experience metrics, system performance, and sustainability along with challenges in its use [42]. Tara J. Brigham et al. compare the use of AR, VR, and MR technologies on the basis of usability concerns, security, and ethical and privacy issues [43]. López et al. review the benefits of AR in education on the basis of existing literature by using documentary analysis, bibliometric analysis, and scientific mapping of research articles [44]. Matt Bowera et al. analyse the use of AR in education and discuss the pedagogical potential of the AR as educational technology [45]. Kesima et al. discuss the potential of AR in education in order to provide 3D-based visualizing experience to the learners [46]. Mustafa Sırakaya et al., in the systematic review, identify the characteristics, advantages, and challenges of AR in K–12 STEM education and indicate growing trend in this area of education [47]. Danakorn Nincarean et al., in the review, discuss the possibilities or AR and MAR for providing joyful, useful learning experience to learners and need to take learning theory and pedagogical principles into consideration to make them useful [48].

Robert Godwin-Jones et al. explores the advantages and challenges for use of AR-based games in learning languages [25]. Slavica et al. compare the effectiveness traditional learning and MAR in vocational education and skill development and finds MAR effective and easy in achieving learning outcomes [49]. Maas et al., in a review, compare the use of AR, VR, and MR in K–12 learning environments and their challenges. The results show that AR is easy to access and use in education [50]. Mubarak et al. discuss merits of AR and proposes conceptual model for use of AR in education to attain increased learning outcomes [51]. Rafal Wojciechowski et al. study the use of AR environments for learning chemistry in secondary level students. The study aims at analysing the perception of usefulness, enjoyment, and satisfaction in traditional and AR environments [52]. Phil Diegmann et al., in systematic review, highlight the benefits of ARLE for language learning and building mechanical skills and spatial skills [53]. Muhammad Saleem et al. provide a conceptual model of AR-based application in higher education [54]. Neha Tuli et al. list 23 usability principles that can be used by researchers as a reference for development of MAR applications for effective kindergarten education [55]. Deepti Prit Kaur et al. designed and developed the framework of ARLE for teaching of linear time-invariant control systems in engineering education [56]. Cecilia et al. provide a review of publications published in the AR technology in education and advocates benefits of using AR for its use at different stages of education [57]. Nikolaos Pellas et al. review the current trends for use of AR with game-based learning approach in education [58]. Desi Dwistratanti et al. discuss user expectations, interaction issues, and design factors that can help in increasing user acceptances of AR prototypes for use in education [59]. Francisco et al. explore the potential of AR as a tool for inclusive education (SDG 4) with the help of mobile devices [60].

11.11 RESULTS

11.11.1 RQ1. What Are Recent Emerging Trends of AR in Education?

The research findings highlight the positive impact of AR tools and applications on learning experience of students in terms of increased learner motivation, engagement, and knowledge retention [1]. The results also advocate increased effectiveness of AR training tools in skill development of learners over traditional training methods. Various studies indicate the need of developing age suitable AR tools with pedagogical value. The studies also analyse the potential benefits of ARGBL (AR game-based learning environment) as an extension of ARLE for development of positive attitude towards learning, acquisition of skills, and transfer and retention of knowledge in learners [2].

Regarding the recent trends in AR research, the review study indicated that there is significant rise in the number of research articles contributing towards the area of AR in education. The systematic review of literature indicates that the main countries contributing towards AR in education and research publications are United States, Spain, and Taiwan. Regarding the emerging trends for AR in education, the most trending research topics are educational needs of special children, Industry 4.0, digital storytelling, 3D printing, and the use of AR in medical and higher education [3].

11.11.2 RQ2. What Are the Major Challenges in Mobile Augmented Reality?

Although this technology is easy to use, technological inhibitions or negative attitudes can have an impact on users [61]. Privacy or security concerns like sharing of sensitive information, age suitability, usability issues can avert users [62]. If not designed properly, the AR applications may confuse students between the real and virtual experiences. There must be focus on the design and development of simple, realistic, and easy-to-use apps to reduce cognitive overload, particularly in younger children. There are technical challenges in MAR, like power usage of mobile devices and tracking and registration problems in outdoor use, that need to be addressed. There must be development in AR-supporting devices that are portable, unobtrusive, lightweight, and fashionable and standardization of mobile technology to support AR technologies in every device [5]. Collaborative MAR interfacing by combining different mobile devices together can help to overcome bandwidth limitations and help in achievement of service availability, security, heterogeneity, and mobile computing problems [4]. The collaborative approach can help students learn easily and decrease cognitive overload by scaffolding [68].

11.11.3 RQ3. What Pedagogical and Design Principles Are Required to Develop Effective AR Tools?

Lack of pedagogical value and curricular relevance may limit usefulness of the technology. Addressing issues like threatening or unproductive, unplanned use of technology in education without yielding learning outcomes for which it has been designed.

The studies suggest the need of collaboration between the researchers, developers, and educators to bridge pedagogical and technological gaps in the AR use [63]. AR-based learning environments should be more interactive and flexible to facilitate customization of learning environment by the users [4]. The review points out certain issues in current applications, like lack of usability standards and pedagogical approach and significant gap between pedagogy and technology [64]. This limits the effectiveness of AR learning applications because teachers and parents do not find them much useful in achieving learning outcomes [65]. The designers are not much aware about pedagogies and teaching experts are not familiar with the technology [66]. There is need to bridge this gap by involving educational experts in design process of AR applications and providing technical assistance to teachers in using AR [67]. The game-based learning or collaborative learning approaches can be experimented with use of AR to address issues of cognitive overload [68]. There is need to develop interactive AR applications that can provide higher degree of interaction and flexibility so that children don't lose interact in educationally flexible applications [69].

11.11.4 RQ4. What Are the Areas/Subjects in Which AR Is Being Currently Used?

AR is being used for all levels of education (kindergarten, higher education, etc.), and studies indicate significant improvement in learning effectiveness with its use as a teaching tool. The review of literature also reveals that its use has been explored in language learning, STEM, social science, engineering and medical education, teacher education, skill development, and trainings in informal and formal educational settings. The review also indicates use of AR is less popular in early childhood education (ECE) [70]. The reason behind this could be challenges in use of technology with younger children and lack of research to study long-term effects with use in early stages of education [5].

11.11.5 RQ5. Which Areas in Education Can Make Use of AR in the Future for Increased Learning Outcomes?

AR's potential is emerging in learning foreign and native languages [6,7]. But it seems it has not been used yet in learning of Indian languages, Sanskrit, Hindi, Urdu, Shina, Persian, Arabic, and so on. When comparing the use of AR at different stages, it seems that its use is significantly less in early stages of education as compared to other stages [8]. The use of AR can be explored in early stages of education like pre-primary and early-childhood education (ECE) [70]. The study identifies the future scope of research to work on technological issues like reducing registration and tracking time, improving battery life of mobile devices, and using affordable and wearable and unobtrusive fashionable devices for implementation [9].

11.12 CONCLUSION

The results of the analysis indicate that AR has a positive impact on achievement of learning outcomes. AR can help in increasing learner motivation and engagement by overlaying virtual content on a real-life environment. AR can promote authentic learning opportunities in children with the help of 3D visualizations of difficult concepts or phenomena. The increasing popularity of mobile technology has provided opportunity for growth of MAR that uses PDAs or handheld mobile devices for the use of AR, particularly in education. This has resulted in the rapid evolution of the MAR technology, owing to benefits of portability, ease of access, and affordability. But despite all these benefits, there are some limitations and challenges that need to be overcome for the successful implementation of MAR in education. These challenges could be the reason for widespread use of conventional methods in modern-age classrooms in spite of availability of digital infrastructure in schools. Some major challenges include accessibility, interactivity, lack of collaboration opportunities, usability issues, cognitive overload, lack of flexibility to author AR content by teachers, gap between technology and pedagogy, and lack of suitable pedagogical approach in implementing AR for education. There is a need of further research in these areas to help in effective design and use of AR learning tools with the help of carefully designed instructional media.

AR applications can make use of game-based learning approaches or other collaborative mechanism that involves learning in groups to reduce cognitive overload. There is need to develop highly interactive AR applications that can provide opportunity to resize, zoom in and out, and scroll to keep students engaged and motivated while using in educationally flexible applications. There is a scope of in-depth research for the use of AR in ECE to address learning gaps at the foundation stage of education. There is need to investigate the long-term effects of AR on cognitive development of children using AR in formal education. Upcoming studies can increase the popularity and interactivity of AR in classrooms by incorporating technology into existing infrastructure, such as smart boards. The forthcoming research asks for qualitative and quantitative cross-sectional studies on larger heterogenous samples of data to study the factors in detail that impede and support the utilization of AR apps in delivering education across different levels and subjects. AR can be combined with other innovative technologies like robotics to increase its learning potential. Future studies must take into consideration the in-depth analysis of various features and factors responsible for design of effective AR learning applications that help in achieving learning outcomes effectively and easily.

REFERENCES

1. R. T. Azuma, "A survey of augmented reality," pp. 355–385, 1997. https://www.cs.unc.edu/~azuma/ARpresence.pdf (Acedido em 24/11/2020).
2. L. Kerawalla et al., "'Making it real': Exploring the potential of augmented reality for teaching primary school science," *Virtual Real.*, vol. 10, pp. 163–174, 2006.
3. J. J. Martin et al., "Social cognitive theory," *Routledge Handb. Adapt. Phys. Educ.*, vol. 6, pp. 280–295, 2020.
4. Waleed et al, "Effect of augmented reality and simulation on the achievement of mathematics and visual thinking among students," *Int. J. Emerg. Technol. Learn.*, vol. 10748, pp. 164–185.
5. A. M. Bodzin et al., "A mixed methods assessment of students' flow experiences during a mobile augmented reality science game," *J. Comput. Assist. Learn.*, vol. 29, 2013.

6. M. Sırakaya, D. Alsancak Sırakaya, "Augmented reality in STEM education: A systematic review," *Int. Learn. Environ.*, vol. 30, no. 8, pp. 1556–1569, 2022.

7. N. I. Nabila Ahmad et al., "Augmented reality for learning mathematics: A systematic literature review," *Int. J. Emerg. Technol. Learn.*, vol. 15, no. 16, pp. 106–122, 2020.

8. S. Singhal et al., "Augmented chemistry: Interactive education system," *Int. J. Comput. Appl.*, vol. 49, no. 15, pp. 1–5, 2012.

9. A. D. Serio et al., "Impact of an augmented reality system on students' motivation for a visual art course," *Comput. Educ.*, vol. 68, pp. 586–596, 2013.

10. J. Garzón et al., "Meta-analysis of the impact of augmented reality on students' learning gains," *Educ. Res. Rev.*, vol. 27, no. April, pp. 244–260, 2019.

11. M. Billinghurst et al., "The magicbook: A transitional AR interface," *Comput. Graph.*, vol. 25, no. 5, pp. 745–753, 2001.

12. G. Koutromanos et al., "The use of augmented reality games in education: A review of the literature," *Educ. Media Int.*, vol. 3987, no. January, 2016.

13. E. Johnson et al., "Augmented reality: An overview and five directions for AR in education," *J. Educ. Tech. Dev. Exch.*, vol. 4, no. 1, 2011.

14. A. J. Vullamparthi, "Assistive learning for children with autism using augmented reality," *Proc.—2013 IEEE 5th Int. Conf. Technol. Educ. T4E 2013*, IEEE Xplore, pp. 43–46, 2013.

15. S. Deb, "Augmented Sign Language Modeling design on smartphone—An assistive learning and communication tool for inclusive classrooms," *Procedia Comput. Sci.*, vol. 125, pp. 492–500, 2018.

16. P. Fraga-Lamas et al., A review on industrial augmented reality systems for the industry 4.0 shipyard," *IEEE Access*, vol. 6, pp. 13358–13375, 2018.

17. D. Parsons, "Cur perspective on augmented reality in medical education: Applications, affordances and limitations," *Adv Med Educ Pract.*, vol. 12, 2021.

18. D. R. A. Rambli, "An interactive mobile augmented reality playbook: Learning numbers with the thirsty crow," *Procedia—Procedia Comput. Sci.*, vol. 25, pp. 123–130, 2013.

19. K. E. Chang, "Development and behavioral pattern analysis of a mobile guide system with AR for painting application instruction in an art museum," *Comput. Educ.*, vol. 71, 2014.

20. A. Clark et al., "An interactive augmented reality coloring book," *In SIGGRAPH Asia 2011 Emerg. Technol.*, pp. 1–1, 2011.

21. P. Vate-U-Lan, "An augmented reality 3d pop-up book: The development of a multimedia project for English language teaching," *In 2012 IEEE Int. Conf. Multimedia and Expo*, July, pp. 890–895, 2012.

22. L. Y. Midak et al., "Augmented reality in process of studying astronomic concepts in primary school," *CEUR Workshop Proc.*, November, 2020.

23. J. Buchner, "Augmented reality in teacher education. Framework to support teachers' technological content knowledge," *Ital. J. Educ. Technol.*, vol. 28, no. 2, pp. 106–120, 2020.

24. P. Sommerauer et al., "AR in informal learning environments: A field experiment in a mathematics exhibition," *Comput. Educ.*, vol. 79, no. 2014, pp. 59–68, 2015.

25. R. Godwin-Jones, "Emerging technology augmented reality and language learning: From vocabulary to mobile games," *Lang. Learn. Technol.*, vol. 20, no. 3, pp. 9–19, 2016.

26. C. S. C. Dalim et al., "TeachAR: An interactive augmented reality tool for teaching basic English to non-native children," *In 2016 IEEE Int. Symp. Mixed and Augmented Reality (ISMAR-Adjunct)*, pp. 82–86, 2016.

27. Z. Turan et al., "The impact of mobile augmented reality in geography education: Achievements, cognitive loads and views of university students," *J. Geogr. High. Educ.*, vol. 42, no. 3, pp. 427–441, 2018.
28. C. H. Chen et al., "An augmented-reality-based concept map to support mobile learning for science," *Asia-Pac. Educ. Res.*, vol. 25, 2016.
29. C. Arth et al., "The history of mobile augmented reality," *arXiv* preprint arXiv:1505.01319, 2015.
30. Chiang, T. H. C. et al., "An augmented reality-based mobile learning system to improve students' learning achievements and motivations in natural science inquiry activities," *Educ. Technol. Soc.*, vol. 17, no. 4, pp. 352–365, 2014.
31. R. Development, "National education policy 2020," *Econ. Polit. Wkly.*, vol. 55, no. 31, p. 4L, 2020.
32. K. Nesenbergs et al., "Use of augmented and virtual reality in remote higher education: A systematic umbrella review," *Educ. Sci.*, vol. 2021, no. 11, p. 8, 2020.
33. D. W. F. Krevelen et al., "A survey of augmented reality technologies, applications and limitations," *Int. J. Virtual Real.*, vol. 9, no. 2, pp. 1–20, 2010.
34. J. C. Liang, "Current status, opportunities and challenges of augmented reality in education," *Comput. Educ.*, vol. 62, pp. 41–49, 2013.
35. J. Carmigniani, "Augmented reality technologies, systems and applications," *Multimed. Tools Appl.*, vol. 51, no. 1, pp. 341–377, 2011.
36. I. Sural, "Mobile augmented reality applications in education," *Virtual Augment. Real. Concepts, Methodol. Tools, Appl.*, vol. 2, no. 1, pp. 954–969, 2018.
37. R. S. Baragash et al., "Augmented reality and functional skills acquisition among individuals with special needs: A meta-analysis of group design studies," *J. Spec. Educ. Technol.*, vol. 37, no. 1, pp. 74–81, 2022.
38. K. Lee, "Augmented reality in education & training," *Techtrends*, vol. 56, no. 2, pp. 13–21, 2012.
39. H. Safar et al., "The effectiveness of using augmented reality apps in teaching the english alphabet in kindergarten children: A case study in the state of Kuwait," *Eurasia J. Math. Sci. Technol. Educ.*, vol. 13, no. 2, pp. 417–440, 2017.
40. M. Mekni, "Augmented reality: Applications, challenges and future trends," *Appl. Comput. Sci. Anywhere*, pp. 205–214, 2014.
41. T. Khan, "The impact of an A R application on learning motivation of students," *Adv. Hum.-Comput. Interact.*, vol. 2019, 2019.
42. D. Chatzopoulos, "Mobile augmented reality survey: From where we are to where we go," *IEEE Access*, vol. 5, pp. 6917–6950, 2017.
43. T. J. Brigham et al., "Reality check: Basics of augmented, virtual, and mixed reality," *Med. Ref. Serv. Q.*, vol. 3869, 2017.
44. J. López-belmonte, "Augmented reality in education: A scientific mapping in Web of Science," *Interact. Learn. Environ.*, pp. 1–15, 2020.
45. M. Bower, "Augmented reality in education—cases, places and potentials," *EMI. Educ. Media Int.*, vol. 51, no. 1, pp. 1–15, 2014.
46. M. Kesim, "Augmented reality in education: Current technologies and the potential for education," *Procedia—Soc. Behav. Sci.*, vol. 47, no. 222, pp. 297–302, 2012.
47. M. Sırakaya, D. Alsancak Sırakaya, "Augmented reality in STEM education: A systematic review," *Interact. Learn. Environ.*, vol. 30, no. 8, pp. 1556–1569, 2022.
48. D. Nincarean et al., "Mobile augmented reality: The potential for education," *Procedia-Soc. Behav. Sci.*, vol. 103, pp. 657–664, 2013.
49. S. Radosavljevic, "The potential of implementing augmented reality into vocational higher education through mobile learning," *Interact. Learn. Environ.*, pp. 1–15, 2018.

50. M. J. Maas, "Virtual, mixed and augmented reality in K—12 education: A review of the literature of the literature," *Technol. Pedagog. Educ.*, pp. 1–19, 2020.

51. M. Ghare, "Augmented reality for educational enhancement," *IJARCCE*, vol. 6, no. 3, pp. 232–235, 2017.

52. R. Wojciechowski et al., "Evaluation of learners' attitude toward learning in ARIES augmented reality environments," *Comput. Educ.*, vol. 68, pp. 570–585, 2013.

53. P. Diegmann et al., "Benefits of augmented reality in educational environments—A systematic literature review," *12th Int. Conf. Wirtschaftsinformatik*, March 4–6, 2015, Osnabrück, Ger., pp. 1542–1556, 2015.

54. M. Saleem, "Influence of augmented reality app on intention towards e-learning amidst COVID-19 pandemic," *Interact. Learn. Environ.*, pp. 1–15, 2021, doi: 10.1080/10494820.2021.1919147

55. N. Tuli, "Evaluating usability of mobile-based augmented reality learning environments for early childhood," *Int. J. Hum. Comput. Interact.*, vol. 37, no. 9, pp. 815–827, 2021.

56. D. P. Kaur, A. Mantri, "A framework utilizing augmented reality to enhance the teaching–learning experience of linear control systems," *IETE J. Res.*, vol. 67, no. 2, pp. 155–164, 2021.

57. C. Garzon "Augmented reality in education: An overview of twenty-five years of research," *Contemp. Educ. Technol.*, vol. 13, no. 3, p. Ep302, 2021.

58. N. Pellas et al., "Augmenting the learning experience in primary and secondary school education: A systematic review of recent trends in augmented reality game-based learning," *Virtual Real.*, vol. 23, no. 4, pp. 329–346, 2019.

59. Rambli et al., "Preliminary evaluation on user acceptance of the augmented reality use for education," *2010 2nd Int. Conf. Comput. Eng. Appl. ICCEA 2010*, vol. 2, pp. 461–465, 2010.

60. G. M. Méndez et al., "Augmented reality and mobile devices: A binominal methodological resource for inclusive education [SDG 4]. An example in secondary education," *Sustainability*, vol. 10, 2018.

61. M. Alkhattabi et al., "Augmented reality as e-learning tool in primary schools 'education: Barriers to teachers' adoption," *Int. J. Emerg. Technol. Learn.*, vol. 12, no. 2, pp. 91–100, 2017.

62. M. Ohkubo et al., "RFID privacy issues and technical challenges," *IEEE Eng. Manag. Rev.*, vol. 35, no. 2, pp. 51–51, 2007.

63. M. Wang et al., "Augmented reality in education and training: Pedagogical approaches and illustrative case studies," *J. Ambient Intell. Humaniz. Comput.*, vol. 9, pp. 1391–1402, 2018.

64. N. Elmqaddem, "Augmented reality and virtual reality in education. Myth or reality?" *Int. J. Emerg. Technol. Learn.*, vol. 14, no. 3, 2019.

65. M. Billinghurst, "A survey of augmented reality," *Found. Trends Hum-Comput Interact.*, vol. 8, no. 2–3, pp. 73–272, 2014.

66. T. Höllerer et al., "User interface management techniques for collaborative mobile augmented reality," *Comp. Graph.*, vol. 25, no. 5, pp. 799–810, 2001.

67. Zilong Pan et al., "Introducing augmented reality in early childhood literacy learning," *Res. Learn. Technol.*, vol. 29, 2021.

68. H. Y. Chang et al., "Ten years of augmented reality in education: A meta-analysis of (quasi-) experimental studies to investigate the impact," *Comp. Educ.*, vol. 191, p. 104641, 2022.

69. S. Albayrak, R. M. Yilmaz, "An investigation of pre-school children's interactions with augmented reality applications," *Int. J. Hum.–Comput. Interact.*, vol. 38, no. 2, pp. 165–184, 2022.

70. S. A. Hassan et al., "Childar: An augmented reality-based interactive game for assisting children in their education," *Univ. Access. Inf. Soc.*, vol. 21, pp. 545–556, 2022.

12 Discussing Methods of Treatment for Obsessive-Compulsive Disorder by Using Sustainable Technology

Neeraj Singla

12.1 INTRODUCTION

Obsessive-compulsive disorder (OCD) is a mental condition that causes undesirable thoughts and feelings to recur lastly (obsession) or to overdo an action (compulsion) regardless of the circumstances. Industry is commercial activity, and engineers are experienced in determining the economic worth of manufacturing engineering solutions. Measuring social and environmental outcomes is a more difficult engineering and commercial challenge. Activities and procedures used by industrial systems and processes to transform input energy and materials into commercial goods have an influence on sustainability. Waste products and emissions, which are frequently categorized as outcomes, are, in turn, inputs towards other natural and industrial systems, where the influence is felt socially, ecologically, and economically [1].

OCD is more than simply nail-biting and worrying. Obsession may stem from the assumption that numbers or colours may be classified as good or harmful. Another possible root is the need to wash one's hands seven times after touching something or someone. These activities become uncontrolled, and the patient does the unintentionally. Everyone has a habit of thinking the same things again and over. OCD patients are plagued by obsessive thoughts or behaviours. However, these thoughts and acts consume at least an hour every day, interfere with social life, and distort judgement [2].

OCD can cause anxiety, stress, frustration, self-criticism, and problems in one or more important social areas. The World Health Organization (WHO) used to rank it as the tenth most crippling illness in the general population. From 2001 to 2003, 50% of adults with OCD had significant impairment. In the years 2001–2003, more than a third of adults with OCD (34.8%) showed signs of moderate impairment. According to a 2001–2003 study, only 15% of adults with OCD were mildly impaired [2]. Nearly 2% of the world's population is affected by OCD, making it the most common mental illness. As a general rule, the lifetime prevalence of OCD is 1.5% among women and 1% among men. Adults in the United States are estimated

DOI: 10.1201/9781003367161-12

to have a lifetime prevalence of 2.3%. OCD frequently coexists with a number of other conditions, including panic disorders, phobias, PTSD, and all types of anxiety disorders (75.8%); depression, manic-depressive illness, bipolar disorder, and other mood disorders (63.3%); ADHD and other disorders of impulse control (55.9%); and disorders involving the use of drugs and alcohol (38.6%).

12.2 TYPES OF OCD

Checking simple things like locks, alarms, the oven, the stove, or the lights on a regular basis or obsessing over the possibility of a medical problem; fear of contamination and the urge to clean something because it is unclean; an obsession with order and symmetry; the desire to have everything in the right place at all times; an obsession with a particular line of thought, leading to ruminations and intrusions—these are some potentially violent or threatening thought. Following are the different types of OCD [3].

Checking OCD: Repetitive checking behaviours, such as regularly checking locks, appliances, or personal possessions, are caused by persistent uncertainties and worries about potential injury or risk.

Contamination OCD: Excessive fear of germs, filth, or toxins that results in compulsions to clean, avoid specific locations or things, or repeatedly wash one's hands.

Symmetry and Ordering OCD: Strong demand for symmetry, exactness, or order that results in obsessive organising, aligning items, or engaging in ritualistic behaviours to create a sense of equilibrium [4].

Hoarding OCD: Obsessive-compulsive disorder (OCD) known as hoarding: persistent difficulty getting rid of or parting with items, regardless of their real worth, leading to congested living conditions and severe unhappiness.

12.3 SYMPTOMS OF OCD

Some symptoms of OCD include concerns about yourself or others being injured; awareness of one's own bodily movements, such as blinking and breathing; lack of evidence to support the belief that a partner is being unfaithful; performing actions in a specific order or a predetermined number of "good" times; counting things like steps; knocking on doors; and fear of using public washrooms.

While afflicted with OCD, sufferers may notice a variety of changes. These modifications fall into the following categories: feeling accomplished at doing the same thing over and over again, or doing the same thing over and over again; anxiety, apprehension, trepidation, remorse, or even a panic attack; food aversions; nightmares; and a habit of thinking about the past [5].

12.4 TREATMENTS OF OCD

Cognitive behavioural therapy (CBT) is a therapy based on cognitive-behavioural principles that helps people deal with and change the negative thoughts, behaviours, and emotions.

Two types of OCD-related associations will be broken as a result of this therapy. Objects, situations, or thoughts that cause distress can be linked to the sensation of distress. There is also a link between ritualistic behaviour and a reduction in anxiety. Anxiety and its associated ritualistic behaviour can be broken with the help of the therapy we provide.

Symptoms for which CBT is suggested are as follows: when even the most insignificant events cause you anxiety; you panic if you have a phobia of germs; you find yourself constantly rechecking something to know for sure; you worry that you might inadvertently harm your family members [6].

Patients with OCD can benefit from the following CBT techniques: anxiety can be reduced through the use of relaxation techniques, like progressive muscle relaxation and deep breathing.

12.5   COMMONLY USED CBT TECHNIQUES TO TREAT OCD

12.5.1   Exposure and Response Prevention Therapy (ERP)

The most effective CBT technique for treating OCD is called ERP, or exposure and response prevention. This type of therapy allows you to face our fears and allow obsessions to occur without the need to correct them. In order to help patients avoid compulsions in the future, exposing them to a variety of situations helps them develop some coping mechanisms and skills. ERP begins with confronting anxiety-inducing situations and objects, but only those that you can tolerate [7].

12.5.2   The Usefulness of ERP

There are times in which professionals and patients alike are interested in uncovering the root cause of all symptoms and times when the patient is mentally strong enough to face his or her fears [7].

12.5.3   Cognitive Therapy

Cognitive therapy assists an individual in identifying and changing cognitive patterns that produce worry, suffering, or undesirable behaviour. That is to say, a CT scan reveals to patients that their brains are sending "error" signals. Through cognitive therapy, an individual can recognize these errors and confront their obsessions in new ways. Cognitive therapy can be useful when the patient has not been suffering from OCD for a very long time, has mild symptoms, and can control his obsessions [8].

12.5.4   Cognitive Restructuring

Cognitive restructuring is a treatment that teaches people to recognize and replace harmful thought patterns with more beneficial ones. Cognitive restructuring is used in OCD to reduce fear-inducing assumptions about obsessive thoughts or compulsive behaviours. Unhelpful thoughts are dismantled and rebuilt using cognitive restructuring techniques. Symptoms for which cognitive restructuring is suggested are as follows: feeling like a violent person all the time, thinking about your loved one's safety, obsessively thinking about how you are going to murder your family members, having the belief that you should not be angry, and having a fear of watching certain films and television shows because of their violence [9].

12.5.5 SELECTIVE SEROTONIN REUPTAKE INHIBITORS

There is a class of medications known as SSRIs that are used to treat depression and other mental illnesses because they are effective, safe, and well-tolerated. There are antidepressants in the form of SSRIs. They affect the levels of a chemical in the brain known as serotonin, which aids in the efficient transmission of nerve impulses. Nerves may be unable to communicate effectively if serotonin levels drop. SSRIs raise serotonin levels by inhibiting the transporters that transport it away from the brain. It is just that these medications do not work immediately and can take up to 8–12 weeks before they start to have an effect. When low doses of these drugs fail to relieve symptoms, a higher dose is prescribed. Symptoms for which SSRIs are suggested include getting anxious over the smallest situations [10].

12.5.6 DEEP BRAIN STIMULATION

During a DBS procedure, electrical impulses are delivered to specific regions of the brain (the brain nucleus) using an implanted medical device called a neurostimulator. In turn, these electrodes transmit information to the corresponding areas of the brain. DBS is recommended when symptoms are severely impairing quality of life, when symptoms remain uncontrolled despite taking the appropriate dose of medication, and when the side effects of current medications are intolerable [11].

DBS is a surgical procedure used to treat patients whose OCD has gotten out of control or for whom no other options have been effective. As many as 60% of patients reported immediate relief after the procedure was completed, and the effects lasted for months or years. As the patient's OCD worsens, he or she begins to avoid things that might set off his or her intrusive thoughts. Unwelcome thoughts may lead to a rash of violent behaviour. Both the frequency and intensity of panic attacks increase. Suicidal and self-harming thoughts are more prevalent and intense [12].

12.5.7 TRANSCRANIAL MAGNETIC STIMULATION

Unlike DBS, TMS utilizes electromagnetic induction via an insulated coil placed over the head; it is a non-invasive means of stimulating the brain. An area of the brain that governs mood is the focus of this treatment. An electromagnetic coil generates short magnetic pulses (similar to those generated by an MRI) that pass easily and painlessly through the skull and into the brain. TBS is used in conjunction with other OCD treatments [13].

TMS can be differentiated into two types: repetitive transcranial magnetic stimulation and deep transcranial magnetic stimulation. In repetitive TMS, a small device is implanted directly into the skull during this procedure. A wire coil is used in this device to carry electricity and create a magnetic field, which is then detected by a sensor. For this reason, it is referred to as repetitive music. Neurons in the brain are stimulated by the flow of electricity through the device, which alters their activity levels. Neuronal activity has been linked to OCD symptoms. The treatment protocol dictates how many TMS sessions you will need [14].

Deep TMS works by creating an intense magnetic field around a coil implanted directly in the skull. Deep TMS uses an H-coil, which allows the pulse to penetrate deeper into the brain, making it more effective. TMS is suggested in cases where standard antidepressant treatments have failed to provide adequate relief, the symptoms of OCD have not been sufficiently alleviated by psychotherapy or medication, and chronic migraine has not been alleviated by other treatments [15].

TMS can be used to treat the following conditions: doing things you do not really want to do over and over again in your mind; images of taboo subject matters that are unwelcome and intrusive; guilt, shame, or self-blame engulfing you; rising anxiety levels as you try to control them; ignoring an urge, which only makes it stronger; and contemplating on harming yourself or taking your own life.

12.5.8 Aversion Therapy

Aversion therapy, a discrete form of treatment, uses harsh treatments to help someone stop their behaviour or habit by making them associate it with something annoying. The vast majority of studies have focused on the positive effects it has on substance abuse.

There are three types of aversion therapies: electric shocks, chemical stimuli, and olfactory stimuli. Electric shock aversion therapy is one of the most widely used and divisive treatment approaches. An electrical shock can be administered using a device that is attached to a part of the body. An electric shock is delivered to the individual each time he or she takes part in something unwanted or shows signs of unacceptable behaviour, such as drug abuse [16].

In chemical stimuli aversion therapy, an unpleasant substance or medication with distasteful properties or side effects is used. It is used to treat alcoholism. Disulfiram, a drug that is sometimes used as a second-line treatment under medical supervision, is used in aversion therapy. Disulfiram causes sensitivity to even small amounts of ethanol, resulting in irksome reactions that can be clinically severe. In this therapy, three types of drugs are used: emetine, apomorphine, and lithium.

In olfactory stimuli aversion therapy, each time a person engages in or imagines doing the unwanted behaviour, they are subjected to an intensely foul odour, such as that of ammonia. Children and adolescents who were considered sexually deviant were treated with this method of treatment.

12.5.9 Rational Emotive Behaviour Therapy (REBT)

In the 1950s, Albert Ell is developed REBT, a form of cognitive behaviour therapy that has dominated the field of psychological treatment. For Albert Ellis, it was a philosophy of life, and its foundation was a belief that our emotions are not caused by events in our lives but rather by our beliefs about those events. The use of REBT is recommended when a person is suffering from depression or anxiety is coping with the effects of sexual abuse, choose between rational or irrational thoughts [17].

12.5.10 SYSTEMATIC DESENSITIZATION

Exposure therapy that is based on classical conditioning is called systematic desensitization. Wolpe developed it in the 1950s. Counter-conditioning therapy aims to gradually replace a conditioned response to fear with a response to relaxation. It is a method in which a person is gradually exposed to ever-increasing levels of stimuli that they normally avoid. Anxiety-inducing levels are gradually increased throughout the treatment, with the goal of decreasing the client's level of anxiety as they progress. Systematic desensitization may lessen the fear response with time and repetition. Three stages are involved in systematic desensitization.

Anxiety-inducing stimuli must be identified in order to begin the process of systematic desensitization. The next step covers techniques for relaxation and coping. When a person has been taught these skills, the next step is for him or her to put them into practice in order to respond to and overcome situations in the hierarchy of fears that has been set up for them. The individual's goal is to learn how to deal with and overcome fear at each step of the hierarchy. When a patient has a specific phobia, like a fear of heights, dogs, snakes, or small spaces, it is recommended that they go through systematic desensitization so that their anxiety can be measured and they can remember painful things [18].

OCD exists in multiple forms, but most cases belong to at least one of the four broad categories. The first category involves developing a habit of checking things frequently, such as lights witches, alarm systems, ovens, and locks, or imagining that one is pregnant or has schizophrenia. The second category involves feeling like you have been treated like dirt (mental contamination) and having a compulsion to maintain cleanliness or having the fear of things getting dirty. The third category involves maintaining the order and symmetry of things.

One can judge oneself if one is suffering from OCD by checking for the symptoms like having constant awareness of any kind of body sensations like heartbeats, breathing, and blinking; having an unreasonable suspicion of one's partner being unfaithful; worrying about getting hurt or hurting other people; repeating a task numerous times; feeling an urge to count things; and having a fear of using public toilets, handshakes, or touching doorknobs. Patients can experience many changes while going through OCD. Second, you may experience mood-related changes, like anxiety, guilt, and panic attacks. Third, you may experience psychological changes, like fear and depression. Apart from these three categories, some other types of common changes include food aversion, nightmares, and repeated thoughts [9].

12.6 TREATMENTS

The treatment breaks the automatic link between anxiety and ritualistic behaviour. When you start to feel anxious about minimalistic situations, you get afraid of germs, you feel that you need to constantly reaffirm your actions, or you start worrying that you might accidentally hurt your loved ones, CBT is the cure to go. The major CBT techniques practised to cure OCD are ERP, cognitive therapy, cognitive restructuring, aversion therapy, REBT, and systematic desensitization [19].

ERP is the most effective CBT technique for curing OCD. This therapy encourages the patient to face their fears and confronting obsessions without forcing to correct or neutralize them. Exposing patients to different situations helps them develop coping mechanisms and skills, and they can use these skills to avoid compulsions. ERP begins with confronting fear-causing things and situations but the ones that you can endure. ERP has three components: in vivo exposure (realistic exposure where someone is constantly exposed to the stimulus of fear over a long period of time), imaginal exposure (mental visualization of feared stimuli and consequences of exposure to stimuli), and ritual or response prevention (refraining from ritualistic behaviour after being exposed to feared stimuli). The ERP technique is useful when an expert wants to know the basis behind all the symptoms when the patient wants to know more about his obsession when the patient's obsession is fearful rather than obsessive.

The second major technique of CBT is cognitive therapy, which helps identify and modify thought patterns that cause anxiety, stress, or negative behaviours. In other words, CT helps the patient understand that the brain sends "error messages." Through CT, the patient learns to distinguish these mistakes and confront obsessions by responding to them in new ways. CT is beneficial when the patient has not been suffering from OCD [20].

Cognitive restructuring stands at number three in CBT techniques. It is about identifying the OCD. CR technique breaks down useless thoughts and reconstructs them in a more balanced and accurate way. If you have symptoms like constantly feeling that you are a violent person, fearing that you can harm your loved ones, having unwanted thoughts about how you are killing your loved ones, believing that whatever you think might come true, feeling that you should not experience anger, and having fear of watching certain movies/shows because they have some violent scenes, then cognitive restructuring is the technique to go, and SSRIs can act as antidepressant medications. They work by affecting the levels of a chemical named serotonin present in the brain, which helps in effective communication through nerves. When the serotonin level drops, nerves can effectively stop communicating. SSRIs raise serotonin levels by blocking transporters that lower the levels of these chemicals. However, these medicines do not work immediately. Getting anxious over the smallest situations, fear of germs, and developing a habit of rechecking something for a repeated number of times even though you are sure about it are the symptoms for which SSRIs are suggested by the doctor. When symptoms significantly impair quality of life and become uncontrollable despite appropriate dosages, then these side effects may be due to current dosage's nontolerance [21].

Deep brain stimulation is the surgical option. DBS is a neurosurgical procedure in. These electrodes then send signals to these parts of the brain. DBS is a surgical method used when a patient's OCD has become severe or other treatments have not worked. The effectiveness of this treatment is as high as that 60% of patients have found relief as soon as the procedure is completed. Also, the effects are long-lasting. The symptoms for which DBS is suggested include increased severity of OCD, which makes patients begin to avoid what can cause obsessive-compulsive anxiety, having unwanted thoughts that can cause aggression, and panic attacks becoming more frequent and intense. Another method of treating OCD is transcranial magnetic stimulation. When medications and psychotherapies have not given much relief to the symptoms of OCD, TMS is the next option. Repetitive thoughts about doing what you really do not want to do, unwanted and intrusive images of taboo subjects, guilt, shame, or self-blame, the more you try to control this obsession more it keeps on increasing [22].

12.7 CONCLUSION

OCD is a curable illness, and with proper attention and treatment, the patient can recover from this. There are a number of methods for treating OCD, such as cognitive-behavioural therapy, exposure and response prevention therapy, cognitive restructuring, rational emotive behaviour therapy, and aversion therapy. In the future, we can explore the possibilities of curing OCD with the help of AI and creating a more personal program for the patient and increasing the recovery speed.

REFERENCES

[1] Moritz, S., Birkner, C., Kloss, M., Jahn, H., Hand, I., Haasen, C., et al. (2002). Executive functioning in obsessive-compulsive disorder, unipolar depression, and schizophrenia. *Archives of Clinical Neuropsychology*, 17(5), pp. 477–483.

[2] Movahedi, Y., Khodadadi, M., & Mohammad Zadegan, R. (2014). The comparison of cognitive function and theory of mind in people with symptoms of obsessive-compulsive disorder and normal people (Persian). *Journal of Cognitive Psychology*, 2(3), pp. 28–36.

[3] Nejati, V., Kamari, S., & Jafari, S. (2018). Construction and examine the psychometric characteristic of student social cognition questionnaire (SHAD) (Persian). *Social Cognition*, 7(2), pp. 123–144.

[4] Okasha, A., Rafaat, M., Mahallawy, N., El Nahas, G., El Dawla, A. S., Sayed, M., et al. (2000). Cognitive days function in obsessive-compulsive disorder. *Act as Psychiatrica Scandinavica*, 101(4), pp. 281–285.

[5] Ozcan, H., Ozer, S., & Yagcioglu, S. (2016). Neuro-psychological, electro physiological and neurological impairments in patients with obsessive-compulsive disorder, their healthy siblings and healthy controls: Identifying potential endophenotype (s). *Psychiatry Research*, 240, pp. 110–117.

[6] Riggs, D. S., & Foa, E. B. (1993). Obsessive-compulsive disorder. In D. H. Barlow (Ed.), *Clinical handbook of psychological disorders: A step-by-step treatment manual* (pp. 189–239).

[7] Rosa-Alcázar, Á., Olivares-Olivares, P. J., Martínez-Esparza, I. C., Parada-Navas, J. L., Rosa-Alcázar, A. I., & Olivares-Rodríguez, J. (2020) Cognitive flexibility and response inhibition inpatients with obsessive-compulsive disorder and generalized anxiety disorder. *International Journal of Clinical and Health Psychology*, 20(1), pp. 20–28.

[8] Sharp, C., Fonagy, P., & Goodyer, I. (2008). Introduction. In P. Fonagy & I. Goodyer (Eds.), *Social cognition and developmental psychopathology* (pp. 2–6). Oxford: Oxford University Press.

[9] Simpson, H. B., Rosen, W., Huppert, J. D., Lin, S. H., Foa, E. B., & Liebowitz, M. R. (2006). Are there reliable neuropsychological deficits in obsessive-compulsive disorder? *Journal of Psychiatric Research*, 40(3), pp. 247–257.

[10] Snyder, H. R., Kaiser, R. H., Warren, S. L., & Heller, W. (2015). Obsessive-compulsive disorder is associated with broad impairments in executive function: A meta-analysis. *Clinical Psychological Science*, 3(2), pp. 301–330.

[11] Soltani, E., Shareh, H., Bahrainian, S. A., & Farmani, A. (2013). The mediating role of cognitive flexibility in correlation of coping styles and resilience with depression (Persian). *Pajoohandeh*, 18(2), pp. 88–96.

[12] Soriano-Mas, C., & Harrison, B. J. (2019). Structural brain imaging of obsessive-compulsive and related disorders. In L. F. Fontenelle & M. Yücel (Eds.), *A transdiagnostic approach to obsessions, compulsions and related phenomena* (pp. 74–84). Cambridge: Cambridge University Press.

[13] Todorov, A., Fiske, S., & Prentice, D. (2011). *Social neuroscience: Toward understanding the under pinnings of the social mind* (pp. 20–22). Oxford: Oxford University Press.

[14] Vaghi, M. M., Vértes, P. E., Kitzbichler, M. G., Apergis-Schoute, A. M., van der Flier, F. E., Fineberg, N. A., et al. (2017). Specific frontostriatal circuits for impaired cognitive flexibility and goal-directed planning in obsessive-compulsive disorder: Evidence from resting-state functional connectivity. *Biological Psychiatry*, 81(8), pp. 708–717.

[15] Okasha, A., Rafaat, M., Mahallawy, N., El Nahas, G., El Dawla, A. S., Sayed, M., et al. (2000). Cognitive disfunction in obsessive—compulsive disorder. *Acta Psychiatrica Scandinavica*, 101(4), pp. 281–285.

[16] Ozcan, H., Ozer, S., & Yagcioglu, S. (2016). Neuro psychological, electro physiological and neuro logical impairments in patients with obsessive compulsive disorder, their healthy siblings and healthy controls: Identifying potential endopheno type (s). *Psychiatry Research*, 240, pp. 110–117.

[17] Pinkham, A. E., Penn, D. L., Green, M. F., Buck, B., Healey, K., & Harvey, P. D. (2014). The social cognition psycho metric evaluation study: Results of the expert survey and RAND panel. *Schizophrenia Bulletin*, 40(4), pp. 813–823.

[18] Purdon, C., & Clark, D. A. (1993). Obsessive intrusive thoughts in nonclinical subjects. Part I. Content and relation with depressive, anxious and obsessional symptoms. *Behaviour Research and Therapy*, 31(8), pp. 713–720.

[19] Rehni, A. K., Singh, T. G., Jaggi, A. S., & Singh, N. (2008). Pharmacological preconditioning of the brain: A possible interplay between opioid and calcitonin gene related peptide transduction systems. *Pharmacological Reports*, 60(6), p. 904.

[20] Aggarwal, A., Dhaliwal, R. S., & Nobi, K. (2018). Impact of structural empowerment on organizational commitment: The mediating role of women's psychological empowerment. *Vision*, 22(3), pp. 284–294.

[21] Sharma, A., Kukreja, V., Bansal, A., & Mahajan, M. (2022). Multi classification of Tomato leaf diseases: A convolutional neural network model. *2022 10th International Conference on Reliability, Infocom Technologies and Optimization (Trends and Future Directions) (ICRITO)*, Noida, India, 2022, pp. 1–5.

[22] Sakshi, S., Lodhi, S., Kukreja, V., & Mahajan, M. (2022). DenseNet-based attention network to recognize handwritten mathematical expressions. *2022 10th International Conference on Reliability, Infocom Technologies and Optimization (Trends and Future Directions) (ICRITO)*, Noida, India, 2022, pp. 1–5.

13 A Mechanistic Review of Clinical Applications of Microchip Electrophoresis and Recent Advancements

*Ritchu Babbar, Rashmi Arora,
Ramanpreet Kaur, Arashmeet Kaur*

13.1 INTRODUCTION

Microchip electrophoresis (MCE) is a shrunken arrangement of capillary electrophoresis (CE). Electrophoresis is a common method to segregate larger molecules such as proteins, nucleic acids (RNA, DNA), and other biological and chemical species. This is a regular method for DNA size shattering and unscrambling protein blends at a quicker rate. The achievement of efficient electrophoretic separations is due to the application of higher voltages in MCE [1, 2]. The beginning of electrophoresis can be dated back to the year 1937 by a Nobel Prize winner, Tiselius, where narrow tubes were used for the placement of mixtures of proteins along with the buffer with the addition of an electric field. This electric field helps in determining the rate and mobility of the charged particles. The sole basis of this segregation is the variation in the mobilities of the particles, cataphoresis being defined as the positively charged particles while the negatively charged particles are referred to as anaphoresis. Factors on which this separation depends are intensity and the mobility of the individual species [3]. This work of Tiselius proved to be a beginning, though it had various drawbacks, such as low efficiency, a longer time for analysis, and lesser facilities for the detection. Therefore, an alternative of using slabs for performing electrophoresis operation in bore tubes was discovered by Hjérten in 1967. He used a one-millimetre bore tube for minimization of convection effect by rotating them in the longitudinal axis. Another advancement was traced by Virtanen followed by Mikker where electrophoresis was carried out in a 200-micrometre diameter capillary. Another advancement, the 75-micrometre capillary was used by Lukacs and

TABLE 13.1

Classification of Electrophoresis

S. No.	Gel Electrophoresis	Affinity Electrophoresis	Capillary Electrophoresis
1.	This type of electrophoresis involves the usage of gel, which is used over for the separation of particles.	This type of electrophoresis is based upon the electrophoretic changes of charges via complex.	Capillary electrophoresis is a type of electrophoresis is utilized for the separation of ions with respect to friction and charge as well as hydrodynamic radius.
2.	This technique is generally used for the separation of DNA and RNA molecules.	This technique is used for estimating binding constants.	Capillary electrophoresis is used as an analytical procedure postulated on an ion separation focusing on individual mobility of ions upon the applied voltage.

Classification of electrophoresis

Jorgenson around the 1980s. He also clarified all the theoretical relationships and parameters related to the quality and efficiency of CE [4]. The process of electrophoresis can be classified as shown in Table 13.1.

13.1.1 Microchip Electrophoresis

This is a miniaturized form of CE based on the same principles. MCE is an advanced technique of CE. This widens the horizons of miniaturized devices with greater efficacy and resolution. MCE has also achieved recognition in regards to its prospective applications and comprehensive device performance [5]. This lab-on-chip concept has expansive utilization in various segments, such as forensics, proteomics, genomics, biomarkers, and metabolics [6]. In addition, the quantitative estimation has proved to be beneficial for the samples, such as inorganic ions, cations, and anions in tap water analysis; determining nitrates in water; and lithium in blood samples. Organic ions or compounds, such as oxalates, levoglucosan, thiols, and phenols, have also been estimated [7–9].

13.1.1.1 Principle of Microchip Capillary Electrophoresis

The working principle of MCE involves the movement of charged particles under the influence of an electric field. When equal positive and negative charges are present, there will be no net charge produced as the charges get neutralized by one another. On the other hand, when there is presence of unequal charges, the charged particles will move to the oppositely charged regions (for instance, an electrode). The anions move toward the anode, whereas the cations will move towards the cathode. Hence, the process of electrophoresis depends on the molecular size of the particles as well as its mobility in the solution or medium [10].

13.1.1.2 Designing of MCE

The general outlook of the microchip device (Figure 13.1) consists of a series of channels being placed in an elongated tube often referred to as channels of separation. At the site separation channel, the process of separation occurs. In a typical microchip device, these channels are about 50–200 micrometres in width and about 15–50 micrometres in depth. The length of the channel generally 1–10 centimetres. So at base of the channel separation, there are approximately four fluid reservoirs in addition to the separation channels. In addition to separation channels, there are about four fluid reservoirs situated at the base of the channel separation. Out of the four fluid reservoirs, two are for introducing samples and the other two are for holding waste. The reservoirs are surrounded by the electrodes connected at high voltage [8, 30]. The case of the commercial Agilent Bioanalyzer TM chip contains a total of 16 reservoirs, out of which 12 are used for the detection of samples and four of them are for reagents and references [11]. The manufacture of several reservoirs and channels helps in better analysis of the diagnostic procedures. The electrode materials that are used range from different metals, like gold (Ag), platinum (Pt), or palladium (Pd) to types of carbon, like ink, fibre, paste, and pyrolyzed photo-resist film, or PPF [12].

13.1.1.3 Channels of Separation

Separation channels in microchip devices are generally T-shaped or cross-channels being perpendicular to the sample channels, as well as buffer channels. The length of the channels of separation is usually reduced than its diameter, being compared to the capillary micro electrophoretic device. In addition, a greater electric field is required when the path length is reduced; the strength of the electric field is directly proportional to the process of electrophoresis. Therefore, with an increase in electric field by decreasing the path length, the process of electrolysis increases.

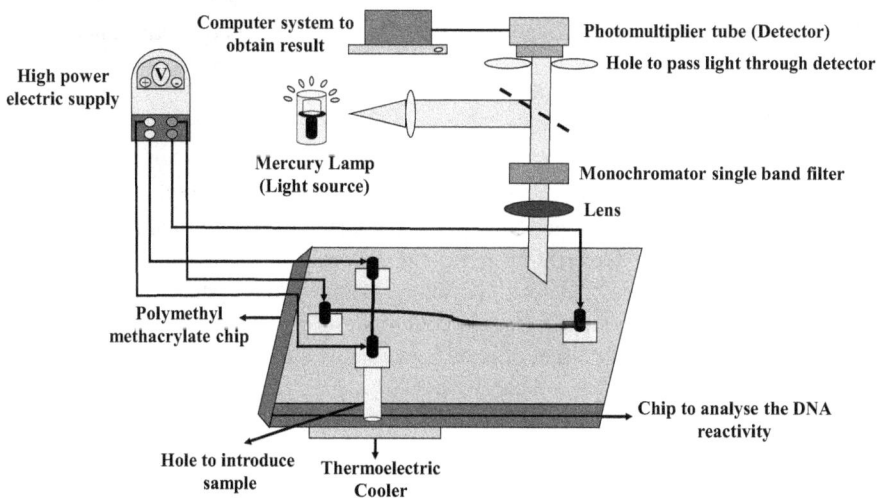

FIGURE 13.1 Mechanistic overview of MCE.

13.1.1.4   Injection of Sample through Channels

Samples are injected into the channels in the CE either with application of high pressure or with high voltage on the end of the capillary tube immersed in the sample. It is difficult to manipulate channels in MCE at each stage. Therefore, peristaltic pumps are used for injecting that either at high pressure or voltage. Two injection modes are generally adopted: electro-kinetic injection or hydrodynamic injection mode [13].

13.1.1.5   Detection

Detection is one of the crucial steps for the MCE. Detectors are generally placed at the channels of separation in the opposite direction. There are usually various types of detection methods used, out of which laser-induced fluorescence is one of the important methods. Other modes include chemiluminescence, mass spectrometry, and electrochemical methods. When the sample is separated, the detection of the sample takes place [13, 14].

13.2   CLINICAL APPLICATIONS OF MICROCHIP ELECTROPHORESIS

The applications of MCE are not only restricted to the industrial field but have a large perspective towards the clinical aspect. This method has emerged as a real advantage when gene detection, polymorphism analysis, detection of blood glucose level, and also analyses of biopharmaceutical drugs etc has to be performed. Separation of DNA can be used in the field of research, forensics, clinical studies, and so on [15]. Analysis of small analyte molecules and analysis of proteins is carried out with the aid of various detection methods, such as mass spectrometry, laser-induced fluorescence, and electrochemical detection. In addition, ion detection can be performed with the aid of dual-channel injection.

In the following sections, listed are some clinical areas where this novel technique has been implemented to enhance the diagnostic and therapeutic functions in these challenging times.

13.2.1   Cancer

The MCE is used in the diagnostic applications of cancer by studying the activity of the telomerase enzyme, detection of carcinoembryonic antigen, and the mutation, that are analysed in the deoxyribonucleic acid (DNA) of p53 oncogenes [16]. In lung cancer and blood cancer, it detects the alterations in EGFR gene and determines neuron-specific enolase [17, 18]. In colorectal cancer, oral cancer, and prostate cancer, it helps to study the alteration of KRAS, evaluates the circulating-cell-free DNA, and assessing oral squamous cell carcinoma [19].

13.2.2   Immune Disorders

These include inflammatory diseases, allergic and hypersensitivity reactions, gout, and arthritis.

For inflammatory conditions and allergic reactions, the common markers C-reactive protein (CRP) and cytokines are detected [20].

13.2.3 NEUROLOGICAL DISEASES

This includes the detection of albumins, globulin in multiple sclerosis, lithium concentration in bipolar disorder, and protein aggregation. Mohamadi and co-workers accomplished the ME reporting of dementia patients; cerebrospinal fluid contains amyloid beta plaque [21].

13.2.4 GENETIC DISORDERS

The following genetic disorders have been analysed in the diagnostic procedures with the help of MCE: it includes the detection of deficiency of glucose-6 phosphate dehydrogenase enzyme, a CAG repeat expansion in muscular and spinal atrophies, and fetal β-globin mutations in the mother's blood with in β-thalassemia [22].

Cardiovascular diseases—deoxyribonuclease I is detected in acute myocardial infarction, detection of mutation angiotensin-converting enzyme (ACE) gene [23].

Diabetes mellitus—in this, we detect the haemoglobin A1c proteins in the blood by chip [24].

13.2.5 INFECTIOUS DISEASES

Virus-induced infections and bacterial infections were studied using genotyping strategies, immunoassays, and zone electrophoresis. For example, hepatitis B is detected by the presence of enzyme alanine aminotransferase [24].

13.2.6 ORGAN DISEASES

The organ diseases and their dysfunctions can also be easily analysed using this chip. Liver disease, kidney disease, and lacrimal gland disease are detected by examining free bilirubin, creatinine concentration, and inorganic ion composition of tears, respectively.

13.2.7 MICROCHIP ELECTROPHORESIS FOR SINGLE AND
BULK CELLS EXAMINATION

In the last 20 years, there is an overabundance of ME to demonstrate biological proceedings and drug evaluation in single cell along with bulk cells. Shi et al. successfully done the screening of NT in PC-12 cells using a ME system connected by a fluorescence microscope. Further, Sin et al. performed the speedy determination of glutathione (GSH) in single rat liver cells utilizing ME coupled with chemiluminescence detection instrument. Also, Li and co-workers described automated single cell (PC-12) study utilizing a ME-MS platform [25].

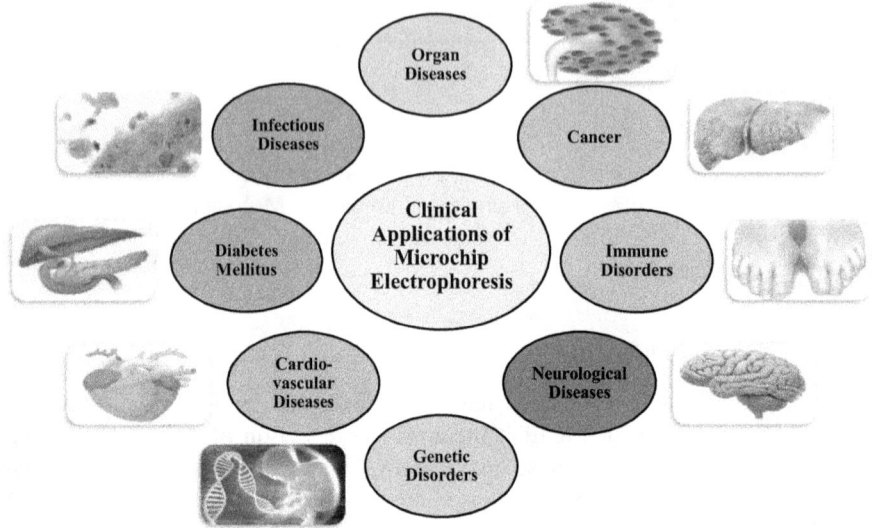

FIGURE 13.2   Clinical applications of microchip electrophoresis.

13.2.8 ORGANS-ON-A-CHIP

The progress of ME-based organs-on-a-chip, mirroring the whole human physique, has become very popular technique in the past decade; for example, a microfluidic-based model was used to replicate the lung, eye, brain, skin, gut, liver, pancreas, heart, and kidney [26].

The summarized content of clinical applications of ME is given in Figure 13.2.

13.2.9 DRUG EXPLORATION AND DEVELOPMENT AND IN PRECLINICAL TOXICOLOGICAL STUDIES

ME is considered as an auspicious tool in drug exploration and development and description of pharmaceutical ingredients. It also includes ME-based approaches, such as genotyping, recognition of toxins in biological samples, foods, and drugs, and analysis of samples concerning clinical toxicology [27].

13.3 RECENT ADVANCEMENTS IN THE FIELD OF MICROCHIP ELECTROPHORESIS

There are many recent progressions in the arena of MCE, which are summarized in Table 13.2.

13.4 PATENTS

There are many recent patents in the arena of MCE. Few are enlisted in Table 13.3.

TABLE 13.2
Current Developments in Microchip Electrophoresis

S. No.	Application	Detection Method	Reference
1	RNA/DNA analysis	Laser-induced fluorescence	[28]
2	*Ginseng* species authentication	Laser-induced fluorescence	[29]
3	Increased amplification for microchip	Surface passivation	[30]
4	Nitrate and chloride determination from drinking water	Conductivity	[31]
5	Creatinine determination from urine	Fluorescence induced by LED	[32]
6	Detection of mutations with the aid of ligase reaction chain	Ligase chain detection	[33]
7	Separation of uric and ascorbic acid using dimethylsiloxane-coated microchip	Amperometric detection	[34]
8	Sirtuin 5 inhibitors identification using Microchip electrophoresis	Fluorescence method	[35]
9	Monitorization of aminophenols with the aid of dual channel microchip-265 electrophoresis	Amperometric method	[36]
10	Assay of food sample to identify *E. coli* using microchip electrophoresis on the basis of aptamer binding	Laser-induced fluoroscence	[37]

TABLE 13.3
Published Patents in the Field of Microchip Electrophoresis

Patent Number	Title	Explanation	Year Granted	Reference
US6969452B2	Method for two-dimensional protein separation and protein-separating device	The invention relates to the two-dimensional protein separation method with the steps (a) ion exchange chromatography and (b) capillary electrophoretic separation of the fraction collective.	2005	[38]
CA2740113C	Hybrid digital and channel microfluidic devices and methods of use thereof	The current innovation offers, in the form of an integrated structure, a hybrid digital and channel microfluidic device in which a droplet can be transported by a digital microfluidic array and transferred to a microfluidic channel.	2019	[39]

(*Continued*)

TABLE 13.3 *(Continued)*
Published Patents in the Field of Microchip Electrophoresis

Patent Number	Title	Explanation	Year Granted	Reference
US8137512B2	Process for analysing sample by capillary electrophoresis method	A method used in the analysis of the sample is done by capillary electrophoresis, which allows the equipment to be reduced to micro scale, allows high analytical accuracy to be achieved and can be easily performed.	2012	[40]
US20070039822A1	Undeviating quantification of antimicrobials, carbohydrates, proteins, and antibiotics by ME	This invention includes a pulsed amperometric electrophoresis microchip for the partition and identification of non-derivatized carbohydrates, proteins, and antibiotics having sulphur atom.	2011	[41]

13.5 FUTURE PERSPECTIVES

With the emerging use of science and technology in the pharmaceutical and medical line, the future of MCE is expanded to biological science, production of drugs, understanding the mechanism of disease or in studying plant stress, and research into marine life. Improvement in the effectiveness of both *in vitro* and *in vivo* models, reproduction of the *in vivo* environment and estimation of the therapeutic efficacy of clinical trials, and a wide array of combined systems pairing 2D and 3D cell culture experimentation with microfluidics have been generated. A significant consideration that has been made in designing the chip is the minimization of cost, which is taken into account in order to facilitate the microchip CE as an ideal approach for the future generation in the analysis of milk. Therefore, low-cost components are used to implement the fluorescence spectroscopy method, which includes UV-light-emitting diode for the excitation of fluorescence. An electrical low-pass filter passes the output signal from the transimpedance circuit, and the resulting signal is stored on a computer. An unconventional micro-sized analytical approach has been used to determine the presence of carminic acid, which a red-coloured food natural is MCE, and it follows detection of the acid in food and pharmacy related materials. Looking into the innumerable advantages of speed and lower consumption in the use of reagents, high-throughput, and amount of sample injection used, MCE has gained a lot of attention in the technique used at microscale level for separation.

13.6 CONCLUSION

MCE has now become a considerable landmark for the separation phenomenon. Its high efficiency and high resolution have proven to be one of the major factors for its acceptance as the best electrophoresis technique, especially for DNA analysis. Rapid separation and lesser time for the process contributes to its advantages. Various methods are being employed by the researchers among the world to minimize its demerits and miniaturization in order to explore more for its utility in various fields of research. This advancement of CE working on the same principle as the former has led to the minimization of the CE. The microchip device system has proven to have its utilization in various aspects, though the technique is still under the analytical testing. Various applications of the MCE in the DNA analysis, proteins, and even the small molecules have proven to be beneficial. Though there is a long way for the proper establishment of the procedures for the MCE, this method is still a boon for the analytical and biochemical industry.

13.7 CONSENT FOR PUBLICATION

Not applicable.

13.8 FUNDING

Not applicable.

13.9 CONFLICT OF INTEREST

The authors declare no conflict of interest, financial or otherwise.

REFERENCES

1. Caruso, G.; Musso, N.; Grasso, M.; Costantino, A.; Lazzarino, G.; Tascedda, F.; Gulisano, M.; Lunte, SM.; Caraci, F. Microfluidics as a novel tool for biological and toxicological assays in drug discovery processes: Focus on microchip electrophoresis. *Micromachines*, **2020**, *11*(6), 593.
2. Kheyrodin, H. Introduction of electrophoresis process. *World J. Clin. Pharmacol. Micrbiol. Toxicol.*, **2015**, *1*(3), 14–21.
3. Dolník, V. Wall coating for capillary electrophoresis on microchips. *Electrophoresis*, **2004**, *25*, 3589–3601.
4. de Castro Costa, B. M.; Griveau, S.; d'Orlye, F.; Bedioui, F.; da Silva, J. A. F. Microchip electrophoresis and electrochemical detection: A review on a growing synergistic implementation. *Electrochim. Acta*, **2021**, *391*, 138928.
5. Mackintosh, J.A.; Choi, H.Y.; Bae, S.H.; Veal, D.A.; Bell, P.J.; Ferrari, B.C.; Van Dyk, D.D.; Verrills, N.M.; Paik, Y.K.; Karuso, P. A fluorescent natural product for ultrasensitive detection of proteins in one-dimensional and two-dimensional gel electrophoresis. *Proteomics*, **2003**, *3*, 2273–2288.
6. Castro, E.R.; Manz, A. Present state of microchip electrophoresis: State of the art and routine applications. *J. Chromatogr. A*, **2015**, *1382*, 66–85.

7. Hassan, S.-U. Microchip electrophoresis. *Encyclopedia*, **2021**, *1*, 30–41. https://doi. org/10.3390/encyclopedia1010006.

8. Zuborova, M.; Masar, M.; Kaniansky, D.; Johnck, M.; Stanislawski, B. Determination of oxalate in urine by zone electrophoresis on a chip with conductivity detection. *Electrophoresis,* **2002**, *23*(5), 774–781.

9. Stepanova, S.; Kasicka, V. Analysis of proteins and peptides by electromigration methods in microchips. *J Sep. Sci.*, **2017**, *40*(1), 228–250.

10. Masar, M.; Hradski, J.; Vargova, E.; Miskovcíková, A.; Bozek, P.; Sevcík, J.; Szucs, R. Determination of carminic acid in foodstuffs and pharmaceuticals by microchip Electrophoresis with photometric detection. *Separations*, **2020**, *7*(4), 72.

11. Wuethrich, A.; Quirino, J.P. A decade of microchip electrophoresis for clinical diagnostics–a review of 2008–2017. *Anal. Chim. Acta.*, **2019**, *1045*, 42–66.

12. Gonzalez-Bellido, P.T.; Wardill, T.J.; Kostyleva, R.; Meinertzhagen, IA.; Juusola, M. Overexpressing temperature-sensitive dynamin decelerates phototransduction and bundles microtubules in Drosophila photoreceptors. *J. Neurosci.*, **2009**, *29*(45), 14199–141210.

13. Dispas, A.; Emonts, P.; Fillet, M. Microchip electrophoresis: A suitable analytical technique for pharmaceuticals quality control? A critical review. *TrAC Trends Anal. Chem.*, **2021**, *139*, 116266. https://doi.org/10.1016/j.trac.2021.116266.

14. Jia, X.; Yang, X.; Lou, G.; Liang, Q. Recent progress of microfluidic technology for pharmaceutical analysis. *J. Pharm. Biomed. Anal.*, **2022**, *209*, 114534. https://doi. org/10.1016/j.jpba.2021.114534.

15. Li, S.F.; Kricka, L.J. Clinical analysis by microchip capillary electrophoresis. *Clinic. Chem.*, **2006**, *52*(1), 37–45.

16. Karasawa, K.; Arakawa, H. Detection of telomerase activity using microchip electrophoresis. *J. Chromat. B.*, **2015**, *993*, 14–19.

17. Matsumoto, N.; Mori, S.; Hasegawa, H.; Sasaki, D.; Mori, H.; Tsuruda, K.; Imanishi, D.; Imaizumi, Y.; Kaku, N.; Kosai, K. Simultaneous screening for JAK2 and calreticulin gene mutations in myeloproliferative neoplasms with high resolution melting. *Clin. Chim. Acta*, **2016**, *462*, 166–173.

18. Zhang, H.; Song, J.; Ren, H.; Xu, Z.; Wang, X.; Shan, L.; Fang, J. Detection of low-abundance KRAS mutations in colorectal cancer using microfluidic capillary electrophoresis-based restriction fragment length polymorphism method with optimized assay conditions. *PLoS ONE*, **2013**, *8*(1), e54510.

19. Odenthal, M.; Barta, N.; Lohfink, D.; Drebber, U.; Schulze, F.; Dienes, H.P.; Baldus, S.E. Analysis of microsatellite instability in colorectal carcinoma by microfluidic-based chip electrophoresis. *J. Clin. Path.*, **2009**, *62*(9), 850–852.

20. Herwig, E.; Marchetti-Deschmann, M.; Wenz, C.; Rüfer, A.; Redl, H.; Bahrami, S.; Allmaier, G. Sensitive detection of C-reactive protein in serum by immuno-precipitation–microchip capillary gel electrophoresis. *Anal. Biochem.*, **2015**, *478*, 102–106.

21. Mohamadi, M.R.; Svobodova, Z.; Verpillot, R.; Esselmann, H.; Wiltfang, J.; Otto, M.; Taverna, M.; Bilkova, Z.; Viovy, J.L. Microchip electrophoresis profiling of abeta peptides in the cerebrospinal fluid of patients with Alzheimer's disease. *Anal. Chem.*, **2010**, *82*, 7611–7617.

22. Maruyama, H.; Morino, H.; Izumi, Y.; Noda, K.; Kawakami, H. Convenient diagnosis of spinal and bulbar muscular atrophy using a microchip electrophoresis system. *A.J. of Neurodegener. Dis.*, **2013**, *2*(1), 35–39.

23. Fujihara, J.; Takinami, Y.; Ueki, M.; Kimura-Kataoka, K.; Yasuda, T.; Takeshita, H. Circulating cell-free DNA fragment analysis by microchip electrophoresis and its relationship with DNase I in cardiac diseases. *Clin. Chim. Acta*, **2019**, *497*, 61–66. http:// doi.org/10.1016/j.cca.2019.07.014.

24. Buyuktuncel, E. Microchip electrophoresis and bioanalytical applications. *C. P'ceutical Anal.*, **2019**, *15*(2), 109–120.
25. Li, X.; Zhao, S.; Hu, H.; Liu, Y.M. A microchip electrophoresis-mass spectrometric platform with double cell lysis nano-electrodes for automated single cell analysis. *J. Chromatogr. A*, **2016**, *1451*, 156–163.
26. Danku, A.E.; Dulf, E.H.; Braicuz, C.; Jurj, A.; Berindan-Neagoe, I. Organ-on-a-chip: A survey of technical results and problems. *Front. Bioeng. Biotechnol.*, **2022**, *10*, Article ID 840674. http://doi.org/10.3389/fbioe.2022.840674.
27. Ouimet, C.M.; Amico, C.I.; Kennedy, R.T. Advances in cappilary electrophoresis and the implications for drug discovery. *Expert. Opin. Drug Discov.*, **2017**, *12*(2), 213–224. http://doi.org/10.1080/17460441.2017.1268121.
28. Nishikawa, F.; Murakami, K.; Matsugami, A.; Katahira, M.; Nishikawa, S. Structural studies of an RNA aptamer containing GGA repeats under ionic conditions using microchip electrophoresis, circular dichroism, and 1D-NMR. *Oligonucleotides*, **2009**, *19*(2), 179–190.
29. Bosma, R.; Devasagayam, J.; Singh, A.; Collier, C.M. Microchip capillary electrophoresis dairy device using fluorescence spectroscopy for detection of ciprofloxacin in milk samples. *Sci. Rep.*, **2020**, *10*(1), 1–8.
30. Lou, X.J.; Panaro, N.J.; Wilding, P.; Fortina, P.; Kricka, L.J. Increased amplification efficiency of microchip-based PCR by dynamic surface passivation. *Biotechniques*, **2004**, *36*(2), 248–252.
31. Masar, M.; Bomastyk, B.; Bodor, R.; Horciciak, M.; Danc, L.; Troska, P.; Kuss, H.M. Determination of chloride, sulfate and nitrate in drinking water by microchip electrophoresis. *Microchimica Acta.*, **2012**, *177*(3–4), 309–316.
32. Wang, S.; Li, X.; Yang, J.; Yang, X.; Hou, F.; Chen, Z. Rapid determination of creatinine in human urine by microchip electrophoresis with LED induced fluorescence detection. *Chromatographia*, **2012**, *75*(21–22), 1287–1293.
33. Lou, X.J.; Panaro, N.J.; Wilding, P.; Fortina, P.; Kricka, L.J. Mutation detection using ligase chain reaction in passivated silicon-glass microchips and microchip capillary electrophoresis. *Biotechniques*, **2004**, *37*(3), 392–398.
34. Hang, Q.L.; Xu, J.J.; Lian, H.Z.; Li, X.Y.; Chen, H.Y. Polycation coating poly (dimethylsiloxane) capillary electrophoresis microchip for rapid separation of ascorbic acid and uric acid. *Anal. Bioanal. Chem.*, **2007**, *387*(8), 2699–2704.
35. Guetschow, E.D.; Kumar, S.; Lombard, D.B.; Kennedy, R.T. Identification of Sirtuin 5 inhibitors by ultrafast microchip electrophoresis using nanoliter volume samples. *Anal. Bioanal. Chem.*, **2016**, *408*(3), 721–731.
36. Chen, C.; Hahn, J.H. Enhanced aminophenols monitoring using in-channel amperometric detection with dual-channel microchip capillary electrophoresis. *Env. Chem. Letters.*, **2011**, *9*(4), 491–497.
37. Zhang, Y.; Zhu, L.; He, P.; Zi, F.; Hu, X.; Wang, Q. Sensitive assay of Escherichia coli in food samples by microchip capillary electrophoresis based on specific aptamer binding strategy. *Talanta.*, **2019**, *197*, 284–290.
38. He, Y.; Pang, H.M.; Luo, S.; Han, F. Method for multiplexed capillary electrophoresis signal cross-talk correction. *US Patent 6969452*, **2005**.
39. Watson, M.W.; Abdelgawad, M.; Jebrail, M.; Yang, H.; Wheeler, A.R. Hybrid digital and channel microfluidic devices and methods of use thereof. *US Patent 9039973*, **2015**.
40. Tanaka, Y.; Wakida, S.; Nakayama, Y.; Yonehara, S. Process for analyzing sample by capillary electrophoresis method. *US Patent 8137512*, **2012**.
41. Henry, C.; Garcia, C. Direct determination of carbohydrates amino acids and antibiotics by microchip electrophoresis with pulsed amperometric detection. *US Patent 10/568975*, **2007**.

14 An Insight into the Potential of Natural Products as Anti-inflammatory Agents
A Systematic Review

Rashmi Arora, Ritchu Babbar,
Ramanpreet Kaur, Parteek Rana

14.1 INTRODUCTION

An anti-inflammatory agent is the characteristics of a material or action that diminishes redness or puffiness. It prevents certain substances in the human body that instigate inflammation. Inflammation is the human body's initial response to infection or wound and is life-threatening for both innate and adaptive immunity. These expelled atoms act as antigens to excite a broad-spectrum immune response and to begin the proliferation of leukocytes [1]. The exploration for natural complexes and phytoconstituents that can hinder with these mechanisms by stopping a prolonged inflammation could be useful for human health. Natural products have assumed a significant role all throughout the world in treating and forestalling human illnesses for millennia, and in recent many years, critical endeavours have been made to look at present-day uses of natural items with more prominent viability and lower adequacy [2]. Many exploration and review articles on plant mitigating properties have been distributed in ongoing many years. In this review, we present a few features from the writing, for the most part from the last 30 years, with a couple of references to prior examinations. Diseases such as arthritis [3], obesity [4], diabetes [5], and cancer [6] are correlated with inflammation in Figure 14.1.

DOI: 10.1201/9781003367161-14

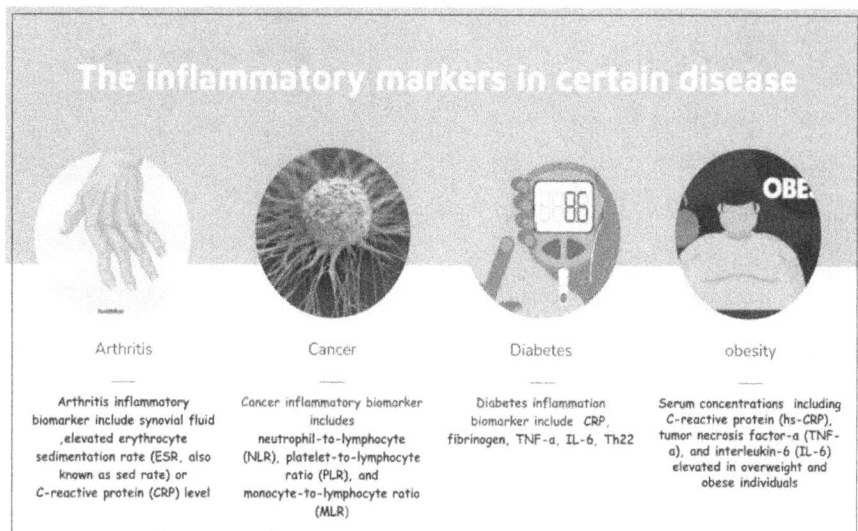

FIGURE 14.1   Depiction of the inflammatory markers in inflammatory diseases.

14.2   NATURAL PLANT PRODUCTS WITH ANTI-INFLAMMATORY PROPERTIES

14.2.1   GARLIC (*ALLIUM SATIVUM*)

Garlic (*Allium sativum* L.; Amaryllidaceae) is a fragrant herb that contains hundreds of phytochemicals, including sulphur-containing compounds like allicin, E/Z ajoene, allyl sulphide, 3-methylsulphanyl prop-1-ene, S-allylcysteine, 3-(allylsulphinyl)-L-alanine, and diallyl thiosulphinate) (Figure 14.2) which were used in human volunteers for pre-clinical examinations [7]. These blends curbed LPS-impelled disturbance by reducing levels of PG, NO, IL-1β, IL6, and TNF-alpha, by extending levels of IL-10, by controlling activity of COX-2, iNOS, and NF-alpha [8].

14.2.2   ALOE (*ALOE VERA*)

Aloe vera is one of the succulent plant species present in the *Aloe* genus, with lupeol as main phytoconstituent (Figure 14.3). It originates in the Arabian peninsula and spreads untamed in tropical, semitropical, and arid regions all over the world. The activity portion of disengaged blends in LPS-fortified mice and murine mononuclear phagocyte system RAW264.7 cells has been the focus of ongoing research [9]. As an

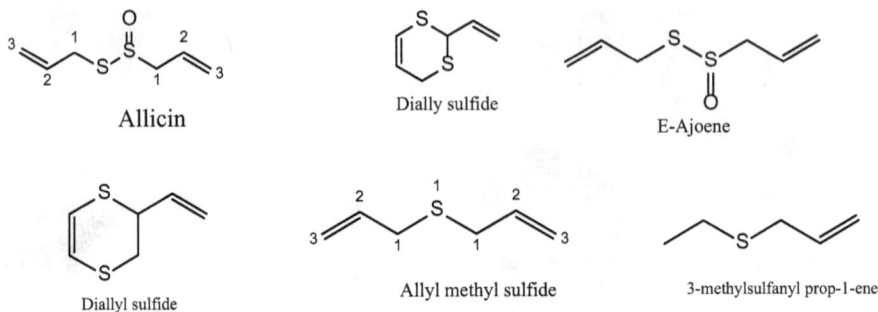

FIGURE 14.1 Garlic phytochemicals.

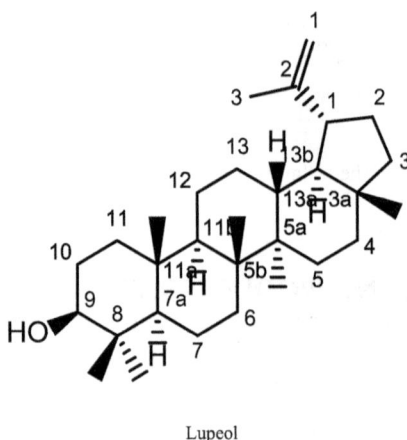

Lupeol

FIGURE 14.3 *Aloe vera* phytoconstituents.

outcome, aloe's calming properties are ascribed to its capacity to repress cytokines, responsive oxygen species (ROS), and the JAK1-STAT1/3 flagging pathway [10].

14.2.3 PAPAYA (CARICA PAPAYA)

Papaya, an individual from the Caricaceae family, is developed in various pieces of the world, including India. Papaya's proteolytic catalysts (papain and chymopapain) have been shown to have immunomodulatory and calming properties (Figure 14.4) [11]. In another examination, a methanol concentrates of papaya leaf hindered the development of nitric oxide (NO) in IFN (100 U/ml)—or LPS (5 µg/ml)—invigorated murine monocytic macrophages (IC50: 60.18 µg/ml) (RAW 264.7 cell line) [12].

14.2.4 GINGER (ZINGIBER OFFICINALE)

Ginger, a flowering plant rhizome called as zingiber root or zingiber, is frequently used as a spice and medicine for humans, with shogaol and paradol as the main

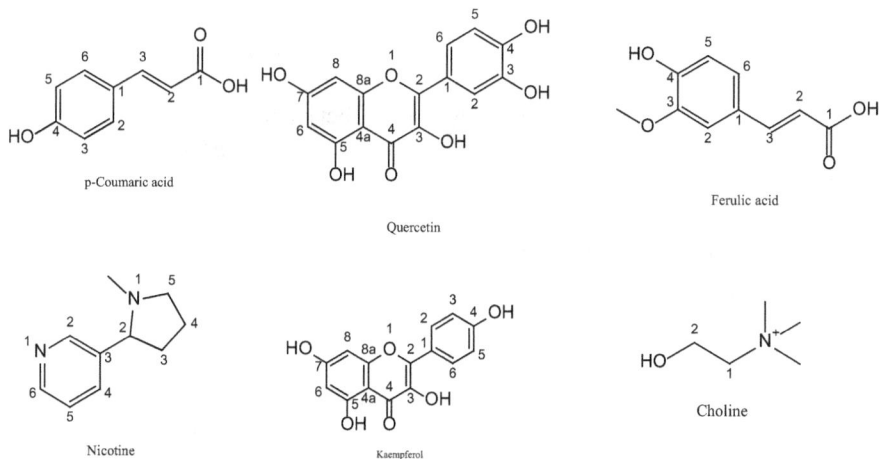

FIGURE 14.4   Papaya's phytoconstituents.

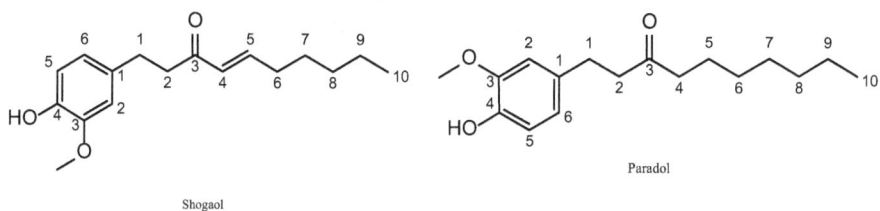

FIGURE 14.5   Ginger phytoconstituents.

phytoconstituents (Figure 14.5). It is a herbaceous perennial that produces yearly pseudostems (false stems created from the repositioned leaf bases) that are about 1 metre tall and have sharp edges. In 2D cell culture conditions, HepG2 cells were treated with three types of ginger root separated from the maceration (G1), Soxhlet (G3), and sonication (G5) strategies to discover the concentrate with the most note-worthy cytotoxicity on HepG2 cells. The discoveries showed that the Soxhlet extri-cate had the most intense cytotoxic impact on HepG2, with an IC50 of 83.3 ± 0.9189 mg/ml, while the IC50 upsides of concentrates arranged by maceration and sonica-tion were 159 ± 7.6 mg/ml and 284 ± 5.116 mg/ml, separately. These discoveries uncovered that the Soxhlet ginger root remove had the best antitumour cytotoxicity against HepG2 [13].

14.2.5   BLUEBERRY (*VACCINIUM CORYMBOSUM*)

Vaccinium corymbosum, the northern highbush blueberry, has become a food prod-uct of huge monetary significance. Studies have shown that the improved degrees of free extremists, qualities, and protein articulation of incendiary cytokines in post-awful pressure problems in rodents have been standardized by blueberry-advanced eating regimens [14]. In 2014, Huang *et al.* directed an examination to explore the inhibitory impact on the provocative reaction in endothelial cells of the two significant blueberry anthocyanins known as malvidin-3-glucoside and malvidin-3-galactoside. They found that the alleviating effect of malvidin-3-glucoside was more conspicuous than that of malvidin-3-galactoside and that its quieting limit was interceded by the NF-3-B pathway.

14.3 ACTIVE CHEMICAL MOIETIES OF NATURAL ORIGIN WITH ANTI-INFLAMMATORY PROPERTIES

14.3.1 Quercetin

Quercetin (Figure 14.6), a typically happening polyphenol, has huge alleviating effect and is accessible in many food products. It represses the advancement of chemicals causing irritation (COX and LOX) [15]. In an examination, the impact of quercetin in combination with vitamin C and omega-3 for oxidative pressure and aggravation on 60 competitors. Quercetin and vitamin C, in combination, lowered C-receptive protein and IL-6 levels when compared with bunches managed with vitamin C and quercetin alone. The discoveries demonstrate that quercetin's inhibitory impact on HL-60 cell development is connected to its inhibitory consequences for PKC and/or TPK *in vitro*, the same as on phosphoinositide production. [16]. TXNIP establishment of the NLRP3 inflammasome is also hindered by quercetin. Recent clinical and preclinical examinations have proposed that quercetin is also a promising common treatment for provocative skin infections, such as atopic dermatitis [17].

14.3.2 Parthenolide

Parthenolide, a sesquiterpene lactone, is obtained from feverfew. In macrophage cell lines, it has NF-kB repressing properties. Proinflammatory pathways like LPS have been demonstrated to be powerfully restrained by parthenolide. In *in vitro*

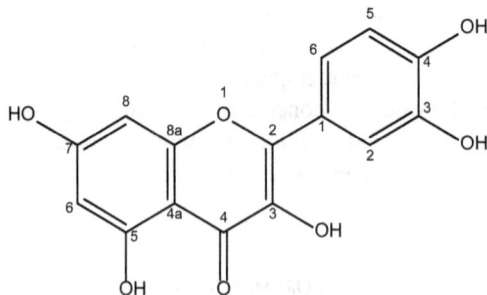

Quercetin

FIGURE 14.6 Quercetin.

(monocytes, macrophages, neutrophils) and *in vivo* (monocytes, macrophages, neutrophils) tests, parthenolide was found to fundamentally lessen IL1, IL2, IL6, IL8, and TNF advancement pathways [18].

14.3.3 RESVERATROL

Resveratrol (Figure 14.7), a stilbenoid, is produced in response to attack by microorganisms like bacteria and fungi. It represses the pathways of COX and NF-kB that are associated with cardiovascular, neurodegenerative, and metabolic issues. Resveratrol, which centres on irritation related issues, has been recorded in a few clinical preliminaries, many of which manage diabetes, weight, cardiovascular diseases, malignant growth, and neurodegenerative infections [19].

14.3.4 CURCUMIN

A pinnacle component obtained from the plant *Curcuma longa* of the ginger family *Zingiberaceae*, curcumin (Figure 14.8) can inhibit lipoxygenase by binding to either lipoxygenase or phosphatidylcholine micelles, demonstrating its broad range of action. In a few clinical preliminaries, curcumin has been shown as a viable rheumatic control specialist [20]. The addition of curcumin to the diet of multidrug-resistant gene-deficient mice that developed colitis spontaneously decreased intestinal inflammation significantly. The IC50 value of curcumin was found to be 10.5 µM [21].

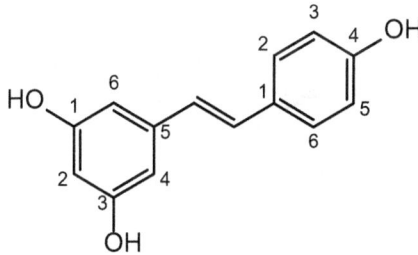

Resveratrol

FIGURE 14.7   Resveratrol.

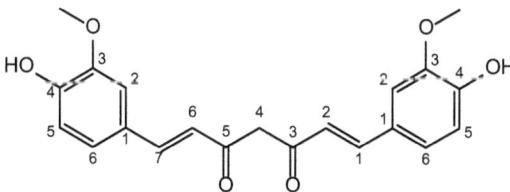

Curcumin

FIGURE 14.8   Curcumin.

14.3.5 Cucurbitacin

Cucurbitacin (Figure 14.9) is class of biochemical mixtures that belonging to the family, Cucurbitaceae. Cucurbitacins, triterpenes, sterols, and alkaloids are among the bioactive compounds present in this family. Carrageenan-induced rat paw oedema was substantially suppressed by intraperitoneal cucurbitacin E (CE) injection. The outcome showed that by inhibiting COX and reactive nitrogen species, CE is potentially useful in treating inflammation. More selectivity towards COX-2 was shown by the compound. The IC50 value of cucurbitacin was found to be 10 μM [22]. The position of the carbonyl group at carbon-11 and the absence of a double bond across C-23 and C-24, according to the scientists, may help to explain why different cucurbitacins have different toxicity profiles.

14.3.6 Capsaicin

Capsaicin (8-methyl-N-vanillyl-6-nonenamide) (Figure 14.10) is a functioning part of bean stew peppers. Studies on capsaicin have shown that it neutralizes neuropathic pain by actuating the vanilloid subfamily part 1 (TRPV1) transient receptor potential channel, a non-selective cation channel primarily located in nociceptive neurons. To induce desensitization and thus decrease the pain sensation, prolonged

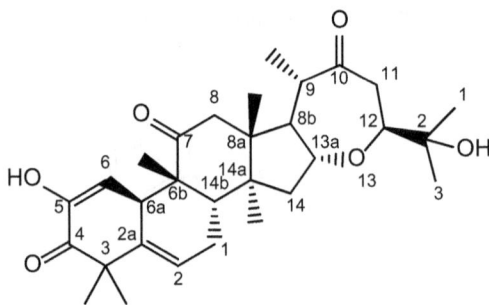

Cucurbitacins

FIGURE 14.9 Cucurbitacins.

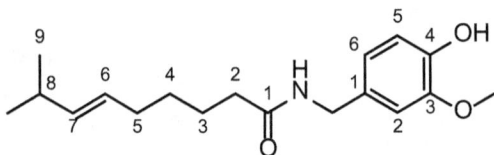

Capsaicin

FIGURE 14.10 Capsaicin.

activation of TRPV1 by capsaicin is addressed. In the treatment of osteoarthritis, capsaicin is effective [23].

14.4 OTHER NATURAL PRODUCTS WITH ANTI-INFLAMMATORY PROPERTIES

14.4.1 MARINE SPONGE

Sponges (phylum Porifera) are sessile marine filter feeders that have created proficient protection components against unfamiliar aggressors, such as infections, microorganisms, or eukaryotic organisms. In 2016, some diterpenoids were disconnected from marine sources and their mitigating movement was appeared to hinder NF-kB initiation and adjust arachidonic corrosive digestion [24].

14.4.2 MUSHROOM

A mushroom or toadstool is the substantial, spore-bearing fruiting body of a life-form, usually conveyed over the ground, on soil, or on its food source [25]. In 2011, Jedinak et al. focused on shellfish mushroom concentrate (OMC) that covered TNF-alpha, IL-6 and IL-12 LPS-subordinate creation in a part response way. The effect was not achieved by OMC cytotoxicity, as RAW264.7 cell reasonably less impacted by OMC.

14.4.3 HONEY

Honey is a sweet, gooey food substance made by honeybees and other related insects. Bumblebees produce honey from the nectar of plants or from releases of various bugs, by spewing, enzymatic activity, and water disappearing. Nectar, moreover, thwarted the improvement of NF-kB, IL-1β, and IL-6. In the provocative model of colitis, nectar was just probably as convincing as prednisolone treatment, without the genuine outcomes related with NSAIDs and corticosteroids [26].

14.4.4 FISH OIL

Fish oil supplements has omega-3 fatty acids, giving them dynamic properties for good health. It also helps in reduction of the inflammation related with illness, such as diabetes, cancer, heart disease, and several other medical situations [27]. There are especially two useful types of omega-3s: eicosapentaenoic acid (also known as EPA) and docosahexaenoic acid (also known as DHA). Precisely, DHA is well-known to have anti-inflammatory effects, which help in the reduction of levels of cytokine and promotion of gut strength. It also prevents the muscle damage that arise after workout and decrease inflammation [28].

14.5 FUTURE PERSPECTIVES

Several studies are being conducted in the field of natural sciences to discover more forward-thinking and safer decisions to fight inflammatory reactions. Recently, the

World Health Organization (WHO) and other regulatory bodies published an annual report that highlighted the protection of local courses of action and trademark itemizing being a more current field of study for investigators. According to the World Health Organization, about 25% of current medications are extracted from plants used in common drug. Normal medicine, as of now, is recognized by WHO for clinical consideration. Thus, key research in the field of herbals is required with respect to their new development and to improve regulatory checks.

14.6 CONCLUSION

The facts summarized in this chapter propose that many compounds derived from natural products wield powerful anti-inflammatory properties. If proven effective and safe, the utilization of natural compounds should be sponsored by policy creators and health systems. Consistent consumption of such amazing products may develop a successful and innocuous strategy to treat chronic inflammatory circumstances. Ongoing tests and clinical trials should be continued to guide and deliver their logically and scientifically based effectiveness to decrease inflammation and endorse wellness.

14.7 CONSENT FOR PUBLICATION

Not applicable.

14.8 FUNDING

Not applicable.

14.9 CONFLICT OF INTEREST

The authors declare no conflict of interest, financial or otherwise.

REFERENCES

1. A. Attiq, J. Jalil, K. Husain, & W. Ahmad, Raging the war against inflammation with natural products, *Frontiers in Pharmacology*, 976 (2018).
2. F. Cao, S.Y. Gui, X. Gao, W. Zhang, Z.Y. Fu, L.M. Tao, & X. Wang, Research progress of natural product-based nanomaterials for the treatment of inflammation-related diseases, *Materials & Design*, 110686 (2022).
3. Y. He, Y. Yue, X. Zheng, K. Zhang, S. Chen, & Z. Du, Curcumin, inflammation, and chronic diseases: How are they linked? *Molecules*, 20(9), 183–213 (2015). http://doi.org/10.1155/2015/608613
4. A.E. Litwic, C. Parsons, M.H. Edwards, D. Jagannath, C. Cooper, & E.M. Dennison, Comment on: Inflammatory mediators in osteoarthritis: A critical review of the state-of-the art, prospects, and future challenges, *Bone*, 106 (2018) 28–29. http://doi.org/10.1016/j.bone.2016.08.001
5. R. Fernández-Rodríguez, S. Monedero-Carrasco, B. Bizzozero-Peroni, M. Garrido-Miguel, A.E. Mesas, & V. Martínez-Vizcaíno, Effectiveness of resistance exercise on

inflammatory biomarkers in patients with type 2 diabetes mellitus: A systematic review with meta-analysis, *Diabetes & Metabolism Journal*, 47(1), 118–134 (2022).

6. G. Grosso, D. Laudisio, E. Frias-Toral, L. Barrea, G. Muscogiuri, S. Savastano, & A. Colao, Anti-inflammatory nutrients and obesity-associated metabolic-inflammation: State of the art and future direction, *Nutrients*, 14(6), 1137 (2022).

7. G. El-Saber Batiha, A.G. Magdy Beshbishy, L. Wasef, Y.H. Elewa, A. Al-Sagan, M.E. Abd El-Hack, & H. Prasad Devkota, Chemical constituents and pharmacological activities of garlic (Allium sativum L.): A review, *Nutrients*, 12(3), 872 (2020).

8. J. Azelmat, S. Fiorito, VA. Taddeo, S. Genovese, F. Epifano, D. Grenier, Synthesis and evaluation of antibacterial and anti-inflammatory properties of naturally occurring coumarins, *Phytochemistry Letters*, 13, 399–405 (2015).

9. Y. Ma, T. Tang, L. Sheng, Z. Wang, H. Tao, Q. Zhang, Z. Qi, Aloin suppresses lipopolysaccharide-induced inflammation by inhibiting JAK1-STAT1/3 activation and ROS production in RAW264. 7 cells, *International Journal of Molecular Medicine*, 42, 1925–1934 (2018).

10. K. Jiang, S. Guo, C. Yang, J. Yang, Y. Chen, A. Shaukat, G. Deng, Barbaloin protects against lipopolysaccharide (LPS)-induced acute lung injury by inhibiting the ROS-mediated PI3K/AKT/NF-κB pathway, *International Immunopharmacology*, 64, 140–150 (2018). http://doi.org/10.1016/j.intimp.2018.08.023

11. A.B. Wadekar, M.G. Nimbalwar, W.A. Panchale, B.R. Gudalwar, J.V. Manwar, R.L. Bakal, Morphology, phytochemistry and pharmacological aspects of Carica papaya, an review, *GSC Biological and Pharmaceutical Sciences*, 14(3), 234–248 (2018).

12. S. Pandey, P.J. Cabot, P.N. Shaw, A.K. Hewavitharana, Anti-inflammatory and immunemodulatory properties of Carica papaya, *Journal of Immunotoxicology*, 13, 590–602 (2016).

13. A. Lichota, K. Gwozdzinski, Anticancer activity of natural compounds from plant and marine environment, *International Journal of Molecular Sciences*, 19(11), 3533 (2018). http://doi.org/10.3390/ijms19113533

14. W.Y. Huang, H.C. Zhang, W.X. Liu, C.Y. Li, Survey of antioxidant capacity and phenolic composition of blueberry, blackberry, and strawberry in Nanjing, *Journal of Zhejiang University Science B*, 13(2), 94–102 (2012). http://doi.org/10.1631/jzus.B1100137

15. S.J. Desai, B. Prickril, A. Rasooly, Mechanisms of phytonutrient modulation of cyclooxygenase-2 (COX-2) and inflammation related to cancer, *Nutrition and Cancer*, 70(3), 350–375 (2018).

16. N.B. Samad, T. Debnath, M. Ye, M.A. Hasnat, B.O. Lim, In vitro antioxidant and anti–inflammatory activities of Korean blueberry (Vaccinium corymbosum L.) extracts, *Asian Pacific Journal of Tropical Biomedicine*, 4(10), 807–815 (2014).

17. V. Karuppagounder, S. Arumugam, R.A. Thandavarayan, R. Sreedhar, V.V. Giridharan, K. Watanabe, Molecular targets of quercetin with anti-inflammatory properties in atopic dermatitis, *Drug Discovery Today*, 21, 632–639 (2016).

18. P. Magni, M. Ruscica, E. Dozio, E. Rizzi, G. Beretta, R.M. Facino, Parthenolide inhibits the LPS-induced secretion of IL-6 and TNF-α and NF-κB nuclear translocation in BV-2 microglia, *Phytotherapy Research*, 26(9), 1405–1409 (2012). http://doi.org/10.1002/ptr.373210.1002/ptr.3732

19. L.X. Zhang, C.X. Li, M.U. Kakar, M.S. Khan, P.E. Wu, R.M. Amir, & J.H. Li, Resveratrol (RV): A pharmacological review and call for further research, *Biomedicine & Pharmacotherapy*, 143, 112164 (2021).

20. F. Matthew, L. Cecilia, B. Jai, T.L. Henry, Curcumin: An age-old anti-inflammatory and antineoplastic agent, *Journal of Traditional and Complementary Medicine*, 7, 339–346 (2017).

21. N. Samaan, Q. Zhong, J. Fernandez, G. Chen, A.M. Hussain, S. Zheng, G. Wang, Q.-H. Chen, Design, synthesis, and evaluation of novel heteroaromatic analogs of curcumin as anti-cancer agent, *European Journal of Medicinal Chemistry*, 75, 123–131 (2014).

22. I. Touihri-Barakati, O. Kallech-Ziri, W. Ayadi, H. Kovacic, B. Hanchi, K. Hosni, & J. Luis, Cucurbitacin B purified from Ecballium elaterium (L.) A. Rich from Tunisia inhibits α5β1 integrin-mediated adhesion, migration, proliferation of human glioblastoma cell line and angiogenesis, *European Journal of Pharmacology*, 797, 153–161 (2017).

23. M. Haanpaa, RD. Treede, Capsaicin for neuropathic pain: Linking traditional medicine and molecular biology, *European Neurology*, 68, 264–275 (2012).

24. Z. Liang, Y. Xu, X. Wen, H. Nie, T. Hu, X. Yang, Rosmarinic acid attenuates airway inflammation and hyper responsiveness in a murine model of asthma, *Molecules*, 21, 769 (2016).

25. A. Jedinak, S. Dudhgoankar, Q. Wu, J. Simon, D. Sliva, Anti-inflammatory activity of edible oyster mushroom is mediated through the inhibition of NF-B and AP-1 signaling, *Nutrition Journal*, 10, 52 (2011).

26. S.Z. Hussein, K.M. Yusoff, S. Makpol, Y.A. Yusof, Gelam honey inhibits the production of proinflammatory, mediators NO, PGE (2), TNF-a, and IL-6 in carrageenan-induced acute paw edema in rats, *Evidence-Based Complementary and Alternative Medicine*, Article ID: 109636, 1–13 (2012).

27. F. Dangardt, W. Osika, Y. Chen, et al., Omega-3 fatty acid supplementation improves vascular function and reduces inflammation in obese adolescents, *Atherosclerosis*, 212(2), 580–585 (2010). http://doi.org/10.1016/j.atherosclerosis.2010.06.046.

28. M. Martorell, X. Capó, A. Sureda, Effect of DHA on plasma fatty acid availability and oxidative stress during training season and football exercise, *Food & Function*, 5(8), 1920–1931 (2014). http://doi.org/10.1039/c4fo00229f.

15 Deep-Learning-Based Road Surface Classification for Intelligent Vehicles

Rishu Chhabra, Saravjeet Singh, Ravneet Kaur

15.1 INTRODUCTION

The development of transportation infrastructure and automotive technology focuses on the safety and convenience of road users. The premium vehicles with advanced driver assistance systems (ADAS) provide a good driving experience to the drivers and reduce the count of road crashes (Dong, Hu, Uchimura, & Murayama, 2010; Pevec, Babic, & Podobnik, 2019; Sun, Bebis, & Miller, 2006). However, research is still going on to enhance the reliability of the driving assistance systems (DAS) to support intelligent vehicles like autonomous vehicles known as autonomous intelligent vehicles. This requires more information regarding the surrounding context including the environment and road infrastructure. Road surface information is the prevailing factor to enhance the performance and reliability of vehicle control systems (Lee, Kim, Kim, & Lee, 2021).

As shown in Figure 15.1, three main categories of road surfaces are asphalt, concrete, and gravel roads, which differ based on the substance used for the construction of the roads.

1. Asphalt roads: These are regarded as smooth road surfaces produced using petroleum. They are easy to maintain, cost-effective, and recyclable. It is one of the safest driving surfaces but requires maintenance, especially in case of bad weather conditions ("Different Road Construction Surfaces: Pros and Cons : KH Plant," 2020).
2. Concrete roads: These are generally best suited and constructed for highways using the concrete slabs joined together. It is considered a greener alternative for road surfaces and does not use any natural resources. It is instead produced using limestone. Vehicle fuel consumption is also less while driving on concrete roads as compared to asphalt roads ("Different Road Construction Surfaces: Pros and Cons : KH Plant," 2020).

DOI: 10.1201/9781003367161-15

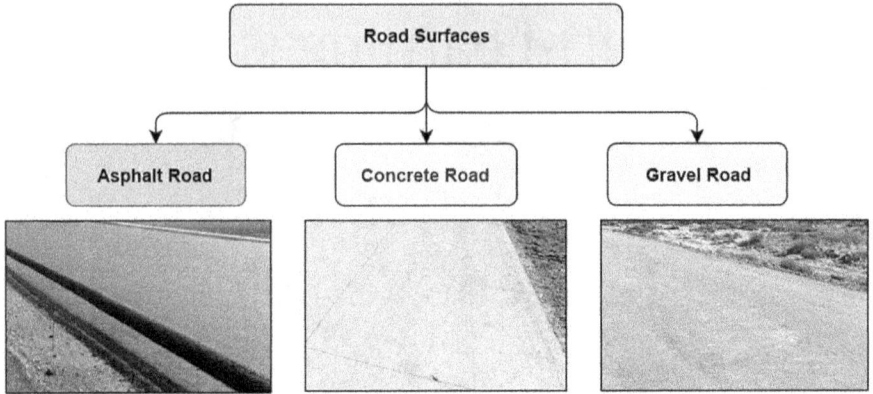

FIGURE 15.1   Different types of road surfaces.

3. Gravel roads: These are generally found in interior rural areas like villages that experience very less traffic. Gravel roads are easy to maintain
and are the least expensive. However, as it is made up of clay; it requires
frequent maintenance and is very difficult for driving in bad weather like
rain ("Different Road Construction Surfaces: Pros and Cons : KH Plant,"
2020).

The US, China, and India are the three countries with the largest road networks
across the globe. A road network is a critical component to evaluate the financial development of a country. Good driving road surface leads to a smooth travel
experience (Chhabra, Krishna, & Verma, 2021). Mainly for intelligent autonomous
vehicles, the road surface information is a crucial control input for adjusting the
vehicle dynamics like steering control and braking behaviour. Intelligent vehicles
play a crucial role in efficient energy consumption, reducing road crashes and traffic
congestion.

Image classification techniques paired with intelligent vehicles improve the
safety of the people on the roads. The application of image classification techniques
to detect and classify road surfaces is a step ahead for autonomous driving. The
performance of the employed classification algorithm is directly dependent on the
quality of the image, and thus, efficient methods for road surface classification are
the need of the hour (Cheng, Zhang, & Shen, 2019).

Deep learning models overcome the limitations of traditional image classification techniques like reduced reliability. Deep learning techniques automate the
process of feature learning, reduce the training time, and are suitable for intricate
classification and identification scenarios. In this chapter, we propose to categorize
the road surfaces into asphalt, concrete, and gravel roads using CNN, LSTM, and
InceptionV3. The model has been trained using the Mendeley-Road Surface Image
dataset, and the road surface has been classified as asphalt road, concrete road, and
gravel road (Zhao & Wei, 2022).

The remainder of the chapter is organized as follows: Section 2 presents the related work. The methodology and experimental results are given in Section 3 and Section 4 concludes the work with future challenges.

15.2 RELATED WORK

In this section the state-of-art literature in the area of road surface detection and classification using different technologies has been discussed. In Pereira, Tamura, Hayamizu, & Fukai (2018), the authors considered the physical properties of road surfaces using different sensors like optical and ultrasonic sensors for road surface classification. This noncontact sensor approach has been employed in various autonomous vehicles with the advantage of capturing the required data even when the vehicle is not being driven. Deep neural networks have been employed in Smolyanskiy, Kamenev, & Birchfield (2018) for road surface image classification. Another set of works employing artificial intelligence has been presented in Tumen, Yildirim, & Ergen (2018) that uses images from Google Street View and classifies the road surfaces. In Roychowdhury, Zhao, Wallin, Ohlsson, & Jonasson (2018), the authors proposed a deep learning model SqueezeNet for road surface condition estimation. The proposed model yields better performance as compared to CNN and other feature-based models. Road surface classification using LSTM followed by ensemble learning has been proposed in Park et al. (2018). The proposed model overcomes the limitation of overfitting and thus yields 94.6% classification accuracy. Another system proposed in Qin, Langari, Wang, Xiang, & Dong (2017) classifies the road surfaces by employing deep learning techniques. The technique employs softmax regression and sparse autoencoder.

In Sabery, Bystrov, Gardner, & Gashinova (2020) and Sabery, Bystrov, Gardner, Stroescu, & Gashinova (2021), the authors implemented CNN for the analysis of radar images for road surface classification and achieved favourable results. The importance of road surface identification in computing the road-tire friction coefficient has been given in Nolte, Kister, & Maurer (2018). The proposed system compared two CNN models for road surface identification considering the friction coefficient as an important parameter. A system to detect asphalt road surfaces using the acoustics of road/tyre noise has been proposed in Kalliris, Kanarachos, Kotsakis, Haas, & Blundell (2019). The system can detect wet, dry, and icy asphalt surfaces efficiently. The system for road surface classification using a deep learning model has been proposed in (Cheng et al., 2019). An improved activation function based on ReLU has been implemented yielding a model classification accuracy of more than 94%. Transfer-learning-based road surface identification system using CNN has been proposed in Xie & Kwon (2022). The system yields an accuracy of 98.21% and detects whether the road is covered with snow or not. In Marianingsih, Widodo, Pieter, Manullang, & Nanlohy (2022), the author prosed a system for classifying road surfaces into gravel, pavement, and asphalt using k-nearest neighbour (KNN) and Bayesian classification. The KNN algorithm gave better results as compared to Bayesian networks.

15.3   METHODOLOGY AND EXPERIMENTAL RESULTS

The proposed methodology implements different deep learning algorithms to classify road surfaces into asphalt roads, concrete roads, and gravel roads. The Mendeley dataset has been used in algorithms for identifying the different road surfaces. Deep learning algorithms overcome the limitation of machine learning algorithms and can do feature extraction on their own from the available dataset. It yields better results as compared to different machine learning algorithms. In this chapter, CNN, LSTM, and InceptionV3 have been implemented and analysed for road surface classification. As shown in Figure 15.2, InceptionV3 yields

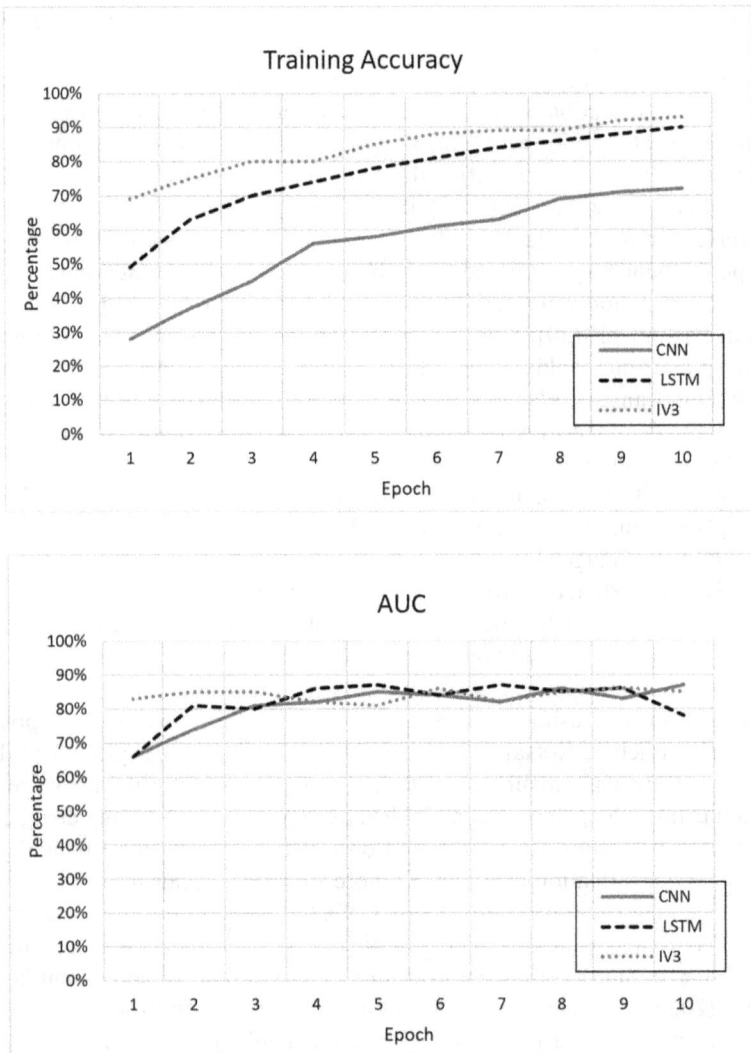

FIGURE 15.2   Deep-learning-based implementation results for road surface classification.

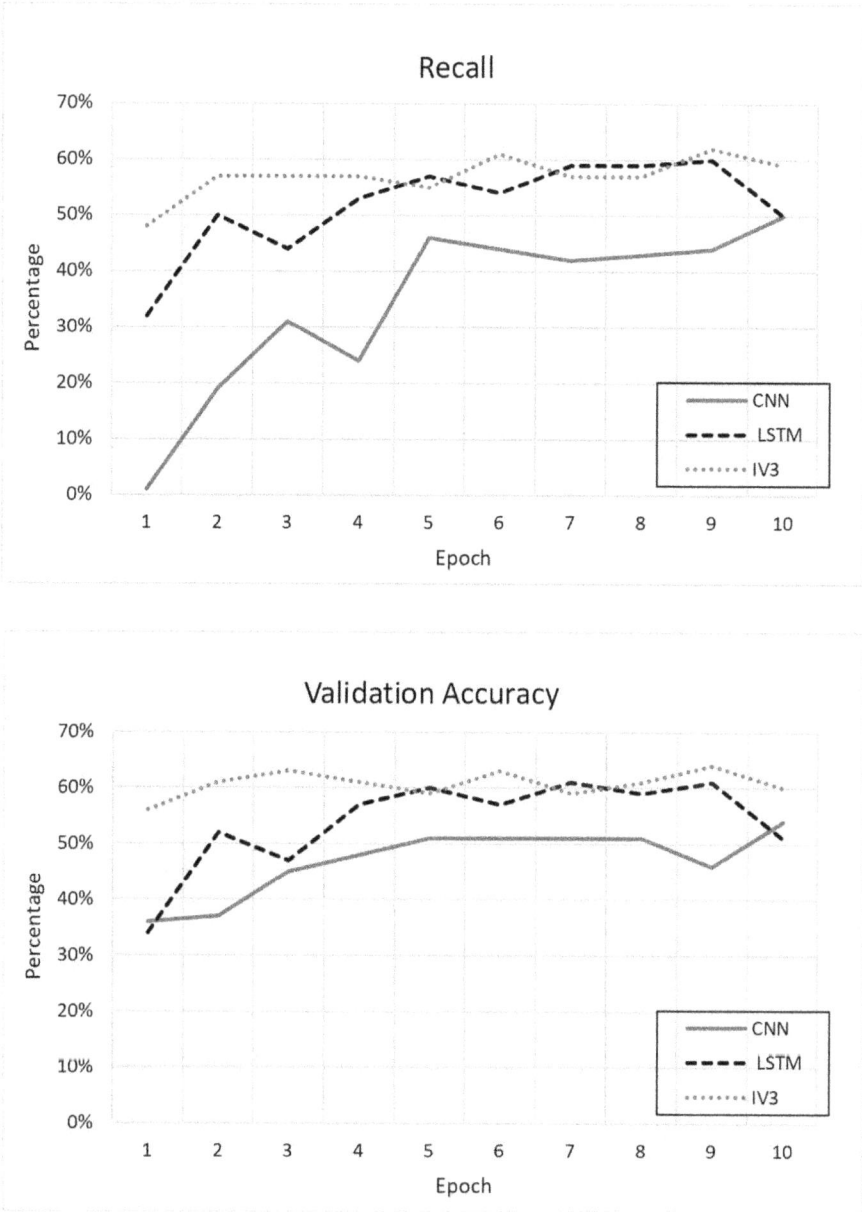

FIGURE 15.2 (Continued)

the highest validation accuracy of 64% and the highest training accuracy of 93%. The validation accuracy measure is supported by the acceptable area under curve (AUC) close to 86%. The numeric results for the implementation have been presented in Table 15.1.

Precision

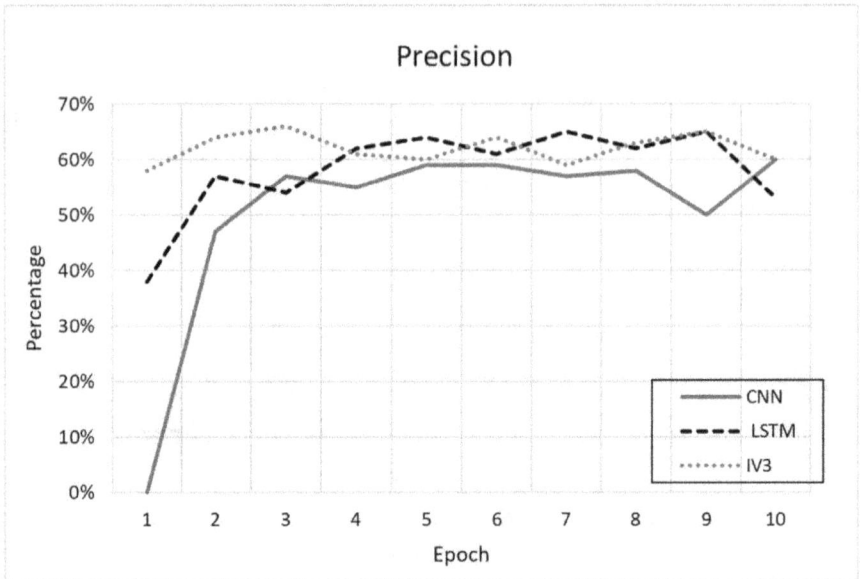

FIGURE 15.2 (Continued)

TABLE 15.1

Comparison of Different Algorithms for Road Surface Classification

Epoch	Training Accuracy			Validation Accuracy			AUC			Precision			Recall		
	CNN	LSTM	IV3	CNN	LSTM	IV3	CNN	LSTM	IV3	CNN	LSTM	IV3	CNN	LSTM	IV3
1	0.28	0.49	0.69	0.36	0.34	0.56	0.66	0.66	0.83	0	0.38	0.58	0.01	0.32	0.48
2	0.37	0.63	0.75	0.37	0.52	0.61	0.74	0.81	0.85	0.47	0.57	0.64	0.19	0.5	0.57
3	0.45	0.7	0.8	0.45	0.47	0.63	0.81	0.8	0.85	0.57	0.54	0.66	0.31	0.44	0.57
4	0.56	0.74	0.8	0.48	0.57	0.61	0.82	0.86	0.82	0.55	0.62	0.61	0.24	0.53	0.57
5	0.58	0.78	0.85	0.51	0.6	0.59	0.85	0.87	0.81	0.59	0.64	0.6	0.46	0.57	0.55
6	0.61	0.81	0.88	0.51	0.57	0.63	0.84	0.84	0.86	0.59	0.61	0.64	0.44	0.54	0.61
7	0.63	0.84	0.89	0.51	0.61	0.59	0.82	0.87	0.82	0.57	0.65	0.59	0.42	0.59	0.57
8	0.69	0.86	0.89	0.51	0.59	0.61	0.86	0.85	0.85	0.58	0.62	0.63	0.43	0.59	0.57
9	0.71	0.88	0.92	0.46	0.61	0.64	0.83	0.86	0.86	0.5	0.65	0.65	0.44	0.6	0.62
10	0.72	0.9	0.93	0.54	0.51	0.6	0.87	0.78	0.85	0.6	0.53	0.6	0.5	0.5	0.59

15.4 CONCLUSIONS AND FUTURE RESEARCH DIRECTIONS

ITS is an amalgamation of technology with road infrastructure for the safety of road users. Road surface conditions play a crucial role and contribute to this aim of ITS. With the increase in intelligent vehicles with DAS and the ongoing research on autonomous vehicles, the information regarding the road surface is an important parameter for controlling the various vehicle dynamics. In this chapter, we have implemented several deep learning algorithms like CNN, LSTM, and InceptionV3 for road surface classification into asphalt, gravel, and concrete. The results clearly show that the InceptionV3 algorithm yields better results as compared to CNN and LSTM. The classification results can be used in controlling the braking behaviour and safe navigation of an autonomous vehicle or speed warning to intelligent vehicles. The secure transfer of information and communication of road surface information to the other vehicles is still a challenge and be worked upon in the future.

REFERENCES

Cheng, L., Zhang, X., & Shen, J. (2019). Road surface condition classification using deep learning. *Journal of Visual Communication and Image Representation*, *64*, 102638.

Chhabra, R., Krishna, C. R., & Verma, S. (2021). A survey on state-of-the-art road surface monitoring techniques for intelligent transportation systems. *International Journal of Sensor Networks*, *37*(2), 81–99.

Different road construction surfaces: Pros and cons. *KH Plant*, https://www.khplant.co.za/blog/different-road-construction-surfaces-pros-and-cons/ (2020)

Dong, Y., Hu, Z., Uchimura, K., & Murayama, N. (2010). Driver inattention monitoring system for intelligent vehicles: A review. *IEEE Transactions on Intelligent Transportation Systems*, *12*(2), 596–614.

Kalliris, M., Kanarachos, S., Kotsakis, R., Haas, O., & Blundell, M. (2019). Machine learning algorithms for wet road surface detection using acoustic measurements. In *2019 IEEE International Conference on Mechatronics (ICM)*, IEEE (Vol. 1, pp. 265–270).

Lee, D., Kim, J.-C., Kim, M., & Lee, H. (2021). Intelligent tire sensor-based real-time road surface classification using an artificial neural network. *Sensors*, *21*(9), 3233.

Marianingsih, S., Widodo, W., Pieter, M. S. S., Manullang, E. V., & Nanlohy, H. Y. (2022). Machine vision for the various road surface type classification based on texture feature. *Journal of Mechanical Engineering Science and Technology (JMEST)*, *6*(1), 40–47.

Nolte, M., Kister, N., & Maurer, M. (2018). Assessment of deep convolutional neural networks for road surface classification. In *2018 21st International Conference on Intelligent Transportation Systems (ITSC)*, IEEE (pp. 381–386).

Park, J., Min, K., Kim, H., Lee, W., Cho, G., & Huh, K. (2018). Road surface classification using a deep ensemble network with sensor feature selection. *Sensors*, *18*(12), 4342.

Pereira, V., Tamura, S., Hayamizu, S., & Fukai, H. (2018). Classification of paved and unpaved road image using convolutional neural network for road condition inspection system. In *2018 5th International Conference on Advanced Informatics: Concept Theory and Applications (ICAICTA)*, IEEE (pp. 165–169).

Pevec, D., Babic, J., & Podobnik, V. (2019). Electric vehicles: A data science perspective review. *Electronics*, *8*(10), 1190.

Qin, Y., Langari, R., Wang, Z., Xiang, C., & Dong, M. (2017). Road excitation classification for semi-active suspension system with deep neural networks. *Journal of Intelligent & Fuzzy Systems*, *33*(3), 1907–1918.

Roychowdhury, S., Zhao, M., Wallin, A., Ohlsson, N., & Jonasson, M. (2018). Machine learning models for road surface and friction estimation using front-camera images. In *2018 International Joint Conference on Neural Networks (IJCNN)*, IEEE (pp. 1–8).

Sabery, S. M., Bystrov, A., Gardner, P., & Gashinova, M. (2020). Surface classification based on low terahertz radar imaging and deep neural network. In *2020 21st International Radar Symposium (IRS)*, IEEE (pp. 24–27).

Sabery, S. M., Bystrov, A., Gardner, P., Stroescu, A., & Gashinova, M. (2021). Road surface classification based on radar imaging using convolutional neural network. *IEEE Sensors Journal*, *21*(17), 18725–18732.

Smolyanskiy, N., Kamenev, A., & Birchfield, S. (2018). On the importance of stereo for accurate depth estimation: An efficient semi-supervised deep neural network approach. In *Proceedings of the IEEE Conference on Computer Vision and Pattern Recognition Workshops*, IEEE (pp. 1007–1015).

Sun, Z., Bebis, G., & Miller, R. (2006). On-road vehicle detection: A review. *IEEE Transactions on Pattern Analysis and Machine Intelligence*, *28*(5), 694–711.

Tumen, V., Yildirim, O., & Ergen, B. (2018). Recognition of road type and quality for advanced driver assistance systems with deep learning. *Elektronika Ir Elektrotechnika*, *24*(6), 67–74.

Xie, Q., & Kwon, T. J. (2022). Development of a highly transferable urban winter road surface classification model: A deep learning approach. *Transportation Research Record*, 03611981221090235.

Zhao, T., & Wei, Y. (2022). A road surface image dataset with detailed annotations for driving assistance applications. *Data in Brief*, *43*, 108483.

16 An Empirical Review of Potholes Classification Using Road Images

Saravjeet Singh, Rishu Chhabra, Rupali Gill

16.1 INTRODUCTION

Roads play a very important role in the economy and gross domestic product (GDP) of the country. Roads provide a physical connection between different stakeholders and open new opportunities for development and growth (Varjan et al., 2017; Singh & Singh, 2017). India being a developing country has the world's second-largest road network. Indian road transportation contributes 3.6% to the GDP of India, and this is 2/3 of the GDP from all transportation mediums. Indian road transportation is majorly used for passenger and freight transfer (*Roads*, n.d.). Transportation medium, cost of transportation, and conditions of transport medium highly affect the economy. The decision to choose road as transportation highly depends on road conditions. In many scenarios, due to bad road conditions, passengers prefer other transportation modes instead of road transportation. Road conditions can be determined based on the presence of cracks, roughness, potholes, and so on, on road surfaces (Chhabra & Singh, 2021a,b). Due to vehicle movement, atmospheric conditions road surface gradually starts damaging and it leads to holes on the surface. These holes are known as potholes (H. Agrawal et al., 2021; R. Agrawal et al., n.d.). Potholes put a negative effect on road transport, and in many cases, it leads to high traffic and accidents. Figure 16.1 shows an example of cracks and potholes. Figure 16.1 also classifies the small and large potholes.

With the advancement in vehicular technology, pothole identification is very important to reduce accidents. Many techniques exist for pothole identification and classification (Kim et al., 2022). These techniques use road images, sensor reading, smartphone data, and video graphic data to identify the potholes (Thiruppathiraj et al., 2020). These techniques have four basic steps to identify and classify potholes. These steps are data collection, data processing, feature extraction, and then pothole classification and identification. The first steps involve data preparation and synthesis. For pothole classification, data is collected in the form of road images and videos, vibration readings using vehicle/device movement, and sensor readings. In the data processing step, collected data is processed to remove the errors and null values. Filtering, outlier detection, masking, and cleaning are the different techniques that are used to pre-process the data. The next step is to extract the features of potholes

DOI: 10.1201/9781003367161-16

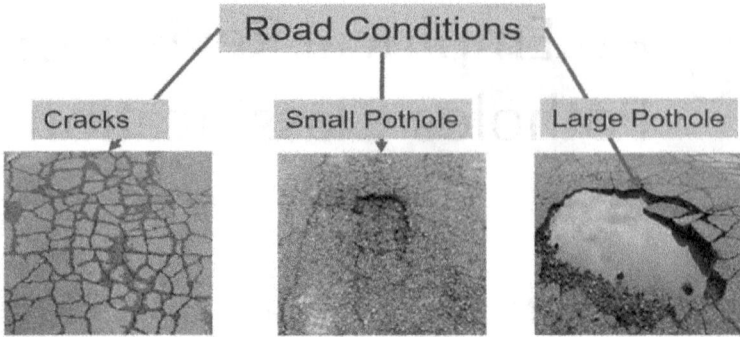

FIGURE 16.1 Classification and sample scenario of bad road conditions (cracks and potholes).

from the collected data. This step is responsible to provide the actual features and specifications of the pothole classifier. Based on the pothole features, the classifier identifies the potholes and non-potholes data (Singh & Singh, 2020a, 2020b).

In the last few years, many techniques have been proposed for pothole classification and identification. This chapter provides pothole classification techniques using the transfer-learning-based convolutional neural network (CNN). In this study, an experiment is performed on the image dataset using Visual Geometry Group (VGG) 16 and 19 CNN models. VGG is developed by the Visual Geometry Group at Oxford with motivation from its predecessor AlexNet. VGG is based on Deep Neural Network and supports a large number of layers. A self-generated road dataset of Indian roads is used to classify the potholes and normal roads using the VGG16 and VGG19. The rest of the chapter provides the details of the performed experiment. Section 2 provides the recent literature related to pothole classification, Section 3 describes the pothole classification using VGG16 and VGG19. Results and analysis are provided in Section 4, and then Section 5 concludes the chapter with future directions.

16.2   LITERATURE STUDIED

With the advancement of technology, the identification of road conditions is very important. Good roads play important role in passenger, goods, and freight transfer and lead to a good contribution to GDP. Earlier road conditions were analysed visually by the human expert, but now many automated techniques and devices are used to identify the road conditions and potholes. Pothole classification techniques are majorly divided into three categories: vibration, three-dimensional reconstruction, and vision-based (Du et al., 2020a). In a vibration-based approach acceleration sensor is used to identify the potholes. The acceleration sensor identifies the z-axis movement of the device, and corresponding to the z-axis reading, potholes are classified. Smartphones and sensors are used to implement this approach. This technique is very cost-effective. The three-dimensional reconstruction approach uses stereo-vision technology to identify the shape and depth of holes on the road surface. Based on shape and depth, these methods are able to identify the pothole. Three-dimensional reconstruction approaches are the most accurate way to identify potholes. Vision-based approaches used road video or images dataset to identify the potholes. These approaches use deep learning and machine learning models for classification and identification. These approaches can be easily applied in real-time applications [11]. Many research studies identify the potholes using these techniques and Table 16.1 provides a brief survey of these research studies.

TABLE 16.1

Survey of Recent Pothole Classification Techniques

Reference	Techniques	Details	Dataset
(Park et al., 2021)	Vision-based	Yolov4-tiny was used to identify the potholes on a system with 12 GB RAM and K80 GPU.	1199 road images
(Baek & Chung, 2020)		YOLO with edge detection feature extraction model is used to classify the pothole. Maximum accuracy is 77% on an I5 processor with 12 GB RAM and GPU.	665 road images
(Ye et al., 2019)		CNN-based feature extraction and classification technique used for pothole classification. Maximum accuracy is 80% on I7 processor with 32 GB RAM and 8 GB GPU.	96,000 pothole images
(Chen et al., 2020)		Localization-based CNN and part-based classification are used. Maximum accuracy is 95% on a system with 11 GB RAM and GeForce GTX 1080Ti GPU.	5,676 road images
(Wang et al., 2017)		Gaussian filter was used for feature extraction. LS-SVM and ANN were used to classify the dataset. The maximum achieved accuracy was 88%.	200 road images
(Yousaf et al., 2018)		A combination of scale-invariant feature transformation, support vector machine, and bag-of-words (BoW) approach was used to classify the porthole. Maximum accuracy is 95% on a Core i3 processor with 4 GB RAM.	120 road images

(Continued)

TABLE 16.1 (*Continued*)
Survey of Recent Pothole Classification Techniques

Reference	Techniques	Details	Dataset
(Du et al., 2020b)	Vibration-based	Improved Gaussian model and k-nearest neighbour classifier was used for pothole classification. Ninety-six per cent accuracy was achieved using an Android mobile phone.	171 potholes captured using smartphone
(Allouch et al., 2017)		Fourier transformation was used for the feature extraction. Three classifiers—decision tree, SVM, and Naïve Bayes—were used with an accuracy of 98%, 95%, and 96%, respectively. An Android app was used for the processing.	2,000 potholes captured using smartphone
(Wu et al., 2020)		Fourier transformation was used for the feature extraction. Three classifiers—linear regression, SVM, and random forest—were used with an accuracy of 95%, 94%, and 95%, respectively. An android-based app with a sensor sampling rate of 50 HZ was used for the processing.	4,088 records considered
(Ul Haq et al., 2019)	3D-reconstruction-based	A combination of a key point, block matching, and stereo triangulation is used for the pothole matching. This technique also used histogram equalization and a high pass filter for the data preprocessing. Eighty-two per cent accuracy was obtained with two stereo webcams on the windows system.	-
(Guan et al., 2021)		Modified U-net with principal component analysis was used to identify the potholes. Ninety-six per cent accuracy as achieved on the GoPro HERO8 device.	-

16.3   EXPERIMENTAL STUDY

This section provides the details of porthole classification techniques using VGG16 and VGG19. The complete experimental process is divided into four steps. The first step is data collection, and for this experiment, 1,719 road images were collected. These images are related to normal roads and potholes. Further, these images were converted into the same size. These images were pre-processed using the rotation, same width and height, shear, zoom, and flip and fill mode factors. After pre-processing image dataset is divided into train and testing samples using the "categorical" class mode. These images belong to three classes: "Large Pothole," "Normal Road," and "Small Pothole." These pre-processed images were converted to machine learning format. A sample example of converted data is shown in Figure 16.2. The next step is to design the model. In this experiment, VGG16 and VGG19 were used for classification. In VGG16 and VGG19, a total of 13 and 16 (respectively), 2D convolutional layers in five blocks were used. Each block had a max pooling layer. After the convolutional blocks, flattened, dense, and dropout layers were used. Details of these layers are provided in Table 16.2. Further, the details of total parameters, trainable and non-trainable parameters are also provided in Table 16.2. Processed dataset is provided to these models for feature extraction and then based on the extracted features model performed the classification.

16.4   RESULT ANALYSIS

The designed model based on details presented in Table 16.2 was implemented using tensor flow and Karas library. I5 processor with 12 GB RAM and 2 GPU was used

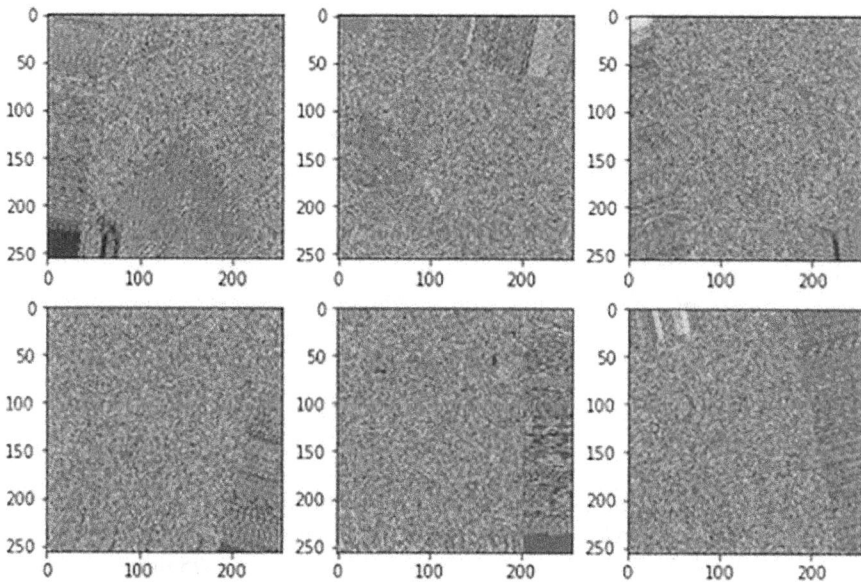

FIGURE 16.2   Image dataset according to VGG input format.

TABLE 16.2
Details of Layers for VGG16 and VGG19

Layers	VGG16	VGG19
Input layer	1	1
5 blocks of 2D convolutional layers	13 CNN + 4 max pooling	16 CNN + 4 max pooling
Flatten	1	1
Dense	1	1
Dropout	1	1
Dense	1	1
Parameters	Total params: 153,331,715	Total params: 158,641,411
	Trainable params: 138,617,027	Trainable params: 138,617,027
	Non-trainable params: 14,714,688	Non-trainable params: 20,024,384

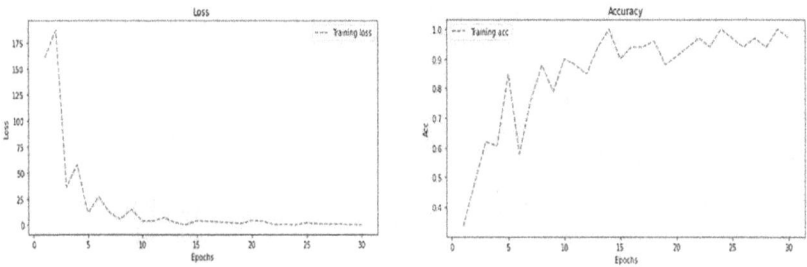

FIGURE 16.3   Training loss and accuracy graph for VGG16.

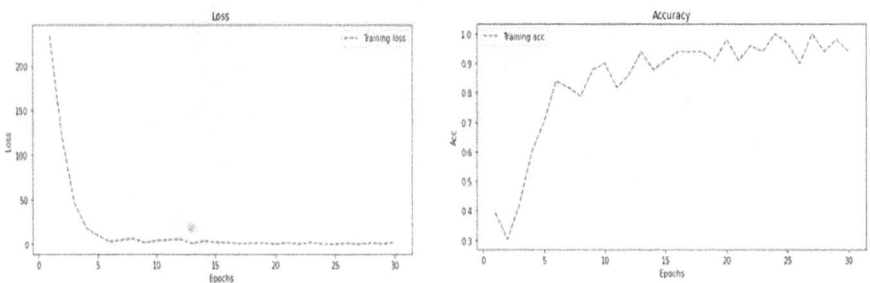

FIGURE 16.4   Training loss and accuracy graph for VGG19.

to implement the model. The complete dataset was divided into 80:20 ratio of training and testing size. Model was executed for the 30 epochs. The batch size for both models was set to 25. VGG16 and VGG19 were trained using 1375 images. Training loss and accuracy loss for VGG16 and VGG19 are shown in Figures 16.3 and 16.4, respectively.

According to performed experiment, the maximum test loss and accuracy for VGG16 were 47.09 and 55.88%, respectively, whereas the maximum monitored value of validation loss and accuracy were 2.18 and 92.30%, respectively. Similarly, the maximum monitored value of test loss and accuracy for VGG19 were 39.27 and 70.05%, respectively, whereas the maximum monitored value of validation loss and accuracy was 5.11 and 93.11%, respectively. Detailed classification reports of VGG16 and VGG19 are shown in Table 16.3. The confusion matrix for VGG16 and VGG19 are shown in Figures 16.5 and 16.6, respectively.

TABLE 16.3
Classification Report of VGG16 and VGG19

	Classification Report—VGG16				Classification Report—VGG19			
	Precision	Recall	F1-score	Support	Precision	Recall	F1-score	Support
Large pothole	0.89	0.53	0.93	15	0.93	0.87	0.94	15
Normal road	0.88	0.92	0.91	12	0.91	0.92	0.93	13
Small pothole	0.91	0.89	0.91	15	0.90	0.91	0.93	15
Accuracy	0.92				0.93			

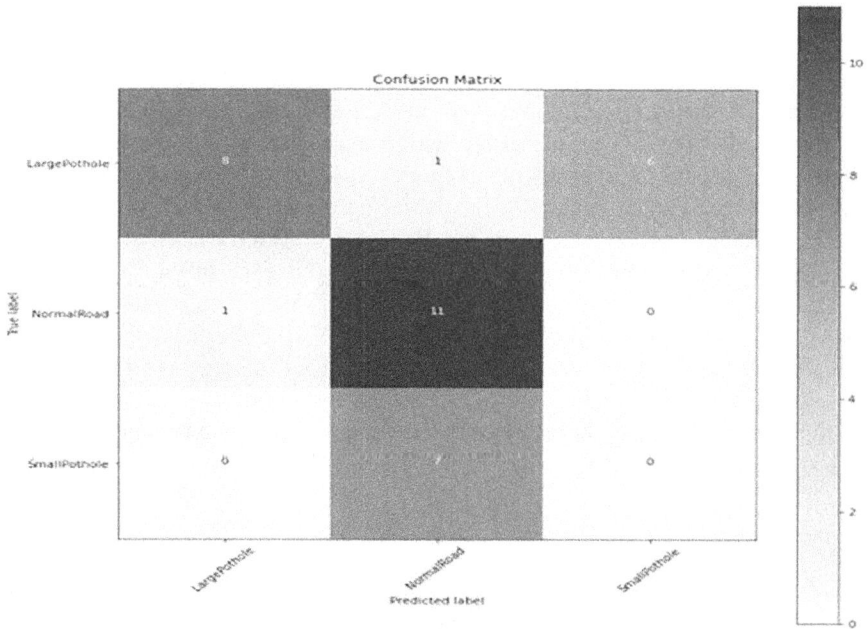

FIGURE 16.5   Confusion matrix for VGG16.

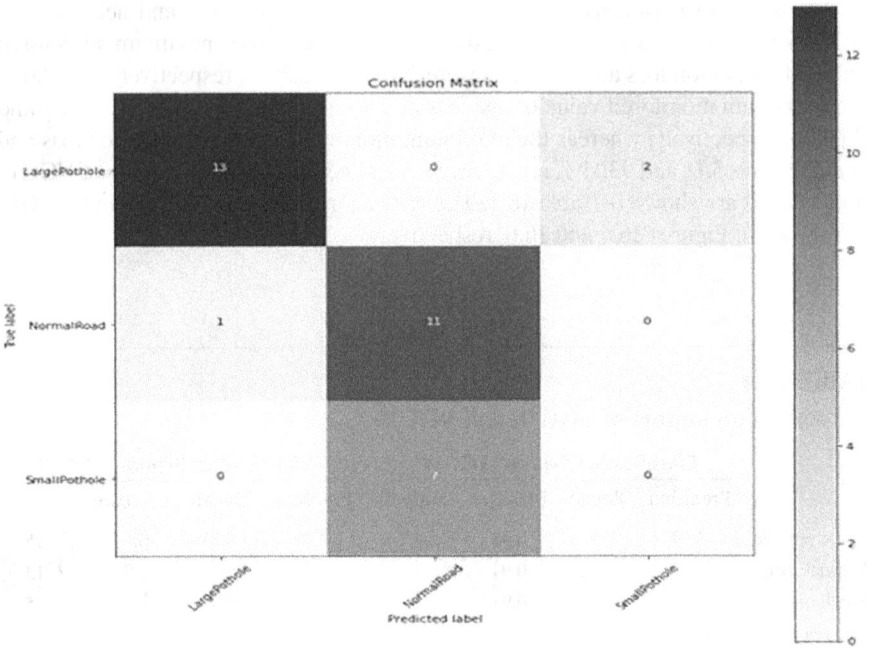

FIGURE 16.6   Confusion matrix for VGG19.

16.5   CONCLUSION

Road conditions are very important for the overall development of the country. Goods and passenger transfer highly depend on the road condition. Roads with bad conditions and potholes are the prime reason for accidents and unsafe journeys. Many techniques were proposed by the research community for pothole detection. This chapter presents pothole classification using the vision-based approaches. In this experiment, transfer-learning-based VGG16 and VGG19 CNN models were used. Road images numbering 1,719 were used in this experiment. Models were trained per the CNN approach, and after training, models were validated using 344 images. According to performed experiment, VGG19 has a 1% higher accuracy than VGG16. For VGG19, the maximum achieved accuracy is 93%, whereas VGG16 provided maximum accuracy of 92%. In the future, these techniques can be implemented with a higher number of images and can be implemented in a real-time situation.

REFERENCES

Agrawal, H., Gupta, A., Sharma, A., & Singh, P. (2021). Road Pothole Detection Mechanism using Mobile Sensors. *Proceedings of International Conference on Technological Advancements and Innovations, ICTAI 2021*, 26–31. https://doi.org/10.1109/ICTAI 53825.2021.9673193

Agrawal, R., Chhadva, Y., Addagarla, S., & Chaudhari, S. (2021). Road Surface Classification and Subsequent Pothole Detection Using Deep Learning. In *2021 2nd International Conference for Emerging Technology (INCET)* (pp. 1–6). IEEE.

Allouch, A., Koubaa, A., Abbes, T., & Ammar, A. (2017). RoadSense: Smartphone Application to Estimate Road Conditions Using Accelerometer and Gyroscope. *IEEE Sensors Journal, 17*(13), 4231–4238. https://doi.org/10.1109/JSEN.2017.2702739

Baek, J. W., & Chung, K. (2020). Pothole Classification Model Using Edge Detection in Road Image. *Applied Sciences, 10*(19), 6662. https://doi.org/10.3390/APP10196662

Chen, H., Yao, M., & Gu, Q. (2020). Pothole Detection Using Location-Aware Convolutional Neural Networks. *International Journal of Machine Learning and Cybernetics, 11*(4), 899–911. https://doi.org/10.1007/S13042-020-01078-7

Chhabra, R., & Singh, S. (2021a). A Survey on Smart Phone-Based Road Condition Detection Systems. *International Conference on Emerging Technologies: AI, IoT, and CPS for Science & Technology Applications.*

Chhabra, R., & Singh, S. (2021b). A Survey on Smart Phone-Based Road Condition Detection Systems. *CEUR Workshop Proceedings, 3058.*

Du, R., Qiu, G., Gao, K., Hu, L., & Liu, L. (2020a). Abnormal Road Surface Recognition Based on Smartphone Acceleration Sensor. *Sensors, 20*(2), 451. https://doi.org/10.3390/S20020451

Du, R., Qiu, G., Gao, K., Hu, L., & Liu, L. (2020b). Abnormal Road Surface Recognition Based on Smartphone Acceleration Sensor. *Sensors, 20*(2), 469. https://doi.org/10.3390/S20020451

Guan, J., Yang, X., Ding, L., Cheng, X., Lee, V. C. S., & Jin, C. (2021). Automated Pixel-Level Pavement Distress Detection Based on Stereo Vision and Deep Learning. *Automation in Construction, 129*, 103788. https://doi.org/10.1016/J.AUTCON.2021.103788

Kim, Y. M., Kim, Y. G., Son, S. Y., Lim, S. Y., Choi, B. Y., & Choi, D. H. (2022). Review of Recent Automated Pothole-Detection Methods. In *Applied Sciences (Switzerland)* (Vol. 12, Issue 11). MDPI. https://doi.org/10.3390/app12115320

Park, S. S., Tran, V. T., & Lee, D. E. (2021). Application of Various YOLO Models for Computer Vision-Based Real-Time Pothole Detection. *Applied Sciences, 11*(23), 11229. https://doi.org/10.3390/APP112311229

Roads. (n.d.). Retrieved September 6, 2022, from www.ibef.org/pages/16539

Singh, S., & Singh, J. (2017). Management of SME's Semi Structured Data Using Semantic Technique. In *Applied Big Data Analytics in Operations Management* (pp. 133–164). IGI Global.

Singh, S., & Singh, J. (2020a). Location Driven Edge Assisted Device and Solutions for Intelligent Transportation. *Fog, Edge, and Pervasive Computing in Intelligent IoT Driven Applications*, 123–147.

Singh, S., & Singh, J. (2020b). Intrinsic Parameters based Quality Assessment of Indian OpenStreetMap Dataset using Supervised Learning Technique. In *2020 Indo–Taiwan 2nd International Conference on Computing, Analytics and Networks (Indo-Taiwan ICAN)* (pp. 52–57). IEEE.

Thiruppathiraj, S., Kumar, U., & Buchke, S. (2020). Automatic Pothole Classification and Segmentation Using Android Smartphone Sensors and Camera Images with Machine Learning Techniques. *IEEE Region 10 Annual International Conference, Proceedings/TENCON, 2020-November*, 1386–1391. https://doi.org/10.1109/TENCON50793.2020.9293883

Ul Haq, M. U., Ashfaque, M., Mathavan, S., Kamal, K., & Ahmed, A. (2019). Stereo-Based 3D Reconstruction of Potholes by a Hybrid, Dense Matching Scheme. *IEEE Sensors Journal, 19*(10), 3807–3817. https://doi.org/10.1109/JSEN.2019.2898375

Varjan, P., Rovňaníková, D., & Gnap, J. (2017). Examining Changes in GDP on the Demand for Road Freight Transport. *Procedia Engineering*, *192*, 911–916. https://doi.org/10.1016/J. PROENG.2017.06.157

Wang, P., Hu, Y., Dai, Y., & Tian, M. (2017). Asphalt Pavement Pothole Detection and Segmentation Based on Wavelet Energy Field. *Mathematical Problems in Engineering*, *2017*. https://doi.org/10.1155/2017/1604130

Wu, C., Wang, Z., Hu, S., Lepine, J., Na, X., Ainalis, D., & Stettler, M. (2020). An Automated Machine-Learning Approach for Road Pothole Detection Using Smartphone Sensor Data. *Sensors*, *20*(19), 5582. https://doi.org/10.3390/S20195564

Ye, W., Jiang, W., Tong, Z., Yuan, D., & Xiao, J. (2019). Convolutional Neural Network for Pothole Detection in Asphalt Pavement. *Road Materials and Pavement Design*, *22*(1), 42–58. https://doi.org/10.1080/14680629.2019.1615533

Yousaf, M. H., Azhar, K., Murtaza, F., & Hussain, F. (2018). Visual Analysis of Asphalt Pavement for Detection and Localization of Potholes. *Advanced Engineering Informatics*, *38*, 527–537. https://doi.org/10.1016/J.AEI.2018.09.002

17 Robust Watermarking for Medicine

Preeti Sharma

17.1 INTRODUCTION

Recently, an unforeseen increase in digital automation has changed the normal life of each person. Most of the data nowadays has to be saved in a digital arrangement [1]. This has escalated the possibility of advancing and meddling of copied information. A strong watermarking technique proves an amicable and convenient option to such problems. This has helped to hide invisible watermarks in other images for justifying the right of ownership [2]. Inserting robust watermarks within images has helped to preserve the proprietary rights also. While a watermark must be strong enough to tolerate the different intentional or unintentional attacks, it must be imperceptible too. A lot of work and efforts have been carried out to achieve a balance between conflicting requirements of robustness and imperceptibility [3–5]. The technique of image-adaptive watermarking grants a profitable approach for countering various impediments [6–8].

17.2 REVIEW OF WORK DONE

All techniques of watermarking must have three optimum characteristics: superior invisibility, higher robustness, and adequate size. Though it is a challenging assignment to obtain all these required features at the same time, the approach of image-adaptive watermarking offers the best solution [9]. Different models [10] utilizing this approach suggested the usage of a constant scaling factor for embedding the watermarks in various sized blocks of the input image [11]. Discrete wavelet transform (DWT) and discrete cosine transform (DCT) were also utilized for hiding a single [12–13] or dual [14–15] watermarks in medical images with varying strength factors.

The present work suggests dividing the given image into various blocks of dimension 8 × 8. Only blocks with high entropy were chosen for inserting the bits of the watermark. DWT is practised upon the selected blocks for obtaining coefficients with lesser frequency. The calculation of STF was executed using the notion of entropy. As the suggested technique is partially blind, hence, useful parallel information related to points of higher entropy blocks, mean value of lower frequency components, and dimensions of the watermark are also transmitted with the watermarked image. It aids in recovering the watermark at the destination.

DOI: 10.1201/9781003367161-17

The chapter is categorized into different sections. Section 17.3 discusses the proposed technique of insertion and extraction of watermark. Section 17.4 details the calculation of STF. Sections 17.5 and 17.6 discuss the variation of IQA for medicinal images and conclusion, respectively.

17.3 PROSPECTIVE WORK

The prospective technique for insertion and extraction of the watermark is given in Sections 17.3.1 and 17.3.2 correspondingly. The block representations are shown in Figure 17.1.

17.3.1 INSERTION OF WATERMARK

The given image is decomposed into various sized (8 × 8) blocks. Average entropy is determined from the higher entropy blocks, which are further utilized for insertion of binary bits 1 and 0. The numeral of blocks is usually equal or higher than the number of watermark bit. This helps to hide a single bit in every block.

DWT is appealed on each preferred block for obtaining the four bands. The lowest frequency measures (assumed as M) are preferred for watermark bit insertion,

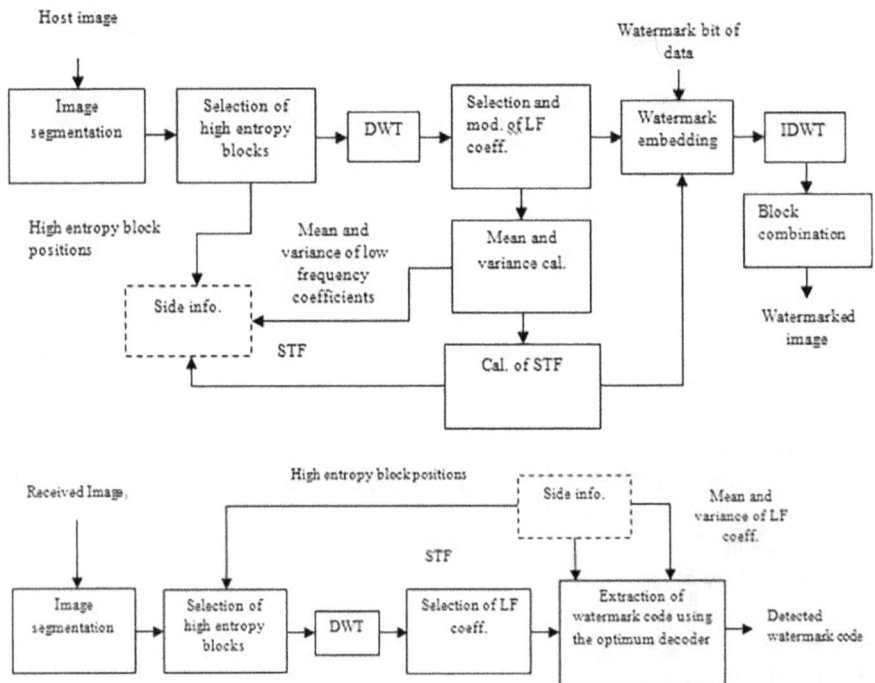

FIGURE 17.1 Representation for the proposed watermarking methodology: (A) insertion and (B) extraction of watermark.

since a change in these coefficients is not easily reflected to the naked eye. The altered measures are presumed as M'.

The bits 1 and 0 of watermark are inserted as

$$M' = M (1 + STF) \text{ and } M' = M (1 - STF). \tag{17.1}$$

Lastly, inverse DWT is endured on the blocks for achieving the watermarked image. The transmission of the parallel information with the watermarked image is done simultaneously.

17.3.2 WATERMARK DECODING

Similar to the previous process, the received image was decomposed in various sized (8 × 8) blocks. Again, DWT is appealed on each preferred block of higher entropy. The mean for the lowest recurring coefficients of watermarked figure was computed. Parallel information was also utilized for providing the mean of the communicated image from the transmitter.

The watermark was decoded using a statistical approach based upon maximum likelihood estimation (MLE). The selected coefficients were modelled using Gaussian distribution to calculate mean (μ) and variance (σ^2) with the probability density function (pdf) expressed as follows:

$$f\left(b|\mu,\sigma^2\right) = \frac{1}{\sqrt{2\pi\sigma^2}} e^{\frac{-(b-\mu)^2}{2\sigma^2}} \tag{17.2}$$

Quantums (z) were oppressed by null-average additive white Gaussian noise (AWGN) at destination, written like $\sigma_{z|1}^2 = \left(1+\beta_d\right)^2 \sigma^2 + \sigma_q^2$ and $\sigma_{z|0}^2 = \left(1-\beta_d\right)^2 \sigma^2 + \sigma_q^2$, where β_d was STF utilized for hiding the bits.

As the values were independently and identically distributed (i.id) upon every degree of desolation, hence, the dispensation in every unit with "q" coefficients z_1, z_2, z_3, ... z_q for lodging 0 is written as

$$P\left(z_1,z_2,z_3,...z_n \mid 0\right) = \prod_{i=1}^{n}\sqrt{2\pi\sigma_{z|0}^2}\frac{1}{e^{-(z_i-(1-\beta_d)\mu)^2}/2\sigma_{z|0}^2}, \tag{17.3}$$

The dispensation of lodging 1 is written as

$$P\left(z_1,z_2,z_3,...z_N \mid 1\right) = \prod_{i=1}^{n}\sqrt{2\pi\sigma_{z|1}^2}\frac{1}{e^{-(z-(1+\beta_d)\mu)^2}/2\sigma_{z|1}^2}. \tag{17.4}$$

The recommendation of ML is

$$P\left(z_1,z_2,z_3,...\{zn|1\}\right)\underset{<0}{\overset{>1}{}}P\left(z_1,z_2,z_3,...zn\mid0\right). \tag{17.5}$$

The rule of extraction is implied upon the comparison of the probabilities of inserting bits 0 and 1. Also, for higher values of signal-to-noise ratio (SNR), β_d approached zero, thus making the maximum likelihood (ML) decoder unconventional of STF. This helped to extract the watermark victoriously.

17.4 STRENGTH FACTOR (STF)

The strength factor was calculated using entropy (H). It is clarified as mean instruction accommodated for every unified production of the origin as

$$H = -\sum_{j=1}^{J} P\left(y_j\right).logP\left(y_j\right)$$ (17.6)

where y_j acts for a discrete deposition of feasible matters where chances are specified as $P(y_j)$.

The STF was calculated from H as

$$STF_d = H_i / \left(H_{max}\right)^{\rho}.$$ (17.7)

Here, H_i is the entropy of minor recurrent estimates for every block(s) within representation.

H_{max} is the largest entropy in all chunks where ρ is the pliable substructure in the denomination whose value is constrained betwixt 1 and 2 for preserving the watermarks imperceptibility over least BER merit for extracted watermarks.

17.5 EXPERIMENTATION

In the presented work, "peppers," "baboon," "knee," and "lungs X-ray" of size 512×512 were utilized as natural and medical images for watermarking. Two watermarks, "CAMERAMAN" and "COPYRIGHT" of proportions 18×25 and 32×32, respectively, were also used in the study. These are shown in Figures 17.2 and 17.3, respectively. STF was utilized to insert the watermarks into the input images. ρ was altered between 1 and 2 to choose 1, 1.5, and 2.

Table 17.1 presents the watermarked images and the extracted watermarks for varied values of ρ. The quality of these images deteriorates as ρ approaches extreme values of the given range.

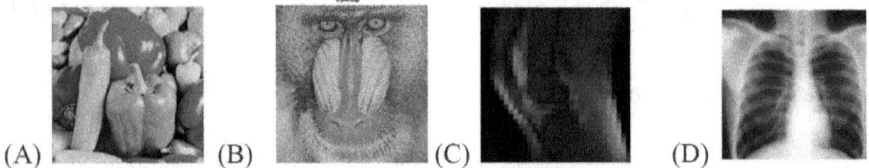

(A)   (B)   (C)   (D)

FIGURE 17.2 Natural and medical figures of proportion 512×512: (A) "peppers," (B) "baboon," (C) "knee hinge," and (D) "lungs X-ray" [16–17].

FIGURE 17.3   Watermarks utilized (A) "CAMERAMAN" (18 × 25) and (B) "COPYRIGHT" (32 × 32).

TABLE 17.1

Outcome of PSNR for Watermarked Figures and BER of Extradited Watermark alongside Invisibility

Original Figure and Watermark Used	Watermarked Received Figure	Recovered Watermark(s)	PSNR for Received Watermarked image	BER (%) of Discovered Watermark
			25.08,22.11 45.91,39.34 67.27,66.12	0,0 0.22,0.097 47.77,18.35
			21.62.19.48 40.73,47.52 58.64,57.68	0,0 0,0 3.11,4.1
			21.25,23.58 44.09,41.44 58.86,55.011	0,0 0,0.2705 41.23,15.33

(Continued)

TABLE 17.1 (*Continued*)

Outcome of PSNR for Watermarked Figures and BER of Extradited Watermark alongside Invisibility

Original Figure and Watermark Used	Watermarked Received Figure	Recovered Watermark(s)	PSNR for Received Watermarked image	BER (%) of Discovered Watermark
			31.29,29.57	0,0.1
			40.07,41.15	0,0.29
			55.66,58.02	57.53,17.13

17.6 CONCLUSION

The effectiveness of the watermarking technique is justified through the invisibility of the watermarked images. Higher merits of PSNR for natural images in range 43–49 dB prove the optimum requirements of imperceptibility with least BER of extracted watermarks. As an observation, as PSNR is increased a distortion is observed in the quality of the extracted watermarks. The technique has also contributed successfully in the field of telemedicine for watermarking the medical images. Further, more combinations of transforms may be opted for obtaining higher robustness of watermarking techniques.

REFERENCES

[1] Mittal, K., Singh, K., Jindal, N. (2017) Image compression algorithm with reduced blocking artifacts. *Turkish Journal of Electrical Engineering & Computer Sciences*, 25(5), 1946–1962.

[2] Singh, H., Kaur, L., Singh, K. (2014) Fractional M-band dual tree complex wavelet transform for digital watermarking. *Sadhana*, 39, 345–361.

[3] Cox, I.J., Kilian, J., Leighton, F.T., Shamoon, T. (1997) Secure spread spectrum watermarking for multimedia. *IEEE Transactions on Image Processing*, 6(12), 1673–1687 (1997).

[4] Wolfgang, R.B., Podilchuk, C.I. (1999) Perceptual watermarks for digital images and video. *Proceedings of the IEEE*, 87(7), 1108–1126.

[5] Watson, A.B. (1993) DCT quantization matrices visually optimized for individual images. In *Human Vision, Visual Processing, and Digital Display IV*, SPIE, vol. 1913, pp. 202–216. IS&T/SPIE's Symposium on Electronic Imaging: Science and Technology, 1993, San Jose, CA, United States.

[6] Watson, A.B., Yang, G.Y., Solomon, J.A., Villasenor, J. (1997) Visibility of wavelet quantization noise. *IEEE Transactions on Image Processing*, 6(8), 1164–1175.

[7] Watson, A.B., Borthwick, R., Taylor, M. (1997) Image quality and entropy masking. In *Human Vision and Electronic Imaging II*, SPIE, vol. 3016, pp. 2–12.

[8] Barni, M., Bartolini, F., Pive, A. (2001) Improved wavelet-based watermarking through pixel-wise masking. *IEEE Transactions on Image Processing*, 5(10), 783–791.

[9] Podilchuk, C.I., Zeng, W. (1998) Image-adaptive watermarking using visual models. *IEEE Journal on Selected Areas in Communication*, 16, 525–539.

[10] Akhaee, M.A., Sahraeian, S.M.E., Sankur, B., Marvasti, F. (2009) Robust scaling based image watermarking using maximum-likelihood decoder with optimum strength factor. *IEEE Transactions on Multimedia*, 11, 822–833.

[11] Yadav, N., Singh, K. (2014) *Robust image-adaptive watermarking using an adjustable dynamic strength factor.* Springer Verlag, London.

[12] Bhinder, P., Singh, K., Jindal, N. (2018) Image-adaptive watermarking using maximum likelihood decoder for medical images. *Multimedia Tools and Applications*, 77, 1030310328.

[13] Bhinder, P., Jindal, N., Singh, K. (2020) An improved robust image-adaptive watermarking with two watermarks using statistical decoder. *Multimedia Tools and Applications*, 79, 183–217. https://doi.org/10.1007/s11042-019-07941-2.

[14] Yan, Y., Cao, W., Li, S. (2009) Block-based adaptive image watermarking scheme using Just Noticeable Difference. In *2009 IEEE International Workshop on Imaging Systems and Techniques*, Shenzhen, IEEE, pp. 377–380. http://doi.org/10.1109/IST.2009.5071669

[15] Peng, H., Wang, J., Wang, W. (2010) Image watermarking method in multiwavelet domain based on support vector machines. *Journal of Systems and Software*, 83, 1470–1477.

[16] https://radiopaedia.org/images/14804579.jpg

[17] http://cdn.innovativelanguage.com/wordlists/media/thumb/8444_fit512.jpg

[18] Kansal, I., Popli, R., Sethi, M. (2022) Effect of non uniform illumination compensation on dehazing/de-fogging techniques. In *AIP Conference Proceedings*, vol. 2357, no. 1. AIP Publishing LLC, Melville, NY.

[19] Kaur, H., Koundal, D., Kadyan, V. (2021) Image fusion techniques: A survey. *Archives of Computational Methods in Engineering*, 28(7), 4425–4447.

[20] Koundal, D., Gupta, S., Singh, S. (2018). Computer aided thyroid nodule detection system using medical ultrasound images. *Biomedical Signal Processing and Control*, 40, 117–130.

18 Modern Deep Learning Networks for Medical Image Segmentation
An Overview

Mohit Pandey, Abhishek Gupta

18.1 INTRODUCTION

The medical images are acquired by various modalities like computed tomography (CT), X-ray, ultrasound, and magnetic resonance imaging [1]. The information contained in these images is helpful for experts. They can utilize the information for clinical diagnosis and treatment planning [2]. Image segmentation is vital in examining specific organs (like kidneys, liver, and heart). It is also helpful for tumour or cancer detection even at an early stage[3, 4].

Image segmentation is a task that involves the partitioning of an image according to the region of interest (RoI). For example, segmenting kidney (RoI) in the abdominal CT scan is shown in Figure 18.1.

Image segmentation can be done manually or automatically. Manual segmentation of clinical images is a tedious and time-consuming task. Manual segmentation is difficult in rural areas due to a lack of medical imaging experts. Also, the quality of segmentation is expert-dependent [4–9]. Various researchers proposed automatic segmentation techniques to overcome the limitations of manual procedures. These methods used thresholding, deformable model, atlas model, edge-based model, image-based model, shape prior information [10], classification, region-based model, and deep-learning model [11, 12]. In recent years, deep learning has turned into well-liked in medical image processing [13, 14]. In literature, many deep-learning-based methods were proposed and showed improved accuracy.

This chapter aims to provide an overview of the different modern deep learning networks developed by various researchers for medical image segmentation.

18.2 LITERATURE REVIEW

Deep networks have evolved renowned in clinical image segmentation-related works in recent years. Many works based on convolutional neural networks (CNNs) related to the segmentation of kidney, liver, thoracic organ-at-risk [15], and tumours are found in the literature.

DOI: 10.1201/9781003367161-18

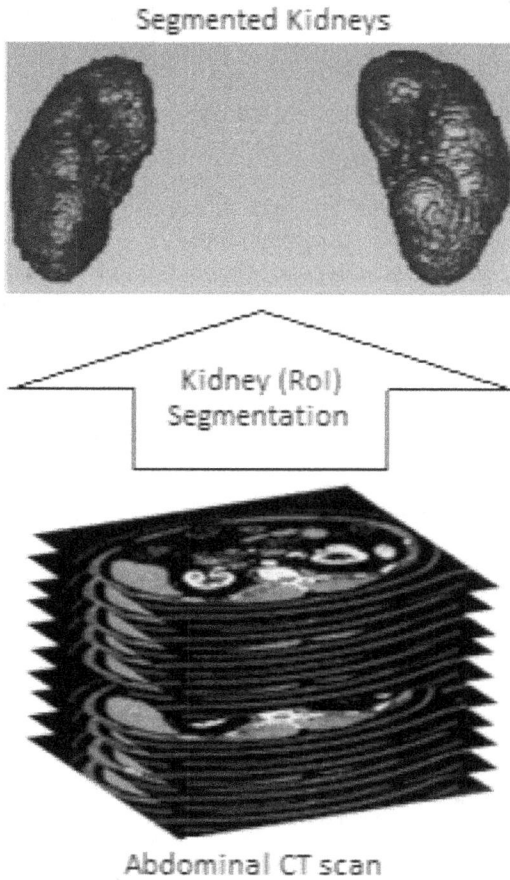

FIGURE 18.1   Demonstrating kidney (RoI) segmentation in abdominal CT scan [4].

Sharma, Rupprecht, et al. [16], 2017, proposed a fully CNN-based procedure for computer-based kidney segmentation. They used 244 CT for the training and testing. They achieved 86% of the dice score.

Yang, Li, et al. [9], 2018, also proposed a CNN-based technique for kidney segmentation in CT images. The proposed method used a pyramid pooling module, which helps to segment multiclass. They attained 93% of the dice score.

In 2020, da Cruz, Araújo, et al. [17] used AlexNet and U-Net in his work of kidney segmentation in CTs. They used the Kits19 dataset for the experiment. First, this method extracts all slices of the CT scan containing the kidney region using AlexNet. Further, another network, U-Net, has been employed to segment the kidneys. They attained a 95.7% of dice score. In the same year, Xie, Li, et al. [7], and Fatemeh, Nicola, et al. [18] also suggested a method based on deep learning. Xie, Li, et al. [7] used three different networks, SE-Net (squeeze and excitation), ResNeXT, and U-Net and achieved a 96.7% dice score. Fatemeh, Nicola, et al. [18] used only

U-Net and achieved a 96.25% dice score for kidney boundary and 87.9% for kidney mass. In 2020, Türk, Lüy, et al. [19] used a network V-Net for kidney and tumour segmentation. They have modified the original V-Net network and attained a 97.7% dice score. The structure of V-Net is very similar to the U-Net.

In 2021, Lin, Cui, et al. [20] suggested an algorithm based on 3D U-Net. They used a sequence of 3D U-Net for coarse-to-fine kidney segmentation in CT scans and achieved a 97.3% dice score. The developed model also segments kidney tumour.

Many works related to segmenting renal tumours based on modern deep blocks with promising accuracy exist in the literature. In 2018 Jackson et al. [21] suggested a 3D CNN-based technique to segment renal tumours and reach 88.5% of the dice score on non-public data. In 2020, Qayyum, Lalande, and Meriaudeau [22] combined 3D ResNet (residual network) and SE (squeeze and excitation) blocks to segment renal tumours and achieved 86.8% of the dice score. In 2021, Cruz et al. [6], for a tumour segmentation task, used DeepLabv3 + 2.5D model with the DPN-131 encoder [23].

18.3 MODERN DEEP LEARNING BLOCKS

This section provides a description of modern networks and their mathematical model.

18.3.1 U-Net

Ronneberger et al. [24] designed a network using an encoder-decoder mechanism named U-Net for clinical image segmentation. The architecture of deep network U-Net has two parts downsampling and upsampling in a symmetric manner. The original U-Net architecture is depicted in Figure 18.2. The downsampling or encoder part is responsible for extracting features by using 3 × 3 convolution operation. At the same time, the upsampling or decoder part is responsible for segmenting the object with the correct location of the object. The convolution layer of the encoder part is connected with the decoder part for inputting the features produced in each encoder part layer to the decoder part's corresponding layer.

1. *Advantages of the U-Net:* The U-Net was used by many researchers for segmenting organs and tumours. Because of the ability of the U-Net to segment medical images, the researchers taking advantage of the U-Net:
   - For training, any deep learning model needs a large dataset, which is very challenging in the case of medical images. The U-Net model performs well with small training dataset.
   - U-Net model can predict a fair segmentation map because of its symmetric architecture, which allows it to keep contextual information and location information.
2. *Limitations of the U-Net model*: While many researchers believe in the U-Net model for clinical image segmentation. On another side there are a few limitations of the U-Net-based model which the researchers face:
   - Originally the U-Net model accepted the input image of size 572×572. However, the U-Net architecture can be modified to suit the requirements.

- Learning slows down in the intermediate levels of U-Net.
- Training is time-consuming because of the deepness of the U-Net model.

However, many variants of the original U-Net deep network have been developed in the past: U-Net++ [25], RA-U-Net [26], SD-U-Net [27], and 3D-U-Net [28].

18.3.2 V-NET

It is a fully CNN-based model used for medical image segmentation [29]. The V-Net is also a symmetric architecture like U-Net. The left half of the architecture is called the compression network, and the other symmetric part is called the decompression network. At each level of the compression network, convolution layers with

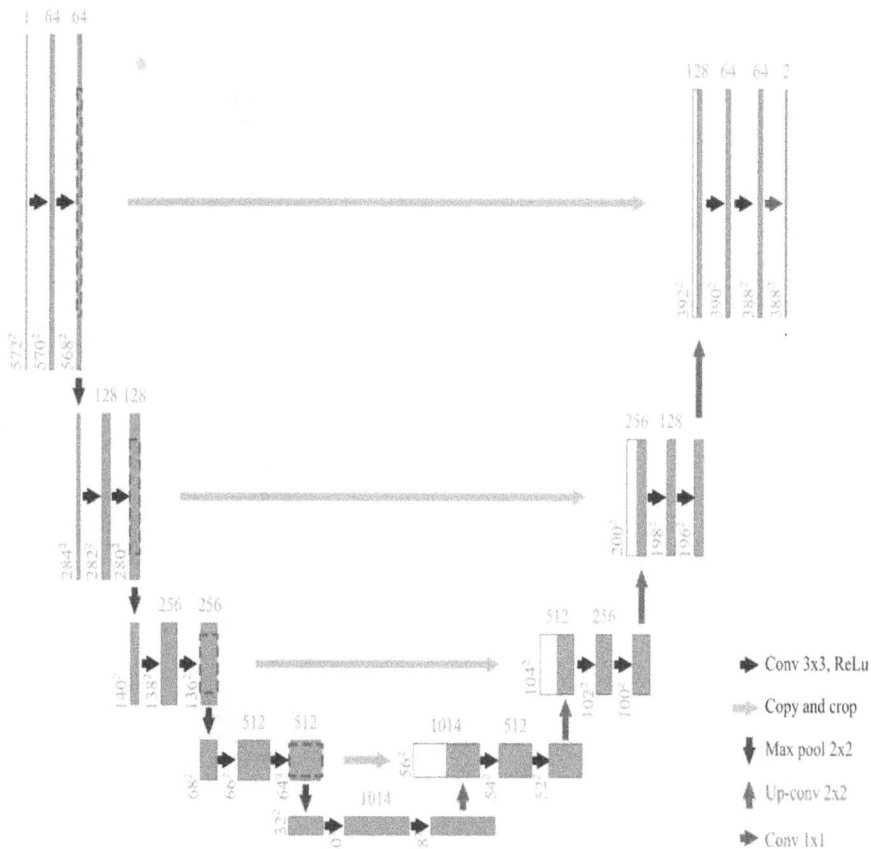

FIGURE 18.2 Original U-Net architecture; the figure is taken from reference [24]. Black boxes show feature set maps. The numbers on the left downside of each box show the feature map's size. The numbers at the top of the box show the number of maps. From the left, the number above the first box shows input image channels, and the number above the last box shows segmentation classes.

residual functions are used. Volumetric kernels were used in these convolution layers. The decompression network collects features from low-resolution feature maps and increases their spatial representation. It provides probabilistic segmentation in two channels for both foreground and background areas.

18.3.3 DeepLab

For extracting quality features from images, the DeepLab model leverages a pretrained CNN ResNet-101 or VGG16 and atrous convolution [30]. The DeepLab utilizes benefits of the atrous convolution:

- Power of handling resolution of features generated in CNN
- Transforming an image classification model into a dense feature vector producer besides requiring further parameter learning
- Usage of a conditional random field to provide finely segmented results

DeepLab passes the input image via a deep convolution neural network layers along with 1 or 2 atrous-convolution layers, as shown in Figure 18.3; this produces a rough

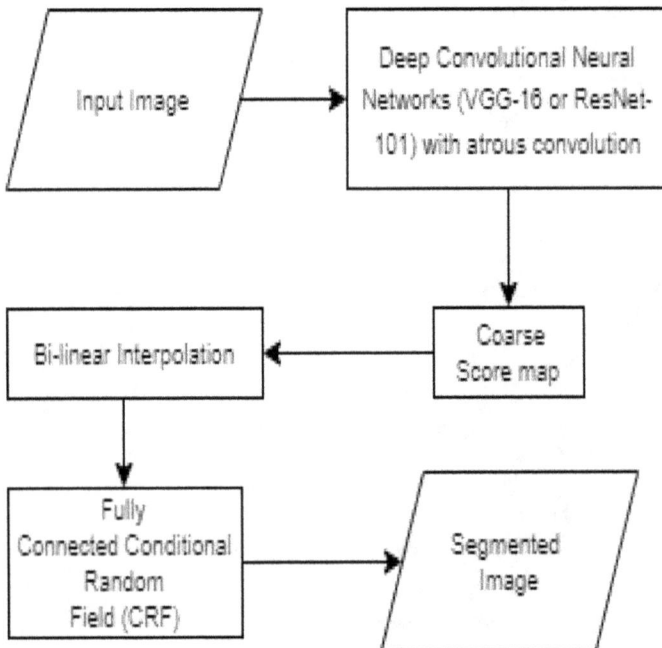

FIGURE 18.3 DeepLab model flow. A fully convolutional deep convolutional neural network (DCNN), such as VGG16/ResNet-101, is used, with atrous convolution used to lower the degree of signal downsampling. The feature maps are enlarged to the input image resolution using a bilinear interpolation step. The result is wholly linked with a connected conditional random field, which enhances the segmentation outcome and identifies the object's edges [30].

feature map. After that, the feature map is undergone the upsampling procedure to get the image's original size by bilinear interpolation. The bilinear interpolated image is used fully connected conditional-random-field (CRF) to get the final result (segmented image).

18.3.4 VGG19

The VGG19 network is similar to the AlexNet design in that it has successive convolutional layers with increasing filters as you progress further into the network. The model consists of 16 convolutional layers, three fully connected layers, and five pooling layers using the maximum pooling approach with 2×2 windows (see Figure 18.4). The choice of smaller filters was inspired by the design since the perceptual field was proved to be equally as efficient with larger filters. Furthermore, a reduced filter size minimizes the number of training parameters [31].

18.3.5 RESNET

When designing a deep learning model, it is evident that a larger number of layers enables more abstract information to be extracted and, as a result, increasingly difficult issues to handle. However, gradient fading restricts network training since the gradient propagates backward across distinct layers, and recurring multiplication over several layers renders the gradient too tiny. He et al. [32] developed the notion of residual connection (as shown in Figure 18.5) to address this issue. The link essentially establishes parallel paths to the convolutional (conv) layer sequences, enabling the gradient to flow across the lattice and preventing it from disappearing. Various version of ResNet has been proposed in the literature, like ResNet-34, ResNet-50, and ResNet-101. Many researchers use ResNet and VGG16/VGG19 as a backbone of the U-Net model for clinical image segmentation.

18.3.6 EFFICIENTNET

The EfficientNet is a CNN with stacked layers in sequence, but it is scaled uniformly. For example, unlike other designs, which have variable dimensions, the EfficientNet adjusts the depth, breadth, and resolution parameters consistently (see Figure 18.6). The method's logic is that if the input image is huge, the network should have more layers to expand the receptive fields and more channels to catch the image's minute patterns [34].

18.4 APPLICATIONS OF DEEP NEURAL NETWORKS IN CLINICAL IMAGE SEGMENTATION

Deep learning can be used in many areas of medical image analysis:

- Object identification
- Classification
- Object detection
- RoI segmentation

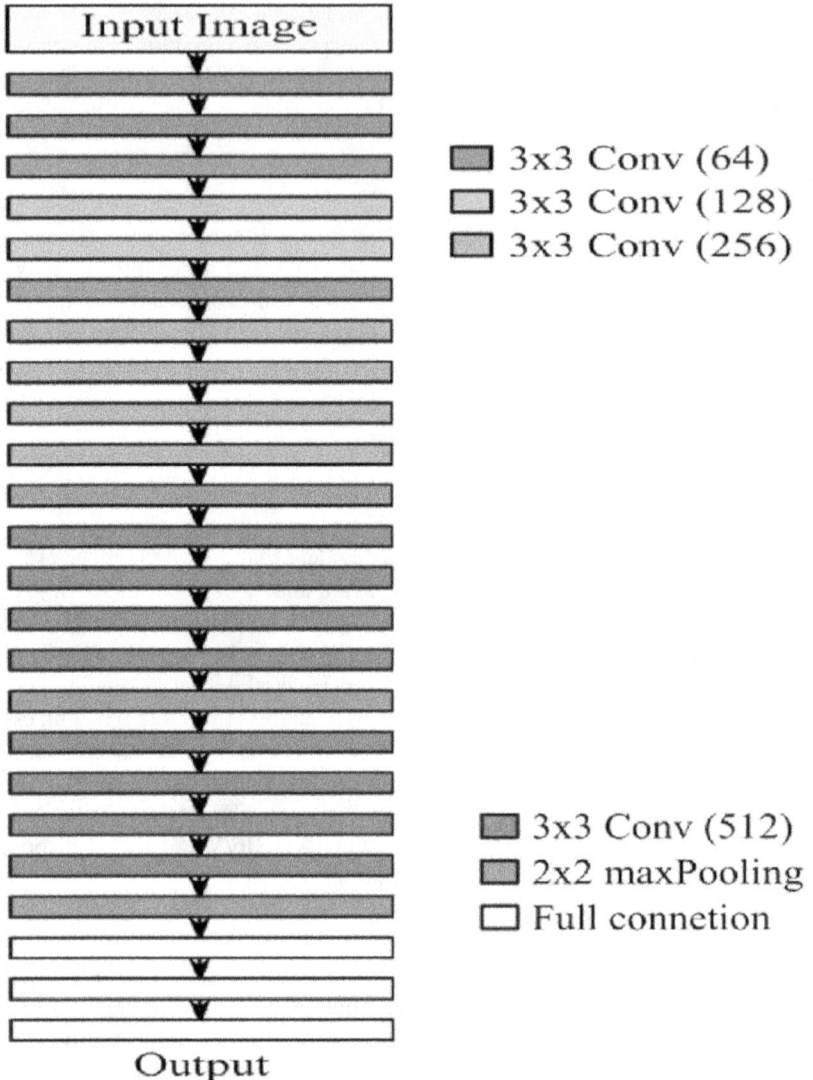

FIGURE 18.4   The basic architecture of the VGG19 deep network [33].

This fundamental deep learning model configuration, as shown in Figure 18.7, is used in a variety of medical applications [36], including image segmentation. Segmentation of clinical images aims to find regions of interest (RoI), such as tumours, lesions, or any organ. Computer-aided segmentation of clinical images is a tough problem since clinical images are typically complicated due to various factors like artifacts, inhomogeneity intensity, low-contrast images, and variability in the organ shape and size in the different images. In the literature, many deep learning models have been suggested. The selection of a specific deep learning model is influenced by a variety of criteria,

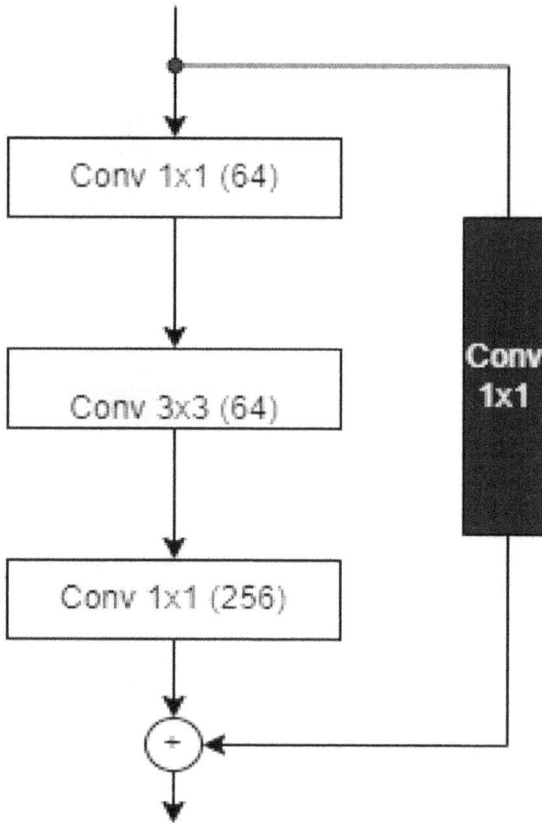

FIGURE 18.5   Residual block used in ResNet deep network [33].

such as what organ to be segmented, the imaging modalities used, and the kind of condition, as different body parts and disorders have distinct needs.

Besides the success of deep learning in medical image analysis, there are various challenges encountered by the researchers:

1. Model overfitting: It occurs when the model gives high accuracy on the training dataset compared to untouched data instances. It generally happens when the model is trained with a modest training dataset [37]. It can be mitigated by the following:
   - Expanding the size of the dataset through the use of augmentation techniques [38].
   - Dropout approaches [39, 40] also aid in the management of overfitting by rejecting the outcome of part of randomly selected neurons of the CNN throughout every cycle.
2. Model efficiency in terms of memory: High memory is required for segmenting clinical images [40]. These models must be streamlined in terms of compatibility with specific devices, such as mobile phones.

a)   b)

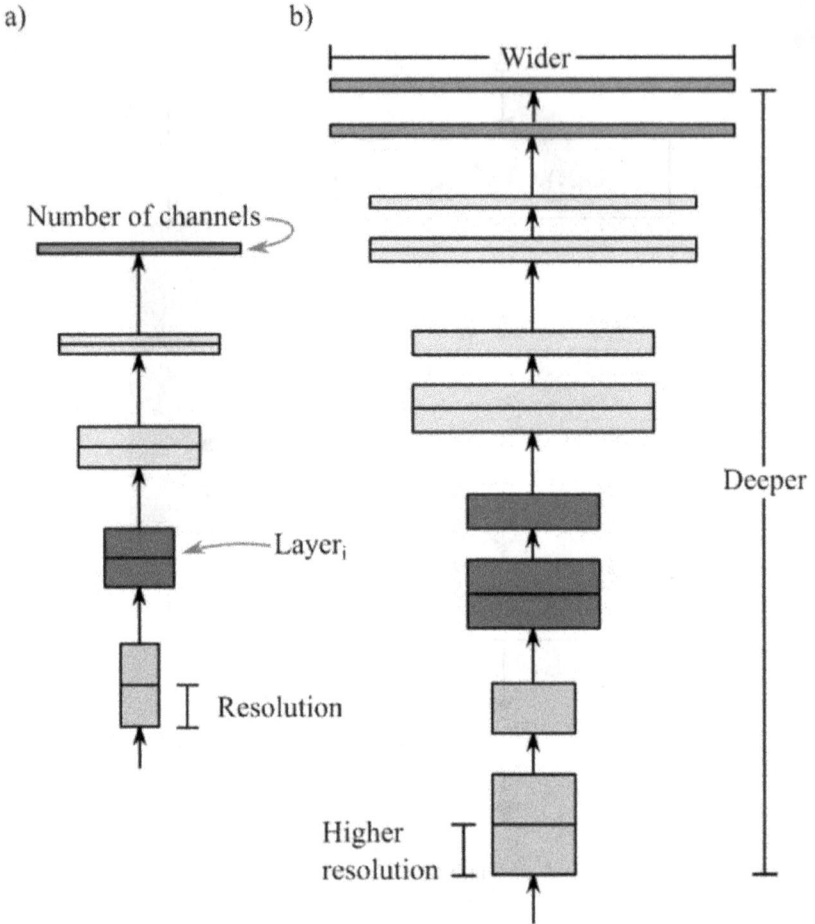

FIGURE 18.6 Graphical representation of not uniformly scaled method: (a) example and (b) architecture of not uniform scaled in depth, width, and resolution. This graph was taken from reference [34].

3. Training time: Deep network training is time-consuming. Fast convergence of training time is needed in the segmentation of clinical images.

    This issue can be solved by the following:
    - Using batch normalization [41]. It determines the pixel values close to zero by subtracting them from the image's mean value. It is effective in achieving rapid convergence.
    - Adding extra pooling layers for minimizing parameter dimension can also result in faster convergence.

4. Vanishing gradient: The vanishing gradient problem confronts deep neural networks [41]. It happens when the ultimate gradient loss cannot be back-propagated. In 3D models, this problem is more prominent. There are various solutions found in the literature to handle this problem:

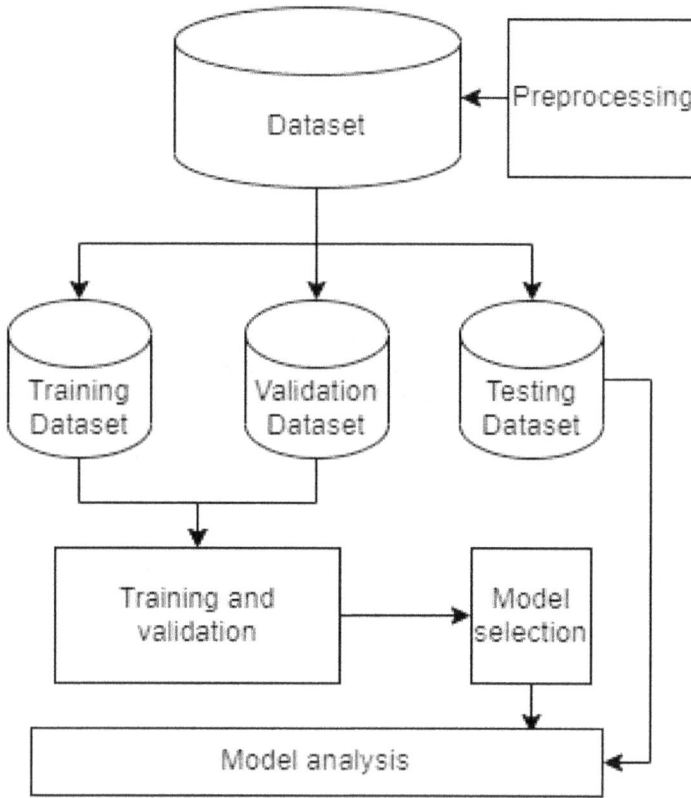

FIGURE 18.7 General scheme for developing a deep-learning-based system.

- The supplementary loss and the primary loss of the hidden layer are merged to enhance the gradient units by upsampling the intermediary hidden layer outputs using de-convolutional operation and activation functions like softmax [42].
- It can also be avoided by using optimal initial network weights [43].

5. Computational complexity: In deep learning, feature analysis algorithms must be highly computationally efficient. These methods necessitate high-performance computer equipment and GPUs [3, 4]. Some of the best techniques may need supercomputers to train the model, which may not be accessible. To address these challenges, the researcher must evaluate a restricted set of factors to achieve a limited degree of precision [44].

18.5 CONCLUSION

This chapter consists of an overview of several modern deep learning networks for analysing medical images. In recent research, deep networks have been widely employed in clinical image segmentation and classification. However, the researchers face many challenges besides the benefits of deep-learning-based algorithms. This chapter gives an overview for researchers who are just starting in the field and want to address the prominent challenges in using deep learning in medical image processing.

REFERENCES

1. Gupta, A., *Current research opportunities of image processing and computer vision.* Computer Science, 2019. **20**(4).
2. Gupta, A., *Challenges for computer aided diagnostics using X-ray and tomographic reconstruction images in craniofacial applications.* International Journal of Computational Vision and Robotics, 2020. **10**(4): p. 360–371.
3. Ashok, M. and A. Gupta, *A systematic review of the techniques for the automatic segmentation of organs-at-risk in thoracic computed tomography images.* Archives of Computational Methods in Engineering, 2021. **28**(4): p. 3245–3267.
4. Pandey, M. and A. Gupta, *A systematic review of the automatic kidney segmentation methods in abdominal images.* Biocybernetics and Biomedical Engineering, 2021. **41**(4): p. 1601–1628.
5. da Cruz, L.B., J.D.L. Araújo, J.L. Ferreira, J.O.B. Diniz, A.C. Silva, J.D.S. de Almeida, A.C. de Paiva, and M. Gattass, *Kidney segmentation from computed tomography images using deep neural network.* Computers in Biology and Medicine, 2020. **123**: p. 103906.
6. da Cruz, L.B., D.A.D. Júnior, J.O.B. Diniz, A.C. Silva, J.D.S. de Almeida, A.C. de Paiva, and M. Gattass, *Kidney tumor segmentation from computed tomography images using DeepLabv3+ 2.5 D model.* Expert Systems with Applications, 2022. **192**: p. 116270.
7. Xie, X., L. Li, S. Lian, S. Chen, and Z. Luo, *SERU: A cascaded SE-ResNeXT U-Net for kidney and tumor segmentation.* Concurrency and Computation: Practice and Experience, 2020. **32**(14): p. e5738.
8. Yang, G., J. Gu, Y. Chen, W. Liu, L. Tang, H. Shu, and C. Toumoulin, *Automatic kidney segmentation in CT images based on multi-atlas image registration.* in 2014 36th Annual International Conference of the IEEE Engineering in Medicine and Biology Society, 2014. IEEE.
9. Yang, G., G. Li, T. Pan, Y. Kong, J. Wu, H. Shu, L. Luo, J.-L. Dillenseger, J.-L. Coatrieux, and L. Tang. *Automatic segmentation of kidney and renal tumor in ct images based on 3d fully convolutional neural network with pyramid pooling module.* in 2018 24th International Conference on Pattern Recognition (ICPR). 2018. IEEE.
10. Maken, P. and A. Gupta, *2D-to-3D: A review for computational 3D image reconstruction from X-ray images.* Archives of Computational Methods in Engineering, 2022: p. 1–30.
11. Trivedi, M. and A. Gupta, *Automatic monitoring of the growth of plants using deep learning-based leaf segmentation.* International Journal of Applied Science and Engineering, 2021. **18**(2): p. 1–9.
12. Trivedi, M. and A. Gupta, *A lightweight deep learning architecture for the automatic detection of pneumonia using chest X-ray images.* Multimedia Tools and Applications, 2022. **81**(4): p. 5515–5536.
13. Suzuki, K., *Overview of deep learning in medical imaging.* Radiological Physics and Technology, 2017. **10**(3): p. 257–273.
14. Lundervold, A.S. and A. Lundervold, *An overview of deep learning in medical imaging focusing on MRI.* Zeitschrift für Medizinische Physik, 2019. **29**(2): p. 102–127.

15. Ashok, M. and A. Gupta. *Deep learning-based techniques for the automatic segmentation of organs in thoracic computed tomography images: A Comparative study.* in 2021 International Conference on Artificial Intelligence and Smart Systems (ICAIS). 2021. IEEE.

16. Sharma, K., C. Rupprecht, A. Caroli, M.C. Aparicio, A. Remuzzi, M. Baust, and N. Navab, *Automatic segmentation of kidneys using deep learning for total kidney volume quantification in autosomal dominant polycystic kidney disease.* Scientific Reports, 2017. **7**(1): p. 2049.

17. da Cruz, L.B., J.D.L. Araújo, J.L. Ferreira, J.O.B. Diniz, A.C. Silva, J.D.S. de Almeida, A.C. de Paiva, and M. Gattass, *Kidney segmentation from computed tomography images using deep neural network.* Computers in Biology and Medicine, 2020. **123**: p. 103906.

18. Fatemeh, Z., S. Nicola, K. Satheesh, and U. Eranga, *Ensemble U-net-based method for fully automated detection and segmentation of renal masses on computed tomography images.* Medical Physics, 2020. p. 4032–4044. John M. Boone I University of California at Davis.

19. Türk, F., M. Lüy, and N. Barışçı, *Kidney and renal tumor segmentation using a hybrid v-net-based model.* Mathematics, 2020. **8**(10): p. 1–17.

20. Lin, Z., Y. Cui, J. Liu, J. Sun, S. Ma, X. Zhang, and X. Wang, *Automated segmentation of kidney and renal mass and automated detection of renal mass in CT urography using 3D U-Net-based deep convolutional neural network.* European Radiology, 2021. **31**(7): p. 5021–5031.

21. Jackson, P., N. Hardcastle, N. Dawe, T. Kron, M.S. Hofman, and R.J. Hicks, *Deep learning renal segmentation for fully automated radiation dose estimation in unsealed source therapy.* Frontiers in Oncology, 2018. **8**: p. 215.

22. Qayyum, A., A. Lalande, and F. Meriaudeau, *Automatic segmentation of tumors and affected organs in the abdomen using a 3D hybrid model for computed tomography imaging.* Computers in Biology and Medicine, 2020. **127**: p. 104097.

23. Chen, Y., J. Li, H. Xiao, X. Jin, S. Yan, and J. Feng, *Dual path networks.* Advances in Neural Information Processing Systems, 2017. **30**.

24. Ronneberger, O., P. Fischer, and T. Brox. *U-net: Convolutional networks for biomedical image segmentation.* in International Conference on Medical Image Computing and Computer-Assisted Intervention. 2015. Springer.

25. Cui, H., X. Liu, and N. Huang. *Pulmonary vessel segmentation based on orthogonal fused U-Net++ of chest CT images.* in International Conference on Medical Image Computing and Computer-Assisted Intervention. 2019. Springer.

26. Jin, Q., Z. Meng, C. Sun, H. Cui, and R. Su, *RA-UNet: A hybrid deep attention-aware network to extract liver and tumor in CT scans.* Frontiers in Bioengineering and Biotechnology, 2020: p. 1471.

27. Guo, C., M. Szemenyei, Y. Pei, Y. Yi, and W. Zhou. *SD-UNet: A structured dropout U-Net for retinal vessel segmentation.* in 2019 IEEE 19th International Conference on Bioinformatics and Bioengineering (BIBE). 2019. IEEE.

28. Çiçek, Ö., A. Abdulkadir, S.S. Lienkamp, T. Brox, and O. Ronneberger. *3D U-Net: Learning dense volumetric segmentation from sparse annotation.* in International Conference on Medical Image Computing and Computer-Assisted Intervention. 2016. Springer.

29. Milletari, F., N. Navab, and S.-A. Ahmadi. *V-net: Fully convolutional neural networks for volumetric medical image segmentation.* in 2016 Fourth International Conference on 3D Vision (3DV). 2016. IEEE.

30. Chen, L.-C., G. Papandreou, I. Kokkinos, K. Murphy, and A.L. Yuille, *Deeplab: Semantic image segmentation with deep convolutional nets, atrous convolution,*

and fully connected CRFs. IEEE Transactions on Pattern Analysis and Machine Intelligence, 2017. **40**(4): p. 834–848.

31. Simonyan, K. and A. Zisserman, *Very deep convolutional networks for large-scale image recognition.* arXiv preprint arXiv:1409.1556, 2014.

32. He, K., X. Zhang, S. Ren, and J. Sun. *Deep residual learning for image recognition.* in Proceedings of the IEEE Conference on Computer Vision and Pattern Recognition. 2016. IEEE.

33. Anaya-Isaza, A., L. Mera-Jiménez, and M. Zequera-Diaz, *An overview of deep learning in medical imaging.* Informatics in Medicine Unlocked, 2021. **26**: p. 100723.

34. Tan, M. and Q. Le. *Efficientnet: Rethinking model scaling for convolutional neural networks.* in International Conference on Machine Learning. 2019. PMLR.

35. Khan, A., A. Sohail, U. Zahoora, and A.S. Qureshi, *A survey of the recent architectures of deep convolutional neural networks.* Artificial Intelligence Review, 2020. **53**(8): p. 5455–5516.

36. Kermany, D.S., M. Goldbaum, W. Cai, C.C. Valentim, H. Liang, S.L. Baxter, A. McKeown, G. Yang, X. Wu, and F. Yan, *Identifying medical diagnoses and treatable diseases by image-based deep learning.* Cell, 2018. **172**(5): p. 1122–1131. e9.

37. Shen, D., G. Wu, and H.-I. Suk, *Deep learning in medical image analysis.* Annual Review of Biomedical Engineering, 2017. **19**: p. 221.

38. Merkow, J., A. Marsden, D. Kriegman, and Z. Tu. *Dense volume-to-volume vascular boundary detection.* in International Conference on Medical Image Computing and Computer-Assisted Intervention. 2016. Springer.

39. Cui, H., H. Zhang, G.R. Ganger, P.B. Gibbons, and E.P. Xing. *Geeps: Scalable deep learning on distributed gpus with a gpu-specialized parameter server.* in Proceedings of the Eleventh European Conference on Computer Systems, 2016. ACM Digital Library. https://doi.org/10.1145/2901318.2901323.

40. Srivastava, N., G. Hinton, A. Krizhevsky, I. Sutskever, and R. Salakhutdinov. *Dropout: A simple way to prevent neural networks from overfitting.* The Journal of Machine Learning Research, 2014. **15**(1): p. 1929–1958.

41. Ioffe, S. and C. Szegedy. *Batch normalization: Accelerating deep network training by reducing internal covariate shift.* in International Conference on Machine Learning. 2015. PMLR.

42. Guo, Y. and A.S. Ashour, *Neutrosophic sets in dermoscopic medical image segmentation,* in Neutrosophic set in medical image analysis. 2019, Elsevier: p. 229–243.

43. Kamnitsas, K., C. Ledig, V.F. Newcombe, J.P. Simpson, A.D. Kane, D.K. Menon, D. Rueckert, and B. Glocker, *Efficient multi-scale 3D CNN with fully connected CRF for accurate brain lesion segmentation.* Medical Image Analysis, 2017. **36**: p. 61–78.

44. Panday, M. and A. Gupta, *Tumorous kidney segmentation in abdominal CT images using active contour and 3D-UNet.* Irish Journal of Medical Science (1971–), 2022: p. (In-press).

19 Cloud and IoT for the Future Internet

Neeraj Singla

19.1 INTRODUCTION

The future internet (FI) can be used for a variety of cutting-edge purposes, including linking all physical objects together. In order to supply a wide range of services for everyone, improve and monitor our lives, and give access to many tools in various disciplines, these applications and devices can cooperate and communicate. The term "future internet" refers to a group of upcoming technologies for data transmission networks. The internet of devices offers a standard worldwide Information Technology (IT) [1, 2].

A platform for fusing networked things and seamless networks is the most well-known and potent notion of the FI. People could access the internet, using any service and any network in a proper manner. The internet of things (IoT) also take saholistic approach to connecting people and things through communication, computation, convergence, collections, and information. Most of the time, an abun dancetal., Kansai University 2020 Technology Reports 2180 IoT smart sensors will collect data from the environment and ecosystem and deliver it to a cloud service for decision-making [3, 4].

The foundation of the "things" paradigm is an active, global network architecture with intelligent, self-configuring inter sections that are linked together and catenated. The most common characteristics of IoT devices are their small size, limited storage capacity, and computing power, which have implications for privacy, security, performance, and dependability. But cloud computing has practically limitless possibilities regarding processing and storage capacity. Additionally, it has s significantly more evolved technology, and it has at least partially or entirely overcome the IoT difficulties [5, 6]. Therefore, cloud computing and the IoT can be seen as two complementary technologies that have been joined simultaneously in a new IT architecture that is intended to consume both existing and future internet. The foundation of global computer networking was expanded in the proposed platform to include every object (thing) that is allowed by communication. The definition of "things" is constantly growing, from automobile actuators and sensors to wearables, the industrial internet, smart buildings, traffic and environment sensors, tablets, and smartphones, as well as home security and appliance sensors. These things can also influence one another and communicate with one another. The development of a

DOI: 10.1201/9781003367161-19

comparable network can then take place thanks to quick advancements in mobile computing, wireless communication, smart sensing, and controlling, as shown by the IoT example. IoT builds a larger network with a variety of physical things around the globe [7, 8.]

It is important to first define cloud computing and the specific benefits it provides. Cloud computing refers to the usage of a network remote server kept on the internet in place of a personal computer or a local server to host, manage, and process data. Where it is possible to provide per-request, appropriate, and scalable network access, enabling the contribution of computing resources. Dynamic data fusion from a number of data sources is authorized and licensed successfully. In cloud computing, services are provided that enable the distribution of computing resources across the internet [9–11]. The basic structure of cloud computing IOT is shown is shown in Figure 19.1.

Since physical items and things are connected to the internet, they are no longer separated from the virtual world. The IoT is the revolution of the following internet-related generation. Numerous interconnected and communicating devices exchange data and information with one another and improve the quality of our daily lives thanks to the IoT As the internet has accomplished so much, the IoT can revolutionize the world [12]. The various applications of IoT in different domains is visible in Figure 19.2.

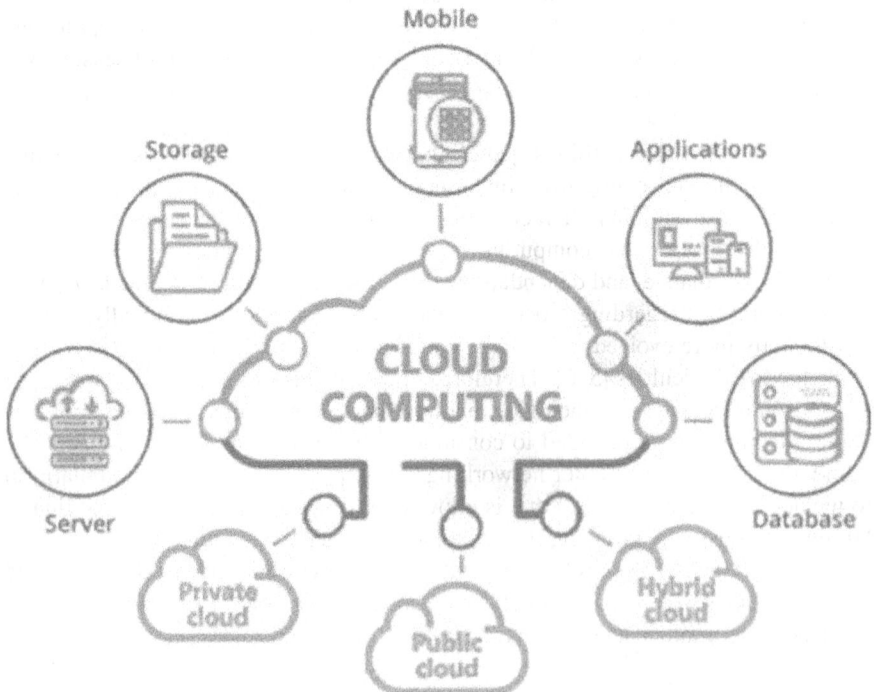

FIGURE 19.1   Basic structure of cloud computing IOT.

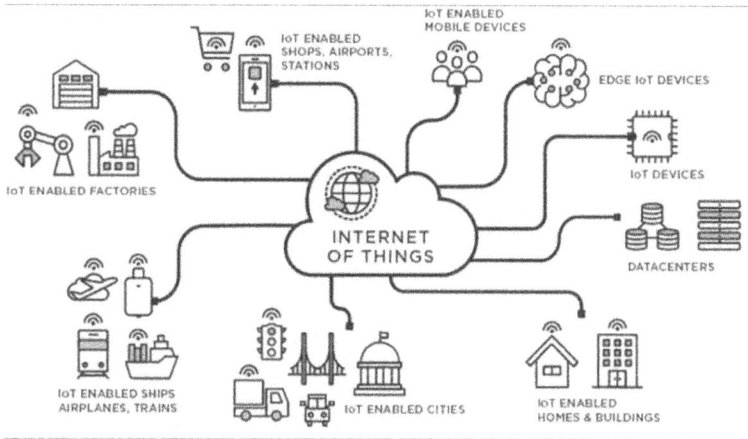

FIGURE 19.2   Applications of IoT in different domains.

19.2   RECENT WORK

1. The IoT is made up of actual physical things, or objects, with embedded sensors, software, electronics, and communication components that enable data collection and exchange. Since physical items and things are connected to the internet, they are no longer separated from the virtual world [25]. The IoT is the revolution of the following internet-related generation. Numerous interconnected and communicating devices exchange data and information with one another and improve the quality of our daily lives thanks to the IoT As the internet has accomplished so much, the IoT can revolutionize the world [13, 14].

2. The phrase "cloud of things" refers to the combination of cloud computing with IoT. This paradigm addressed issues with data reachability, analysis, and computation in the IoT. Additionally, it has opened up new possibilities like "smart things" and "things as a service." Dáz et al. declared various existing integration levels and examined the proposed solutions with an emphasis on embedded IoT devices. But subsequent publications from 2017 to 2019 that was published have not been taken into account in this chapter. Additionally, they have not gone into great depth about the paper selection approach [15, 16].

3. IoT and cloud computing definitions, reference architectures, security issues, and potential solutions have all been investigated individually by Amairah et al. They reviewed the most popular research and then suggested future directions. Future research will be necessary to develop the platforms for the linked apps and devices, which rely on cloud computing and IoT. Additionally, more study is needed to solve the security flaws and to determine the best ways to integrate these technologies [17, 18].

19.2.1   IoT AND CLOUD COMPUTING

Cloud computing has developed and grown to offer a platform that makes it easier for apps to store and analyse data. Cloud computing is typically used by IoT devices to store and process data. Cloud offers fresh business models and opportunities to show off created solutions for enhancing existing information systems. However, some concerns, such as security and privacy, are made possible by cloud computing. The combination of IoT with cloud computing aims to provide clients with trustworthy and adaptable decision-making tools by transforming some common resources like sensors, work processes, and machines into smart things [19–21]. The different utilities of IoT and cloud computing is shown in Figure 19.3.

A logical foundation for the Health Level 7 protocol has been given by Plathong and Surakratanasakul, using cloud computing and the IoT for real-time healthcare monitoring. A novel architecture with real-time communication was presented by Dáz et al. as CoAP. In many industries, the combination of cloud computing with IoT can be quite profitable. This will enable the cloud to connect to actual physical items since IoT makes it possible to connect heterogeneous objects. IoT can be made to overcome its technological limits, such as those related to energy, processing, and storage, with the help of cloud computing and its limitless virtual capabilities and resources. Cloud computing and IoT service composition are specifically used to handle applications that use devices and the data they produce [22–24].

Nevertheless, the cloud can benefit from IoT's worldwide reach and cope with physical sensors thanks to its dynamic and dispersed ways of providing new services in a variety of real-world circumstances [25].

Together, these functional building blocks, as shown in Figure 19.4, comprise an effective IoT system, which is necessary for top performance. Despite the fact that a

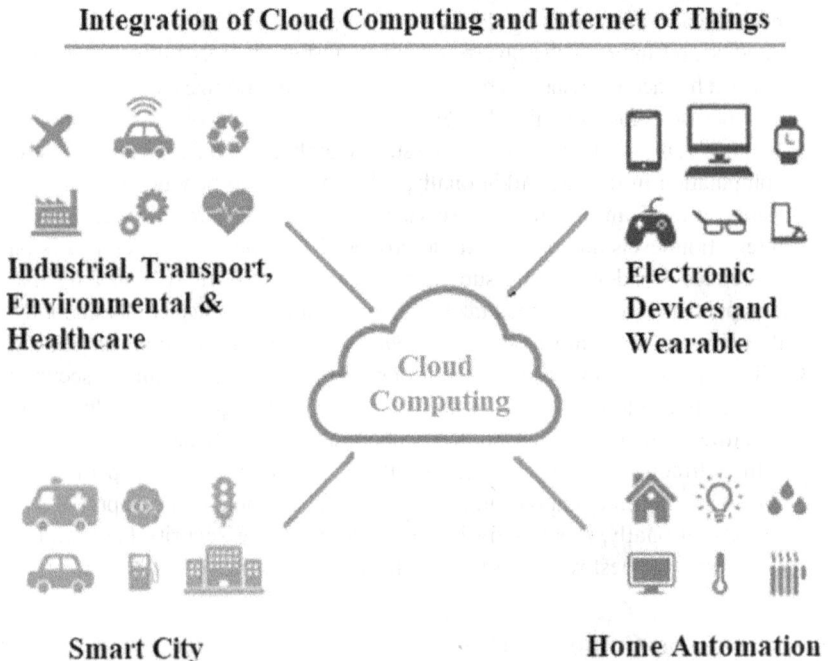

FIGURE 19.3   Utilities of IoT and cloud computing: IoT architecture and technologies.

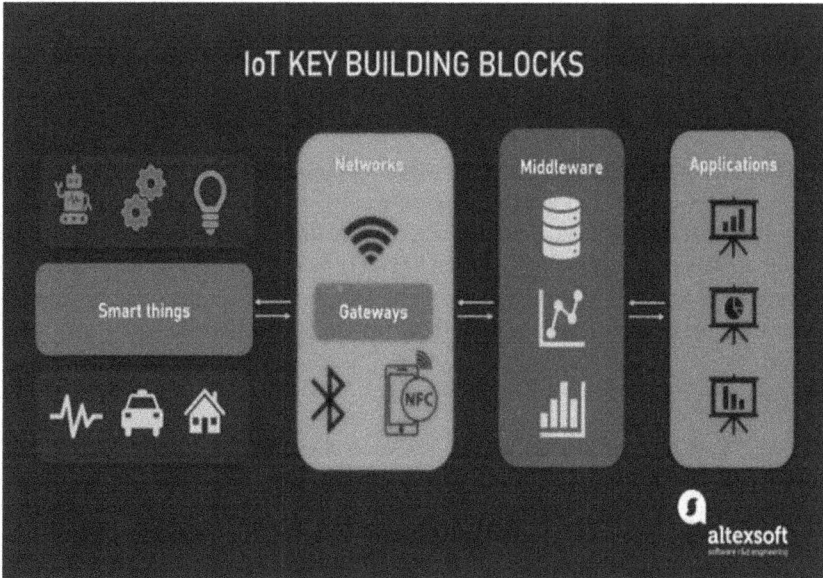

FIGURE 19.4   Building blocks of IoT.

number of reference architectures have been presented along with technical details, these remain far from the standardized architecture that is appropriate for the global IoT. In order to meet the needs of the IoT globally, a suitable architecture still has to be built to depict the basic IoT system operational structure. Researchers have emphasized the utilization of IoT in many tools, including mobile computing, data processing to improve for visualization applications [26, 27].

19.3   OBJECTIVE OF CLOUD COMPUTING USING IOT

It is important to first define cloud computing and the specific benefits it provides. Cloud computing refers to the usage of a network remote server kept on the internet in place of a personal computer or a local server to host, manage, and process data. Where it is possible to provide per-request, appropriate, and scalable network access, enabling the contribution of computing resources. Dynamic data fusion from a number of data sources is authorized and licensed successfully. In cloud computing, services are provided that enable the distribution of computing resources across the internet. The cloud is made up of a number of per-request compute services that are available and present on the internet [28, 29].

19.4   METHODOLOGY

The IoT and cloud computing areas complement one another to give a better IoT service. Despite the fact that they have significant disparities that force them to operate both independently and collaboratively. Cloud computing can manage vast amounts of data generated by IoT. This procedure takes advantage of big data. The combination of IoT and cloud computing automates the system in a cost-effective manner, providing real-time control and data monitoring [30, 31].

19.5   CONCLUSION

IoT and cloud computing definitions, reference architectures, security issues, and potential solutions have all been investigated individually by Amairah et al. They reviewed the most popular research and then suggested future directions. Future research will be necessary to develop the platforms for the linked apps and devices, which rely on cloud computing and IoT. Additionally, more study is needed to solve the security flaws and to determine the best ways to integrate these technologies [32, 33].

REFERENCES

[1] Luigi Atzori et al., "The Internet of Things: A Survey," *Computer Networks*, no. 54, pp. 2787–2805, 2010.
[2] Sandip Roy et al., "A Fog-Based DSS Model for Driving Rule Violation Monitoring Framework on the Internet of Things," *International Journal of Advanced Science and Technology*, pp. 23–32, 2015.
[3] Melanie Swan, "Sensor Mania! The Internet of Things, We Are Able Computing, Objective Metrics, and the Quantified Self 2.0," *Sensor and Actuator Networks*, vol. 1, no. 3, pp. 217–253, 2012. doi:10.3390/jsan1030217.
[4] Mohammad A. Alsmirat, Yaser Jararweh, Islam Obidat, and Brij B. Gupta, "Internet of Surveillance: A Cloud Supported Large Scale Wireless Surveillance System," *The Journal of Super Computing*, vol. 3, pp. 2–9, 2016.
[5] J. Mongay Batalla and P. Krawiec, "Conception of ID Layer Performance at the Network Level for Internet of Things," *Springer Journal Personal and Ubiquitous Computing*, vol. 18, no. 2, pp. 465–480, 2014.
[6] Y. Kryftis, G. Mastorakis, C. Mavromoustakis, J. Mongay Batalla, E. Pallisand, and G. Kormentzas, "Efficient Entertainment Services Provision Over a Novel Network Architecture," *To be published in IEEE Wireless Communications Magazine*, 2016.
[7] M. R. Rahimi et al., "Mobile Cloud Computing: A Survey, State of Art and Future Directions," *Mobile Networks and Applications*, vol. 19, no. 2, pp. 133–143, 2014.
[8] Zou Quan, *IEEE International Conference (ICCCBDA)*, Chengdu, Chinapp, pp. 29–33, 2017.
[9] AriI, Olme Zogullari E and Celebi of 2012, *IEEE International Conference on Cloud Computing Technology and Science Taipei Taiwan*, IEEE Xplore, pp. 857–862.
[10] V. C. Emeakaroha, N. Cafferkey, P. Healy, and J. P. Morrison, *3rd International Conference on Future Internet of Things and Cloud*, IEEE Xplore: Rome, Italy, pp. 50–57, 2015.
[11] A. R. Maria, P. Sever, and V. Carlos, *Conference Grid Cloud & High-Performance Computing in Science (ROLCG)*, IOP Conference Series: Materials Science and Engineering, Cluj-Napoca, Romaniapp, pp. 1–4, 2015. ISSN: 1757-899X.
[12] J. R. Dinakar and S. Vagdevi, *International Conference on Electrical, Electronics, Communication, Computer, and Optimization Techniques (ICEECCOT)*, IEEE Xplore: Mysuru, pp. 342–345, 2017.
[13] Zhu Chunsheng, Li Xi, Ji Hong, and Victor M. Leung, *IEEE 7th International Conference on Cloud Computing Technology and Science*, IEEE Xplore, pp. 101–110, 2015.
[14] R. Singh, M. V. Gaonkar, S. Sharma, P. Grover, and A. Khatri, *IEEE International Conference on Machine Learning, Big Data, Cloud and Parallel Computing*, IEEE Xplore, pp. 90–94, India, 2019.

[15] S. Chen, Xu Huihanzhi, D. Liu, B. Hu, and H. Wang, IEEE Internet of Things Journal, vol. 1, *India Computer International Conference on Computing for Sustainable Global Development*, no. 4, Ast2014, 2014.

[16] J. Shenoy and Y. Pingle, *India Computing International Conference on Computing for Sustainable Global Development*, pp. 24–25, New Delhi, India, 2016.

[17] Y. Pingle, S. R. Chaudhari, S. N. Dalvi, and P. Bhatkar, *India Computer International Conference on Computing for Sustainable Global Development*, no. 4, pp. 54–61, New Delhi, India, 2016.

[18] S. Banka, I. Madan, and S. S. Saranya, "Smart healthcare monitoring using IoT," *International Journal of Applied Engineering Research*, vol. 13, no. 15, pp. 11984–11989, 2018, ISSN 0973-4562.

[19] V. Sharma and R. Tiwari, "Combination of Spectral and Texture Features for Remote Sensing Image Segmentation," *International Journal of Science Engineering and Technology Research (IJSETR)*, vol. 5, no. 2, pp. 472–476, 2016.

[20] M. T. Reddy and R. K. Mohan, "Applications of IoT: A Study," *International Journal of Trend in Research and Development (IJTRD)*, pp. 86–87, 2017.

[21] M. S. Dawood, J. Margaret, and R. Devika, *International Journal of Advanced Research in Computer Engineering & Technology (IJARCET)*, vol. 7, no. 12, pp. 841–845, 2018.

[22] Miao Wu et al., "Research on the Archite Cture of Internet of Things," in *The Proceedings of 3rd International Conference on Advanced Computer Theory and Engineering*, 20–22 August, 2012, Beijing, China.

[23] Gerd Kortuem, Fahim Kawsar, Daniel Fitton, and Vasughi Sundra Moorthi, "Smart Objects and Building Blocks of Internet of Things," *IEEE Internet Computing Journal*, vol. 1, no. 1, pp. 44–51, 2010.

[24] Shuai Zhang et al., "Cloud Computing Research and Development Trend," in *The Proceedings of International Conferenceon Future Networks*, 22–24 January, 2010, Sanya, China.

[25] W. Ma et al., "The Survey and Research on Application of Cloud Computing," in *The Proceedings of 7th International Conference on Computer Science and Education*, 02–04 November, 2012, Wuyishan.

[26] Y. Jadeja et al., "Cloud Computing-Concepts, Architecture and Challenges," in *The Proceedings of International Conference on Computing Electronics and Electrical Technologies*, IEEE Xplore, 21–22 March, 2012.

[27] Raj Kumar Buyya, Christian Vecchiola and S. Thamarai Selvi, "Mastering Cloud Computing Foundations and Applications Programming," Morgan Kaufmann, Elsevier, USA, vol. 3, pp. 23–29, 2013. Elsevier Inc.

[28] Dan C. Marinescu, "Cloud Computing Theory and Practice," *2013 Elsevier Ileana Castrillo, Derrick Rountree and Hai Ji angas Technical Editor, The Basics of Cloud Computing: Understanding the Fundamentals of Cloud Computing in Theory and Practice*, 2013. ElsevierInc.

[29] J. R. Dinakar and S. Vagdevi, *International Conference on Electrical, Electronics, Communication, Computer, and Optimization Techniques (ICEECCOT)*, IEEE Xplore: Mysuru, pp. 342–345, 2017.

[30] Y. Zhang, Q. Chen, and S. Zhong, "Privacy-Preserving Data Aggregation in Mobile Phone Sensing," *IEEE Transactions on Information Forensics and Security*, vol. 11, no. 5, pp. 980–992, 2016.

[31] L. Wang, J. Yang, and W. Liu, "Leveraging Participatory Extraction to Mobility Sensing for Individual Discovery in Crowded Environments," *International Journal of Distributed Sensor Networks*, vol. 9, no. 10, pp. 246–916, 2013.

[32] A. Sharma, V. Kukreja, A. Bansal, and M. Mahajan, "Multi Classification of Tomato Leaf Diseases: A Convolutional Neural Network Model," *2022 10th*

International Conference on Reliability, Infocom Technologies and Optimization (Trends and Future Directions) (ICRITO), Noida, India, 2022, pp. 1–5, doi:10.1109/ICRITO56286.2022.9964884.

[33] S. Sakshi, S. Lodhi, V. Kukreja, and M. Mahajan, "DenseNet-based Attention Network to Recognize Handwritten Mathematical Expressions," *2022 10th International Conference on Reliability, Infocom Technologies and Optimization (Trends and Future Directions) (ICRITO)*, Noida, India, 2022, pp. 1–5, doi:10.1109/ICRITO56286.2022.9964619.

20 Increasing Crop Yield with Minimum Resource Utilization

Neeraj Singla

20.1 INTRODUCTION

The world's population is increasing day by day an expected to grow to almost ten billion by 2050, that will increase in demand for food and the pressure would increase on fruits, vegetables, and other products, which will add pressure on the natural resources even more than it is today. People today are leaving the agriculture field as an occupation and moving more towards the urban area for better living conditions and a good lifestyle [1]. Because of this, production is also taking a hit. The growth of yields has been decreasing, and that is not viable in the long run. Food wastage and loss due to improper storage is also an area where we need to improve. However, the necessary increase in food production would result in the degradation of natural resources, ecosystems, and the development of pests and illnesses in both animals and plants that have developed [2]. The farming practices that require resource-intensive farming or that can release massive greenhouse gasses that can deplete resources like water and soil or that will increase the levels of chemicals cannot deliver sustainable food and agriculture production. So there is a need for innovative farming practices that address such issues. And we need to move towards a holistic approach so that we can improve the conditions of soil depletion, water table decrease, and global warming [3]. We need to use technologies that will improve the agriculture economy and reduce the dependency on fossil fuels [4]. Increased cultivating area, inappropriate use of agrochemicals, and irrigation all played important roles in agricultural production increases during the Green Revolution. Though now it is observed that seen that these methods have bought in effects including land degradation, depletion of groundwater, and buildup of pest resistance [5]. It has also damaged the environment through deforestation, and the emission of greenhouse gases [6]. The water crisis is arriving sooner than we expect due to extensive water use for human needs, such as drinking, washing, cooking, and sanitation, as well as agriculture from rivers, lakes. In dry climates, we have to supply more water for irrigation but in wet areas, these are fulfilled through rain and soil moisture [7]. With the increasing population, the need for irrigating more area [8] will be required, and this will demand more water, which can be only obtained by irrigation efficiency or through reusing and recycling more water [9].

DOI: 10.1201/9781003367161-20

Cotton is the most used material in the clothing industry worldwide. One of the major problems with growing cotton is the requirement for massive amounts of water and pesticides [10]. The number of city dwellers has increased significantly in the past years, and with the increased earnings of people, the consumption of meat has also increased [11]. With this, the requirement for water for the purpose of rearing livestock, and growing crops for their food is increasing gradually [12]. Water tables are already dropping by as fast as ten metres in many areas, which is an alarming situation for many countries [13]. This is leading to less reliable resources of water supply [14]. One of the crops that can be a good alternative in those areas with water scarcity is millets [15]. It is grown as food and fodder around the world [16]. They are flood and drought-proof crops, and they are resistant to pests, so we would have to use less or nearly no pesticides to grow which not only reduces labour but also decreases natural resource utilization [17]. They are healthy and perfect for the daily diet [18].

20.2 LITERATURE SURVEY

The production of food grains has increased nearly five times in the past 60 years. And among these, rice has increased by ~435%. Wheat has shown an increase of ~1,533%. Cotton increased by nearly ~1,000%. Sugarcane has increased by over 436%. However, the population increased by only ~271% from 1950 to 2017. The production data of major crops during the years 1950–2017 is shown in Figure 20.1. And the major states that are producing these water-intensive crops, such as rice, wheat, sugarcane, and cotton in Punjab, Arunachal Pradesh, Haryana, Telangana, and Uttar Pradesh [19]. The state-wide coverage data of major crops in the years 2014–2015 is shown in Figure 20.2. When the water level was studied before monsoon 2018, 38% of all wells experienced an increase in water level, and 58% experienced a decline in water level. The remaining 3% were investigated a change in water level. The water level changed from January 2020 to January 2019, with 69% indicating an increase and 29% showing a decrease. The remaining 2% of studied stations showed no change in water level in contrast to the depth to the water level before monsoon 2019 with the ten-year mean be for monsoon water level since 12% of wells have dropped by 2–4 metres 42% of wells exhibited a fall in the water in the range of 0–2 metres, while the remaining 7% have dropped by more than 4 metres [20]. Figure 20.3 represents the India states showcasing depth of water table in the month of January 2020.

20.3 PROBLEM FORMULATION

The main reason we decided to work on this topic is the future problems that are bound to happen if we do not change the way things are happening. The decreasing water table all around the world is a major issue because of the drought that is coming faster than we think. And we are growing crops that are water-intensive in areas that have low levels of groundwater [21]. The potency of soil is decreasing gradually due to overexploitation and excessive use of pesticides that also seep into the ground and contaminate the water, which makes it unfit for consumption. Also, the demand for meat is increasing due to overconsumption among the wealthy urban class, so this also leads to the problem of water consumption where water is used in food crops for animals and also for their consumption. Also due to the overuse of pesticides,

(Million Tonnes)

S.No (1)	Crops (2)	1950-51 (3)	1960-61 (4)	1970-71 (5)	1980-81 (6)	1990-91 (7)	2000-01 (8)	2010-11 (9)	2011-12 (10)	2012-13 (11)	2013-14 (12)	2014-15 (13)	2015-16 (14)	2016-17* (15)
1	Foodgrains	50.82	82.02	108.42	129.59	176.39	196.81	244.49	259.29	257.13	265.04	252.02	251.57	275.68
	Rice	20.58	34.58	42.22	53.63	74.29	84.98	95.98	105.30	105.23	106.65	105.48	104.41	110.15
	Wheat	6.46	11.00	23.83	36.31	55.14	69.68	86.87	94.88	93.51	95.85	86.53	92.29	98.38
	Maize	1.73	4.08	7.49	6.96	8.96	12.04	21.73	21.76	22.26	24.26	24.17	22.57	26.26
	Nutri Cereals	15.38	23.74	30.55	29.02	32.70	31.08	43.40	42.01	40.04	43.29	42.96	38.52	44.19
2	Pulses	8.41	12.70	11.82	10.63	14.26	11.08	18.24	17.09	18.34	19.25	17.15	16.35	22.95
	Gram	3.65	6.25	5.20	4.33	5.38	3.86	8.22	7.70	8.83	9.53	7.33	7.06	9.33
	Tur (Arhar)	1.72	2.07	1.88	1.96	2.41	2.25	2.86	2.65	3.02	3.17	2.81	2.56	4.78
	Lentil (Masur)	–	–	0.37	0.47	0.85	0.92	0.94	1.06	1.13	1.02	1.04	0.98	–
3	Oilseeds	5.16	6.98	9.63	9.37	18.61	18.44	32.48	29.80	30.94	32.75	27.51	25.25	32.10
	Groundnut	3.48	4.81	6.11	5.01	7.51	6.41	8.26	6.96	4.70	9.71	7.40	6.73	7.585
	Rapeseed & Mustard	0.76	1.35	1.98	2.30	5.23	4.19	8.18	6.60	8.03	7.88	6.28	6.80	7.98
	Soy Bean	–	–	0.01	0.44	2.60	5.28	12.74	12.21	14.67	11.86	10.37	8.57	13.79
	Sunflower	–	–	0.08	0.07	0.87	0.65	0.65	0.52	0.54	0.50	0.43	0.30	0.24
4	Cotton #	3.04	5.60	4.76	7.01	9.84	9.52	33.00	35.20	34.22	35.90	34.80	30.01	33.09
5	Jute & Mesta @	3.31	5.28	6.19	8.16	9.23	10.56	10.62	11.40	10.93	11.68	11.13	10.52	10.60
6	Sugarcane	57.05	110.00	126.37	154.25	241.05	295.96	342.38	361.04	341.20	352.14	362.33	348.45	306.72
7	Tobacco	0.26	0.31	0.36	0.48	0.56	0.34	0.80	0.75	0.66	0.74	0.84	0.80	–

* 4th Advance Estimates # Million bales of 170 kg. each @ Million bales of 180 kg. each
Source: Directorate of Economics and Statistics, DAC&FW

FIGURE 20.1 Production of major crops during the years 1950–2017.

(Percentage)

State	Rice	Wheat	Total Cereals	Total Pulses	Total Foodgrains	Sugarcane	Fruits & Vegetables	Total Oilseeds	Cotton	Total Irrigated area under all crops
(1)	(2)	(3)	(4)	(5)	(6)	(7)	(8)	(9)	(10)	(11)
Andhra Pradesh	97.1	15.0	89.5	2.0	66.5	92.4	49.5	18.7	19.9	50.5
Arunachal Pradesh	20.6	52.9	20.9	-	20.9	0.0	112.6	0.0	-	18.7
Assam	11.0	10.9	10.9	1.7	10.4	1.5	0.0	2.5	0.0	9.2
Bihar	65.0	94.9	75.2	4.5	69.8	79.3	68.2	57.2	-	68.7
Chhattisgarh	35.7	75.7	34.9	15.0	31.6	98.9	55.6	4.5	57.6	31.2
Goa	33.7	-	33.7	97.6	44.5	100.0	10.1	16.5	-	24.6
Gujarat*	61.5	90.8	52.5	13.1	45.4	94.5	91.9	29.8	58.7	47.1
Haryana	99.9	99.5	94.0	20.7	92.7	100.0	96.8	83.1	99.8	89.1
Himachal Pradesh*	66.1	21.8	21.4	12.3	21.0	54.1	17.3	16.4	100.0	21.0
Jammu & Kashmir*	90.5	28.5	40.9	13.5	40.2	40.9	40.5	69.7	-	42.8
Jharkhand	5.0	90.9	9.4	3.9	8.8	68.3	71.4	23.2	150.0	14.3
Karnataka	76.0	60.9	36.2	8.6	27.3	99.4	49.5	29.9	27.2	34.2
Kerala	76.1	0.0	75.9	0.0	74.6	99.9	14.2	20.8	0.0	17.9
Madhya Pradesh	34.2	93.4	67.2	42.8	59.7	99.9	88.4	5.3	60.4	43.3
Maharashtra*	26.1	73.9	21.0	10.9	18.0	70.5	81.6	3.2	2.7	18.2
Manipur*	30.7	0.0	27.3	0.0	24.4	0.0	0.0	0.0	-	18.0
Meghalaya	90.7	100.0	76.3	0.0	71.9	0.0	15.9	45.5	0.0	37.1
Mizoram	72.5	-	58.2	94.3	63.0	0.0	0.0	0.0	0.0	14.5
Nagaland	51.7	17.2	36.5	2.6	32.5	0.0	0.0	4.6	0.0	21.2
Odisha	33.3	100.0	32.0	2.2	29.0	100.0	100.0	12.7	-	28.7
Punjab	99.7	99.1	99.0	87.5	99.0	97.7	99.7	86.8	100.0	98.7
Rajasthan	68.6	99.6	41.3	20.7	35.9	98.4	97.0	63.8	95.2	42.0
Sikkim*	98.8	0.4	18.8	0.2	17.0	-	3.4	0.2	-	8.9
Tamil Nadu	94.4	100.0	71.8	10.9	56.8	100.0	54.5	66.5	26.5	56.6
Telangana	98.1	99.2	75.3	5.1	64.3	100.0	80.2	40.5	12.5	47.6
Tripura*	33.9	100.0	33.5	28.3	33.3	50.3	23.7	18.8	0.0	24.0
Uttarakhand	70.0	59.5	47.3	7.9	44.8	98.7	39.7	25.0	-	49.5
Uttar Pradesh	86.7	98.8	87.0	27.4	80.4	95.4	88.6	42.5	97.2	80.2
West Bengal*	46.9	98.7	49.7	15.0	48.4	84.6	89.5	79.1	80.0	58.8
All India	60.1	94.2	60.1	19.9	53.1	90.2	65.6	27.4	33.7	48.6

Table 5.8: State-wise Coverage of Irrigated Area under Major Crops during 2014-15

(Percentage)

(1) State	(2) Rice	(3) Wheat	(4) Total Cereals	(5) Total Pulses	(6) Total Foodgrains	(7) Sugarcane	(8) Fruits & Vegetables	(9) Total Oilseeds	(10) Cotton	(11) Total Irrigated area under all crops
Andhra Pradesh	97.1	15.0	89.5	2.0	66.5	92.4	49.5	18.7	19.9	50.5
Arunachal Pradesh	20.6	52.9	20.9	-	20.9	0.0	112.6	0.0	-	18.7
Assam	11.0	10.9	10.9	1.7	10.4	1.5	0.0	2.5	0.0	9.2
Bihar	65.0	94.9	75.2	4.5	69.8	79.3	68.2	57.2	-	68.7
Chhattisgarh	35.7	75.7	34.9	15.0	31.6	98.9	55.6	4.5	57.6	31.2
Goa	33.7	-	33.7	97.6	44.5	100.0	10.1	16.5	-	24.6
Gujarat*	61.5	90.8	52.5	13.1	45.4	94.5	91.9	29.8	58.7	47.1
Haryana	99.9	99.5	94.0	20.7	92.7	100.0	96.8	83.1	99.8	89.1
Himachal Pradesh*	66.1	21.8	21.4	12.3	21.0	54.1	17.3	16.4	100.0	21.0
Jammu & Kashmir*	90.5	28.5	40.9	13.5	40.2	40.9	40.5	69.7	-	42.8
Jharkhand	5.0	90.9	9.4	3.9	8.8	68.3	71.4	23.2	150.0	14.3
Karnataka	76.0	60.9	36.2	8.6	27.3	99.4	49.5	29.9	27.2	34.2
Kerala	76.1	0.0	75.9	0.0	74.6	99.9	14.2	20.8	0.0	17.9
Madhya Pradesh	34.2	93.4	67.2	42.8	59.7	99.9	88.4	5.3	60.4	43.3
Maharashtra*	26.1	73.9	21.0	10.9	18.0	70.5	81.6	3.2	2.7	18.2
Manipur*	30.7	0.0	27.3	0.0	24.4	0.0	0.0	0.0	-	18.0
Meghalaya	90.7	100.0	76.3	0.0	71.9	0.0	15.9	45.5	0.0	37.1
Mizoram	72.5	-	58.2	94.3	63.0	0.0	0.0	0.0	0.0	14.5
Nagaland	51.7	17.2	36.5	2.6	32.5	0.0	0.0	4.6	0.0	21.2
Odisha	33.3	100.0	32.0	2.2	29.0	100.0	100.0	12.7	-	28.7
Punjab	99.7	99.1	99.0	87.7	99.0	97.7	99.7	86.8	100.0	98.7
Rajasthan	68.6	99.6	41.3	20.7	35.9	98.4	97.0	63.8	95.2	42.0
Sikkim*	98.8	0.4	18.8	0.2	17.0	-	3.4	0.2	-	8.9
Tamil Nadu	94.4	100.0	71.8	10.9	56.8	100.0	54.5	66.5	26.5	56.6
Telangana	98.1	99.2	75.3	5.1	64.3	100.0	80.2	40.5	12.5	47.6
Tripura*	33.9	100.0	33.5	28.3	33.3	50.3	23.7	18.8	0.0	24.0
Uttarakhand	70.0	59.5	47.3	7.9	44.8	98.7	39.7	25.0	-	49.5
Uttar Pradesh	86.7	98.8	87.0	27.4	80.4	95.4	88.6	42.5	97.2	80.2
West Bengal*	46.9	98.7	49.7	15.0	48.4	84.6	89.5	79.1	80.0	58.8
All India	60.1	94.2	60.1	19.9	53.1	90.2	65.6	27.4	33.7	48.6

* The figures related to irrigated area (Part-II) are either estimated based on the data for the latest available year received from the State/UT or are estimated/taken from Agriculture Census.

Source: Directorate of Economics and Statistics, DAC&FW.

FIGURE 20.2 State-wide coverage of major crops in the years 2014–2015.

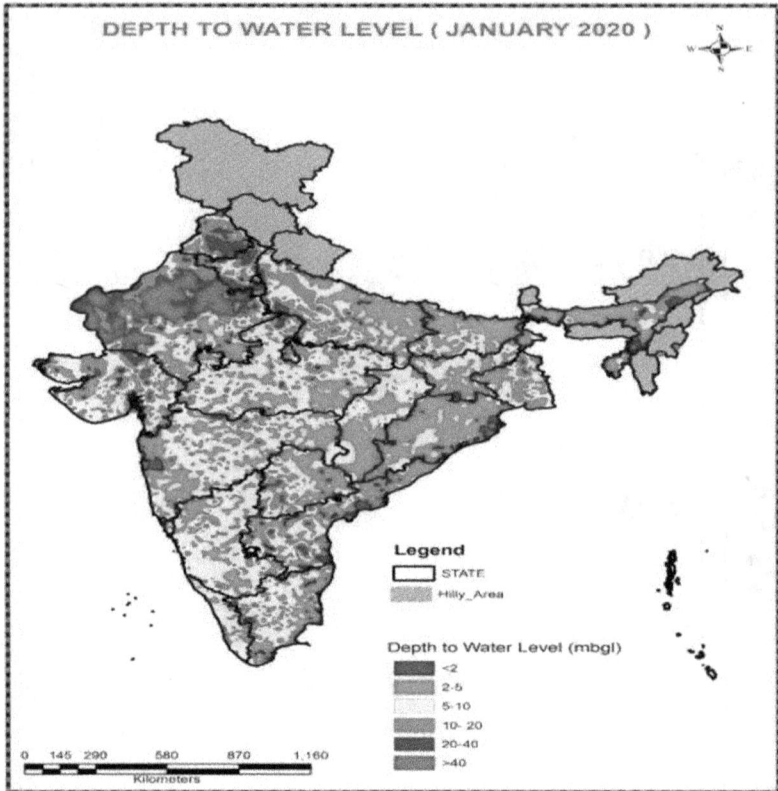

FIGURE 20.3 Depth of water table in the month of January 2020.

the quality of the crop is decreasing, and they are containing more harmful chemical that enter the food chain and can causes ever problems including cancer. The vast life-forms living inside oil that helps the farmer to grow his crops is also affected by using pesticides in a major way [22].

Objectives

- Improve crop yield and nutritional value.
- Reduce overconsumption of groundwater.
- Prevent the use of pesticides to improve soil health.
- Remove the presence of harmful chemicals in crops.
- Rest or water table level.
- Reduce the presence of hard metals in soil.

20.4 TOOL IMPLEMENTATION

While judging the quality of soil, we have to take certain parameters into consideration, such as physical and chemical properties. Physical properties include texture, water retention, colour, and amount of organic content present in the soil [23]. Chemical properties include pH, nutritional content, and presence of any hard metals in the soil. In crops, we have to consider seed type, seed quality, germination rate, soil requirements, water requirements, pesticides, climate conditions, and other parameters. These parameters can be used to arrange the data and improve the outcomes using tools like MATLAB and some Python libraries like Matplotlib, Num.py, and Pandas [24].

20.5 CONCLUSION

At a point where profit per hectare is the largest, any move that is in the direction of using the external resources at an efficient rate will come at the price of the profits [25]. Regulations must be put in place that would control the way the crops are being grown without any checks on resource utilization. And we need to work more in the area to provide more agricultural growth by utilizing minimum space. Some of the solutions could be vertical farming techniques, such as hydroponics, aquaponics, and aeroponics [26]. With the increase in population in the coming years, the demand for crop yield is going to increase, and in order to cope with the demand, higher yield should be produced, but with the decreasing water table and other resources, it is not possible to continue the methods of agriculture with the same rate [27].

REFERENCES

[1] Shinoj, P. & Mathur, V.C. (2008). Comparative advantage of India in agricultural exports vis-á-vis Asia: A post-reforms analysis. *Agricultural Economics Research Review*, 21, 60–66.

[2] Limbore, N.V. & Mane, B.S. (2014). A study of banking sector in India and overview of performance of Indian banks with reference to net interest margin and market capitalization of banks. *Review of Research*, 3(6), 1–11.

[3] Limbore, N.V. & Chandgude, A.S. (2013). A review of the current scenario of corporate social responsibility in it business sector with the special reference to Infosys. *Golden Research Thought*, 2(9), 1–10.

[4] APEDA. *Annual reports and statistics on exports of agricultural commodities from India import & export data base.* Ministry of Commerce and Industry, Government of India (accessed 1 January 2015). https://apeda.gov.in/apedawebsite/Annual_Reports/Apeda_Annual_Report_English_2014-15.pdf

[5] Sinha, N., Abraham, B. & Chaudhary, T. (2012). Effectiveness of contract farming: A case of Nadukkara Agro Processing Company Ltd., *BVIMR Management Edge*, 5(1), 94–106.

[6] Alston, J.M., Norton, G.W. & Pardey, P.G. (1995). *Science under scarcity: Principles and practice for agricultural research evaluation and priority setting.* Cornell University Press, Ithaca, NY.

[7] Todkar, R.S., Limbore, N.V. & Zargad, B.B. (2013). To study the current scenario of cloud computing in business with special reference to the mobile phone industry. *Indian Streams Research Journal (ISRJ)*, 2(12), 1–13.

[8] Renkow, M. (2000). Poverty, productivity and production environment: Are view of the evidence. *Food Policy*, 25(4), 463–478.

[9] Austin, R.B. (2019). Physiological limitations to cereal yields and ways of reducing them by breeding. In *Opportunities for increasing crop yields*, eds. R.G. Hurdetal. Pitman, Boston, pp. 1–11.

[10] Kerr, J. & Kovalli, S. (1999). Impact of agricultural research on poverty all eviation: Conceptual frame work with illustrations from the literature. *Environment and production technology division discussion paper 58.* International Food Policy Research Institute, Washington, DC.

[11] Fan, S. & Chan-Kang, C. (2004). Returns to investment in less-favore dare as in developing countries: A synthesis of evidence and implications for Africa. *Food Policy,* 29(4), 431–444.

[12] Almekinders, C.J.M. & Elings, A. (2001). Collaboration of farmers and breeders: Participatory crop improvement in perspective. *Euphytica,* 122(3), 425–438.

[13] Atlin, G.N., Cooper, M. & Bjørnstad, A. (2001). A comparison of formal and participatory breeding approaches using selection theory. *Euphytica,* 122(3), 463–475.

[14] Singh, M., Ceccarelli, S. & Grando, S. (1999). Genotype environment interaction of cross overtype: Detecting its presence and estimating the crossover point. *Theoretical and Applied Genetics,* 99(6), 988–995.

[15] Bellon, M.R. & Taylor, J.E. (1993). 'Folk' soil taxonomy and the partial adoption of new seed varieties. *Economic Development and Cultural Change,* 41(4), 763–786.

[16] Bänziger, M. & Cooper, M. (2001). Breeding for low input conditions and consequences for participatory plant breeding examples from tropical maize and wheat. *Euphytica,* 122(3), 503–519.

[17] Ceccarelli, S., Grando, S., Bailey, E., Amri, A., El-Felah, M., Nassif, F., et al. (2001). Farmer participation in barley breeding in Syria, Morocco and Tunisia. *Euphytica,* 122(3), 521–536.

[18] Baan Hofman, T. (2019). *Effecten van stikst of gift en maaifrequentieop de drogest of opbrengst van Engels dieverschilleninpersistentie.* CABO-verslag 86, Wageningen, pp. 1–10.

[19] Bellon, M.R., Hodson, D., Bergvinson, D., Beck, D., Martinez-Romero, E. & Montoya, J. (2005). Targeting agricultural research to benefit poor farmers: Relating poverty mapping to maize environments in Mexico. *Food Policy,* 30(5/6), 476–492.

[20] Baan Hofman, T. & van der Meer, H.G. (2019). *Schatting van deopbrengstderving door ziekten en plagen in grasland unit pro evenmetbiociden.* CABO-verslag 64, Wageningen.

[21] Dercon, S. (2002). Incomerisk, coping strategies and safety nets. *The World Bank Research Observer,* 17(2), 141–166.

[22] Buringh, P., VanHeemst, H.D.J. & Staringh, G.J. (2020). Computation of the absolute maximum food production for the world. In *MOIRA, model of international relations in agriculture,* eds. H. Linneman et al. North-Holland Publishing Company, Amsterdam.

[23] Fafchamps, M. (1999). Rural poverty, risk and development. *FAOE Conomic and Social Development Paper 144.* Food and Agriculture Organization (FAO), Rome.

[24] Chen Caihong, C. & Chenliu, Liu. (2020). Studies on the nutritional character of high yielding hybrid rice Shantou63. *Gricultural University,* 1, 1–9.

[25] Anderson, J.R. & Hazell, P.B.R. (1989). *Variability in Grain Yields: Implications for agricultural research and policy in developing countries.* Johns Hopkins Press, Baltimore, MD.

[26] Sharma, A., Kukreja, V., Bansal, A. & Mahajan, M. (2022). Multi classification of tomato leaf diseases: A convolutional neural network model. *2022 10th International Conference on Reliability, Infocom Technologies and Optimization (Trends and Future Directions) (ICRITO),* Noida, India, pp. 1–5, doi:10.1109/ICRITO56286.2022.9964884.

[27] Sakshi, S., Lodhi, S., Kukreja, V. & Mahajan, M. (2022). DenseNet-based attention network to recognize handwritten mathematical expressions. *2022 10th International Conference on Reliability, Infocom Technologies and Optimization (Trends and Future Directions) (ICRITO),* Noida, India, pp. 1–5, doi:10.1109/ICRITO56286.2022.9964619.

21 A Comparison of the Effectiveness of Unreal and Unity Engines in Constructing Virtual Environments

Neeraj Singla

21.1 INTRODUCTION

Everyday newness is the basis of invention. The more interactive thing with fewer efforts was the biggest urge in game development in industries. That is when the open source started, which allowed developers all over the world to build progressive content, a two-way experience, and a hypnotic world. In no time it gained fame and became the most popular and widely used system. One such greatest invention of all times is creating a world that does not exist actually called the virtual world. A virtual world is a three-dimensional virtual space where people, things, and places can live. It can be merged into the real world in different ways, and it also reacts to the actions and interactions of users who experience it. It follows the various characteristics of three dimensions, it is occupied by an entity that serves the purpose determined by the creators of the world, it reacts to the user's presence and behaviour by transforming itself in some way. With such advancement, a lot of real-life approaches have been taken, which is the major reason for its boom. The game development industry is one such industry whose backbones lie in artificial reality. Games built using either Unreal Engine or Unity follow the concept of false reality. Figure 21.1 shows the comparison data on the use of Unreal Engine with other popular gaming engines. Both the engines have different approaches to building games Unity was simple to use and required little compilation time, however, the Unreal Engine editor allows for the creation of scripts in a visual Blueprint system and had more up-to-date capabilities for producing materials or generating terrain and vegetation [1–3]. The growth of usage of Unity 3D in years has been shown in Figure 21.2 and the comparison on significant parameters for the Unity 3D and Unreal Engines is compared in Table 21.1.

21.1.1 INPUT

An actor can bind several types of input events to delegated functions using an input component, which is a temporary component. Rendering of input components takes

DOI: 10.1201/9781003367161-21

217

Unity 3D Engine(Unity Technologies) — 62%
Internal proprietary Engine — 47%
Other — 23%
Unreal Engine(Epic Games) — 12%
Open Source — 10%
Cocos2d-x — 9%
CryEngine (Crytek) — 5%
Marmalade SDK (Marmalade) — 5%
Torque Engine (Garage Games) — 2%
None of the above — 2%
Phyre Engine (Sony) — 1%

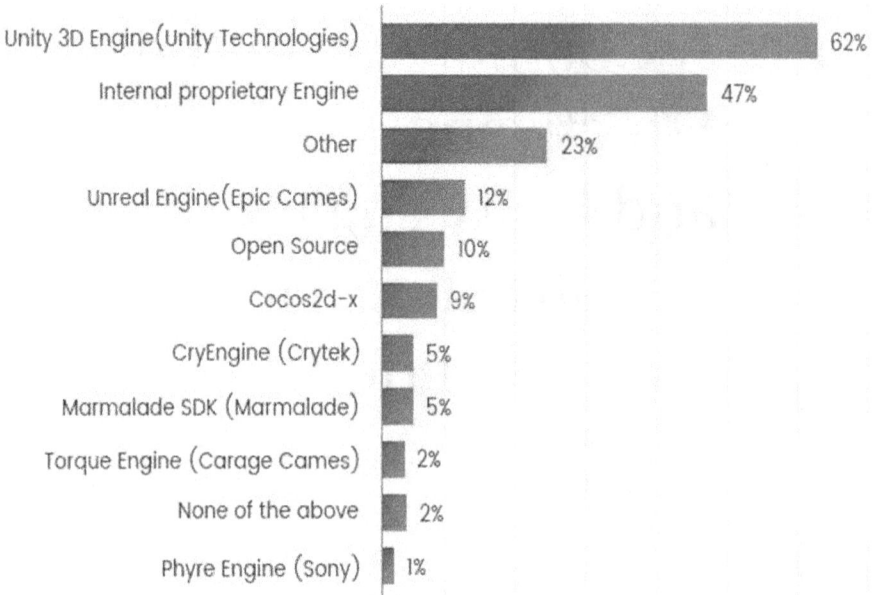

FIGURE 21.1 Percentage of people using Unreal Engine.

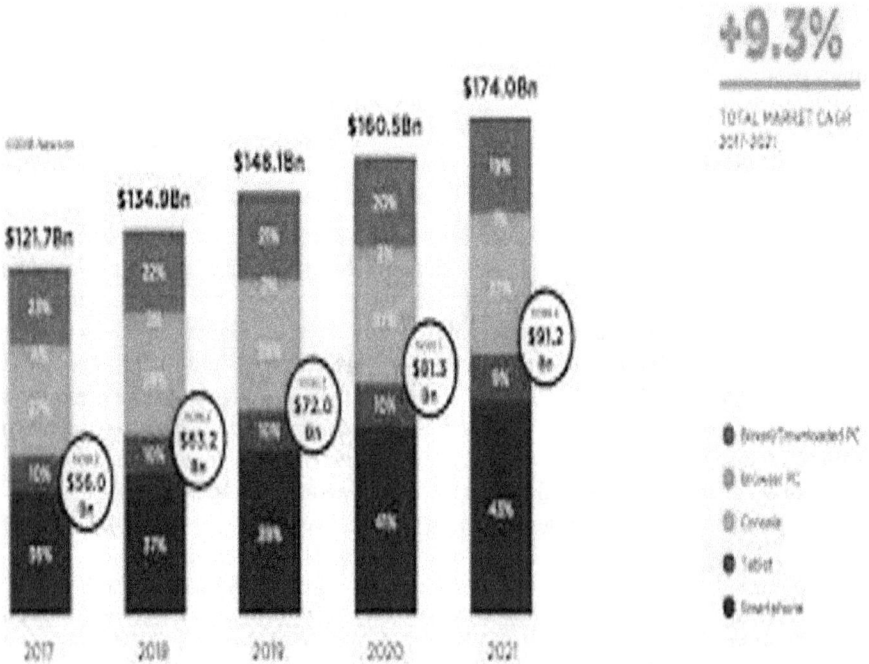

+9.3%

TOTAL MARKET CAGR 2017-2021

$121.7Bn $134.9Bn $148.1Bn $160.5Bn $174.0Bn

$56.0Bn $63.2Bn $72.0Bn $81.3Bn $91.2Bn

2017   2018   2019   2020   2021

● Direct/Downloaded PC
● Browser PC
● Console
● Tablet
● Smartphone

FIGURE 21.2 It demonstrates these sequential performance growths of Unity 3D over the years and how its utility continues to grow.

TABLE 21.1

Literature Review about Unity 3D and Unreal Engine

Sr. No.	Parameters	Unreal	Unity
1	Versions released	5 versions released so far: Unreal Engine1 in 1998 Unreal Engine2 in 2002 Unreal Engine3 in 2007 Unreal Engine4 in 2014 Unreal Engine5 in 2021 The languages used are Unreal Script and C++.	5 versions released: Unity2.0 in 2007 Unity3.0 in 2010 Unity4.0 in 2012 Unity5.0 in 2015 Unity6.0 in 2017 The language used is C# [5].
2	Graphics	Physically based rendering, global illumination volumetrics lights out of box, material editor	Physically based rendering, global illumination volumetrics lights after plugins installed [6]
3	Unique features	AI, network support	Rich 2D support
4	Target audience	AAA-game studios, indies, artists	Mostly indies, coders [7]
5	Coding	C++, Blueprints	C#, Prefab, Bolt
6	Community	About 100K members	More than 200K members [8]
7	Prototypes	Fast, no line of code	No line of code but purchase license [9]
8	Assets	1,000 assets in total	31K assets in total
9	Artificial reality	Unreal in preferred in 3D because of improved graphics quality	Unity is preferred in 2D because of larger focus and toolset [10]. It is best suited for augmented reality and virtual reality because the plugins are very durable and versatile [11].
10	Multiplayer	Unity is best suited for and can do multiplayer because it is the only on with integrated support.	Unity requires a third-party tool support and hence is not preferred [12].

place from a stack that player controller and player input maintain. By block in go their elements on the input stack from handling the input [13, 14].

21.1.2 GAMEPLAY FRAMEWORK

The Unreal Engine gameplay framework provides a powerful set of classes to create your game. The framework is very flexible and does some hard work and sets some standards. It has a pretty deep integration with the engine [15]. The important components included in a game engine are shown in Figure 21.3.

21.1.3 PHYSICS ENGINE

To beautify participant immersion, Unreal Engine makes use of PhysX through default to pressure its bodily simulation calculations and to carry out all collision calculations. The PhysX engine subsystem plays correct collision detection and simulates physical interactions among gadgets in the world [16, 17].

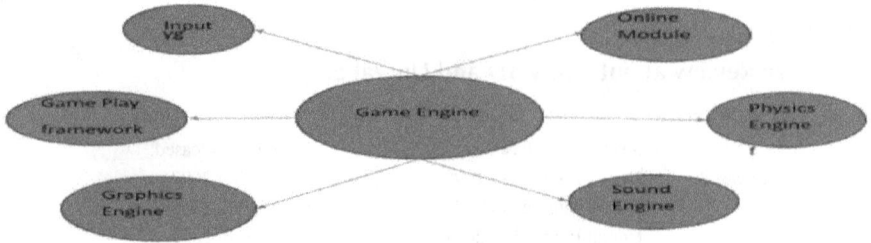

FIGURE 21.3   The components included in a game engine.

21.1.4 SOUND ENGINE

Sound layout is a vital part of every and every game and having exceptional sound consequences and track play a large position inside the game [18].

21.2 GAME COMPONENT USING UNITY

21.2.1 LIGHTS

They are an important aspect of any scenario. The atmosphere and colour of the surroundings are defined by light. In most cases, we deal with more than one light in a scene [19, 20].

21.2.2 GRAPHICS

It has incredible rendering capabilities, with a variety of characteristics like shadows [21].

21.2.3 ANIMATION

Retargetable animations, full control over animation weights at runtime, event calling from within the animation playback, advanced state machine hierarchies and transitions, blending shapes for face animations, and many more capabilities are available in Unity Animation [22].

Unity is a cross-platform engine for creating two- and three-dimensional video games for computers, consoles, and mobile devices. It allows for three-dimensional manipulation through the use of functions specified in programming languages like C# [23, 24].

The input mechanism of the Unreal Engine translates player keystrokes and button pushes into actions carried out by the character in the game. The gameplay framework, which includes capabilities to monitor game progress and manage game rules, can be used to configure this input method [26]. Additionally, it has a variety of organized tools and editors that let users maintain and update their properties in order to produce artwork for their games [27]. It comes pre-installed with Unreal Editor. The user can view and interact with various sub-platforms and editors via this

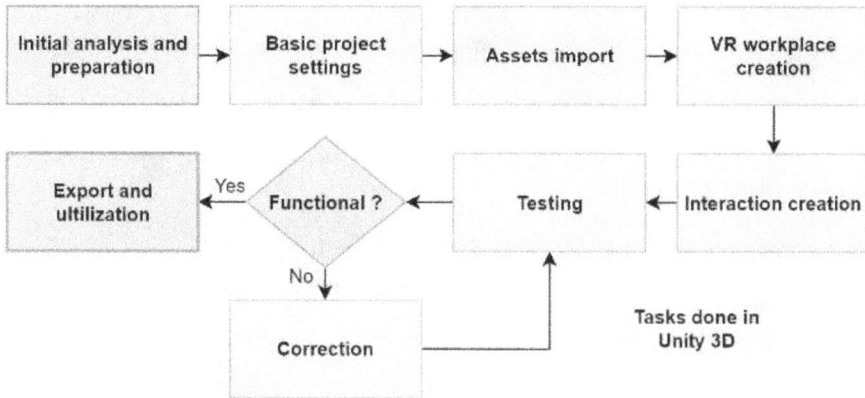

```
Initial analysis and  →  Basic project  →  Assets import  →  VR workplace
   preparation              settings                            creation
                                                                    ↓
  Export and        Yes
  ultilization    ←  Functional ?  ←  Testing  ←  Interaction creation
                         ↓ No                              ↑
                                                    Tasks done in
                     Correction  ────────────────   Unity 3D
```

FIGURE 21.4   Shows the tasks in the Unity 3D used while the development of a project methodology for Unreal Engine.

editor, which is the primary editor. In the flowchart, shown in Figure 21.4, initializing the engine, creating and initializing a game instance, loading a level, and then beginning play is the typical order of events [28].

21.3   CONCLUSION

A virtual world is a three-dimensional virtual space where people, things, and places can live. It can be merged into the real world in different ways and reacts to the actions and interactions of users who experience it. It follows the various characteristics of three dimensions, it is occupied by an entity that serves the purpose determined by the creators of the world, and it reacts to the user's presence and behaviour by transforming itself in some way. With such advancement, many real-life approaches have been taken, which is the major reason for its boom. The game development industry is one such industry whose backbones lie in artificial reality. Games built using either Unreal Engine or Unity follow the concept of false reality. Both the engines have different approaches to building games. Unity is simple to use and requires little compilation time, but the Unreal Engine editor allows for the creation of scripts in a visual Blueprint system and has more up-to-date capabilities for producing materials or generating terrain and vegetation [29–32].

REFERENCES

[1] Antinucci, F. (2007). The virtual museum. *Archeologiae Calcolatori*, 1, 7986. Retrieved July 2, 2021, from www.archcalc.cnr.it/indice/Suppl_1/6_Antinucci.pdf.
[2] Kęsik, J., Montusiewicz, J., & Kayumov, R. (2017). An approach to computer-aided reconstruction of museum exhibits. *Advances in Science and Technology Research Journal*, 11, 8794. Retrieved July 2, 2021, from http://yadda.icm.edu.pl/yadda/element/bwmeta1.element. Baztech-73 ba2929-b9ff-4741-978bbf780441dd3f/c/kesik.pdf.

[3] Šmíd, A. (2017). *Comparison of Unity and Unreal Engine*, Bachelor Thesis, Czech Technical University, Prague. Retrieved July 2, 2021, from https://dcgi.fel.cvut.cz/theses/2017/smidanto.

[4] Siarkowski, K., Sprawka, P., & Plechawska-Wójcik, M. (2017). Metody Optymalizacji Wydajności Silnika Unity 3D woparciu o grę z widokiemperspektywytrzeciejosoby. *Journal of Computer Sciences Institute*, 3, 46–53. Retrieved July 2, 2021, from https://doi.org/10.35784/jcsi.592.

[5] Puławski, E., & Tokarski, M. (2019). Wykorzystanie post processing uijegow pływunawy dajnośćrenderowania w silniku Unreal Engine 4. *Journal of Computer Sciences Institute*, 10, 54–61. Retrieved July 2, 2021, from https://doi.org/10.35784/jcsi.206.

[6] Haas, J. (2014). *A History of the Unity Game Engine, An Interactive Qualifying Project*, Worcester Polytechnic Institute, Worcester. Retrieved July 2, 2021, from https://digital.wpi.edu/pdfviewer/2f75r821k.

[7] Sankhyan, A., & Pawar, P. K. (2013). Metformin loaded non-ionic surfactant vesicles: Optimization of formulation, effect of process variables and characterization. *DARU Journal of Pharmaceutical Sciences*, 21(1), 1–8.

[8] Christopoulou, E., & Xinogalos, S. (2017). Overview and comparative analysis of game engines for desktop andmobile devices. *International Journal of Serious Games*, 4(4), 21–36.

[9] Pattrasitidecha, A. (2014). Comparison and evaluation of 3D mobile game engines. *Master of Science Thesis in the Program International Design*.

[10] Petridis, P., Dunwell, I., Panzoli, D., Arnab, S., Protopsaltis, A., Hendrix, M., & Freitas, S. (2012). Game enginesse lection frame work for high fidelity serious applications. *International Journal of Interactive Worlds*, 2012, 19, 45–50.

[11] How do game engines work? (2016). *Interesting Engineering Website*. Retrieved 21 November, from https://interestingengineering.com/how-game-engines-work.

[12] Concepts and feature of game engines. (2017). *UK Essays Website*. Retrieved 23 November, from www.ukessays.com/essays/computer-science/concepts-features-game-engines-6438.php.

[13] 5 years of unity vs. unreal: Who is leading the war of giants. (2019). *Programace Website*. Retrieved 25 November, from https://program-ace.com/blog/5-years-of-unity-vs-unreal/.

[14] Kumar, P., Sharma, V., Jaggi, C., Malik, P., & Raina, K. K. (2017). Orientational control of liquid crystal molecules via carbon nanotubes and dichroic dye in polymer dispersed liquid crystal. *Liquid Crystals*, 44(5), 843–853.

[15] Petty, J. (n.d.). What is unity 3D & what is it used for? *Concept at Empire Website*. Retrieved 25 November, from https://conceptartempire.com/what-is-unity/

[16] Unity Guruz. (2019, April 10). *Unity vs Unreal: Graphics Comparison 2019 (HDRPDEMO)*. Retrieved from www.youtube.com/watch?v=sMIi_Z33pkw

[17] Autodesk Knowledge Network. (n.d.). *Procedural Textures*. Retrieved from https://knowledge.autodesk.com/support/maya/learexplore/caas/CloudHelp/cloudhelp/2016/ENU/[19]Maya/files/GUID-B2C969C0-48CD-45AB-8C7B-E6FC9E34AD19-htm; Html, A.W.S. (n.d.). *Working with Gloss Maps*.

[18] Blevins, N. (2014). *Anisotropic Reflections in the Real World*. Retrieved from www.neilblevins.com/cg_education/aniso_ref_real_world/aniso_ref_real_world.htm.

[19] Unity Manual. (2018). *Light Mapping: Getting Started*. Retrieved from https://docs.unity3d.com/Manual/Lightmapping.html.

[20] Williams, G. (2017). *What is the Difference between Shadow Volume and Shadow Mapping in Computer Graphics*? Retrieved from www.quora.com/What-is-the-difference-betweenShadow-Volume-and-Shadow-Mapping-in-Computer-Graphics.

[21] Kersten, T. P., Buyiiksalih, G., Tschirschwitz, F., Kan, T., Deggim, S., Kaya, Y., et al. (2017). The selmiye mosque of edirne Turkey-an immersive and interactive virtual reality experience using HTC vive. *ISPRS-Internationa Archives of the Photo Grammetry Remote Sensing and Spatial Information Sciences*, XLII-5/W1, 403–409.

[22] Wei-chao, Q., & Yuille, A. (2016). Unreal CV: Connecting computer vision to unreal engine. *Lecture Notes in Computer Science*, 16, 109–116.

[23] Junxiong, H. (2016). The application and development of virtual reality technology in game entertainment. *Heilongjiang Science and Technology Information*, 29, 23–24.

[24] Qi-ming, R. (2012). *Small Team Lots Process Next-Gen Game Engine Virtual Reality Application Research Project*. Guangzhou: South China University of Technology.

[25] Desai, P. R., Desai, P. N., Ajmera, K. D., & Mehta, K. (2014). A review paper on oculus rift—a virtual reality headset. *International Journal of Engineering Trends and Technology (IJETT)*, 13, 175–179.

[26] Ting, D., Yan, S., & Cheng-hu, C. (2016). The development of virtual reality based on UE4. *China Science and Technology Review*, 5, 230–230.

[27] Biffi, E., Beretta, E., Cesareo, A., Maghini, C., Turconi, A. C., Reni, G., et al. (2016). An immersive virtual reality plat form to enhance walking ability of children with acquired brain injuries. *Methods of Information Inmedicine*, 23, 119–126.

[28] Bock, M., & Schreiber, A. (2018). Visualization of neural networks in virtual reality using Unreal Engine. In *Proceedings of the 24th ACM Symposium on Virtual Reality Software and Technology*, VRST '18, 132:1–132:2. New York, NY: ACM. https://doi.org/10.1145/3281505.3281605.

[29] Bondi, E., & Kapoor, A., Dey, D. D., Piavis, J., Shah, S., Hannaford, R., Iyer, A., Joppa, L., & Tambe, M. (2018). Near real-time detection of poachers from drones in Air Sim. In *Proceedings of the Twenty-Seventh International Joint Conference on Artificial Intelligence*, IJCAI-18, 5814–5816. International Joint Conferences on Artificial Intelligence Organization, 7, pp. 5814–5816. https://doi.org/10.24963/ijcai.2018/847.

[30] Cover, A., Posser, R. D., Campos, J. P., & Rieder, R. (2018). Methodology of communication between a criminal data base and a virtual reality environment for forensic study. In *2017 19th Symposium on Virtual and Augmented Reality (SVR)*, pp. 215–222. https://doi.org/10.1109/SVR.2017.35.

[31] Sharma, A., Kukreja, V., Bansal, A., & Mahajan, M. (2022). Multi classification of tomato leaf diseases: A convolutional neural network model. In *2022 10th International Conference on Reliability, Infocom Technologies and Optimization (Trends and Future Directions) (ICRITO)*, Noida, India, pp. 1–5. https://doi.org/10.1109/ICRITO56286.2022.9964884.

[32] Sakshi, S., Lodhi, S., Kukreja, V., & Mahajan, M. (2022). Dense net-based attention network to recognize handwritten mathematical expressions. In *2022 10th International Conference on Reliability, Infocom Technologies and Optimization (Trends and Future Directions) (ICRITO)*, Noida, India, 2022, pp. 1–5. https://doi.org/10.1109/ICRITO56286.2022.9964619.

22 Implementation of Augmented-Reality-Based English and Vocabulary Learning

Bhavika Singla, Navnoor Kaur, Bhawna Chetan, Shubham Gargrish

22.1 INTRODUCTION

In the era of 21ˢᵗ century, we all are well aware about the benefits of technology like computersand mobile phones. COVID-19-era technology has helped us in best possible way, and it got enhanced in lot of fields like medicine, education, artificial intelligence, and communication. Further after the enhancement and development of technology there are still some problems arising as technology is very vast, and hence has so many fields associated with it. Now through this chapter we are going to discuss that how education can be associated with technology [1]. There is lot of advancement in technology, but students are still facing the problems as Indian education system is still adhere to traditional methods of teaching.

Due to that, students are unable to understand the topic in best possible way—that is, they have to cram it, when instead they can understand a topic in better way, which can help them to retain it for longer period of time. While in traditional method of teaching there are books of a particular subject that a student is entrusted to study a topic, most of time they are cramming the topic, but in this way, they cannot understand the concepts which are difficult to visualize topics, like 3D geometry, trigonometry, laws of physics, and so on [2].

But we give them an alternative approach which is a new technology that invokes the overlay of computer graphics in the real world.

With the help of this, the concepts would be easier to understand and fun to learn, which will make students understand them completely without losing interest in them. It will also help them to visualize things that they are not able to do by simply learning through the traditional methods. Imagine if learning and technology are mixed to create a learning system that is more attractive, engaging, and motivating.

Here comes the concept of an AR. It uses technology to add digital information to an image of something being viewed through a gadget to produce an augmented representation of reality. Now we will get to know about the three characteristics of this technology to understand it in a better sense:

DOI: 10.1201/9781003367161-22

- Superimposing of the digital and the real world
- Real-time interaction
- Alignment and registration in 3D

22.2 LITERATURE

Research Works	Learning	Methodology	Domain and AR Devices Used	Conclusion
"AR in Healthcare," by Ms. Khushboo Sethiya and Prof. M. Guruprasad	To bring new therapies to life and examine how new drugs and medical devices interact with human body	The study is qualitative as the aim is to understand the impact of AR technology, which is still used by a handful of people having adequate awareness. Twelve publications were studied, where experimentation method was used as a mode of analysis [3].	AR on mobile device is the way to go.	After studying 11 journals and numerous books, it can be concluded that augmented reality has a transformational effect on healthcare. This era has diverse benefits and holds the ability to make healthcare greater powerful and green.
"Using Augmented Reality for Entertainment," by Blair MacIntyre and Brendan Hannigan	We focus on AR techniques that use head-worn displays, and for localized 3D sound headphones are utilized.	There were two dramatic AR experiences: they immerse users in the scene, and people were allowed to be in a world with independent, physically located characters, which brings them into a story [4].	We focus on AR techniques that use head-worn displays, and for localized 3D sound headphones are utilized.	From this we were able to gather knowledge that AR could refine a person's experience while experiencing the story, and we could see the potential of AR in entertainment.
"Augmented Reality in Retail," by R. Chen, P. Perry, R. Boardman, and H. McCormick (2021)	This report synthesizes peer-reviewed journal publications on augmented reality in retail contexts to determine the current research priorities in education in this emerging field and to create a conceptual framework for future research agendas.	A thorough search of high-quality peer-reviewed studies over the years (1997 to 2020) yielded a sample of 76 papers for schematic analysis [5].	Implemented via outlets looking to give customers an enhanced buying experience and make AR products affordable.	It was the first comprehensive review of the literature on AR in retail contexts, encompassing more than one disciplinary perspective (HCI and advertising and marketing/ control).

(Continued)

Research Works	Learning	Methodology	Domain and AR Devices Used	Conclusion
"Virtual and Augmented Reality on the 5G Highway," by Jason Orlosky, Kiyoshi Kiyokawa, and Haruo Takemura	The 5G network will create a ton of brand-new possibilities for telepresence, streaming media, healthcare, and education. Soon we will be able to transfer 3D model or video stream data that will enable users to be entirely transported to another planet, without cables or tethering, even though there will still be issues with latency and bandwidth in more remote or populated areas.	As a result of improvements in elevated mobile networks and a growth in the number of smart devices, it is predicted that the global market for digital content would climb from $477 billion in 2018 to $1,175 billion in 2025. Future digital content markets are predicted to be driven by virtual and augmented realities, with the market for VR/AR content alone expected to rise from $1934 million in 2018 to $3,264 million in 2025 [6].	Applications in the near and distant future will be significantly impacted by 5G infrastructure. Examples include entertainment, assistance with industry, and interactive gaming, among others.	Virtual workplaces will probably become more popular for many purposes, including medical, remote machine operation, education, and general conferencing. Other fields, such as mobile gaming, vision improvement, and collaborative art, make promises of providing rich settings for the creation of new content and improved performance via fascinating new human-machine interfaces.
"Augmented Reality to Help Public Safety Workers," by K. Joy Santhra and Dr M. N. Vijayalakshmi	Public safety personnel can benefit from AR by being trained to perform complex tasks like mid-air refuelling and flying formations, which would be very challenging to perform in the field without any prior training. AR may also assist these safety officers by offering effective maintenance for the vehicles, airplanes, tools, and other equipment they	This project works on marker-based augmented reality method. A static image that a user can scan with their mobile device using an augmented reality app is necessary for marker-based augmented reality. The additional content (video, animation, and 3D objects) that was previously produced will then start to surface on top of the marker as a result of the mobile scan.	The Unity game engine will be used to create the proposed AR application.	In this manuscript, we looked at how well policing can employ augmented reality technologies. Its various applications support public safety personnel by offering simulated training, equipment maintenance, geo-enabled augmented reality applications, and so on.

Research Works	Learning	Methodology	Domain and AR Devices Used	Conclusion
	utilize. First responders can utilize this technology to show them in real time who might require their assistance by just donning AR glasses when they arrive at a fire or earthquake scene and are trying to determine who needs help and the best route to transport residents of that region to safety.	The 3D area of the battlefield will be the triggering 3D model for this project [17]. The database will thereafter be downloaded online from Vufor ia portal and used in the creation of an augmented reality application. You can create a UI for creating and joining the room different scene in Unity Editor.		Additionally, it suggested an augmented reality application that would enable military forces to view a 3D map of their battleground before the actual fight starts.
"A Review of Augmented Reality Research for Design Practice: Looking to the Future," by Lorenzo Giunta, Elies Dekoninck, James Gopsill, Jamie O'Hare	This study seeks to (1) link recent research on augmented reality (AR) technology to design practice, (2) pinpoint areas where significant technological advancement has taken place, and (3) specify topics that might profit from more study.	Google Scholar searches on augmented reality and augmented reality for design turned in relevant papers for the review. The matrix was then filled with these documents. "Design," "mixed reality," "augmented reality," "assembly," and "creativity" were among the search terms used. The top page of results was the only ones taken into account. After gathering the references, they were categorized according to the technology employed and the stage of the design process.	Van Krevelen and Poelman (2010) provide a clear categorization of AR technologies, dividing the field into three major categories: handheld, head-mounted displays, and spatial augmented reality.	Table 21.2 includes the papers that have been prepared for the mapping effort after the review identified 21 studies. The summaries provide an overview of the research that was done for each resource. This gives a fundamental grasp of the rationale for the classification of each resource, taking into account both the type of technology utilized and the stage of the design process.

1. Game Design
   According to the research we have done some games that are helping students to learn their curriculum subjects are related to science and math, but there are less or no such serious education games (SEG) developed for literature. So, we need to keep in mind some factors/things while designing that are as follows:
   a. There must be interactivity, enjoyment, effectiveness, usability
   b. game should attract the students in such a way that they should try it repeatedly.
   c. initially there should not be any complex problem so that the student might not feel frustrated, irritated or upset that he/she is not able to engage himself/herself.
   d. The game should be level based so that the students come to know about his performance on daily basis and student will learn that their success is determined by every decision they make.
   e. The development process should be interactive process, with repeated phases of testing and revision.
2. Game Description

As augmented reality (AR) is a technology that make human work easy by a technology that make human work easy by analysing the situation at the situation time or simultaneously at the same time [7]. As we know that students are facing problems in literature and hence in speaking also. For those for whom English is not their first language, learning the language is crucial. They need to learn, understand, and speak English well in order to function in society. English is a common language to use while speaking and exchanging ideas with people from different geographical areas who speak other languages. For the sake of surviving in foreign nations, students who wish to travel abroad for academic purposes must be able to understand and speak English. The following are issues that pupils have speaking English:

a. Common grammatical errors made when speaking English:
   Grammar instruction in English is particularly challenging for students. Students frequently make grammatical, active/passive, and vocabulary errors when speaking English. They err in their tenses. They speak in the present tense rather than the past tense when they want to speak about the past. They struggle to understand the distinctions between the proper usage of the past, present, and future tenses.
b. Lack of confidence in their ability to speak the language:
   They lack self-assurance when speaking in public in the English language. The majority of the time, their teachers discourage students from using English in class or in front of others.
c. English language shyness:
   This is another major reason pupils struggle to communicate in the language. They feel uneasy and anxious around people who speak English because they are bashful.

But what if there is a game which could help the students learn better and fluent English. As students generally get confused between tenses and grammar, it would help if there is a platform which presents all tenses that a student must use a sentence. If the student chooses the wrong option, then the instructor must translate even that wrong sentence to the student in his/her native language so that the student could know that a single word can change the whole meaning, and he/she could put the appropriate word in the sentence. Hence, there should be different levels, like beginner, intermediate, and advanced. In the beginner level, the student is asked to put only tenses in between the blanks. In the intermediate level, there is a combination of two and three lines, and in the advanced level, students are asked to write a paragraph in English while the game instructor is speaking in the native language of student and he/she has to translate it. After he/she has written, the instructor must verify it and award the student with small vouchers or some gift in order to maintain the interest of students in the game and attract more and more students also. The difficulty level of the game should change accordingly as the student completes his/her tasks. And also, we could train the instructor in certain topics for debate so that users can argue.

Our prototype also incorporates a parental monitoring mechanism as shown in Figure 22.1.

Parents can set a time restriction so that the application can automatically be stopped or stopped online, and follow their child's development. Parents can use such monitoring tools to limit their child's application usage and assuage their concerns about their child's health. We polled 20 parents and 10 non-parents, and they all agreed that the application is effective, as shown in Table 22.1.

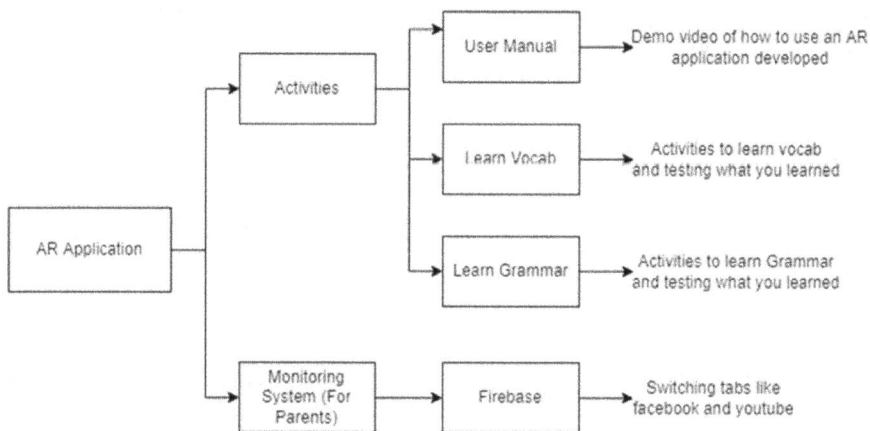

FIGURE 22.1 Methodology of the AR application.

TABLE 22.1
Parents' Survey Sheet

Items	1	2	3	4	5
1. The UI of the AR application is easy to understand for students.	0%	15%	40%	37%	8%
2. Students can learn English (language) grammar successfully through AR app.	0%	10	37	45	8%
3. Monitoring feature can decrease my concern of my children using the mobile.	0%	18	40	35	7%
4. I will prefer this mobile AR application for my students' learning.	0%	7	37	46	10%

Hence, this could benefit a lot of people in learning the language and enhance their skills and thus they would be able to speak fluent English.

22.3 CONCLUSION

By employing a monitoring mechanism, our AR mobile application prototype may efficiently teach kindergarten pupils English vocabulary in an interactive and appealing manner, while also assuaging parents' concerns that long-term use of electronic gadgets may harm their child's health. We believe that sharing our experience developing other professionals can learn how augmented reality can be used to enhance learning effectiveness in early childhood education from AR mobile applications with educational games. They can also learn how to manage ICT usage time to lessen negative effects on children's health while still gaining educational benefits.

In light of the feedback from the parents and the foregoing limitations, future work should focus on adding learning resources for more difficult English language vocabulary, providing child facts and figures for the tracking system, improving the UI to make it more intuitive for kindergarten students, and optimizing the application code for better AR user experience.

REFERENCES

1. Cascales, A., Pérez-López, D., & Contero, M. (2013). Study on parent's acceptance of the augmented reality use for preschool education. *Procedia Computer Science, 25*, 420–427.
2. Christou, C. (2010). Virtual reality in education. *Affective, Interactive and Cognitive Methods for E-learning Design*, 228–243.
3. American Academy of Pediatrics (2016). *Kids & tech: Tips for parents in the digital age* [Online]. Available: www.healthychildren.org/English/family-life/Media/Pages/Tips-for-Parents-Digital-Age.aspx [Accessed Feb 4, 2017].
4. Billinghurst, M., Clark, A., & Lee, G. (2015). A survey of augmented reality. *Foundations and Trends in Human-Computer Interaction, 8*(2–3), 73–272.
5. Gargrish, S., Mantri, A., & Kaur, D. P. (2020). Augmented reality-based learning environment to enhance teaching-learning experience in geometry education. *Procedia Computer Science, 172*, 1039–1046.

6. Gargrish, S., Kaur, D. P., Mantri, A., Singh, G., & Sharma, B. (2021). Measuring effectiveness of augmented reality-based geometry learning assistant on memory retention abilities of the students in 3D geometry. *Computer Applications in Engineering Education*, *29*(6), 1811–1824.
7. Gargrish, S., Mantri, A., & Singh, G. (2020, February). Measuring students' motivation towards virtual reality game-like learning environments. In *2020 Indo–Taiwan 2nd International Conference on Computing, Analytics and Networks (Indo-Taiwan ICAN)* (pp. 164–169). IEEE.

23 Study and Analysis of Low-Cost Wall Finishes for Affordable Housing

Harleen Kaur, Asmita Sharma, Atul Dutta

23.1 INTRODUCTION

Having an affordable house for all individuals is an emerging task being faced by developing nations like India (Janani et al., 2018). To make a house cost-effective, it is not only construction materials that contribute but materials used in interiors too add a delight to one's space. The house is a place which mirrors an individual's interest, love for art, sense towards life's beauty, and selection of its inhabitants (Patel & Sayed, 2018). In modern India, as real estate has become highly costly, it has become a challenge to afford a house that is not only economically viable but also aesthetically beautiful (Gangani et al., 2016; Patel & Sayed, 2018). A residence can be made cost-effective without sacrificing its strength and construction quality, with the use of locally available interior materials that are durable and tough in the context of their maintenance (Tam, 2011). Affordability has different meanings for people based on different income levels (Khan, 2012). Although largely, affordable housing is meant for people belonging to the median income levels, who also have a dream to have their own house that may be a small but attractive one (Shinde & Karankal, 2013).

As budget is of main concern, a person can choose design elements that are low in cost but can give amazing results making a house look perfect. Today, there are many interior design solutions available at a low cost, specially designed to meet the needs of people who want low-cost interiors. Designing a house within a given budget is an exigent task for a designer (Patel & Sayed, 2018). A careful choice of interior materials can reduce the price, while wall treatments come in a variety of styles and are an important aspect to consider when constructing a home. Interior wall finishes include wallpapers, wall tiles, paints, and many kinds of wall panellings. Before finalizing any finish, one should look at its impact, longevity, maintenance, and aesthetic appearance. In today's market, there are many such materials available that are low in cost yet may work as a decorative feature, making it economically beneficial to its user. The intent of this chapter is to suggest different types of wall finishes based on their cost and durability. A survey of 50 homeowners has been done in tri-city (Chandigarh, Panchkula, and Mohali) based on a questionnaire to study the desire of common middle-class people to decorate their home.

 DOI: 10.1201/9781003367161-23

23.2 RESEARCH METHODOLOGY

The exploratory study is based on a methodology of study and literature analysis in relation to several accessible wall-finishing products. Market research was carried out to determine the cost of the available materials. A well-structured questionnaire was administered and analysed further among 50 homeowners. The questionnaire included survey questions on several key elements in order to collect information for the assessment of wall-finishing materials that are regularly used, economically effective, and easily available for wall finishing.

23.3 LITERATURE STUDY

Residential interiors are one of the most considered sections, when one thinks of having a beautiful house. Wall finishes is the factor that makes a place look delightful and lively. Frequently used wall-covering materials which are used to decorate, enhance, and groom a space are explored in the research study.

23.3.1 Wall Paint

Paint is a thin layer applied as a coating on a wall to protect the surface and decorate it (Binggeli, 2013). Paint is a mixture of four ingredients—pigments, binders, solvents (fluids), and added substances. Pigments are used to give different shade to the paint and help to create coverup, while binders are used to make the paint film by binding the pigments together. Solvents are the fluids that suspend the fixings and permit the paint to stay on the surfaces, and added substances are fixings that give explicit paint properties, such as buildup opposition (Bosveld, 2013). Good quality paint provides protection against moisture, preventing mould and mildew growth and even more extensive damage that can be caused by moisture. A paint has a property that helps in clean home while repelling dust, pollutants, and allergens (Abrams, 2019). It is a smart and cost-effective option for those who want a reasonable solution to make a home look beautiful, warm, and energetic (Earthry Facilities, 2019). Table 23.1 lists the types and corresponding cost of paint as per the market survey.

Paint can also be used to create decorative wall art. Wall painting is able to achieve the harmonious coherence of the wall's purpose and its visual appeal due to the coordination of the wall and the art. Through its composition, structure, colour, and other qualities, the wall painting integrates and unifies with the current surroundings and area (Yuan, 2019). Best paints suited for interiors are follows:

> *Oil-bound distemper*—This paint is a water-based mixture available in a variety of shades, and its application is also easy. This paint is a good option to decorate various inner structures of a house, including walls and ceilings, and keeps up phenomenal porosity (The Traditional Paint Company, n.d.).
> *Emulsion paint*—It is a water-based paint which consists of minute particles of pigments filled in polymer. This paint is difficult to get back off the wall once dried, as these particles fuse together to create a film of paint (Lorch, n.d.).

Plastic emulsion—This paint has similar properties as emulsion paint, but the one exceptional quality is that it is a washable paint. This paint gives a smooth matte look to walls (ColourDrive, 2019).

Easy wash (acrylic emulsion)—As the name suggests, it is an easy washable paint, which gives a rich finished smooth matte finish (Surfa Coats India Pvt. Ltd., n.d.).

23.3.2 WALLPAPERS

Wallpaper is a product available in many qualities. The selection of wallpaper depends upon the entire look of the room or space for which it has to be used. These backdrops can add unique character to a room as they are available in various designs, ranging from nature-inspired themes to contemporary geometric patterns. It also gives an option to have a customized print per personal desire to make a space more personalized. Wallpaper is one of the budget-friendly wall finishes (Ferrao, 2019). The wallpaper costing details as per the market survey is given in Table 23.2.

Installation of wallpaper requires a flat smooth surface, free from cracks or any other defect. Wallpaper can also be hung on the wall that already has wallpaper over it, but in a good condition. Otherwise, existing wallpaper can be removed using hot water and scrubber. Water takes about 15–30 minutes to relieve the glue from old pasted paper. The different varieties of available wallpapers generally available are as follows (Binggeli, 2013):

Paper—On a paper base, a decorative design is printed; this sheet is without a top coating, which is used to keep design intact. This quality of wallpaper is comparatively cheap.

Paper-backed vinyl—This chapter has a coat of liquid vinyl over which design is printed. This is also a cheaper quality of wallpaper, which can be wiped and damp proof.

Non-woven—This one is a little expensive and is made of natural or synthetic fibres which give a sophisticated appearance to the final design. It is easy to install or remove.

TABLE 23.1
Paint Types and Costing Details (from Market Survey)

Sr. No.	Materials	Capacity	Approx. Price Range (per Bucket)
1	Oil-bound distemper	20 kg (bucket)	Rs 650–700
2	Emulsion		Rs 1,800–2,000
3	Plastic emulsion		Rs 4,000–4,500
4	Easy wash (acrylic emulsion)		Rs 7,500–8,000
5	Exterior apex		Rs 4,000–7,000

TABLE 23.2
Wallpaper Costing Details (from Market Survey)

Material	Sizes	Price Range
Wallpaper	57-foot roll	Rs 500–3,000 per roll

Solid vinyl—It is a more durable and washable wallpaper made by pasting a vinyl sheet over paper or fabric, and design is then printed on that sheet.

PVC or coated vinyl—It is basically a layer of acrylic sprayed on a paper base, making it moisture resistant, and this is a reasonable option.

Fabric—It may have a layer of vinyl with design on it with a fabric-backed base, or it will be a normal design printed on a fabric base.

Wallpaper is one of the most-sought-after interior design elements since it is relatively easy to remove or replace and can be altered easily per the changing taste and trend. They come in a variety of styles, such as floral prints, abstract designs, embroideries, strips, and geometric patterns, and some can even be used to create borders that run all around the four sides of a room. Wallpapers are not just used on walls, but they may also be used on ceilings to cover flaws or as an artistic feature (Hoagland, 2015).

23.3.3 PVC PANELS

A very popular product which is being used these days is panels made of polyvinyl chloride (PVC), which is a very friendly material in terms of cost and durability. It has taken over various traditional materials used in building interiors (Wall Affairs, n.d.). Application of PVC panels is advised because of its hardness and mechanical properties, which make it a flame-resistant product (*Polyvinyl Chloride*, n.d.). As in comparison with other interior wall coverings, installation of these panels is an easy job in terms of sound disturbance and fixation time. Moreover, it does not create much dirt during installation. It is a stain-proof, waterproof, and moisture-proof material which is quite safe and recyclable and involves no pain in cleaning afterwards. These panels are one of the best suitable materials to be used in wet or damp areas of the house, like bathrooms and kitchens (Wall Affairs, n.d.). Available in a variety of shades and textures, as well as other vibrant hues, they are suitable for any interior space, whether it is a bedroom, drawing room, or any other room in the house. It is a one-of-a-kind product that is both lightweight and termite-resistant (Sri Kamakshi Enterprises, n.d.). PVC panels are certainly available in different thickness and quality which might affect their cost a little. Overall, this is a good choice and is one of the best products suited for interiors.

TABLE 23.3
PVC Panels Costing Details (from Market Survey)

Sr. No.	Material	Sizes	Price Range
1	Polyvinyl chloride panels	10″ × 10′	Rs 200–350 per panel Rs 40–100 per square foot

These panels can not only be used on walls but are also a great option for ceilings when it comes to interiors (Table 23.3). This product can even resemble materials like stone, wood, or cork. This material is usually used for decorating interiors since it is easy to install and has features such as protection, insulation, and sound absorption (Sadıklar & Tavşan, 2016).

23.3.4 CEMENTED BOARDS

Cement boards are in market since long time. These boards are ductile in nature, and the main benefits of cement boards are that they have a high level of fire, fungus, and decay resistance. These boards are made by pouring the slurry directly onto the paper covering, which adheres without the use of glue because of cement hydration (Tittelein et al., 2012). Cement boards are characterized by a very high pH level, making it resistant to moisture, which ensures protection against mould growth, making it suitable to be used in interiors, as well as an exterior element, such as finishing attics, windows, and balconies (Bugno & Krampikowska, 2020). Nowadays, designers are using this material as it allows a good amount of thermal insulation, which in turn lowers the buildings inside heat and helps in maintaining good quality of acoustic levels (Bugno & Krampikowska, 2020; Tittelein et al., 2012). The sturdy design of these boards allows them to preserve their original shape and exhibit extremely high mechanical strength even in changing weather conditions (Bugno & Krampikowska, 2020). A typical cemented board composition details are shown in Figure 23.1, and a survey based cost estimation for the cemented board is listed in Table 23.4.

23.3.5 WALL TILES

Wall tiles have long been utilized as a decorative feature in interiors, such as *azulejo*, a type of ceramic tile employed in interiors during the 15th century to increase the aesthetic aspects of Portuguese churches, monasteries, and palaces, as these tiles were linked with higher social groupings. Tiles, on the other hand, have become an affordable interior feature for houses and other public areas over time (Pereira et al., 2009). Ceramic tile is usually made from clay that is quarried out, prepared as required and then developed into a shape using a mould. Mostly ceramic products are made of single type of clay but at times may be combined with mineral modifiers like feldspar and quartz (Awoyera et al., 2017).

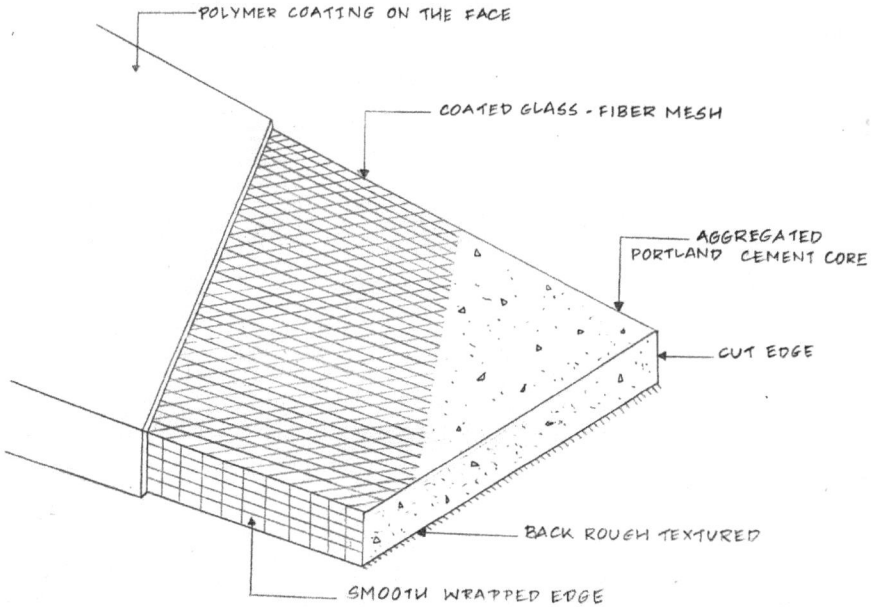

FIGURE 23.1 Cemented board: composition details (Spence, 1998).

TABLE 23.4
Cemented Board Costing Details (from Market Survey)

Sr. No.	Material	Sizes	Price Range
1	Cement board	6 mm, 8 mm, 10 mm, 12 mm, 20 mm, 25 mm	Rs 3 per mm per square foot

Manufacturing of tiles is done through various assorted materials like clay, metal, stone, quartz, and terrazzo. The best-suited material as wall finish is ceramic tiles; usually, they do not lose their colour. These tiles can be easily cleaned, are stain proof and a very durable product. These are available in glazed and unglazed finish. The tiles with glossy or crystalline finish are glazed tiles available in many colours and are made of ceramic materials. When it comes to interiors, floor tiles are a kind of solution that may increase the initial estimate of your home for once but keeping it aesthetically appealing for long life once installed (*11 Different Types of Tile Flooring*, 2015).

Ceramic tiles are used on walls as they have strong adhesive property and never fall off in the condition of dampness on walls. There are variety of colours, textures, materials, and patterns available in tiles with a mix of style and durability. Tiles helps in achieving desired ambience for a house, setting a mood of the space where is it is being used although careful choice is required for desired outputs (The Tile Shop, n.d.). A particular ceramic tiles costing estimation is given in Table 23.5.

TABLE 23.5

Ceramic Tiles Costing Details (from Market Survey)

Sr. No.	Variety of Ceramic Tiles	Sizes	Approx. Price Range
1	Glazed Polished	10″ × 15″	Rs 20–60 per square foot
2	Digital Polished	12″ × 18″	
3	Wooden Ceramic	12″ × 24″	
4	Matt Polished	12″ × 36″, 16″ × 32″	

TABLE 23.6

Wooden Panelling Costing Details (from Market Survey)

Sr. No.	Material	Sizes	Price Range
1	Teak ply	8′ × 4′ (4 mm, 6 mm)	40–45 per sq. ft.
2	Teak wood	5″ × 7′ (19 mm)	1800 per sq. ft.
3	Pine wood	9″ × 16′ (38 mm)	1500 per sq. ft.

23.3.6 WOODEN PANELLING

Wooden panels are installed on interior walls to make them look aesthetically appealing and are good for controlling acoustic levels and improves indoor climate making a space comfortable for living. These panels are available in different types of wood and can be painted or polished. But as shortcoming, it needs to be treated with a special compound to present it from the capricious action of the surroundings (Mekhriniso Murodovna & Gulshan Sergeevna, 2022). Table 23.6 lists and approximate wooden panelling cost data for reference.

Sheets of wood panel products are made of layers (plies) or particles of wood and adhesives. They provide flat surfaces wider than those available with solid wood. Wood panel products were developed as efficient ways of using limited wood resources. Their properties have made them very widely used wherever flat surfaces are needed. Bent plywood developed its own aesthetic for architecture and furniture (Binggeli, 2013).

Wood is a natural material which is a very safe and environmentally friendly to be used in interiors. In current scenario, people are choosing wood as an interior element more as it has many advantages and make a space beautiful and stylish. Apart from adding an aesthetical value to space, it the one product they may serve for long (Mekhriniso Murodovna & Gulshan Sergeevna, 2022).

23.4 DISCUSSIONS AND RESULTS

This survey was conducted within the tri-city region and 50 houseowners were involved. The structured questionnaire, listed in Table 23.7 was administered to all

TABLE 23.7
Questionnaire Survey Response Details

S. No.	Survey Questionnaire	Wall Finishes					
		Wall Paint	Wallpapers	PVC Panels	Wall Tiles	Wooden Panelling	Cement Boards
1	Q.1 At present, what kind of wall finish have been used at your home?	56%	20%	14%	6%	4%	0%
2	Q.2 After 5 years, which finish will you prefer in your home?	28%	18%	12%	16%	18%	8%
3	Q.3 In your opinion, which one of the following is a long-lasting/durable finish?	16%	14%	22%	36%	10%	2%
4	Q.4 In your region, which of the following materials is easily available?	50%	10%	12%	22%	4%	2%
5	Q.5 According to you, which of the following takes less time in execution?	24%	34%	28%	6%	4%	4%
6	Q.6 Which of the following do you think will make your interior space look more beautiful?	12%	26%	14%	6%	38%	4%
7	Q.7 In your opinion, which of the following is an affordable material in relevance to its installation?	36%	34%	8%	12%	8%	2%
8	Q.8 At present, what kind of wall finish have been used at your home?	38%	28%	14%	14%	6%	0%

the survey participants so as to have assessment of the choice of respondents for the effective wall finishes that can be used in affordable housing. The data received was analysed, and the response graphs in the form of pie charts are shown in Figures 23.2 and 23.3. It was found that presently, the majority of people are using wall paint for interiors followed by wallpapers. The choice for future use also being the wall paint for majority. If we compare the long-lasting/durable finish, wall tiles are the first choice followed by PVC panels. The most easily available material is again the wall paint followed by wall tiles. Based on the less time in execution, wallpapers were the first choice followed by PVC panels.

For making the interior space look beautiful, wooden panelling is the first choice of selection followed by wallpapers. In terms of installation, wall paint was again the first choice followed by wallpapers. Considering a cost-effective solution in terms of its appearance, the first choice of respondents was wall paint followed by wallpapers.

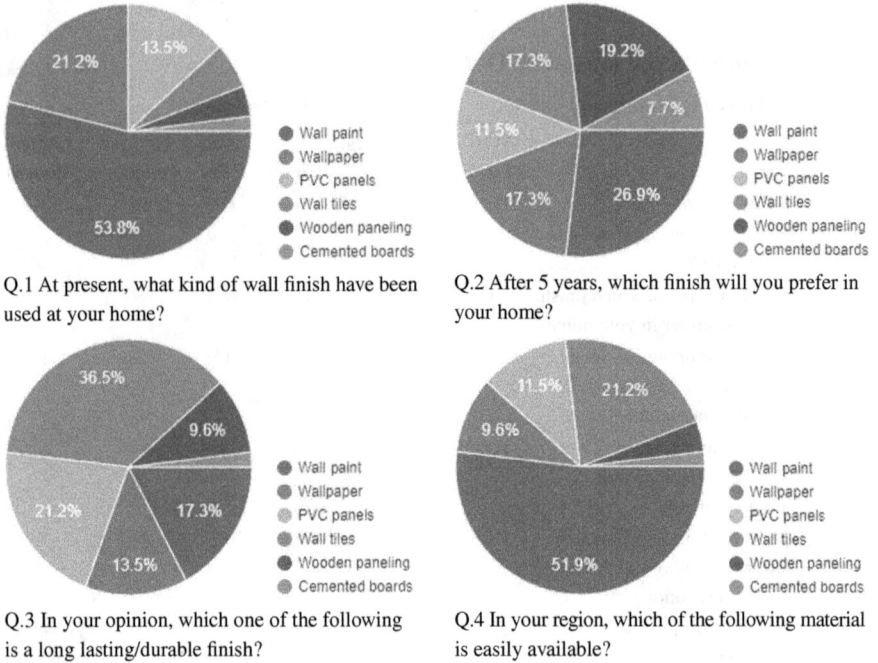

Q.1 At present, what kind of wall finish have been used at your home?

Q.2 After 5 years, which finish will you prefer in your home?

Q.3 In your opinion, which one of the following is a long lasting/durable finish?

Q.4 In your region, which of the following material is easily available?

FIGURE 23.2   Questionnaire survey response analysis (Q1 to Q4).

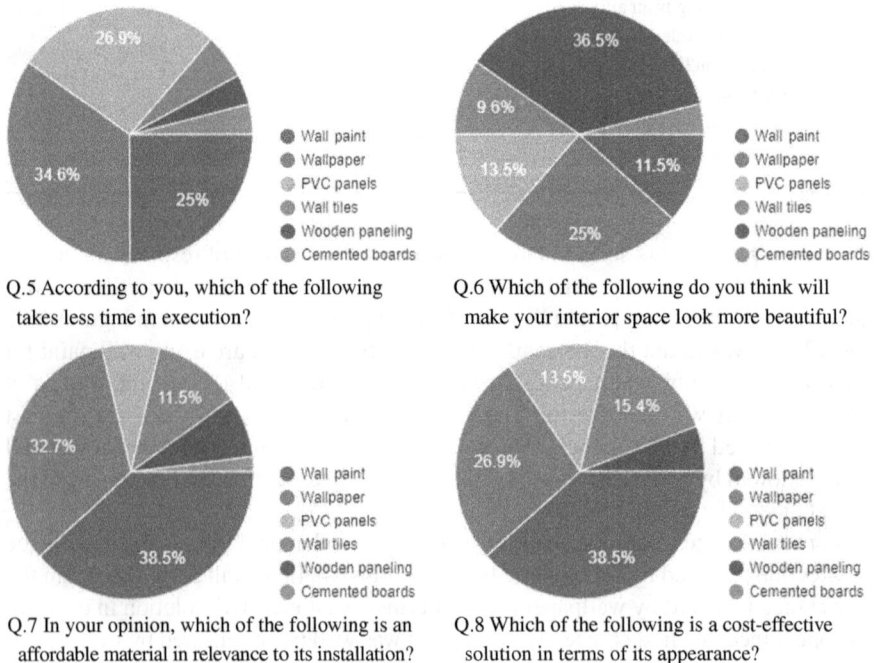

Q.5 According to you, which of the following takes less time in execution?

Q.6 Which of the following do you think will make your interior space look more beautiful?

Q.7 In your opinion, which of the following is an affordable material in relevance to its installation?

Q.8 Which of the following is a cost-effective solution in terms of its appearance?

FIGURE 23.3   Questionnaire survey response analysis (Q5 to Q8).

23.5 CONCLUSION

An adroit-looking home is a basic need of every individual. Walls are the initial segment for making the interiors of an accommodation visually appealing. Wall paint is most frequently used option in terms of easy availability, affordability, and cost-effective solution and hence also being the most used presently. Further, respondents also considered options of different available materials as their choice due to the intrinsic qualities of the distinct materials. There are varieties of materials available for wall finishes and appropriate selection can enhance the ambience of the place with affordable solutions.

REFERENCES

11 Different Types of Tile Flooring. (2015). Endeavour Homes. https://endeavourhomes.com. au/blog/11-different-types-of-tile-flooring/

Abrams, S. (2019). *3 Surprising Benefits of Painting Your Home.* Paintzen. www.paintzen. com/blog/3-benefits-home-painting

Awoyera, P. O., Akinmusuru, J. O., Ndambuki, J. M., & Lucas, S. S. (2017). Benefits of Using Ceramic Tile Waste for Making Sustainable Concrete. *Journal of Solid Waste Technology and Management, 43*(3), 233–241. https://doi.org/10.5276/JSWT.2017.233

Binggeli, C. (2013). *Materials for Interior Environments* (2nd Edition). John Wiley & Sons, Inc. www.wiley.com/enus/Materials+for+Interior+Environments%2C+2nd+Edition-p-9781118306352

Bosveld, T. (2013). *What is Paint Made of?* www.dunnedwards.com/pros/blog/whats-in-your-paint/

Bugno, A. A., & Krampikowska, A. (2020). The basics of a system for evaluation of fiber-cement materials based on acoustic emission and time-frequency analysis. *MBE, 17*(3), 2218–2235. https://doi.org/10.3934/mbe.2020118

ColourDrive. (2019). *What is the Difference Between Plastic Emulsion Paint and Royale Paint?—ColourDrive Home Varsity.* www.colourdrive.in/diy/difference-plastic-emulsion-paint-royale-paint/

Earthry Facilities. (2019). *The Benefits of Painting Your Home's Interior and Exterior.* https:// earthryfacilities.wordpress.com/

Ferrao, C. (2019). *Types of Wallpapers, Why Wallpaper & Everything Else in Between!— The Urban Guide.* Urban Company. www.urbancompany.com/blog/interiors/types-of-wallpaper-india/

Gangani, M. G., Suthar, H. N., Pitroda, D. J., & Singh, A. R. (2016). A Critical Review on Making Low Cost Urban Housing in India. *International Journal of Constructive Research in Civil Engineering, 2*(5), 21–25. https://doi.org/10.20431/2454-8693.0205004

Hoagland, S. M. (2015). *Paper, Pins, and Preservation: The Evolution of Wallpaper Conservation in a "Ruin" Environment.* The Book and Paper Group Annual. https:// scholar.google.com/scholar?hl=en&as_sdt=0%2C5&q=Paper%2C+Pins%2C+and+Pr eservation%3A+The+Evolution+of+Wallpaper+Conservation+in+a+"Ruin"+Environm ent&btnG=

Janani, R., Kalyana Chakravarthy, P., & Ilango, T. (2018). Budget Houses for Low-Income People. *International Journal of Mechanical Engineering and Technology (IJMET), 9*(13), 109–117.

Khan, H. R. (2012). Enabling Affordable Housing for All-Issues & Challenges. *International Conference on Growth with Stability in Affordable Housing Markets,* 1–14. https:// rbidocs.rbi.org.in/rdocs/Speeches/PDFs/EAHAICS130412.pdf

Lorch, M. (n.d.). What is Emulsion Paint? *BBC Science Focus Magazine*. Retrieved September 17, 2022, from www.sciencefocus.com/science/whats-in-emulsion-paint/

Mekhriniso Murodovna, R., & Gulshan Sergeevna, K. (2022). Wood-As a Decorative Material of the Interior. *International Journal on Integrated Education*, 5(3), 310–315. https://scholar.google.com/scholar?hl=en&as_sdt=0%2C5&q=Wood+–+As+a+Decorative+Material+of+the+Interior&btnG=

Patel, S., & Sayed, F. (2018). A Study on Low Budget Interior Design Options. *The Indian Journal of Home Science*, 30(1), 48–55. www.homescienceassociationofindia.com

Pereira, M., Gomes, M. J. M., Cerqueira Alves, L., & Lacerdo-Aroso, T. de. (2009). Ancient Portuguese Ceramic Wall Tiles ("Azulejos"): Characterization of the Glaze and Ceramic Pigments. *Article in Journal of Nano Research*, 8, 79–88. https://doi.org/10.4028/www.scientific.net/JNanoR.8.79

Polyvinyl Chloride. (n.d.). ChemicalSafetyFacts.Org. Retrieved September 17, 2022, from www.chemicalsafetyfacts.org/polyvinyl-chloride/

Sadıklar, Z., & Tavşan, F. (2016). A Study on Selection of Polymer Based Surface Materials in Interior Design. *New Trends and Issues Proceedings on Humanities and Social Sciences*, 2(1), 387–396. https://doi.org/10.18844/GJHSS.V2I1.323

Shinde, S. S., & Karankal, A. B. (2013). Affordable Housing Materials & Techniques for Urban Poor's. *International Journal of Science and Research*, 1(5), 30–36. www.ijsr.net

Spence, W. P., & William, P. (1998). *Construction Methods, Materials, and Techniques* (First). Thomson Delmar Learning.

Sri Kamakshi Enterprises. (n.d.). *PVC Wall Panels*. Retrieved September 17, 2022, from www.indiamart.com/proddetail/pvc-wall-panels-2305278391.html

Surfa Coats India Pvt. Ltd. (n.d.). *Interior Paint*. Retrieved September 17, 2022, from www.surfacoatspaints.in/interior-paint.html

Tam, V. W. Y. (2011). Cost Effectiveness of using Low Cost Housing Technologies in Construction. *Procedia Engineering*, 14, 156–160. https://doi.org/10.1016/J.PROENG.2011.07.018

The Tile Shop. (n.d.). *Wall Tile Design Ideas*. Retrieved September 17, 2022, from www.tileshop.com/inspiration/tile-ideas/walls

Tittelein, P., Cloutier, A., & Bissonnette, B. (2012). Design of a Low-Density Wood–Cement Particleboard for Interior Wall Finish. *Cement and Concrete Composites*, 34(2), 218–222. https://doi.org/10.1016/J.CEMCONCOMP.2011.09.020

The Traditional Paint Company. (n.d.). *Oil Bound Distemper*. Retrieved September 17, 2022, from www.traditionalpaint.co.uk/product/oil-bound-distemper/

Wall Affairs. (n.d.). *PVC Panels FAQs*. Retrieved September 17, 2022, from www.madaanindia.com/pvc-wall-panels-faqs#:~:text=PVC means Polyvinyl Chloride, friendly and easy to install.&text=PVC panels can be used, in bedrooms%2C bathrooms and kitchens

Yuan, L. (2019). The Application Value of Wall Painting Design Art in Interior Space. *International Conference on Humanities, Cultures, Arts and Design (ICHCAD 2019)*, 264–268. https://doi.org/10.25236/ICHCAD.2019.052

24 Design of Image Retrieval Descriptor Based on the Fusion of Colour and Texture Feature Descriptors

Himani Chugh, Meenu Garg, Sheifali Gupta

24.1 INTRODUCTION

The traditional image retrieval techniques worked efficiently in the case of small-scale image databases, however, proved impractical, expensive, and subjective while annotating large-scale image databases [1]. Content-based image retrieval (CBIR), being more practical towards large databases, soon overtook traditional retrieval techniques and offered a reduction in ambiguities related to textual indexation. CBIR is an application of computer vision that offers a less-time-consuming solution to image retrieval problems. Rather than analysing the descriptors, tags, and words associated with the image, it analyses the derivable information of the image, such as shapes, RGB content, or textures [2]. CBIR-based systems process information from an image and abstract it in terms of visual qualities. All the query operations are concerned with the abstracted parameters and not just the image. The technology has several applications that include historical research, digital libraries, medical diagnosis, fingerprint identification, crime prevention, biodiversity information systems, and so on. Selection and extraction of the visual image feature are of high significance while designing an efficient CBIR because the features utilized for discrimination instantly influence the efficacy of the entire system. Without the need for human intervention, attributes can be extracted directly from the image.

The work presented in the article is focused on searching for random images. The chapter presents a novel extractable feature descriptor, and each image has a different colour and texture combination. The significant contributions of the work done are as follows:

1. For the design of an elaborative feature map, a novel approach that combines micro-structure descriptor (MSD) and colour difference histogram (CDH) has been proposed for efficient image retrieval.

DOI: 10.1201/9781003367161-24

243

2. HSV related to the image are utilized by the CDH descriptor for computation of the colour features, while the GLCM is utilized by the MSD descriptor for computing the texture features.
3. The new feature map that is created by concatenating features from features derived from MSD and CDH descriptors improves the accuracy of the image retrieval.
4. The Euclidean distance metric has been used to calculate the degree of similarity between the database image and the query image.
5. Using F-measure, recall, and precision, the proposed framework's performance is examined.

The following is how the chapter is organized: The literature review pertinent to the subject is in Section 24.2. The proposed model and approach are thoroughly discussed in Section 24.3. The concluding remarks are highlighted in Section 24.4 and Section 24.5, which provide an experimental analysis of the work.

24.2 LITERATURE REVIEW

Different researchers work on the colour and shape features of an image. The colour and shape invariants were combined to index and retrieve images [3]. Different colour models, independent of object geometry, object pose, and illumination, were proposed. Shape-invariant features were computed from colour-invariant edges derived from these colour models. The colour feature was represented by a cumulative histogram similar to the colour co-occurrence matrix (CCM) or GLCM and was used to extract texture features (CCM) [4]. The two primary design features of an image are colour and texture [5]. The colour of every sub-block was obtained by resampling the HSV colour space into non-equal ranges, and the colour feature was expressed by a cumulative histogram. GLCM was used to determine the texture of each sub-block. A study [6] represented an image by its feature vectors referred to as descriptors. The authors suggested a method for obtaining an effective colour-based descriptor that combines the colour correlogram (CC) and dominant colour (DC) [7]. A colour histogram, colour moments, and colour coherence vector (CCV) feature as colour descriptors and co-occurrence matrices and discrete wavelet transform features as texture descriptors are applied to check the effectiveness of the system [8].

24.3 PROPOSED METHODOLOGY

A model is designed to retrieve images using colour and texture features. The block diagram of the proposed methodology is shown in Figure 24.1.

24.3.1 IMPLEMENTATION ON DATASET AND QUERY IMAGE

A generic Corel 5K dataset is used having with variety of colours and shapes for the simulation purposes. Figure 24.2 displays sample images for each category in the dataset. For testing, a dataset of 100 images is produced. Each image belonged to one of four groups, including various flower types, buses, dinosaurs, and horses.

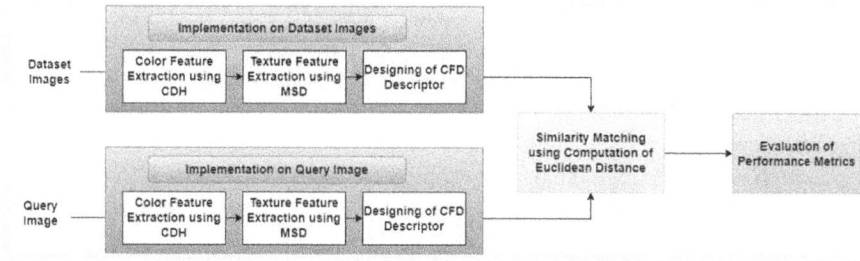

FIGURE 24.1   Proposed methodology to retrieve images with the combination of CDH and MSD descriptor

FIGURE 24.2   Classifications of random images in a dataset.

Each category had 25 pictures. The images are each 128 by 192 pixels (or 192 by 128 pixels) in size and are in the JPEG format [9], [10]. For simulation purposes, the flower's query image was captured.

The images gathered from the dataset had various hues, dimensions, and shapes. Image resizing is required because uniformity must be achieved by having all of the images the same size. This uniformity helps to produce accurate computing features. To boost the number to unique pixel information in the image, the images are scaled down to 384 by 256 pixels [11].

24.3.1.1   Colour Feature Extraction Using CDH Descriptor

It is a global image based on the difference of edge orientation and colour. It combines colour information and edge information of colour images. Colour features that were retrieved [12] include the HSV histogram, colour moments, autocorrelogram, and wavelet moments with feature size of 1×32, 1×64, 1×6, and 1×40, respectively.

24.3.1.2   Texture Feature Extraction Using MSD Descriptor

A variety of ways to assess texture, including the co-occurrence matrix, fractals, Gabor filters, versions of the wavelet transform, and other methods are used to extract texture features [13]. Using the texture spectrum, researchers have improved their techniques for expressing local patterns. Other techniques for analysing textures include using edge information and co-occurrence matrix attributes. Finally, representing an image as a two-dimensional array is required to identify a particular texture in it [14]. A method of examining a picture by counting its pixels and dots is statistical texture analysis. The second-order statistics known as feature extraction based on GLCM is applied to evaluate texture characteristics.

GLCM is a feature extraction technique that uses pair-wise distance measures to extract features from all pixels in the image matrix [15]. It determines, for every pixel in your image, the grey level that is most often found in couples of pixels having definite values. To regulate the degree of grey, a specific distance (d) and angular orientation (h) are analysed. There are several different distance (d) options available for the GLCM method, including 1, 2, 3, and 4. While 0°, 45°, 90°, and 135° of the angle orientation (h) are frequently employed. The method of distance and angle orientation is described in Figure 24.3. The F-measure (Fm), according to Eq, is a measure that combines recall and precision (8)

When determining the angular orientation (h), 0° is often used, but it is also possible to use 45°, 90°, or 135°. This technique uses distance and angle measurements to locate features on a map as shown in Figure 24.4.

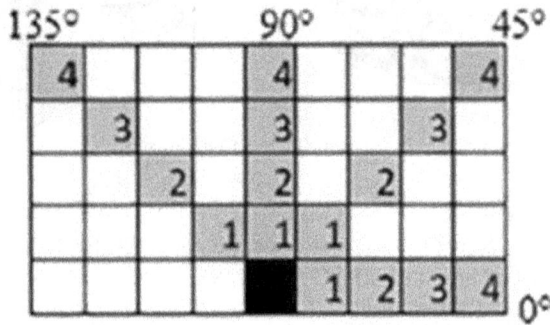

FIGURE 24.3 Method to determine distance and angle orientation.

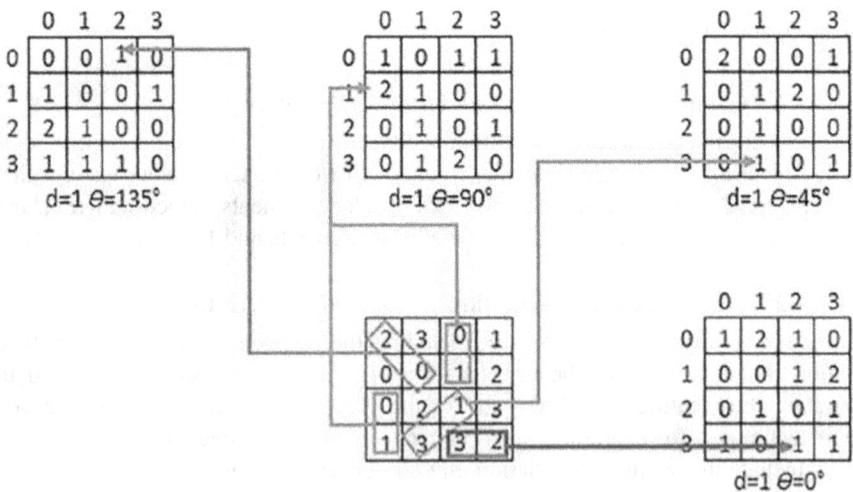

FIGURE 24.4 Process of creation of GLCM.

In this study, four statistical parameters contrast, correlation, energy, and homogeneity are extracted at $\delta = 0°$ as explained and calculated using equations (24.1) to (24.4).

1. Contrast
   The contrast between an image and its surrounding pixels is referred to as spatial frequency. It also shows the grey-level variation. If two neighbouring pixels have the same value, then their contrast will be 0.

$$C = \sum\sum (m-n)^2 P(m,n) \tag{24.1}$$

2. Correlation
   Correlation is a mathematical measure of how two things change together. Digital image correlation is a tool in which changes in images can be measured with accuracy by using tracking and image registration techniques. This method is frequently used in a variety of scientific and engineering fields. The motion of an optical mouse is one of the most prevalent applications.

$$Corr = \frac{\sum\sum m, n P(m-n) - \mu_x \mu_y}{\sigma_x \sigma_y} \tag{24.2}$$

where μx and σx are the mean and standard deviation of P_m and μy, and σy is the mean and standard deviation of P_n. Then Pm is the sum of each row in the co-occurrence matrix and P_n is the sum of each column in the co-occurrence matrix.
3. Energy
   Energy is a quantity of how equally pixels are distributed throughout an image. An image with high energy is a homogeneous image and will have a large value for the uniformity or angular second moment.

$$E = \sum\sum P^2(m,n) \tag{24.3}$$

4. Homogeneity
   The degree to which the distribution of items in the GLCM is homogeneous with respect to the GLCM diagonal is measured by homogeneity [13].

$$H = \sum \frac{P(m,n)}{1+|m-n|} \tag{24.4}$$

The four statistics computed are shown in Figure 24.5.

24.3.1.3 Designing of CFD Descriptor

A combined feature descriptor is made using concatenation of CDH descriptor and SSH descriptor as shown in Figure 24.6.

```
stats =

            Contrast:  1.367289490861618e-01
         Correlation:  9.395083960021600e-01
              Energy:  2.289641199527440e-01
         Homogeneity:  9.455112121953874e-01
```

FIGURE 24.5   Evaluated texture features using MSD descriptor.

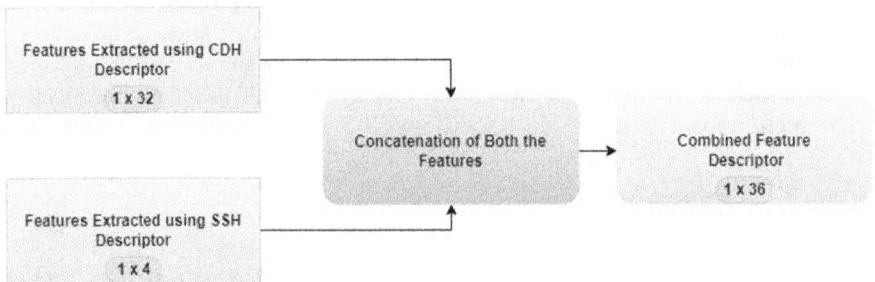

FIGURE 24.6   Process of designing CFD descriptor.

24.3.2   SIMILARITY MATCHING USING COMPUTATION OF EUCLIDEAN DISTANCE

Distance measuring is crucial in image retrieval systems. Finding the connections between the query image and the image dataset, as well as their similarities, is crucial. Therefore, the Euclidean distance between the dataset image and the query image is calculated in order to perform similarity matching. The best way to determine the separation between two points is to use the Euclidean distance. Assuming we have the points (x, y), where x = (m1, n1) and y = (m2, n2), we can calculate the Euclidean distance between them using equation (24.5).

$$E_d = \sqrt{\left(m_1 - m_2\right)^2 + \left(n_1 - n_2\right)^2} \tag{24.5}$$

24.3.3   EVALUATION OF PERFORMANCE METRICS

The proposed descriptor is assessed using a variety of measurements in order to assess the effectiveness of image retrieval systems.

1. According to equation (24.6), precision (Pr) is the proportion of appropriate images that were recovered from entire database.

$$P_r = \frac{No.\,of\ Relevant\ Images\ Extracted}{Total\ No.\,of\ Images\ Extracted} \qquad (24.6)$$

2. By dividing the total number of images in the dataset by the number of relevant images that were extracted, recall (Re) is calculated. Equation (24.7) is a representation of the equation used to calculate this parameter.

$$R_e = \frac{No.\,of\ Relevant\ Images\ Extracted}{Total\ No.\,of\ Images\ in\ the\ Dataset} \qquad (24.7)$$

3. A measure that combines precision and recall is known as the F-measure (Fm), according to equation (24.8)

$$F_m = 2 \times \frac{Pr(I_i) \times Re(I_i)}{Pr(I_i) + Re(I_i)} \qquad (24.8)$$

24.4 RESULT ANALYSIS

The HSV feature map obtained out from CDH is combined well with GLCM feature map obtained again from MSD during the procedure. Overall using CDH, 32 features are computed, and using MSD, four feature maps are computed. Upon concatenation of both features, the total computable features increases to 36, which further increase the system image retrieval efficiency.

Various threshold values of 40%, 50%, 60%, and 70% are applied to the maximum sorted distance. The implementation of CDH and MSD for the flower image at all the threshold values is shown in Table 24.1.

When a flower image is taken as the query image, it is observed that 28 images are retrieved in which 23 match exactly of that query image, which are called as true positive (TP) images, at 50% threshold. The performance parameters of precision, recall, and F-measure are computed to be 82%, 92%, and 87%, respectively, which is the maximum of all the thresholds applied, thus showing that at the threshold of 50%, more accurate results can be retrieved. The implementation of CDH and MSD for the bus image at all the threshold values is shown in Table 24.2.

TABLE 24.1

Implementation of CDH and MSD for the Flower Image at Different Threshold Values

Threshold	Images Retrieved	TP	Actual Images	P_r	R_e	F_m
40%	21	19	25	0.90	0.76	0.83
50%	28	23	25	0.82	0.92	0.87
60%	59	25	25	0.42	1	0.60
70%	84	25	25	0.30	1	0.46

When a bus image is taken as the query image, it is observed that 26 images are retrieved in which 23 match exactly of that query image at 50% threshold. The performance parameters are computed to be 88%, 92%, and 90%, respectively, which is the maximum of all the thresholds applied, thus showing that at the threshold of 50%, more accurate results can be retrieved. The implementation of CDH and MSD for the dinosaur image at all the threshold values is shown in Table 24.3.

When a bus image is taken as the query image, it is observed that 23 images are retrieved in which 23 matches exactly of that query image at 60% threshold. The performance parameters are computed to be 100%, 92%, and 95.8%, respectively, which is the maximum of all the thresholds applied, thus showing that at the threshold of 60%, more accurate results can be retrieved. The implementation of CDH and MSD for the horse image at all the threshold values is shown in Table 24.4.

TABLE 24.2

Implementation of CDH and MSD for the Bus Image at Different Threshold Values

Threshold	Images Retrieved	TP	Actual Images	P_r	R_e	F_m
40%	13	13	25	1	0.52	0.68
50%	26	23	25	0.88	0.92	0.90
60%	47	25	25	0.53	1	0.69
70%	58	25	25	0.43	1	0.60

TABLE 24.3

Implementation of CDH and MSD for the Dinosaur Image at Different Threshold Values

Threshold	Images Retrieved	TP	Actual Images	P_r	R_e	F_m
40%	15	15	25	1	0.6	0.75
50%	20	20	25	1	0.8	0.88
60%	23	23	25	1	0.92	0.95
70%	56	25	25	0.44	1	0.61

TABLE 24.4

Implementation of CDH and MSD for the Horse Image at Different Threshold Values

Threshold	Images Retrieved	TP	Actual Images	P_r	R_e
40%	33	23	0.69697	0.92	0.793103
50%	20	20	1	0.8	0.888889
60%	51	25	0.490196	1	0.657895
70%	67	25	0.373134	1	0.543478

Similarly, when a bus image is taken as the query image, it is observed that 20 images are retrieved in which 20 matches exactly of that query image at 60% threshold. The performance parameters are computed to be 100%, 80%, and 88.8%, respectively, which is the maximum of all the thresholds applied, thus showing that at the threshold of 50%, more accurate results can be retrieved. The performance chart for the computation of F-measure at different thresholds for different categories of images is shown in Figure 24.7. Accordingly, the first 25 retrieved images using CDH and MSD (retrieved: 28; true positive: 23) are shown in Figure 24.8.

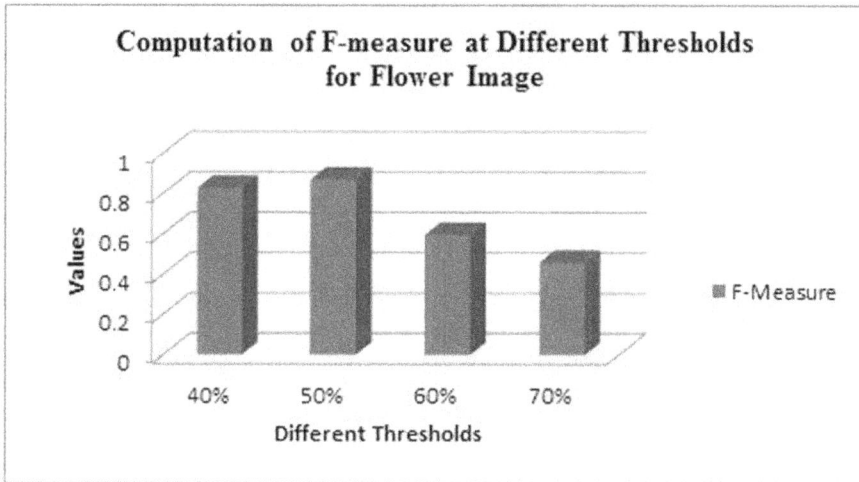

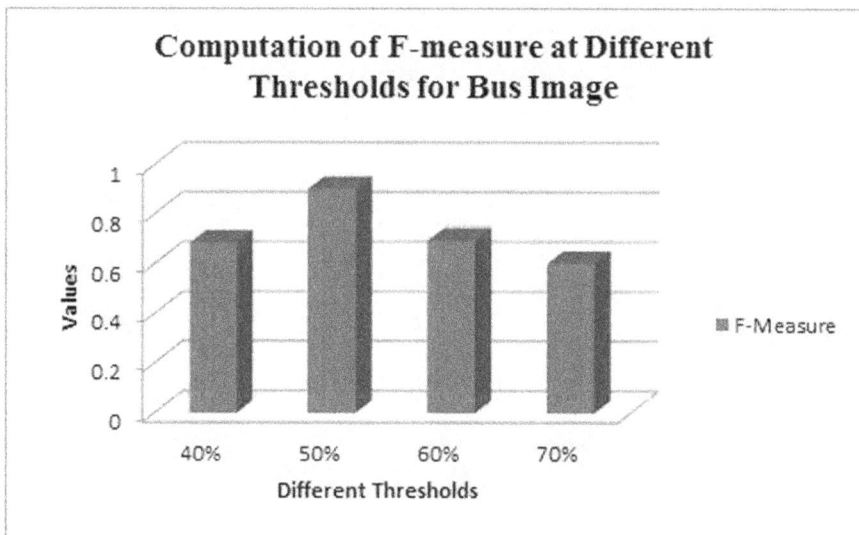

FIGURE 24.7   Performance chart for the computation of F-measure at different thresholds for the image of (a) flower, (b) bus, (c) dinosaur, and (d) horse.

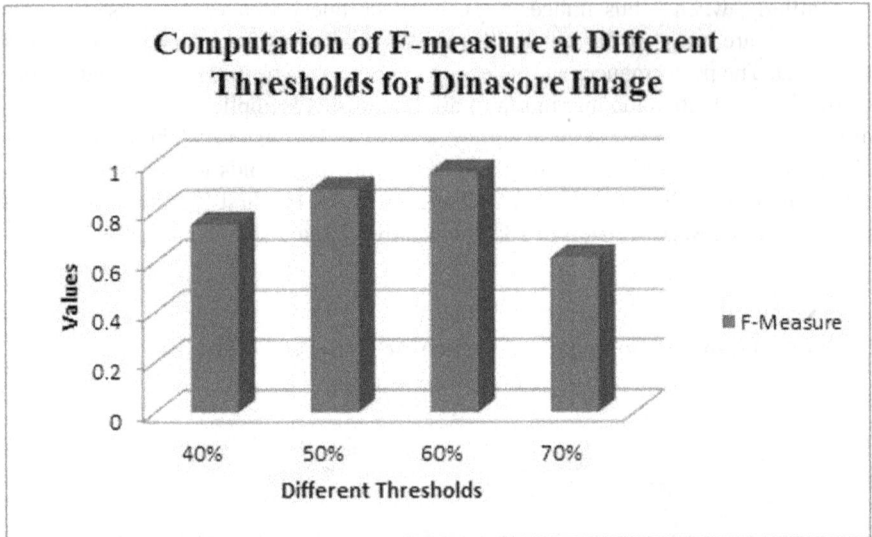

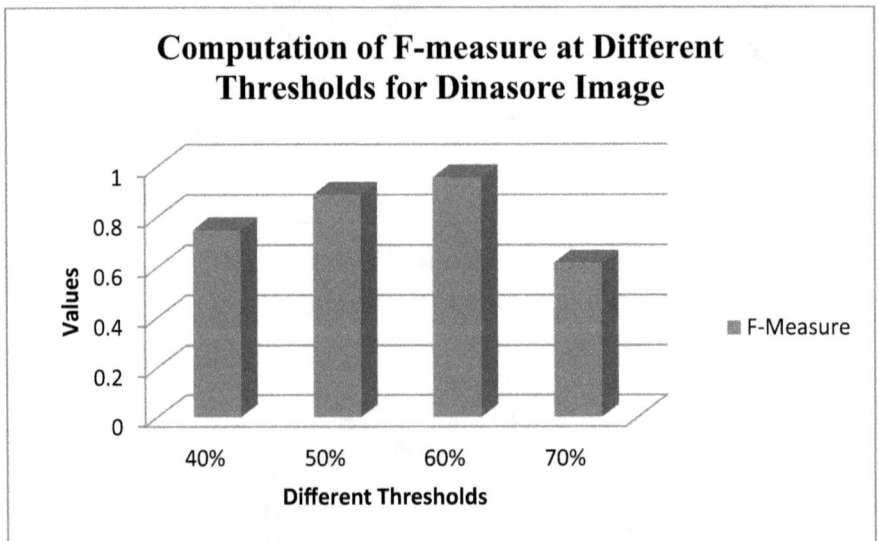

FIGURE 24.7 Continued

FIGURE 24.8 First 25 retrieved images using CDH and MSD (retrieved: 28; true positive: 23).

24.5   CONCLUSION AND FUTURE SCOPE

A new descriptor is proposed for the retrieval of images using the combination of colour and texture features to improve the precision, recall, and F-measure of image retrieval. On the one hand, we implement CDH and MSD as elementary image descriptors, and then we use a framework to combine CDH with MSD. The proposed methodology is implemented on different thresholds, and it was concluded that at a threshold of 50%, 23 images are retrieved as the most similar images compared to the query image. Out of all the images, it is concluded that the threshold value of 50% is most suitable to retrieve similar images as it gives the maximum value of F-measure.

REFERENCES

[1] H. C. Akakin and M. N. Gurcan, "Content-Based Microscopic Image Retrieval System for Multi-Image Queries," *IEEE Transactions on Information Technology in Biomedicine*, vol. 16, no. 4, pp. 758–769, Jul. 2012, doi:10.1109/TITB.2012.2185829.
[2] H. Chugh, S. Gupta, and M. Garg, "Image Retrieval System-An Integrated Approach—IOPscience." https://iopscience.iop.org/article/10.1088/1757-899X/1022/1/012040/meta (accessed Aug. 29, 2022).

[3] T. Gevers and A. W. M. Smeulders, "PicToSeek: Combining Color and Shape Invariant Features for Image Retrieval," *IEEE Transactions on Image Processing*, vol. 9, no. 1, pp. 102–119, Jan. 2000, doi:10.1109/83.817602.

[4] F.-H. Kong, "Image Retrieval Using Both Color and Texture Features," *2009 International Conference on Machine Learning and Cybernetics*, vol. 4, pp. 2228–2232, Jul. 2009, doi:10.1109/ICMLC.2009.5212186.

[5] K. Chaduvula, R. B. Prabhakara, and D. Govardhan, "Image Retrieval Based on Color and Texture Features of the Image Sub-blocks," *International Journal of Computer Applications*, vol. 15, Feb. 2011, doi:10.5120/1958-2619.

[6] A. N. Fierro-Radilla, M. Nakano-Miyatake, H. Pérez-Meana, M. Cedillo-Hernández, and F. Garcia-Ugalde, "An Efficient Color Descriptor Based on Global and Local Color Features for Image Retrieval," in *2013 10th International Conference on Electrical Engineering, Computing Science and Automatic Control (CCE)*, Sep. 2013, pp. 233–238. Doi:10.1109/ICEEE.2013.6676028.

[7] X. Zenggang, T. Zhiwen, C. Xiaowen, Z. Xue-min, Z. Kaibin, and Y. Conghuan, "Research on Image Retrieval Algorithm Based on Combination of Color and Shape Features," *Journal of Signal Processing Systems*, vol. 93, no. 2, pp. 139–146, Mar. 2021, doi:10.1007/s11265-019-01508-y.

[8] A. Shahbahrami, D. Borodin, and B. Juurlink, "Comparison Between Color and Texture Features for Image Retrieval," in *Proceedings of 19th Annual Workshop on Circuits, Systems and Signal Processing* (ProRisc 2008), Veldhoven, The Netherlands, Nov. 2008.

[9] M. J. J. Ghrabat, G. Ma, I. Y. Maolood, S. S. Alresheedi, and Z. A. Abduljabbar, "An Effective Image Retrieval Based on Optimized Genetic Algorithm Utilized a Novel SVM-based Convolutional Neural Network Classifier," *Human-Centric Computing and Information Sciences*, vol. 9, no. 1, p. 31, Dec. 2019, doi:10.1186/s13673-019-0191-8.

[10] W. Li, L. Duan, D. Xu, and I. W.-H. Tsang, "Text-Based Image Retrieval Using Progressive Multi-Instance Learning," in *2011 International Conference on Computer Vision*, Nov. 2011, pp. 2049–2055. doi:10.1109/ICCV.2011.6126478.

[11] A. E. Minarno and N. Suciati, "Batik Image Retrieval Based on Color Difference Histogram and Gray Level Co-Occurrence Matrix," *TELKOMNIKA (Telecommunication Computing Electronics and Control)*, vol. 12, no. 3, Art. no. 3, Sep. 2014, doi:10.12928/telkomnika.v12i3.80.

[12] H. Chugh et al., "Image Retrieval Using Different Distance Methods and Color Difference Histogram Descriptor for Human Healthcare," *Journal of Healthcare Engineering*, vol. 2022, p. e9523009, Mar. 2022, doi:10.1155/2022/9523009.

[13] S. Marianingsih and F. Utaminingrum, "Comparison of Support Vector Machine Classifier and Naïve Bayes Classifier on Road Surface Type Classification," in *2018 International Conference on Sustainable Information Engineering and Technology (SIET)*, Nov. 2018, pp. 48–53. doi:10.1109/SIET.2018.8693113.

[14] H. Chugh, S. Gupta, M. Garg, D. Gupta, H.G. Mohamed, I.D. Noya, A. Singh, and N. Goyal, "An Image Retrieval Framework Design Analysis Using Saliency Structure and Color Difference Histogram," *Sustainability*, vol. 14, no. 16, p. 10357, 2022.

[15] A. Bansal, K. Mehta, and S. Arora, "Face Recognition Using PCA and LDA Algorithm," in *2012 Second International Conference on Advanced Computing & Communication Technologies*. IEEE, Jan. 2012, pp. 251–254.

25 Identification of Facemask in a Hazy Environment
A Review of This Problem with a Solution

Nishant Sharma, Aniran Singh, Ankush Khera, Dev Sayal, Isha Kansal

25.1 INTRODUCTION

A facemask is a protective layer that you wear over your face, to prevent yourself from inhaling contaminated air or to protect your face when you are in a dangerous situation. Facemasks were an essential part of human life, before the spread of COVID-19. There are many reasons, ancient/religious beliefs made facemasks a necessity for human life at different times. The first and foremost reason why people wear facemasks is health. Oftentimes, it is to prevent the spread of one's own germs or illness in public places—a critical point in the world's often over-populated cities. Besides diseases, masks provide good protection against dust and pollen, especially in spring. They filter out most of the pollen particles and this is important for those with asthma or pollen allergies. Religions, such as Jainism, Judaism, and Islam, based on their religious beliefs, instruct their followers to wear a facemask. Jainism prohibits killing, violence, or injury to any living thing; to display this reverence, orthodox Jain monks and nuns wear cloth masks over their faces to prevent them from accidentally inhaling tiny flying insects. The niqab is a piece of cloth worn by women to cover their faces. It is worn by Muslim women across the Indian subcontinent, Saudi Arabia, and parts of the West. Facemasks have always been a part of fashion across Asia. Modern masks, initially seen as embarrassing and should only be worn if absolutely necessary, are now designed in a variety of colours and shaped to fit fashion trends. There are pink masks for the lolita girl subculture, spiked black masks for metal fans, and anime print masks, and the list goes on. Japan has a very reserved culture that is very conscious of the judgements of others. People who are shy and lack self-confidence wear a mask to hide from the bustling world. Some people choose to wear a mask to hide any physical imperfections (pimples, zits, or scars) they may have.

People across the globe are affected gravely due to the COVID-19 pandemic. Every day thousands of people fall prey to the COVID-19 virus. The virus has spread momentarily with lack of awareness, people not wearing a facemask, and not maintaining social distancing (and so on). Elementally, this infection spreads from person-to-person transmission making social distancing critical, to avoid the spread of this disease. However, various vaccines for COVID-19 have been developed, decreasing the intensity of this virus rather than acting as a cure for the virus. Hence, the WHO strongly exhorts wearing a facemask in crowded places and public gatherings to impede the transmission.

Facial recognition is a natural method of recognizing and authenticating people. It is an integral part of our everyday life. The security and authentication of an individual are critical in every industry or institution. Facemasks are one of the most effective ways of preventing the COVID-19 virus. Due to this, facemask detection systems are widely used in different areas. However, the challenging part is inherent diversity in faces, such as face shape, skin colour and texture, presence of a moustache/beard/mask/sunglasses, and so on. Recognizing a facemask is one of the most difficult and intriguing problems in pattern recognition and image processing, an example of what faces look like with and without masks is shown in Figures 25.1 and Figure 25.2. With the aid of such a technology, one can easily detect a person's face by using a dataset of identical matching appearances. The most effective approach for detecting a person's face is to use Python and a convolutional neural network (CNN) in deep learning.

Advances in deep learning, especially CNN, have exhibited surprising results in classifying images. The main idea behind CNN is to construct an artificial replica of the human brain as a visualization field. The biggest advantage of CNN is that more important features can be extracted from the whole image instead of just hand-made features. Researchers have introduced several deep networks based on CNN, and these networks have attained cutting-edge outcomes in computer vision classification, segmentation, object detection, and localization.

Haze is indeed a crucial problem in image processing and computer vision. The image quality deteriorates under haze conditions, as shown in Figures 25.3 and 25.4. It deteriorates the colour and lowers the contrast of pictures, reduces the clarity of scenes, and threatens reliability. Haze usually occurs during the absorption and scattering of smoke and dust particles in comparatively dry air. Reducing haze is a very difficult task in case of image processing.

FIGURE 25.1   Images without facemasks in a clear environment.

FIGURE 25.2 Images with facemasks in a clear environment.

FIGURE 25.3 Images without facemasks in a hazy environment.

FIGURE 25.4 Images with facemasks in a hazy environment.

In this study, deep learning techniques were applied to gather pictures of an individual with and without facemask from a database and then create a classifier to distinguish those captured in clear or foggy environments. This study aims to include the application of CNN to build a facemask sorter and the effect of the number of convolutional neural layers on prediction performance.

25.2 LITERATURE SURVEY

A facemask detection algorithm uses image localization to locate facemasks in a set of images or video streams and draws a bounding box around it. There are mainly two kinds of facemask detection algorithms; one mainly focuses on image classification, and the other focuses on image localization. *Joseph et al.* [1] presented You Only Look Once (YOLO), which became more effective than two-stage detectors for object detection. It has fast detection speed but compromises in positioning accuracy. *Zhao et al.* [2] proposed Mixed YOLOv3-LITE, for mobile smart and embedded devices. It reduced the size to 20.5 MB, which resulted in it being 91.70% smaller than the original model. Using computer vision for enhancement of security of manufacturing manpower in a post-COVID-19 world [3, 24]. This benefitted detecting social distancing and facemask and MobileNetV2 was used for its deployment. Retina Mask is a facemask detector [4] A facemask detector incorporates a context attention module, FPN, and CNNs. The layers of neurons in facemask detector capture all the image data as inputs. Further, the fully connected layers process the inputs to give the final prediction about the image. P. *Gupta et al.* [5] proposed an innovative method of using a DNN in face recognition by providing the extracted facial features as input and not raw pixel values. It results in the process being lighter and faster than a convolutional network and provides no compromise over certainty of the framework, and the average calculated efficiency is 97.05%. *Loey et al.* [6] built a device so people follow facemask regulations. The device makes use of deep learning, AI, and machine learning to implement real-time image processing and face detection. *S. Ge et al.* [7] proposed a model that takes the input image and combines two pre-trained CNNs in order to pull out applicant facial regions and represents them using HD descriptors. *M. Loey et al.* [8] suggested a model to combine extraction feature of Resnet50 with an ensemble algorithm using a support vector machine. It calculated 99.64% accuracy in Real-World Dataset (RMDF), 99.49% in Simulated Masked Face Dataset (SMFD), and 100% in Labelled Faces in the Wild (LFW). [9] Because of its compact fast and compatible design for mobile hardware development, YOLO v3-tiny is best suited for real-time face detection. Based on different datasets, the contrast and colour are awfully degraded due to bad atmospheric conditions like fog and haze particles. The level of degradation decreases with the increase of distance between object and camera. Earlier dehazing algorithms required multiple sets of images to identify the depth of haze and remove it, but recent algos require one single image to identify the levels of haze. New Trends in Image Restoration and Enhancement (NTIRE 2018–2019) has proposed various effective dehazing algorithms given by a number of researchers [10]. The Research of Image Dehazing and Video Dehazing [11] widely used the atmosphere scattering model (ASM). *Schaul et al.* [12] emphasized that the image of the distant object in outdoor photography, there is a decrease in visibility and colour due to the atmospheric haze. They proposed the idea of fusion [23] of near-infrared and visible images of the input to achieve a dehazed image. In *Jiang et al.* [13], a recurrent neural system is utilized to test the preparing procedure that acquires calibrating connection between scene solidity and surface highlights and shading highlights. Further, the scene is calibrated of mist pictures. *Yang et al.* [14] proposed a strategy to remove obscureness from an individual picture. In environmental dispersing model to gauge the air-light, a dim channel is used. *Y. Gao et al.* [15] suggested a model that estimates the transmission by finding the least underlying brightness with small sliding windows and using a filter to soothe out the image to obtain stable outcomes. Researchers [16] proposed a successful and novel defogging algorithm. It analysed the natural light by using haze-free reference pictures of the same site. *Surasak Tangsakul et al.* [17] used the cellular automata model for single image haze removal. Dull channels were refined using the standard of cellular structure which helped to enhance the potential of

DCP. Since dehazing is used in other areas, such as watermarking [18, 19], the solutions are required to be proposed under the area.

25.3 PROPOSED MODEL TO DETECT FACEMASKS IN HAZY WEATHER

This segment contains a proposed model to detect the facemask in hazy weather. First, live foggy videos and images are taken at different times. These images are then fed to an image classifier, which then identifies the ones taken in a clear and foggy environment. The ones taken in a foggy environment are then dehazed. Further, the dataset is passed through various face detection algos to detect the facemasks. The detailed procedure is as follows.

25.3.1 CAPTURING LIVE FOGGY VIDEO

The capturing of live foggy video and images, as presented in Figure 25.5, was performed in different locations at different times of the day. This was done to ensure a rich and diverse dataset. The camera equipment used in the research are a Canon EOS 1500D DSLR camera as the primary image and video capture and a Simpex SL60 5600K studio light to aid the camera in high-contrast and low-light conditions.

25.3.2 DEHAZING FOGGY VIDEO FRAME/IMAGE

Haze is known to be one of the most difficult problems in image processing and computer vision fields. Digital images' quality becomes worse under haze conditions because the colours change and the contrast of images get reduced; it compromises the clarity of the scenes, and it also risks the many applications' dependability such as outdoor monitoring, object identification, photography, and the like [20]. The precision of underwater images and satellite also gets decreased. Haze is formed by

FIGURE 25.5 Capturing a foggy video frame using Canon EOS 1500D DSLR.

absorption and scattering of smoke, dust, and other dry particulates of dry air, the process for which has been described in Figure 25.6. Overcoming haze is one of the most difficult tasks in the case of image processing. Removing haze from the image is called image dehazing [21, 22].

Haze is a natural phenomenon that can be approximately explained by ASM. The ASM gives a reliable theoretical basis for image dehazing research. Its formula is shown in equation (25.1):

$$H(x) = C(x)m(x) + A(1 - m(x)) \qquad (25.1)$$

Here, the coordinate value means the global atmospheric light. In different papers, it may be referred to as air light or ambient light. For ease of understanding, it is noted as atmospheric light in this survey. For the dehazing methods based on ASM, it is usually unknown. H(x) stands for the hazy image and C (x) denotes the clear scene. For most dehazing models, H(x) is the input and the desired output is C(x). The m(x) means the medium transmission map, which is defined in equation (25.2),

$$m(x) = e^{-\beta n(x)} \qquad (25.2)$$

Here n(x) stands for the atmosphere scattering parameter and H(x)'s depth, respectively. Thus, the m(x) is determined by n(x), which can be used for the synthesis of hazy images. If the n(x) can be estimated, C(x), the haze-free image, can be obtained by the following formula shown in equation (25.3),

$$m(x) = e^{-\beta n(x)} \qquad (25.3)$$

It can be seen that the light reaching the camera from the object is changed by the particles in the air. Some works use ASM to describe the formation process of haze,

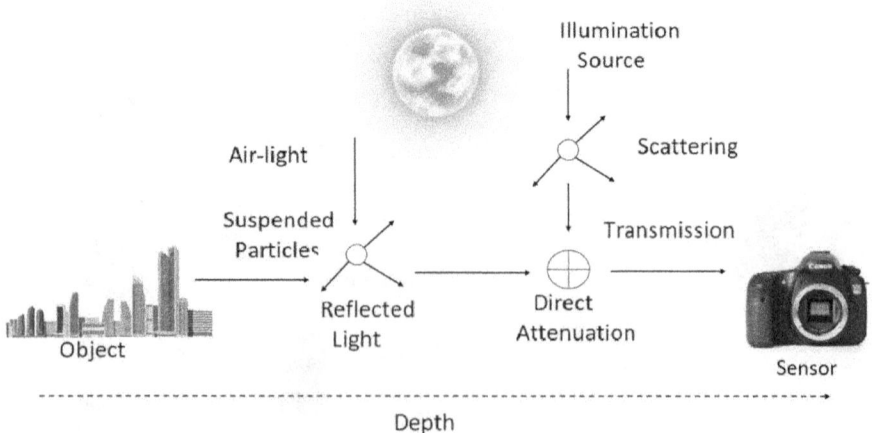

FIGURE 25.6   The effect of haze on image quality.

and the parameters included in the atmospheric scattering model are solved in an explicit or implicit way.

Many researchers have given numerous methods and techniques for dehazing images. With time, we got advancements in different methods and models. Some of the methods include single-image dehazing, which focuses on recovering the clear image from an input hazy or foggy image. Dark channel prior (DCP) is also one of the dehazing methods that observes the key features of the haze-free images. Similarly, equation 25.3 can help meet our purpose of dehazing images.

25.3.3 IDENTIFICATION OF FACEMASKS IN FOGGY VIDEO FRAME

The flowchart in Figure 25.7 explains the mechanism of image dehazing. Here, we have constructed a classifier using deep learning techniques. It will check database and person's images wearing a facemask and not wearing one will be collected and

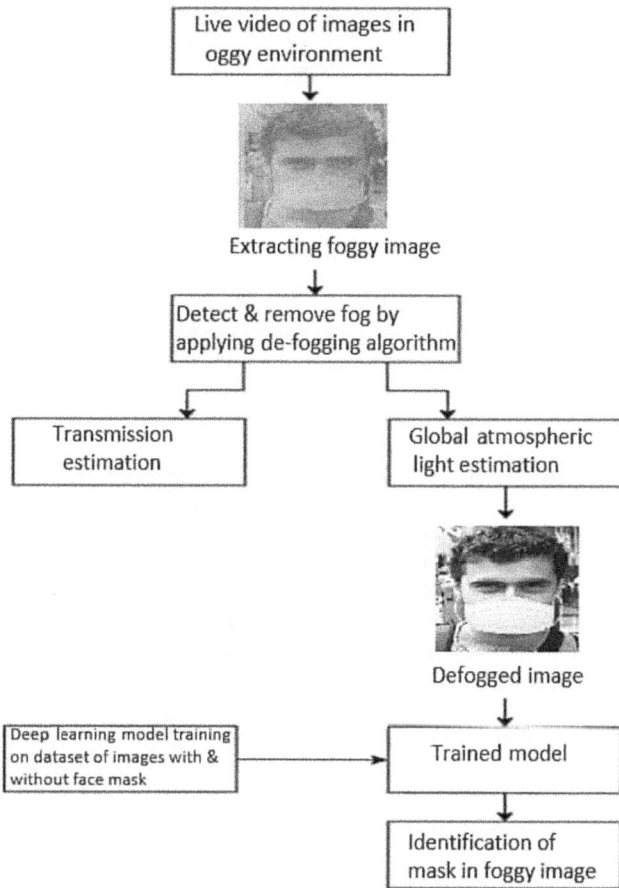

FIGURE 25.7 Flowchart explanation of the proposed model.

then distinguished between the pictures taken in a clear or foggy environment. Then, the images taken in a foggy environment are extracted from the database and defogging algorithms are applied to them to get a clear version of those images.

25.3.4 DATASET

The dataset comprises of two categories of images, ones that are shot in hazy environment and ones that are shot in clear environment. Both of these categories contain images, mainly portraits of two variations, one with people wearing the facemasks and the other in which people are not wearing facemasks [18].

25.4 CONCLUSION AND FUTURE SCOPE

With the growing demand of automatic face-mask detection systems in the world, many researchers gave different techniques and methods to provide accuracy in these systems. This chapter provides a survey on various methods based on deep learning for detecting facemask and their applications. Further we also checked various algorithms designed by researchers for dehazing images to restore the condition of hazy images. The aim of this chapter is to check whether we can apply the face-mask-detection algorithms even in hazy or foggy weather conditions as our system should work with higher accuracy in all weather conditions. In future, we can apply dehazing algorithms in facemask detection model and make use of both by combining these two. This research will increase the trust on deep learning methods as we would then rely more on our machines and would not have to worry about doing all this manually.

REFERENCES

[1] Redmon, J., Divvala, S., Girshick, R. B., & Farhadi, A. (2016). You only look once: Unified, real-time object detection. In *Proceedings of the 2016 IEEE Conference on Computer Vision and Pattern Recognition (CVPR)*, Las Vegas, NV, 27–30 June 2016, pp. 779–788.

[2] Zhao, H., Zhou, Y., Zhang, L., Peng, Y., Hu, X., Peng, H., & Cai, X. (2020). Mixed YOLOv3- LITE: A lightweight real-time object detection method. *Sensors*, 20, 1861.

[3] Khandelwal, P., Khandelwal, A., Agarwal, S., Thomas, D., Xavier, N., & Raghuraman, A. (2020). Using computer vision to enhance safety of workforce in manufacturing in a post COVID world. *arXiv:2005.05287* [Online]. http://arxiv.org/abs/2005.05287

[4] Loey, M., Manogaran, G., Taha, M. H. N., & Khalifa, N. E. M. (2021). A hybrid deep transfer learning model with machine learning methods for face mask detection in the era of the COVID-19 pandemic. *Measurement*, 167, 108288.

[5] Gupta, P., Saxena, N., Sharma, M., & Tripathi, J. (2018). Deep neural network for human face recognition. *International Journal of Engineering and Manufacturing (IJEM)*, 8(1), 63–71.

[6] Loey, M., Manogaran, G., Taha, M. H. N., & Khalifa, N. E. M. (2021). A hybrid deep transfer learningmodel with machine learning methods for face mask detection in the era of the COVID-19 pandemic. *Measurement*, 167, 108288.

[7] Ge, S., Li, J., Ye, Q., & Luo, Z. (2017). Detecting masked faces in the wild with LLE-CNNs. In *Proceedings of the IEEE Conference Computer Vision Pattern Recognition (CVPR)*, Jul. 2017, IEEE, pp. 2682–2690

[8] Loey, M., Manogaran, G., Taha, M. H. N., & Khalifa, N. E. M. (2021). A hybrid deep transfer learning model with machine learning methods for face mask detection in the era of the COVID-19 pandemic. *Measurement*, 167, Art. no. 108288.

[9] Cheng, G., Li, S., Zhang, Y., & Zhou, R. (2020). A mask detection system based on Yolov3-Tiny. *Frontiers in Food Science and Technology*, 2, 33–41.

[10] Codruta, A., Ancuti, C., & Timofte, R. (2020). NH-HAZE: An image dehazing benchmark with non-homogeneous hazy and haze-free images. In *Proceedings of the IEEE/CVF Conference on Computer Vision and Pattern Recognition Workshops*, IEEE Xplore, pp. 444–445.

[11] Cai, B., Xu, X., Jia, K., Qing, C., & Tao, D. (2016). Dehazenet: An end-to-end system for single image haze removal. *IEEE Transactions on Image Processing*, 25(11), 5187–5198.

[12] Schaul, L., Fredembach, C., & Susstrunk, S. (2009). Color image dehazing using the near-infrared. *International Conference on Image Processing (ICIP), 2009 16th IEEE International Conference on*. IEEE.

[13] Jiang, X., Sun, J., Li, C., & Ding, H. (2018). *Video Image De-Fogging Recognition Based on Recurrent Neural Network, 1551–3203 (c)*. IEEE.

[14] Yang, L., Bian, H., Feng, J., Zhao, L., Wang, H., Liu, Q., Wang, H., Wang, H., & Liao, J. (2017). *Optimized Design of Fast Single Image Dehazing Algorithm*. IEEE. 978-1-5386-1010-7/17/$31.00 ©2017.

[15] Gao, Y. Y., Hu, H. M., Wang, S. H., & Li, B. (2014). A fast image dehazing algorithm based on negative correction. *Signal Processing*, 103, 380–398.

[16] Maik, V., Park, H., Park, J., Kim, H., Know, Y., & Paik, J. (2018). Atmospheric light estimation using fog line vector for efficient defogging without color distortion. *IEEE, Conference*. IEEE.

[17] Tangsakul, S., & Wongthanavasu, A. (2018). Improving single image haze removal based on cellular automata model. *15th International Joint Conference on Computer Science and Software Engineering (JCSSE)*, 978-1-5386-5538-2/18/$31.00 ©2018 IEEE.

[18] https://github.com/chandrikadeb7/Face-Mask-Detection/tree/master

[19] Bhinder, P., Singh, K., & Jindal, N. (2018). Image-adaptive watermarking using maximum likelihood decoder for medical images. *Multimedia Tools and Applications*, 77, 1030310328.

[20] Bhinder, P., Jindal, N., & Singh, K. (2020). An improved robust image-adaptive watermarking with two watermarks using statistical decoder. *Multimedia Tools and Applications*, 79, 183–217. https://doi.org/10.1007/s11042-019-07941-2

[21] Kansal, I., & Kasana, S.S. (2020). Improved color attenuation prior based image defogging technique. *Multimedia Tools and Applications*, 79(17), 12069–12091.

[22] Kansal, I., & Kasana, S.S. (2020). Improved color attenuation prior based image defogging technique. *Multimedia Tools and Applications*, 79(17), 12069–12091.

[23] Kaur, H., Koundal, D., & Kadyan, V. (2021). Image fusion techniques: A survey. *Archives of Computational Methods in Engineering*, 28, 4425–4447.

[24] Tiwari, S., Kumar, S., & Guleria, K. (2020). Outbreak trends of coronavirus disease–2019 in India: A prediction. *Disaster Medicine and Public Health Preparedness*, 14(5), e33–e38.

26 A Fusion of U-Net and VGG16 Model for the Automatic Segmentation of Healthy Organs in the GI Tract

Neha Sharma, Sheifali Gupta

26.1 INTRODUCTION

The gastrointestinal (GI) tract is also known as the digestive system or alimentary duct. The whole digestive system of both humans and animals—from the mouth to the anus—is referred to as the GI tract [1]. The most prevalent type of cancer in both men and women is gastrointestinal [2]. The age, health, and stage or kind of cancer a patient has had an impact on their treatment for GI cancer [3]. Radiation therapy, chemotherapy, and surgery are the most often used therapies for GI cancer. Worldwide, an estimated 5 million persons received a diagnosis of GI cancer in 2019. About half of these patients are candidates for radiation treatment, which is typically administered over 10–15 minutes each day for 1–6 weeks [2]. Radiation oncologists attempt to provide strong doses of radiation using X-ray beams aimed at tumours while avoiding the stomach and intestines. Oncologists can now see the tumour and intestines' daily positions, which might change from day to day, thanks to updated technologies like integrated magnetic resonance imaging and linear accelerator systems, or MR-LINACs. In MRI images, the oncologists should physically draw the shape of the stomach and intestines to set the direction of the X-ray beams to improve the dose distribution to cancer while avoiding the healthy organs. Unless deep learning may assist in automating the segmentation process, this is a labour-intensive and time-consuming procedure that can extend treatments from 15 minutes to an hour each day, which can be tough for patients to accept. Treatments might be completed considerably more quickly, and more patients may receive more effective care if the stomach and intestines could be divided.

This problem may be resolved by recent advancements in deep learning, particularly concerning CNN [4]. The automated identification of problems in several human organs, including the brain [5], cervical cancer [6], eye diseases [7], and skin cancer [8], has produced positive results in CNN in recent decades. The CNN

 DOI: 10.1201/9781003367161-26

model has an advantage since it extracts characteristics in a hierarchy, starting with the most fundamental and progressing to the most abstract. Organ segmentation in abdominal imaging can be useful for clinical processes including diagnosis, therapy planning, and administration. For computer-assisted diagnosis and biomarker monitoring systems, organ segmentation is crucial [9]. Organ and treatment volume segmentation is another step in the planning process for radiation therapy [10]. Segmented patient-specific anatomical models that are utilized in combination with intraoperative image-guidance techniques can enhance surgical planning and execution [4]. (Semi-)automated segmentation methods [9] have been developed to streamline clinical procedures since manual segmentation of 3D abdominal pictures is labour-intensive and ineffective. In this case, a deep-learning-based approach to segmentation may be advantageous for the GI tract, which might speed up treatments and enable more patients to receive better care. In this work, we'll develop a deep-learning-based technique for automatically segmenting the digestive tract's stomach and intestines using MRI images. It has been suggested to combine VGG16 with U-Net in this method. The U-Net model uses the VGG16 as an encoder. Three performance metrics, including dice coefficient, IoU, and model loss, have been used to assess the performance of the model. The remainder of this essay is structured as follows: Section 26.2 discusses recent research on GI tract segmentation that is linked to this topic. The dataset that was utilized for this study is shown in Section 26.3. The methods employed in this study is covered in Section 26.4. The outcome following the application of the suggested model will be represented and discussed in Section 26.4. Section 26.5 finishes the chapter and presents the research's findings.

26.2   RELATED WORK

The segmentation of the GI tract has been the subject of extensive research in recent years [1–3]. To characterize WCE pictures, Li, B., et al. (2012) presented a colour texture feature that incorporates a wavelet and a uniform local binary pattern. The suggested attributes explain the multiresolution properties of WCE pictures and are insensitive to changes in illumination [11]. A worldwide statistical technique was published by Zhou, M., et al. (2014) that could automatically identify polyps in WCE frames and calculate their radii. The suggested approach uses available RGB channels to acquire statistical data. The existence and radii of polyps are then determined using a support vector machine (SVM) using the statistical data [12]. To help the endoscopist locate polyps during colonoscopy, Wang, Y., et al. (2015) proposed a software system dubbed Polyp-Alert [13]. For segmenting colorectal polyps, Li, Q., et al. (2017) suggested a brand-new, end-to-end fully convolutional neural network structure. This technique can provide a prediction map with an output size equal to the size of the input network's original picture [14]. A brand-new automated computer-aided approach to detect polyps for colonoscopy recordings was put out by Yuan, Y., et al. in 2017. First, an image is segmented into multilevel superpixels, and then a sparse autoencoder (SAE) is used to generate discriminative features in an unsupervised manner to identify perceptually and semantically relevant prominent polyp areas [15]. Using colonoscopy pictures, Dijkstra, W., et al. (2019)

demonstrated a one-shot method to define polyps. For semantic segmentation, they employed a fully convolutional neural network model [16]. A brand-new polyp segmentation technique called MED-Net was suggested by Nguyen, N., et al. in 2020. It is based on the architecture of multi-model deep encoder-decoder networks. This architecture not only extracts discriminative features at various effective fields of view and many picture scales to collect multi-level contextual information, but it can also significantly perform upsample more accurately to generate a superior prediction. Additionally, it can capture polyp borders with greater accuracy thanks to the use of multi-scale effective decoders [17]. For the segmentation of polyps, Huang, C., et al. (2021) presented the HarDNet-MSEG convolution neural network. It is made up of a decoder and a backbone. Low-memory traffic CNN with the name of HarDNet68 serves as the foundation and has been successfully used for many CV tasks, including image classification, object identification, multi-object tracking, and semantic segmentation, among others [18].

26.3 DATASET DESCRIPTION

A collection of MRI scans was produced by the University of Wisconsin-Madison, a public land-grant research centre in Madison, Wisconsin. The dataset included 85 people with scans that ranged from one to six days. There are 144 or 80 slices in each scan every day, depending on the patient. The measurement of the image in the dataset is $224 \times 224 \times 3$. Pictures numbering 38,496 make up the collection. RLE encoding is used to encrypt the collection's images. Masks are produced from these RLE-encoded pictures using deep learning techniques. Figure 26.1 exhibits a few representative pictures from the dataset. A sample of images from the dataset is shown in Figures 26.1(a), (b), (c), and (d), whereas Figures 26.1(e), (f), (g), and (h) display the equivalent mask produced by the RLE encoding. Here, the colours

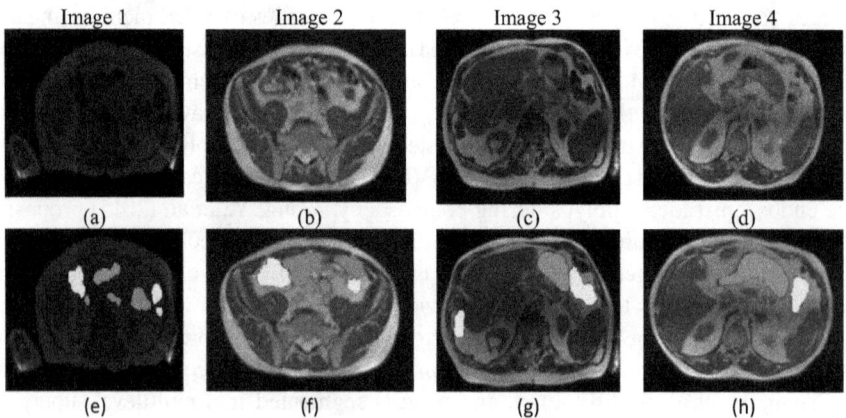

FIGURE 26.1 Sample images from the dataset. Figure 26.1(a), (b), (c), and (d) are some random images from the dataset, and Figure 26.1 (e), (f), (g), and (h) masks were obtained from the RLE encoding.

red, yellow, and green each stand in for one of the bowels: the small intestine, large intestine, or both.

26.4 METHODOLOGY

In this research, a deep-learning-based model for automatic segmentation of the stomach, small bowel and large bowel has been proposed. This model can assist the radio oncologist to deliver a high dose of the X-ray beam to the tumour while avoiding healthy organs in the GI tract using MRI scans. The proposed model is a fusion of the VGG16 [19] and U-Net [20] model. Figure 26.2 shows the block diagram of

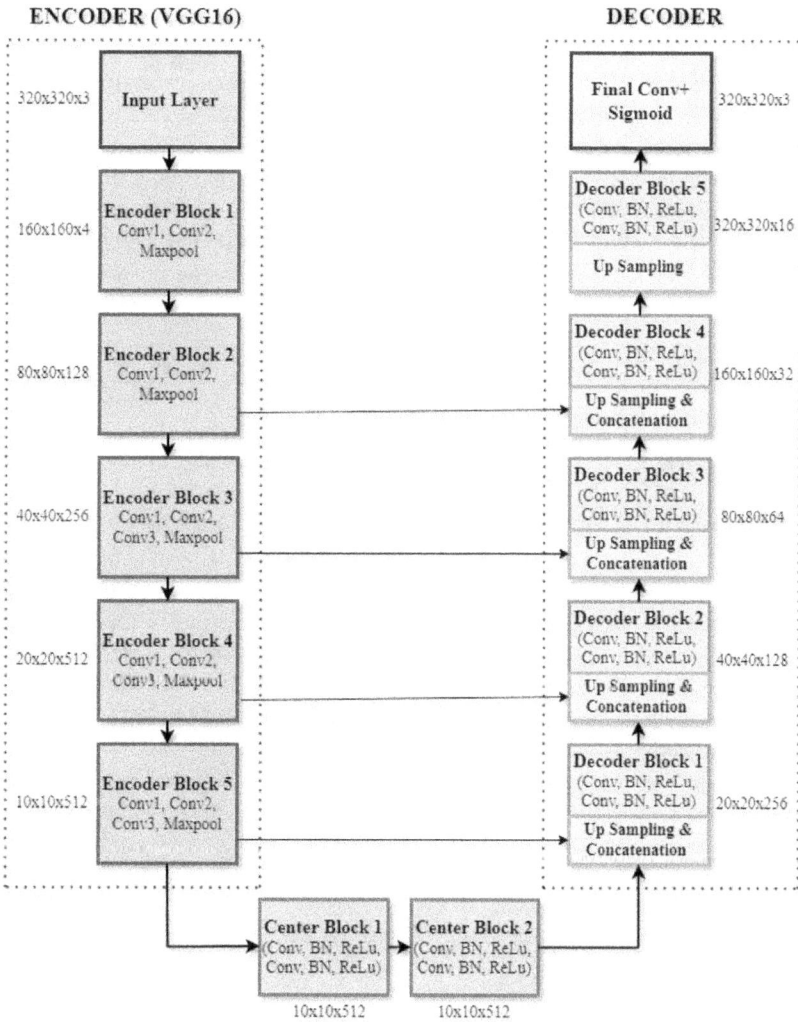

FIGURE 26.2 Proposed model for segmentation of the gastrointestinal tract.

the proposed model which is a fusion of VGG16 and U-Net. U-Net is developed from the conventional CNN. It was implemented in 2015 for medical applications. It is not only able to classify the images; it can also detect the shape and location of the object.

The U-Net mainly consists of two branches one is an encoder and the other is the decoder. The encoder branch includes convolutions and max pool layers. The size of the image is reduced as we go further in the encoder. The decoder branch performs the reverse operation of the encoder. As the number of layers in the decoder increase the size of the image also increases. It performs the transposed convolution operation in every convolution layer. It is a combination of convolution and batch normalization layer; it does not include any max pool layer. The encoder and decoder process of the U-Net makes the shape of the letter "U"; that is why this arrangement is called U-Net. In this proposed model the VGG16 model is used as the encoder of U-Net. The VGG is a transfer learning model which is pre-trained on the ImageNet dataset. The VGG stands for Visual Geometry Group and 16 is for the number of layers it consists of. VGG16 is a type of convolution neural network (CNN). It was implemented in 2014 to win the ImageNet challenge. It has several convolutional layers of filter size 3 × 3 with a stride of 1 and the same padding. The max pool layer has a filter size of 2 × 2. The arrangement of the layers is the same throughout the architecture in the VGG transfer learning model.

Table 26.1 shows the detailed model summary for the proposed network architecture. Table 26.1 includes the detail of all the layers, input size, output size, size of the

TABLE 26.1

Model Summary for the Proposed Model

S. No.	Layers	Input Image Size	Filter Size	Activation Function	Output Image Size	Number of Parameters
1	Input image	320 × 320 × 3	-	-	320 × 320 × 3	0
2	Encoder block 1	320 × 320 × 3	3 × 3	ReLU	160 × 160 × 64	38,720
3	Encoder block 2	160 × 160 × 64	3 × 3	ReLU	80 × 80 × 128	147,584
4	Encoder block 3	80 × 80 × 128	3 × 3	ReLU	40 × 40 × 256	1,475,328
5	Encoder block 4	40 × 40 × 256	3 × 3	ReLU	20 × 20 × 512	5,899,778
6	Encoder block 5	20 × 20 × 512	3 × 3	ReLU	10 × 10 × 512	7,079,424
7	Centre block 1	10 × 10 × 512	3 × 3	ReLU	10 × 10 × 512	2,361,344
8	Centre block 2	10 × 10 × 512	3 × 3	ReLU	10 × 10 × 512	2,361,344
9	Decoder block 1	10 × 10 × 512	3 × 3	ReLU	20 × 20 × 256	2,951,168
10	Decoder block 2	20 × 20 × 256	3 × 3	ReLU	40 × 40 × 128	1,033,216
11	Decoder block 3	40 × 40 × 128	3 × 3	ReLU	80 × 80 × 64	258,560
12	Decoder block 4	80 × 80 × 64	3 × 3	ReLU	160 × 160 × 32	64,768
13	Decoder block 5	160 × 160 × 32	3 × 3	ReLU	320 × 320 × 16	7,040
14	Final conv	320 × 320 × 16	3 × 3	Sigmoid	320 × 320 × 3	435

Total parameters: 23,752,563

Trainable parameters: 23,748,531

Non-trainable parameters: 4,032

filter, and also many parameters in every layer. The input image of size $320 \times 320 \times 3$ is applied to the network. Then it passes through several convolution and max pool layers of the VGG16 model, which acts as an encoder branch of the U-Net model. It contains five encoding blocks, and after the encoder branch, the size of the image is reduced to $10 \times 10 \times 512$. Although the height and width of the image are reduced, its depth is increased. Then two centre blocks were used in the network. These blocks contain the same configuration; it has a combination of convolution, batch normalization and ReLU layer. These blocks do not change the size of the image. Then the image is fed to the decoder branch of the U-Net model. This branch contains five blocks. Before feeding the image to the decoder block it is upsampled and concatenated with the encoder blocks. The decoder block is a combination of convolution and batch normalization layers. The size of the image after passing it through all the decoder blocks is $320 \times 320 \times 16$. Then a final convolution block is used; it contains the convolution layer and sigmoid activation function. The final output size of the image is $320 \times 320 \times 3$, which is the same as the input image.

26.5 RESULTS AND DISCUSSIONS

The performance and implementation of the model, as well as the complexity of the training process, are all crucial considerations in addition to the network's architecture. There were several parameter selections made during the deep neural network training process. The network was constructed with the use of the VGG16 transfer learning model. Experimental data guided the choice of hyperparameters. To improve the weights of the U-Net model, the Adam [21] optimizer was applied. Following extensive testing, the final model was tested using the 32 batch size, 0.0001 learning rate, and 40 epochs. Using the dice coefficient, IoU coefficient, and model loss, the model's performance has been assessed. The hyperparameter values that were utilized to build the suggested model are displayed in Table 26.2.

The model was created using the Keras TensorFlow Package and Python. A free and user-friendly framework for creating neural networks is called Keras. It is a free source and compatible with TensorFlow and Theano. Additionally, deep neural network calculations may be done on top of it. On the Google Colab Platform, using a Colab notebook outfitted with TensorFlow and a GPU, all simulations for this study were executed. The model's final output is shown in Table 26.3. The model produced a model loss of 0.4348, a dice coefficient of 0.5904, and an IoU coefficient of 0.7683.

TABLE 26.2
Values of Hyperparameters

Hyperparameters	Value
Batch size	32
Optimizer	Adam
Learning rate	0.0001
Number of epochs	40

TABLE 26.3

Performance Parameters Results

Performance Parameter	Value
Model loss	0.4348
Dice coefficient	0.5904
Intersection of union (IoU)	0.7683

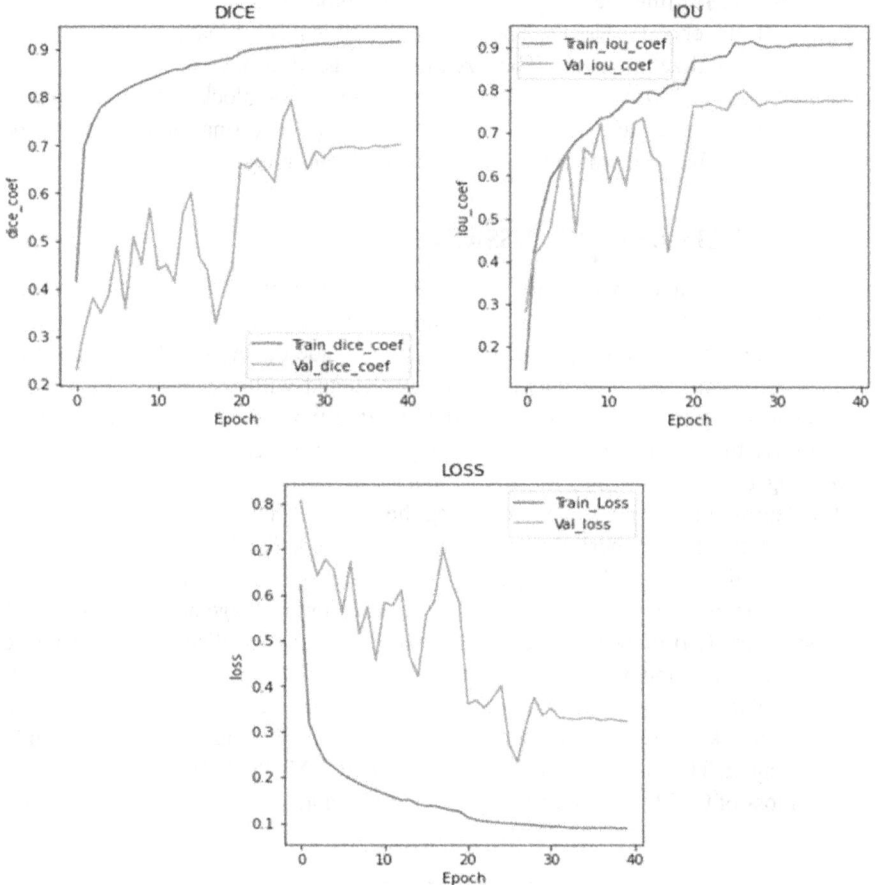

FIGURE 26.3 Plots for proposed model: (a) dice coefficient, (b) IoU coefficient, and (c) model loss.

Figure 26.3 shows different plots after the implementation of the proposed model. Figure 26.3(a) shows the dice coefficient, Figure 26.3(b) is IoU coefficient, and Figure 26.3(c) shows the model loss curve. These curves show the performance of the model during training and validation. It can be concluded from Figure 26.3 that dice and IoU coefficients were gradually increased during training and validation and model loss is gradually decreasing during training and validation.

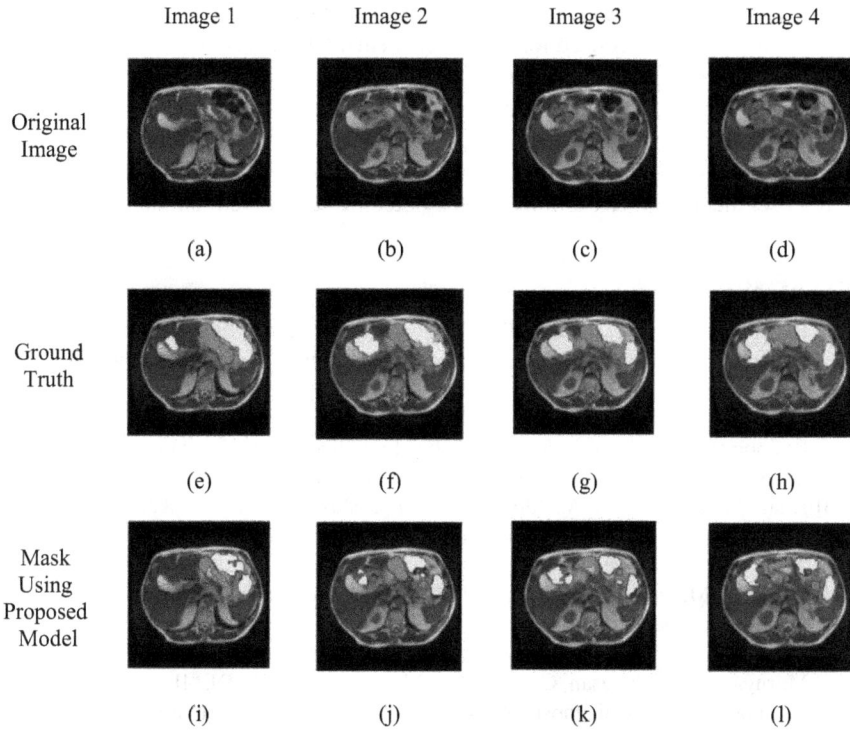

FIGURE 26.4 Output images: (a), (b), (c), and (d), random sample images; (e), (f), (g), and (h), ground truth for same images; and (i), (j), (k), and (l), mask obtained using proposed model.

Figure 26.4 shows the output images after the implementation of the model. Figure 26.4 includes some random original images, their corresponding ground truth and mask obtained using a proposed model for the same images. Here red colour represents the stomach, yellow represents the big bowel, and green represents the small bowel. From the figure, it can be concluded that the predicted mask is very much similar to the ground truth mask, which was obtained from the RLE encoding. The model performed very well for the segmentation of the stomach, large bowel, and small bowel in the GI tract using MRI scans.

26.6 CONCLUSION AND FUTURE WORK

A deep-learning-based strategy for automatically segmenting the stomach, small intestine, and large intestine in the GI tract using MRI images was proposed in this chapter. This work suggests using the VGG16 and U-Net fusion model to help radiation therapy by automatically segmenting GI tract organs. The pre-trained transfer learning VGG16 model, which was pretrained on the ImageNet dataset, serves as the encoder for the U-Net model. For downsampling, the U-Net model utilizes VGG16, and it features several upsampling blocks made up of several convolutions and batch normalization layers. The GI tract image segmentation dataset from the University of Wisconsin-Madison was used for the experiments. Dice coefficient, IoU, and model loss were the three performance measures used to assess the model's effectiveness. The dice coefficient, IOU, and model loss for the suggested model were, respectively, 0.5904, 0.7683, and 0.4348.

In our upcoming study, we will employ a different deep learning method for the GI tract segmentation task to improve the outcomes and expedite radiation therapy.

REFERENCES

[1] Reed, K. K., & Wickham, R. (2009, February). Review of the gastrointestinal tract: From macro to micro. In *Seminars in Oncology Nursing* (Vol. 25, No. 1, pp. 3–14). WB Saunders.

[2] Berzin, T. M., Parasa, S., Wallace, M. B., Gross, S. A., Repici, A., & Sharma, P. (2020). Position statement on priorities for artificial intelligence in GI endoscopy: A report by the ASGE task force. *Gastrointestinal Endoscopy*, 92(4), 951–959.

[3] Khan, M. A., Khan, M. A., Ahmed, F., et al. (2020). Gastrointestinal diseases segmentation and classification based on duo-deep architectures. *Pattern Recognition Letters*, 131, 193–204.

[4] Berzin, T. M., Parasa, S., Wallace, M. B., Gross, S. A., Repici, A., & Sharma, P. (2020). Position statement on priorities for artificial intelligence in GI endoscopy: A report by the ASGE task force. *Gastrointestinal Endoscopy*, 92(4), 951–959.

[5] Murugan, S., Venkatesan, C., Sumithra, M. G., et al. (2021). DEMENT: A deep learning model for early diagnosis of Alzheimer diseases and dementia from MR images. *IEEE Access*, 9, 90319–90329.

[6] Chandran, V., Sumithra, M. G., Karthick, A., et al. (2021). Diagnosis of cervical cancer based on ensemble deep learning network using colposcopy images. *BioMed Research International*, 2021, 1–15.

[7] Khosla, A., Khandnor, P., & Chand, T. (2020). A comparative analysis of signal processing and classification methods for different applications based on EEG signals. *Biocybernetics and Biomedical Engineering*, 40(2), 649–690.

[8] Tang, P., Liang, Q., Yan, X., et al. (2019). Efficient skin lesion segmentation using separable-UNet with stochastic weight averaging. *Computer Methods and Programs in Biomedicine*, 178, 289–301.

[9] van Ginneken, B., Schaefer-Prokop, C. M., & Prokop, M. (2011). Computer-aided diagnosis: How to move from the laboratory to the clinic. *Radiology*, 261(3), 719–732.

[10] Sykes, J. (2014). Reflections on the current status of commercial automated segmentation systems in clinical practice. *Journal of Medical Radiation Sciences*, 61(3), 131–134.

[11] Li, B., & Meng, M. Q. H. (2012). Tumor recognition in wireless capsule endoscopy images using textural features and SVM-based feature selection. *IEEE Transactions on Information Technology in Biomedicine*, 16(3), 323–329.

[12] Zhou, M., Bao, G., Geng, Y., Alkandari, B., & Li, X. (2014, October). Polyp detection and radius measurement in small intestine using video capsule endoscopy. In *2014 7th International Conference on Biomedical Engineering and Informatics* (pp. 237–241). IEEE.

[13] Wang, Y., Tavanapong, W., Wong, J., Oh, J. H., & De Groen, P. C. (2015). Polyp-alert: Near real-time feedback during colonoscopy. *Computer Methods and Programs in Biomedicine*, 120(3), 164–179.

[14] Li, Q., Yang, G., Chen, Z., Huang, B., Chen, L., Xu, D., . . . & Wang, T. (2017, October). Colorectal polyp segmentation using a fully convolutional neural network. In *2017 10th International Congress on Image and Signal Processing, Biomedical Engineering and Informatics (CISP-BMEI)* (pp. 1–5). IEEE.

[15] Yuan, Y., Li, D., & Meng, M. Q. H. (2017). Automatic polyp detection via a novel unified bottom-up and top-down saliency approach. *IEEE Journal of Biomedical and Health Informatics*, 22(4), 1250–1260.

[16] Dijkstra, W., Sobiecki, A., Bernal, J., & Telea, A. C. (2019). Towards a single solution for polyp detection, localization and segmentation in colonoscopy images. In *VISIGRAPP (4: VISAPP)* (pp. 616–625). SCITEPRESS – Science and Technology Publications, Lda.

[17] Nguyen, N. Q., Vo, D. M., & Lee, S. W. (2020). Contour-aware polyp segmentation in colonoscopy images using detailed upsampling encoder-decoder networks. *IEEE Access*, 8, 99495–99508.

[18] Huang, C. H., Wu, H. Y., & Lin, Y. L. (2021). Hardnet-mseg: A simple encoder-decoder polyp segmentation neural network that achieves over 0.9 mean dice and 86 fps. *arXiv preprint arXiv:2101.07172*.

[19] Simonyan, K., & Zisserman, A. (2014). Very deep convolutional networks for large-scale image recognition. *arXiv preprint arXiv:1409.1556*.

[20] Ronneberger, O., Fischer, P., & Brox, T. (2015, October). U-net: Convolutional networks for biomedical image segmentation. In *International Conference on Medical Image Computing and Computer-Assisted Intervention* (pp. 234–241). Springer, Cham.

[21] Zhang, Z. (2018, June). Improved adam optimizer for deep neural networks. In *2018 IEEE/ACM 26th International Symposium on Quality of Service (IWQoS)* (pp. 1–2). IEEE.

27 Systematic Review of Image Computing Under Some Basic and Improvised Techniques

Neelam Kumari, Preeti Sharma, Isha Kansal

27.1 INTRODUCTION

Images have enormous use in human life in various forms—text, video, audio, and so on. Human and computer interaction fully depends on image processing. The field of processing digital photos with an electronic computer is known as DIP. A digital picture is made up of a small number of elements, also known as picture components, image pels, or pixels. Unlike humans, who can catch only objects in the visible spectrum, DIP can works on pictures acquired across any band of the electromagnetic spectrum. So DIP plays an important role in various a broad range of fields, such as medical science, remote sensing, neuroimaging, air imaging, and surveillance. It also has a huge application on the professional front, such as in biomedical, forensic, biometrics, military, photoshop, multimedia, agriculture, traffic control, weather forecasting, digital image steganography, and many more. This chapter is an honest endeavour to bring all these aspects to one place and create a miniature world of image processing. This work tries to cover all possible areas of image processing and uses techniques/algorithms, detailed methods, advantages, limitations, and applications of methods. The upcoming section covers the overview of all areas of DIP and the following subsections cover techniques, literature review, and contributions from image enhancement research of related areas.

27.2 OVERVIEW OF DIP TECHNIQUES

DIP is followed by some fundamental techniques to get the desirable result called DIP techniques, as shown in Figure 27.1. Image acquisition is the very beginning step of image fundamentals. In this process, an image is acquired by a digital device. Image type depends on the light spectrum in which the image is acquired. All the techniques are segregated into three categories: In the first type, output and input are images, also called low-level techniques. In the second type, input is image, and results are attributes extracted from the input image; these are also called medium-level techniques.

DOI: 10.1201/9781003367161-27

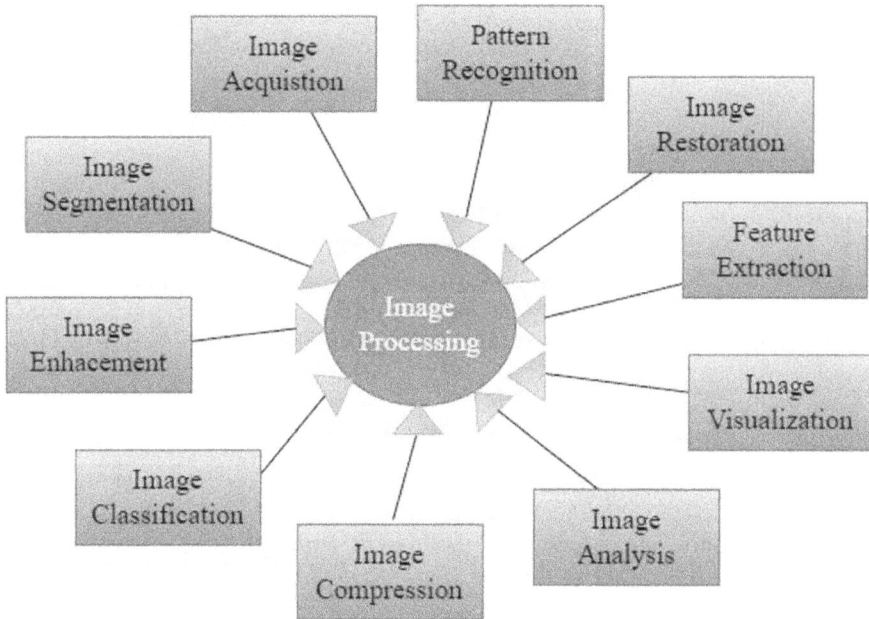

FIGURE 27.1 Broad areas of image processing.

The third type comes under the high-level techniques which has input as image and output understanding and interpretation of the input image, such as pattern recognition: the identification of patterns or objects in an image or frequentness in data. It is widely used in statistical data analysis, signal processing, image analysis, information retrieval, bioinformatics, data compression, computer graphics, and machine learning. In image visualization, the input-output is different from the original image after processing. It is a process of converting (rendering) an image pixel/voxel into two-dimensional (2D) or three-dimensional (3D) graphical representation. Image analysis is a technique to analyse images to read and measure statistics of images or objects that are hard to analyse in physical form (*e.g.* to calculate the radius of a planet). Image restoration, enhancement, compression, segmentation, and classification and feature extraction are some of the techniques that are the centrepoint of study in this chapter because these are the basic and most important techniques used in every area from the last few decades till now.

27.2.1 IMAGE RESTORATION

It is the technique of fixing images by removing noise and fuzziness from the image to restore originality, and it is essential before other image-processing operations. The main reason for noisy and fuzzy images is sensor illumination level, heat, and environmental conditions, such as snow, rain, aerial imaging, and underwater imaging. The image restoration part is decided based on degradation function and additive noise. In the past, various methods have been discussed and proposed in this regard by researchers. This study tries to cover all those filters, methods, and techniques.

27.2.1.1 Techniques of Image Restoration

There are several filters and techniques used in image restoration as shown in Figure 27.2. These are used for different purposes, such as median and contra-harmonic mean filter used to remove impulse noise. Midpoint filter and arithmetic mean remove Gaussian and uniform noise, respectively. Alpha trimmed extracts Gaussian and salt-and-pepper noise. Trimmed filter and geometric mean remove Gaussian and pepper noise. Max filter and min filter remove salt and pepper noise, respectively.

27.2.1.2 Related Research Work on Image Restoration

Conventional image restoration techniques are generally based on image degradation models applied to the latent image. Restoration methods are devoted to the actual reflection feature coverage of the images. However, image degradation is more complex to manifest in the real world. Deep-learning- and machine-learning-based image restoration techniques are in trend to solve complex problems in several domains, such as SAR imaging, medical imaging, agriculture, remote sensing, low-resolution imaging, and underwater image restoration as shown in Table 27.1.

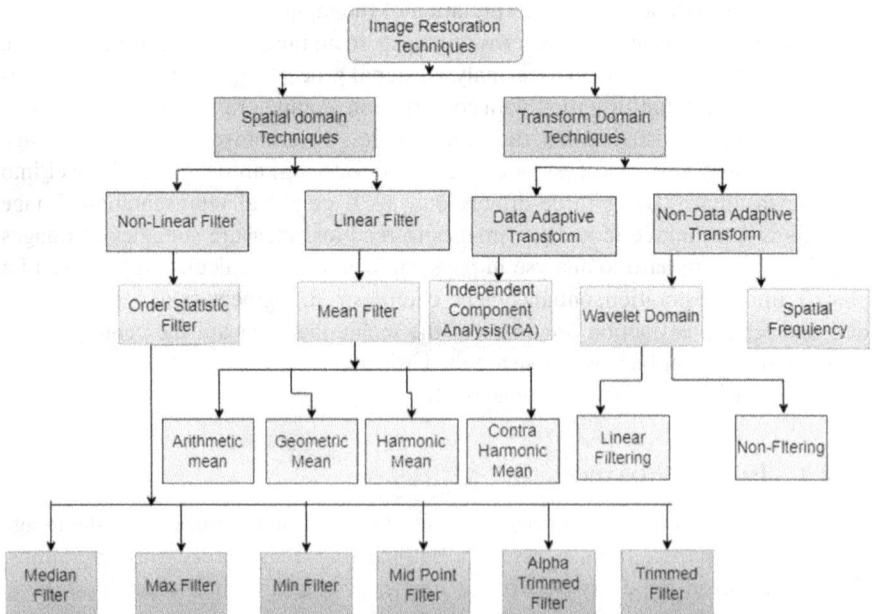

FIGURE 27.2 Image restoration techniques

TABLE 27.1
Related Research on Image Restoration Techniques

Sr. No.	Name	Goals	Methods	Results	Applications
1	Y. Yu *et al.* [1]	To approximate solid depth-map of a single monocular natural image	Split attention multiframe alignment networks (SAMANet), warping repetition detection module, attention fusion module	The superior performance of SAMANet over existing registration algorithms	Real-world noisy photographs, video restoration, *etc.*
2	E. Park *et al.* [2]	To eliminate poor contrast, haze, colour distortion, blur, and light scattering and absorption in underwater imaging	Underwater algorithm for image restoration based on the underwater image formation mode	Superior site radiance restoration performance SAM.	Underwater environment monitoring, fabrication for artificial facilities, underwater imaging to rescue of sunken ships
3	Zhaoxia Wang *et al.* [3]	To successfully restore a damaged image by optimizing the genetic algorithm.	The novel genetic algorithm-based restoration method	Higher peak signal-to-noise ratio (PSNR) values as compared to other contemporary methods	Transportation, medical, aerospace, energy, and other fields
4	Guojia Hou *et al.* [4]	To improve the mentoring system and visibility of driver	Underwater synthesis of image algorithm and synthetic dataset for underwater image	Algorithms are validated and solved from psycho-physical point of view through this approach. Algorithms can be used to develop sophisticated driver visibility descriptors.	Object recognition, underwater geological and biological exploration, AI, *etc.*
5	S. Rishi *et al.* [5]	To find image tamper for self-recovery	Secret-key-inspired pseudo-arbitrary binary sequence for watermarking	This model validated the image and estimated the tampered portion with 99% efficiency.	Digital data, multimedia
6	F. Xuan *et al.* [6]	To restore hyperspectral image (his) affected due to mixed noise, impulse noise, Gaussian noise, and stripe noise	Low-rank prior information and spatial spectral directional total variation (LRSSDTV), with alternating direction method of multipliers	As compared to existing popular approaches, the LRSSDTV method is ahead in terms of objective criteria and visual fidelity.	Agriculture, astronomy, mineralogy, biomedical field, military, environmental monitoring, and urban planning

(Continued)

TABLE 27.1 (*Continued*)

Sr. No.	Name	Goals	Methods	Results	Applications
7	K. Apurva *et al.* [7]	To quickly and efficiently dehaze and defog a single image	Novel thresholding-based restoration approach and gamma transformation.	Low complexity of the calculation, reducing processing time Higher perceptual quality	Intelligent vehicles, terrain classification, surveillance, remote sensing
8	Z. Xianquan *et al.* [8]	To remove noise from embedded images	kNN-bit estimation algorithm	This approach has great feat than some existing filtering techniques.	Military map, stego image used over the internet

27.2.1.3 Contributions from Image Restoration Research Work

In this section, image restoration technique from different research work is discussed for various instances, such as future scope, limitations, and advantages. Y. Yu *et al.* [1] focus their work on image alignment in such cases problems of ghost effect and aperture rise as the component of optical flow can only be measured in the direction of intensity gradient, not the components tangential to the intensity gradient as shown in Figure 27.3. Models based on residual dense network (RDN) have the issue of over-sharpening, which needs to be fixed. E. Park *et al.* [2] used non-reference-based matrix in his work that is why the performance approximation is not always correct. Despite the great result, the synthetic underwater image dataset (SUID) also has some drawbacks. First, the SUID is designed based on the interpreted underwater image formation model (IFM) without seeing the response of forward-scattering. Secret-key-based pseudo-random binary sequence is used by S. Rishi *et al.* [5] as a fragile watermark for tamper detection. The restoration of the tampered region needs to be investigated to get better parametric results dissuades the behaviour of the scheme in future. Motion blur is the reason of forward-scattering exists in some live underwater imaging. Additionally, the effect of non-uniform brightness from unnatural light sources is also not taken into account. Compared with some real-world underwater images captured in extreme deteriorated scenarios, some features such as light sources and plankton light spots are not reproduced in the synthetic underwater image.

27.2.2 Image Enhancement

Image enhancement is a combination of techniques that raise the visual appearance of a picture or to change it to a form better for analysis by a machine or human. This technique enhances image colour, contrast, brightness, saturation, background, and so on; these are enhanced as a prerequisite for further image processing or to get a better image. Image enhancement does not require prior knowledge of image quality

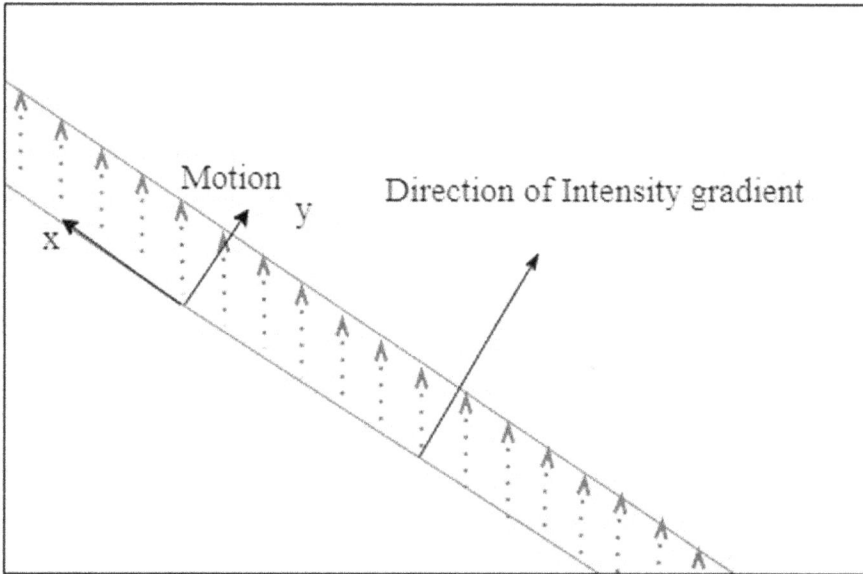

FIGURE 27.3 Optical flow can only measure the component y, not x.

degradation. Picture captured in low light with low-resolution and high-intensity light must always pass this process for better visibility and next-level processing. Histogram equalization, gamma correction, fusion, and retinex-based algorithms are an example of some techniques used during this process. Figure 27.4 shows the process of image enhancement.

27.2.2.1 Techniques of Image Enhancement

There are many techniques to enhance an image. Some of these techniques are shown in Figure 27.5, such as spatial domain and frequency domain, homomorphic filtering, and histogram-based techniques. The spatial domain has two types: point transformation (processing operations applied on pixels) and spatial filtering. The holomorphic filtering techniques normalize the image and increased the contrast globally. Adaptive histogram enhances the contrast locally. To modify pixel brightness in an image histogram, stretching high pass filters is used to suppress the high-frequency band, and low-frequency filters riddle the low-frequency band.

27.2.2.2 Related Research Work on Image Enhancement

Several approaches have been considered in the study to enhance the images aspects from different domains. However, the earliest techniques mainly included contrast expansion [9], grey level sliding, and various histogram processing [10, 11]. Global enhancement approaches may get affected by the loss of detail in parts of local areas due to the inherent distinction present in the image. In this study, as mentioned in Table 27.2, most of the techniques included are gamma correction and histogram equalization

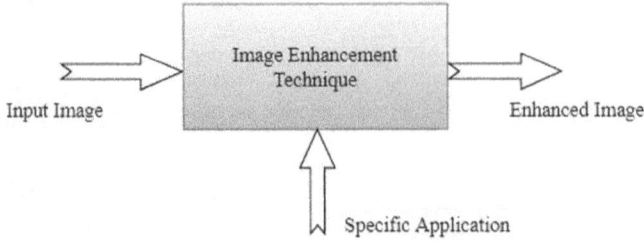

FIGURE 27.4   Image enhancement process.

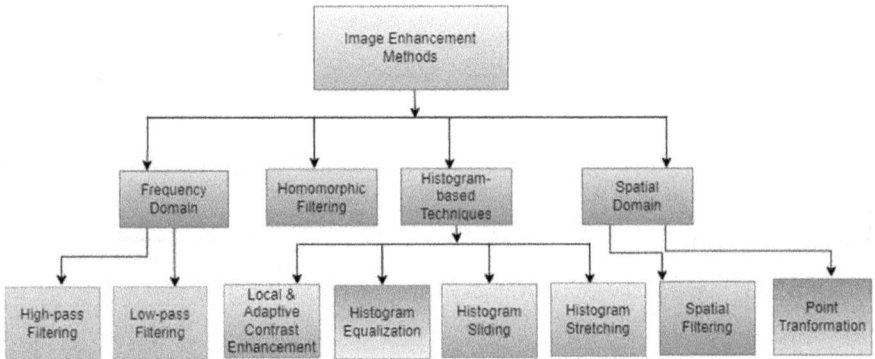

FIGURE 27.5   Image enhancement techniques.

TABLE 27.2
Related Research on Image Enhancement Techniques

Sr. No.	Name	Purpose	Methods	Results	Applications
1	L. Changli *et al.* [12]	To enhance the low-contrast images in the eminent high-intensity light areas	Pair of complementary gamma functions, adaptive gamma correction with weighting distribution	The method can significantly enhance the contrast and the detail of the low-light image.	Dehazing foggy images, enhancement of underwater images and saliency detection.
2	Yu Guo *et al.* [13]	To enhance low visibility and remove impulsive noise from marine images acquired in low light	Brightness-preserving bi-histogram equalization (BBHE), minimum-mean brightness error bi-histogram equalization (MMBEBHE), and background brightness-preserving histogram equalization (BBPHE)	The proposed approach has the scope of generating more natural enhanced images in different low-light surroundings, and this method is easy and simple use.	Maritime traffic supervision and management

Sr. No.	Name	Purpose	Methods	Results	Applications
3	A. Khodabakhsh *et al.* [14]	To remove unavoidable artifacts in the image, such as pulse rate, breathing, *etc.*	SOM as a preprocessor, nearest neighbour algorithm as a classifier, and a high-frequency filter as a high-frequency image elements extractor	In bounded computational means, improved evaluation criteria among the whole traditional state of the art, more steady performance, and resolution on images with different scale factors	Biomedical, *e.g.,* restoring medical diffusion-weighted and MRI images
4	N. Rabia *et al.* [15]	Computed tomography image enhancement as a preprocessing step to highlight the relevant structures	Optimized guided contrast enhancement (OPTGCE).	The method has better enhancement results, reflects better performance, and has higher entropy values.	Medical imaging to better visual appearance of relevant organ structures for improved intervention and diagnosis.
5	P. Tae Hee *et al.* [16]	To eliminate haze and to enhance underwater photos	Innovative colour-balance algorithm for enhancement of sand-dust images	The proposed approach is quick and produces improved results in a wide variety of sand-dust imagery.	Sand-dust images and underwater images
6	P. Qiaoying *et al.* [17]	To enhance the visual effect of unsatisfactory visible and infrared fused images	Multi-level latent low-rank representation combines image enhancement and multiple visual weight information-based infrared and visible image-fusions.	Observations reveal that the multi-visual fusion of detailed images contributes more to bettering the fusion results. Superior as compared to other contemporary methods.	Investigating crime-based scenes, conducting search and rescue operations
7	R. Wang *et al.* [18]	To remove the noise of low-light images during the enhancement process	Mixed-attention guided generative adversarial network (MAGAN)	Mixed-attention methods with GAN for unsupervised low-light image enhancement is superior to even other state-of-the-art models.	Stormy- and foggy-weather images

(HE) based. Both of these approaches work on global enhancement and fusion-based algorithms are used that take care of the local enhancement aspect as well.

27.2.2.3   Contributions from Image Enhancement Research Work

This section covers image enhancement novel techniques from different research work. Researchers used basic techniques in hybrid form and after fusion with other techniques to improve performance and various other parameters to enhance the image quality. Contextual and variational contrast enhancement (CVC) approach is used with histogram-mapping, which emphasizes on large grey-level variation has issues when dealing with very dark regions, due to the limitations of stretching range. Histogram-based methods used by Yu Guo *et al.* [13] are supposed to increase the flexibility of the histogram equalization system by dividing the original histogram to enhance low-light images. MRI imaging is mostly low resolution, concluded from A. Khodabakhsh *et al.*'s [14] work. So the scanning is a very expensive, time-consuming, and cumbersome process. However, the images with optimal resolution and quality can be reconstructed with learning-based super-resolution methods, such as self-organizing map (SOM) used in this work. It is an unsupervised learning-based model and little distinct from other artificial neural networks. SOM does not train by backpropagation with stochastic gradient descent (SGD) to adjust weights in neurons it uses competitive learning. This kind of artificial neural network is used as a dimension reduction model and also discovers the correlation between the data used for medical imaging. Nowadays, the medical field has two main challenges concerned to image enhancement. First, most recent enhancement techniques are applied on only certain types of images. Second, there is no well-established yardstick to estimate the existing enhancement techniques. A 2D histogram has the disadvantage of not considering strong spatial relation of pixels and exploiting it to avoid side effects associated with histogram approaches. The sand-dust image enhancement algorithm proposed by P. Tae Hee *et al.* [16] fully eliminate the red veil when the image's high pixel value's red histogram has a quite narrow peak. The future work of this research is a robust enhancement algorithm that can be deployed well to red-storm images.

27.2.3   FEATURE EXTRACTION

It is the process of converting raw data into numerical features, which may be processed while preserving the feature information in a real dataset with accuracy. There are multiple features based on which object can be extracted for a given set of images for further process. These multiple features may contain the object's shape, colour, texture, or statistical feature, such as entropy, variance, and so on. Feature extraction has many applications in various fields, such as forensics, biometrics, face reorganization, herb reorganization, watermarking [19, 20], and many more.

27.2.3.1   Techniques of Feature Extraction

Feature extraction can be applied to an image by feature extraction and feature selection methods. These techniques are applied to an image in various ways. Some of them are explained in Figure 27.6, like principal component analysis (PCA), grey-level

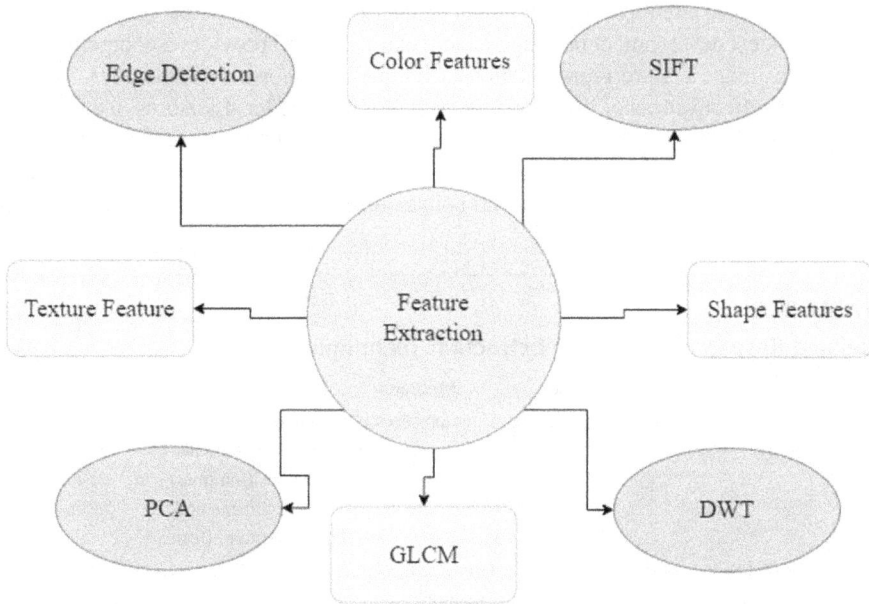

FIGURE 27.6   Feature extraction techniques.

co-occurrence matrix (GLCM) [21], shape features, discrete wavelet transform (DWT), shift invariance-based feature extraction technique (SIFT), colour feature, edge detection, and texture feature [22]. PCA is used for both statistical analyses and dimension reduction. This technique identifies the object and differentiates it on the bases of similarity and dissimilarity, but it is more useful when dealing with higher than 2D data. GLCM is a technique to calculate the texture feature of an image. SIFT locates the local feature in an image. The key points generated using SIFT are used as image attributes at the time of model training. The main advantage of SIFT features over other techniques edge features and histogram of oriented gradients (HOG) is that they are not affected by the size of the image. DWT gives frequency and spatial domain details simultaneously. It is an extensively applied technique in image feature extraction. Colour feature extraction is based on the histogram of hue, saturation, and value (HSV) model. Hue means the effects of colours, such as blue, green, and red. Saturation means the intensity of a particular colour. Value means the brightness of particular colour. They are used in colour pickers, image editing, and image analysis in DIP and computer vision.

27.2.3.2   Related Research Work on Feature Extraction

This study examines the advancement in the field of feature extraction in different areas. Feature extraction is an urgent step for interactive media handling. The detailed instructions required to extricate ideal highlights are indeed an issue in DIP for computer vision. How to extract ideal features that can reflect the intrinsic content of the images as completely is still a challenging problem in computer vision. However, research has paid attention to this problem in the last decades. So in this study, the focus is on the latest development in image feature extraction and provides a comprehensive survey on image feature representation techniques, as shown in Table 27.3. In particular, the effectiveness of the fusion of overall image and local features in automatic image annotation and content-based image retrieval community are analysed, together with some classic models and their illustrations in the literature. Eventually, this study summarizes some vital conclusions and points out future potential analysis directions.

TABLE 27.3

Related Research on Feature Extraction Techniques

Sr. No.	Name	Purpose	Methods	Results	Applications
1	Tao Jiang *et al.* [23]	To extract contour features	Edge-preserving filters (bilateral filter)	Convolution and Filtering together to the positron images in three dimensions have better effects.	Industrial non-destructive testing.
2	Khalid Saeed *et al.* [24]	To create high-resolution images from low-resolution to extract contour features	Granular-level feature-extraction method to create high-resolution images	The overall performance of this high quality resolution image reconstruction methodology is pretty satisfactory.	Forensic, biometrics.
3	Hang Fu, Genyun Sun, *et al.* [25]	To denoise and extract features from (HSI)	Novel 1.5D singular spectrum analysis (SSA) approach for in situ spectral and spatial feature extraction in HSI	1.5D-SSA achieves best results as compared to various state of art spectral-spatial methods.	Real-time pedestrian detection, coastline detection.
4	Mohammad A. Alzubaidi *et al.* [26]	Local feature extraction framework for lung cancer detection using CT scan imagery	SVM with HOG (histogram of oriented gradients)	CT scanning produces a quick test result without pain, it gives information related to the tumour structure, size and location	Radiologists in detecting lung cancer and suspicious areas in CT-scan.
5	Yu Haifie *et al.* [27]	Spectral spatial feature-extraction approach for hyperspectral image	Random forest, GF-3 SAR	The method proposed here is robust and also attains competitive results related to several "state of the art" spectral-spatial HSI classification models when checked on three distinct datasets	Mineral exploration, monitoring the crop, management of resources *etc*.

Sr. No.	Name	Purpose	Methods	Results	Applications
6	C. Azura *et al.* [28]	To distinguish between herb species in the same family based on chromatographic signal patterns	Weighted histogram analysis method (WHAM)	The classification results show the highest accuracy for both SVM and k-NN using weighted histogram analysis method.	Herbs recognition.
7	A. Tusneem *et al.* [29]	Feature extraction from white blood cells	Feature extraction using CMYK moment localization, deep learning and FCL, RF, SVM, and XGBoost	The result was improved detection of white blood cells or leukocyte nuclei compared to other methods and improved detection of leukocytes, especially light-coloured leukocytes.	Different types of leukocyte detection, including both normal and acute myeloid leukaemia cancer cells
8	A. Osamah Mohammed *et al.* [30]	To select features for textual content classification	Metaheuristic algorithms	Increase accuracy, decrease error rate, improve classifier time efficiency, and achieve high performance in terms of text classification accuracy	Data mining, data analysis, and pattern classification
9.	S. A. Alazawi *et al.* [21]	To find features using second-order statistics	Gray-level co-occurrence matrix.	This approach is capable to achieve a high accuracy degree in face image retrieval	Surveillance, crime investigation

27.2.3.3 Contributions from Feature Extraction Research Work

In this section, feature extraction technique from different research work is discussed for different aspects, such as future scope, limitations, and advantages. The detectors' intrinsic noise and scattering within the tested metallic cavity is always an issue. Tao Jiang *et al.*'s [23] work addresses foreign object detection inside dense metal cavities. It provides new research ideas for the non-destructive study of metal cavities, such as pipes and hydraulic components in industrial plants. Performance is limited when using only the spectral features in the IISV data classification. While kurtosis wavelet energy (KWE) and kurtosis curvelet energy (KCE) have good performance for texture recognition and image segmentation in SAR, they can be studied in future research for more efficient feature extraction and data classification in HSV. Mohammad A. Alzubaidi *et al.* [26] investigated a comprehensive and comparative global and local feature extraction framework for lung cancer detection using CT imaging. This study shows that deep learning techniques for feature extraction are more robust in terms of scale, rotation, occlusion, distortion, and so on. Yu Haifie *et al.* [27] included MobileSegNet designed to eliminate train-ing load and improve prediction accuracy, but there are some limitations, such as complicated execution steps, too many hyper-parameters, and large influence of sub-factors. Histogram-based methods have limitation of over-enhanced images. However, in the case of feature extraction, the multiple histograms' weighting tech-nique offers the advantage of extracting all data information at once and reducing the size of the data feature.

27.2.4 IMAGE COMPRESSION

Today's global network consists of a number of different sub-networks that provide various information and communication functions are interconnected using stan-dardized communication protocols. Transmission and storage of such information require good bandwidth and storage space. So utilizing the limited bandwidth and storage compression is the most useful technique. This technique is used to utilize the storage capacity efficiently by compressing images or decreasing the size of images. In this process, large-size images are compressed by some methods in small sizes by lossy and lossless compression techniques. Redundancy and irrelevancy reduction are the two main components of compression [31]. Redundancy is remov-ing duplicates from images/videos. Irrelevant reduction filters out parts of the image/video that are imperceptible or irrelevant.

27.2.4.1 Techniques of Image Compression

Image-compression techniques are usually divided into lossy and lossless com-pression, as shown in Figure 27.7. Lossless compression is useful for applications where the low-quality image is unacceptable, such as in medical imaging. Images

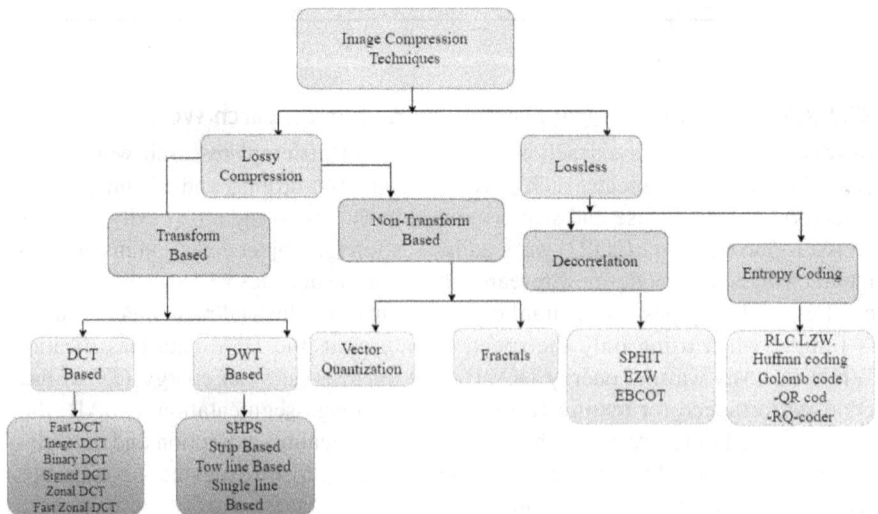

FIGURE 27.7 Image compression techniques.

compressed using lossy techniques are not reconstructed as accurately as the original image. This technique is suitable for applications where the loss of redundant data is acceptable.

Lossy compression techniques are classified into transform-based and non-transform-based techniques. Raw input images are converted into a convenient format that machines can easily access and read. Spatial information present in the image is grouped into shapes based on the frequency of occurrence of the pixel data. This approach is called transform-based compression. Several modified techniques are also introduced, such as discrete cosine transform (DCT), DWT [31], and set partitioning in hierarchical trees (SPIHT) are techniques introduced for compressing multispectral images in the spectral dimension by executing Karhunen Loève transform (KLT), vector quantization (VQ), fractals, embedded zerotree wavelet (EZW), embedded block coding with optimized truncation (EBCOT), run-length coding (RLC), Lempel-Ziv-Welch (LZW), quick response (QR) code, mixed reality (MR), and Golomb coding, proposed by Solomon W. Golomb in 1960s. It is a lossless data compression technique.

27.2.4.2   Related Research Work on Image Compression

Recently, lossless and lossy compression methods have extensively used deep learning, significantly improving image compression performance. Although research has already been done on compressing the original unencrypted binary image, few approaches have focused on compressing the encrypted binary image. Contracts, signatures, and halftone images are still widely used as binary images. Chuntao Wang *et al.* [32] developed a lossy compression technique for encoded or encrypted binary images by using the Markov random field (MRF) approach named sum-product algorithm (SPA). There is some more research work mentioned in Table 27.4

27.2.4.3   Contributions from Image Compression Research Work

This section briefly introduces the image compression techniques mentioned in the Table 27.4 and their future scope, advantage, and disadvantages. MRF-based lossy compression method is a state of art for compression of encrypted binary images. But the main drawback is limited to only binary images, and it is a lossy compression. The DNNs used in real-life critical applications still needs to overcome the trust barrier. To address this issue, a novel end-to-end neural image compression and interpretation method is proposed by Xiang Li *et al.* [34]. In the future, the approach can be expended and reinvented technique in another areas, such as textual and bioinformatics for neural estimation (interpretation). The performance of the compact image schema proposed by Haisheng Fu. *et al.* [35] in the YUV field can be further improved by boosting the performance of the foundation layer and instigating a new arithmetic coding procedure for compact image development. Image-as-protein (IaP) is a novel multi-level lossy type of compression technique for greyscale images only. This can be further modified for colour images. The Compression Artifacts Simulation Network (CAS-Net) used in this study produces images with a better trade-off between image quality and bit rate.

TABLE 27.4

Related Research on Image Compression Techniques

Sr. No.	Name	Purpose	Methods	Results	Applications
1	Chuntao Wang et al. [32]	To compress encrypted binary images	MRF-based lossy compression. Used downsampling and LDPC-based encoding to access encrypted data	Achieves favourable compression efficiency and is comparable or better than the original (Joint Photographic Experts Group) JPEG2	Practical scenarios like a signature, contract, and halftoning
2	Tung Thanh Pham et al. [33]	Experimental image quality assessment solution for image/video with compression artifacts	End-to-end IQA method on high-efficiency video coding (HEVC) compression artifacts	Suitable for images/videos with compression artifacts, not only in HEVC video compression but also with JPEG or JPEG-2000 image compression standards	Distorted images, saliency maps, and reference images
3	Xiang Li et al. [34]	The convolutional neural network (CNN) predictions compress the input images for efficient storage or transmission.	Novel end-to-end Neural Image Explanation and Compression (NICE)	Improved explanation quality and semantic image compression rate	Drones, surveillance cameras, and self-driving cars, where decisions are critical and storage/network bandwidth is less
4	Haishe ng Fu et al. [35]	To make output images compact	Extended hybrid image compression scheme based on soft-to-hard quantification	Performance is excellent over a wide range of bit rates with MS-SSIM metrics and avoids network overfitting.	Visualization of the compact image
5	Mohammad Nassef et al. [36]	With different computational perspectives, compression problems of images	Multilevel lossy-compression method for greyscale images named image-as-protein	Algorithm is promising compared to the famous JPEG lossy image compression standard.	Bandwidth-constrained social media and navigation map applications used to compress medical, satellite, and security surveillance imagery

Sr. No.	Name	Purpose	Methods	Results	Applications
6	Kwang -Hyun Uhm *et al.* [34]	To state the issue that CNN-based ISPs have not explicitly considered conventional-camera image signal processing to be lossy compressed	Compression Artifacts Simulation Network (CAS-Net), Image Signal Processing Network (ISP-Net)	ISP-Net have improved image quality as compared to the compression-agnostic ISP-Net.	High-resolution camera ISP
7	Binghu a Xie *et al.* [37]	To remove compression artifacts and produce photorealistic images	Weakly connected dense generative adversarial network	Outperforms state-of-the-art compression artifact elimination methods in terms of peak signal-to-noise ratio and structural similarity index	Embedding a watermark into the compressed images and to compress medical imagery
8	Bianca Jansen Van Rensburg *et al.* [38]	To compress a large 3D image	Draco 3D object crypto-compression scheme	This approach is effective in respect of time and compression rate.	3D image compression

27.2.5   IMAGE SEGMENTATION

It is a process of subdividing an image into its constituent parts to find the object of interest or partitioning a digital image into multiple regions or sets of pixels on the bases of colour, texture, and intensity gradient called image segmentation. Segmentation applies in many fields, from medicine to the film industry. Medical applications of segmentation include identifying injured tissue and bone and detecting suspicious structures. It is also used to track objects in a sequence of images or classify features, such as oil reserves in satellite imagery.

27.2.5.1   Techniques of Image Segmentation

Many general-purpose techniques have been developed for image segmentation. Since the image segmentation problem has no general solution, these techniques are usually required to be integrated with the domain database to resolve the problem domain segmentation successfully. As shown in Figure 27.8, image segmentation methods are grouped into two main classes: layered segmentation methods

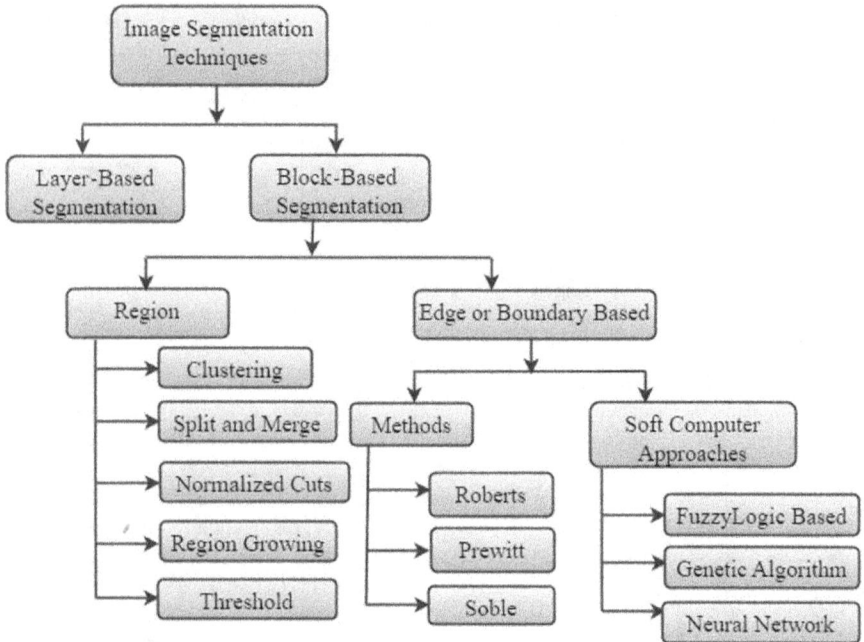

FIGURE 27.8 Image segmentation techniques.

and block-based segmentation. In the first kind of segmentation, information like localization of object, labelled pixel, and masks are combined and explain the looks and deep ordering, which evaluates class and instance segmentation. Block-based segmentation methods are utilized on various features detected in the image, such as colour information or knowledge related to the pixels that specify edges, boundaries, or texture information. Block-based image segmentation methods are based on two things: similarity and discontinuity. Methods to find similarity are clustering, region merges, normalized cuts, region growing, and threshold [39]. Discontinuity is found by edge- or boundary-based methods. This is further classified into methods and soft computer approaches (SCA) [40]. SCA is a class of computational techniques that are based on AI and natural selection; examples are fuzzy logic, genetic algorithm, and neural network. Fuzzy logic is useful for identifying pixels, whether belong to a uniform region or edge. Genetic algorithms represent segmentation as an optimization problem and can easily detect about edges and boundary. An encoder-decoder framework is used in a neural network for segmentation. Encoder work is a kind of barrier, and decoder takes layers of uniform object from that barrier for sampling purpose. Edge- and boundary-based methods are Sobel, Prewitt, and Roberts [41]. In the Sobel method, intensity gradient is calculated pixel-wise in an image. On the basis of intensity, it identifies edges. The Prewitt method changes the image into a high-level noise image, then edge detection filter identifies the high and low contrast. Involution filters apply the negative weight on one edge and positive weight on others. Lawrence Roberts proposed this method in 1963; that is why it is named Roberts. It works on 2D greyscale images for edge detection using intensity gradient.

27.2.5.2   Related Research Work on Image Segmentation

The most trending area of DIP is image segmentation nowadays. In the field of image segmentation, astounding work has been done since the invention of MRI and CT scan, and a lot of work has still to be done. This study in Table 27.5 illustrates the research work done in past few years.

TABLE 27.5

Related Research on Image Segmentation Techniques

Sr. No.	Name	Purpose	Methods	Results	Applications
1	Pengshuai Yin *et al.* [42]	To fully recover the lost information of structure in some tasks, *e.g.* the retinal vessel segmentation	Multiscale network with guided image filter module experiments on datasets (REFUGE, ORIGA, DRIVE, and CHASEDB1) [43]	Recovers spatial information lost by the down sampling operation. This approach enables end-to-end training and rapid inference.	Optic disc, cup and tumour segmentation
2	Hui Liu *et al.* [44]	Efficient and rapid segmentation for imminent four-dimensional heart image	Itti-visual saliency model and GrabCut algorithm, image segmentation technique applied on visual saliency model	The proposed algorithm realizes the automation of positron emission tomography (PET). Picture segmentation boosts segmentation productivity and effectiveness and gives eminent data support in biomedical imagery.	Clinical practice, cardiovascular and neurological disease
3	Liu Weixia *et al.* [45]	Due to the large variety in the human tongues on colour, texture, shape, and weak edges, tongue segmentation is a challenging task. This work introduces new methods to alleviate the issue.	With the representation of sparse matrix, a patch-driven segmentation method	Improved both segmentation accuracy and robustness	TCM (traditional Chinese medicine) and to diagnose various diseases

(Continued)

TABLE 27.5 (*Continued*)

Sr. No.	Name	Purpose	Methods	Results	Applications
4	W. Aichen *et al.* [46]	To achieve pixel-wise semantic segmentation of field picture into crop, weed, and soil	Site-specific weed management with auto contrast algorithm, histogram equalization (HE)	Indicates the colour space conversion and index of vegetation. In presence of near infrared (NIR) information image enhancement significantly improved quality of segmentation accuracy, improvement in robustness	Different lighting conditions in the field, occluded or coinciding leaves of crops and weeds, different levels of plant growth
5	Liu Cheng *et al.* [47]	The main focus is issue of noisy images' segmentation.	Efficient variational-level-set model based on adaptive filtered image.	The proposed method is compared to the general standard histogram distribution and local histogram equalization.	Oil spill and synthetic-aperture-radar images segmentation.

27.2.5.3   Contributions from Image Segmentation Research Work

The purpose of including this section to conclude the paperwork mentioned in Table 27.5 with summary, limitations, and future research directions. The model proposed by Pengshuai Yin *et al.* [42] is similar to U-Net, although U-Net is extensively used for clinical image segmentation. U-Net applies skip connections to recover the spatial data loss because of downsampling technique. Some pieces of information like tiny vessels cannot reconstruct by skip connections or upsampling. An efficient and simple multi-scale guided filter-based network is included in this study to restore the small vessel information. The model used is named deep architecture. By considering the sensitivity of the medical field, the image should be accurately segmented. Neural network and image-saliency-dependent model is proposed to attain the optimum accuracy. By taking into consideration the PET image as poor resolution greyscale, the Itti model and an enhanced GrabCut image segmentation technique with some modifications are proposed by Hui Liu *et al.* [44]. This approach has the demerits, like dependency on expert intervention and poor quality image segmentation with little RGB differences in distinct sphere of life. Previous studies have shown that segmentation results for unenhanced imagery, vegetation index, and colour space transformation did not improve without NIR information. So for that, these three enhancement techniques were used: histogram equalization, Photoshop auto contrast, and deep photo enhancer. In this study, encrypted and decrypted deep learning networks for pixel-by-pixel semantic segmentation of crops and weeds were investigated. Liu Cheng *et al.*'s [47] work is based on the noisy image segmentation problem. Using the adjusted local mean and local entropy, an adaptive local fitting image is suggested and inserted into the information to improve the reliability of the noise model. The whole model is based on a strict convex energy function and has good characteristic in noisy image segmentation. Since the energy function of this model is not convex, the level set function can get stuck in local minima and unexpected segmentation results.

27.2.6 IMAGE CLASSIFICATION

It is a process of categorization of objects on the bases of features such as shape, texture, colour, and some statistical analysis. In this type of processing, first features are extracted and then stored in a database with a label. This predefined dataset is used to classify images on the bases of categorized features. Image classification is widely used from SAR imaging to social security and medical field.

27.2.6.1 Techniques of Image Classification

Images are classified on the basis of two main approaches, unsupervised and supervised [48], as shown in Figure 27.9. In unsupervised classification, on the basis of their properties, pixels are classified. This process is called clustering. This is decided by the user on how much clusters needed. In case of unsupervised classification, untrained pixels are required because prior information is not needed. It does not require human interpretation because it is totally automated. Supervised classification requires prior knowledge in advance for testing process. This model is trained per the expected output. Two datasets are required for this approach: training data

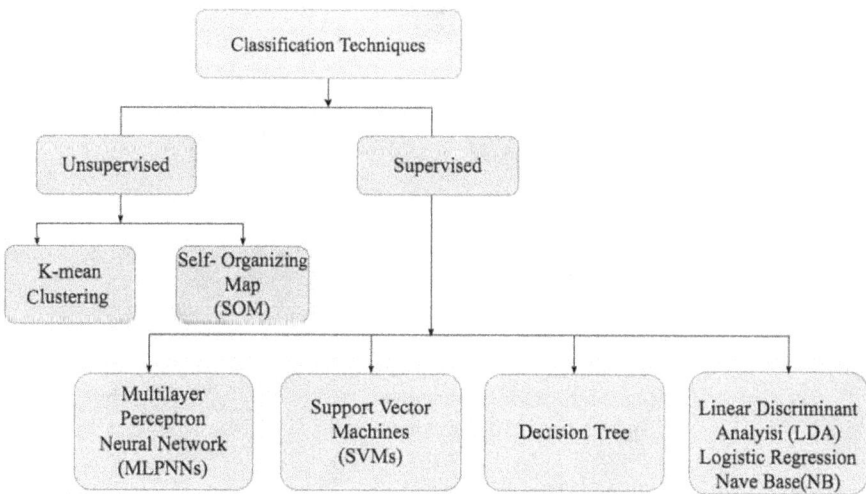

FIGURE 27.9 Techniques of image classification.

and test data. The unsupervised approach is further divided into K-mean clustering and SOM. K-mean clustering is used whenever the data is unlabelled. The important motive of this clustering technique is to find the batches in the data, where the variable K represents a group. Neural networks have different usage in different fields. The SOM is one such variation, also called Kohonen's map.

27.2.6.2   Related Research Work on Image Classification

Objects are somewhat easy to classify, but with the help of a machine, it is a challenging task. The image classification task by machine mainly requires image sensors, object segmentation, picture pre-processing, object-recognition, feature-extraction, and object-classification. An image classification system comprises a database containing predefined patterns that are compared with objects to classify them into the appropriate category. Image classification is a challenging task in multiple application areas, including video surveillance, remote sensing, biomedical imaging [49], biometrics, industrial vision inspection, robot navigation, and vehicle navigation. A study on image classification from some previous paper is elaborated in Table 27.6.

TABLE 27.6

Related Research on Image Classification Techniques

Sr. No.	Name	Purpose	Methods	Results	Applications
1	Yake Zhang, Fang Liu, *et al.* [50]	Features in probabilistic topic model (PTM) do not make full use of high-level structure feature and the feature correlation within similar images to mine discriminative features. Work is concentrated on this problem.	Sketch topic model with structural constraint for SAR image classification (C-SSTM), scale-invariant feature transform	Empirical performance proves that the suggested C-SSTM method performs well in detecting discriminative semantic features from SAR images with high time efficiency.	Regional planning and environmental surveillance.
2	Caihong Mu *et al.* [51]	Spatial data is instigated in supervised or unsupervised ways to get better classification outputs.	Spectral-spatial feature fusion using spatial coordinates based on active learning (SSFFSC), SVM	Better classification results in less time.	Indian Pines dataset and another similar forest

Sr. No.	Name	Purpose	Methods	Results	Applications
3	Zahraa A. Al-Saffar *et al.* [52]	To work on brain image classification	Mutual information-accelerated singular value decomposition local difference (MI-ASVD) in intensity-means, grey-level run-length matrix	MI-ASVD provides accurate and efficient classification results compared to the original feature space and two other standard dimensionality reduction methods, PCA and singular value decomposition.	Biomedical field to detect and classify grades of brain gliomas
4	Hongmin Gao *et al.* [53]	Combinations of shallow and deep spectral-spatial features are ignored. Certain models have been proposed to overcome these problems.	Multi-scale feature extraction (MSFE), dual-branch feature fusion (DBFM), MSDBFA.	The MSDBFA approach outperforms different techniques in phrases of each class's overall performance and computational cost.	Agriculture. The experiment worked well on the Indian Salinas dataset and pines dataset.
5	Hui Bi *et al.* [54]	SAR image classification to attain a high-resolution image in all climate conditions.	Novel target detection and classification framework based on a sparse SAR image dataset, region-based convolutional neural networks (RCNN), and YOLOv3	The method has optimum performance in CNN-based target detection and classification in the extended operating conditions (EOC)	Social security, resource exploration, and military reconnaissance
6	Yue Xi *et al.* [55]	Identification of tiny objects from very-low-resolution (LR) drone-based remote-sensing images.	Dual-stream representation learning generative adversarial network (DRL-GAN).	Perform better with CIFAR-10 than recent modern methods and works much better for very LR images	Drone-based video surveillance, UAV, Earthvision remote-sensing [56] and analysis of privacy-preserving video

(Continued)

TABLE 27.6 (*Continued*)

Sr. No.	Name	Purpose	Methods	Results	Applications
7	Donghang Yu *et al.* [57]	Classification and understanding of challenging remote sensing images.	Semi Supervised Retinal Image Classification method by a Hybrid Graph Convolutional Network (HGCN).	The precision and error curve of HGCN equal to lp = 20% shows that the performance achieved satisfactory result at 160th epoch and became stable.	Remote sensing images.
8	Kang Hyeon Rhee *et al.* [58]	It is not easy to determine which ground truth (GT) image generated by the network model is appropriate, so a superior GT image is requirement in imaging forensics.	Copy a move forgery detection scheme.	The suggested approach can be further utilized in the field of image forensics	Inpainting, cut-copy-paste, and forgery-feature-extraction
9	Chien-Chou Lin *et al.* [59]	Categorization of various objects in a point cloud captured by light detection and ranging (LIDAR) on streets of urban areas.	Bear angle convolutional neural network (BA-CNN) using conventional CNN models	Works well with simple data structures using less computation and memory than other methods	Auto-driving, classifying objects on the road, such as pedestrians, cars, bicycles, and buildings

27.2.6.3   Contributions from Image Classification Research Work

The main contributions from image enhancement research of this section can be summarized the techniques and their future directions, advantage, disadvantage, and so on. Classification is a crucial part of SAR images. Probabilistic theme model (PTM) features do not fully utilize high-level structural features and correlations of objects in similar images to exploit discriminant features. Top-level structure features and the feature correlation do not utilize by most of the available feature learning methods within identical images to mine discriminative features. Hence, the discriminative sketch topic model with structural constraint (C-SSTM) proposed by Yake Zhang, Fang Liu, *et al.* [50] for SAR image classification. This has a large impact on the classification results. A combination of the spectral and spatial knowledge base is an efficient source of getting good results in hyperspectral image classifications as investigated from Caihong Mu *et al.* [51] work. Hence, the spectral and spatial classification algorithms are called as hyperspectral image classification based on spectral-spatial feature fusion using spatial coordinates. HSI images are split in many smaller pictures based on spatial dimensions called samples, and these samples are randomly selected for training samples. These samples are categorized by SVM machine. In future, the sampling weight parameter can be designed as an adaptive parameter that can adaptively alter as the learning pattern set changes, to improve the classifier's performance more effectively. In Zahraa A. Al-Saffar *et al.*'s [52] work, a new method employed called mutual information-accelerated singular value decomposition (MI-ASVD) to select a significant subset of features as input to a classifier is proposed. Using this new algorithm an intelligent system that classifies MRI brain images into three classes, healthy, high-grade glioma and low-grade glioma is developed.

Most of the hyperspectral-image classification methods concentrate on extracting features utilizing layer-wise representations and fixed convolution kernels, resulting in feature extraction details. The feature fusion process is also simple and crude. By stacking modules hierarchically, different methods are established for different levels of feature fusion, ignoring combinations of deep and shallow spectral and spatial features. A novel multi-scale dual-branch functional fusion and attention network is proposed by Hongmin Gao *et al.* [53] to overcome the aforementioned problems. The proposed procedure is slightly longer than the previous procedure. To further improve the expressive power of this model, more nonlinear properties can be obtained by concatenating all feature groups and relaying them to a $(1 \times 1 \times 1)$ convolution using ReLU activation function.

27.2.7 CONCLUSION

The whole study mentioned concluded that there is lots of work already done in the field of DIP. Research groups are continuing work in the field of DIP in different domains. Some of the disciplines completely rely on DIP, such as underwater imaging, biomedical field, SAR imaging, and surveillance. The underwater image suffers degradation due to backward scattering, forward scattering, and absorption. To overcome this problem white balance is used. Retinex and grey-world approaches are used to overcome the white balance issue. Other additional approaches which are very important and used in the literature frequently for image restoration and enhancement are histogram-based and fusion-based [10, 11, 19, 60]. Because of the extensive use of image processing in the medical field, the new field with name digital imaging and communications in medicine (DICOM) has been introduced. Since the evolution of magnetic resonance imaging (MRI) and computerized tomography (CT), a significant amount of work has been done in image segmentation. In the medical field, the big challenge is the unavailability of an appropriate database. This study observed that U-net and GrabCut are used for segmentation of clinical images, as the U-Net model uses skip connection approach, so the information loss issue is common. Segmentation also has vast applications in agriculture and to vegetation index calculation. Nowadays, a global network consists of a number of different subnetwork that provide various information and communication functions and are interconnected using standardized communication protocols. So for storage and transmission of this huge data, compression techniques are very important. Similarly, image classification and feature extraction are the leading technology used in surveillance cameras, drones, self-driving car, agriculture, unmanned aerial vehicles (UAVs), and many more. Therefore, deep study of all these techniques reveals that further research is needed on which features to extract, which feature selection methods to apply and how to build a database of target environment properties.

REFERENCES

[1] Y. Yu, M. Liu, H. Feng, Z. Xu and Q. Li, "Split-Attention Multiframe Alignment Network for Image Restoration," *IEEE Access*, vol. 8, pp. 39254–39272, 2020, doi:10.1109/ACCESS.2020.2967028.

[2] E. Park and J. Y. Sim, "Underwater image restoration using geodesic color distance and complete image formation model," *IEEE Access*, vol. 8, pp. 157918–157930, 2020, doi:10.1109/ACCESS.2020.3019767.

[3] Z. Wang, H. Pen, T. Yang and Q. Wang, "Structure-Priority Image Restoration through Genetic Algorithm Optimization," *IEEE Access*, vol. 8, pp. 90698–90708, 2020, doi:10.1109/ACCESS.2020.2994127.

[4] G. Hou, X. Zhao, Z. Pan, H. Yang, L. Tan and J. Li, "Benchmarking Underwater Image Enhancement and Restoration and Beyond," *IEEE Access*, vol. 8, pp. 122078–122091, 2020, doi:10.1109/ACCESS.2020.3006359.

[5] R. Sinhal, I. A. Ansari and C. W. Ahn, "Blind Image Watermarking for Localization and Restoration of Color Images," *IEEE Access*, vol. 8, pp. 200157–200169, 2020, doi:10.1109/ACCESS.2020.3035428.

[6] X. Fei, J. Miao, Y. Zhao, W. Huang and R. Yu, "Total Variation Regularized Low-Rank Model with Directional Information for Hyperspectral Image Restoration," *IEEE Access*, vol. 9, pp. 84156–84169, 2021, doi:10.1109/ACCESS.2021.3087916.

[7] A. Kumari, S. K. Sahoo and M. C. Chinnaiah, "Fast and Efficient Visibility Restoration Technique for Single Image Dehazing and Defogging," *IEEE Access*, vol. 9, pp. 48131–48146, 2021, doi:10.1109/ACCESS.2021.3068446.

[8] X. Zhang, X. Li, Z. Tang, S. Zhang and S. Xie, "Noise Removal in Embedded Image with Bit Approximation," *IEEE Trans. Knowl. Data Eng.*, vol. 34, no. 3, pp. 1359–1369, 2022, doi:10.1109/TKDE.2020.2992572.

[9] A. Kaur, "A Review on Image Enhancement with Deep Learning Approach," *ACCENTS Trans. Image Process. Comp. Vision*, vol. 4, no. 11, 2018.

[10] I. Kansal and S. S. Kasana, "Improved Color Attenuation Prior Based Image De-Fogging Technique," *Multimed. Tools Appl.*, vol. 79, no. 17, pp. 12069–12091, 2020.

[11] I. Kansal and S. S. Kasana, "Minimum Preserving Subsampling-Based Fast Image De-Fogging," *J. Mod. Opt.*, vol. 65, no. 18, pp. 2103–2123, 2018.

[12] C. Li, S. Tang, J. Yan and T. Zhou, "Low-Light Image Enhancement Via Pair of Complementary Gamma Functions by Fusion," *IEEE Access*, vol. 8, pp. 169887–169896, 2020, doi:10.1109/ACCESS.2020.3023485.

[13] Y. Guo, Y. Lu, R. W. Liu, M. Yang and K. T. Chui, "Low-Light Image Enhancement with Regularized Illumination Optimization and Deep Noise Suppression," *IEEE Access*, vol. 8, pp. 145297–145315, 2020, doi:10.1109/ACCESS.2020.3015217.

[14] K. Ahmadian and H. R. Reza-Alikhani, "Self-Organized Maps and High-Frequency Image Detail for MRI Image Enhancement," *IEEE Access*, vol. 9, pp. 145662–145682, 2021, doi:10.1109/ACCESS.2021.3123119.

[15] R. Naseem, Z. A. Khan, N. Satpute, A. Beghdadi, F. A. Cheikh and J. Olivares, "Cross-Modality Guided Contrast Enhancement for Improved Liver Tumor Image Segmentation," *IEEE Access*, vol. 9, pp. 118154–118167, 2021, doi:10.1109/ACCESS.2021.3107473.

[16] T. H. Park and I. K. Eom, "Sand-Dust Image Enhancement Using Successive Color Balance with Coincident Chromatic Histogram," *IEEE Access*, vol. 9, pp. 19749–19760, 2021, doi:10.1109/ACCESS.2021.3054899.

[17] Q. Pan, L. Zhao, S. Chen and X. Li, "Fusion of Low-Quality Visible and Infrared Images Based on Multi-Level Latent Low-Rank Representation Joint with Retinex Enhancement and Multi-Visual Weight Information," *IEEE Access*, vol. 10, pp. 2140–2153, 2022, doi:10.1109/ACCESS.2021.3139670.

[18] R. Wang, B. Jiang, C. Yang, Q. Li and B. Zhang, "MAGAN: Unsupervised Low-Light Image Enhancement Guided by Mixed-Attention," *Big Data Min. Anal.*, vol. 5, no. 2, pp. 110–119, 2022, doi:10.26599/BDMA.2021.9020020.

[19] P. Bhinder, K. Singh and N. Jindal, "Image-Adaptive Watermarking Using Maximum Likelihood Decoder for Medical Images," *Multimed. Tools Appl.*, vol. 77, p. 1030310328, 2018.

[20] P. Bhinder, N. Jindal and K. Singh, "An Improved Robust Image-Adaptive Watermarking with Two Watermarks Using Statistical Decoder," *Multimed. Tools Appl.*, vol. 79, pp. 183–217, 2020, doi:10.1007/s11042-019-07941-2.

[21] S. A. Alazawi, N. M. Shati and A. H. Abbas, "Texture Features Extraction Based on GLCM for Face Retrieval System," *Period. Eng. Nat. Sci.*, vol. 7, no. 3, pp. 1459–1467, 2019, doi:10.21533/pen.v7i3.787.

[22] P. Kupidura, "The Comparison of Different Methods of Texture Analysis for Their Efficacy for Land Use Classification in Satellite Imagery," *Remote Sens.*, vol. 11, no. 10, 2019, doi:10.3390/rs11101233.

[23] T. Jiang et al., "Outline Feature Extraction of Positron Image Based on a 3D Anisotropic Convolution Operator," *IEEE Access*, vol. 8, pp. 150586–150598, 2020, doi:10.1109/ACCESS.2020.3016674.

[24] K. Saeed, S. Datta and N. Chaki, "A Granular Level Feature Extraction Approach to Construct HR Image for Forensic Biometrics Using Small Training DataSet," *IEEE Access*, vol. 8, pp. 123556–123570, 2020, doi:10.1109/ACCESS.2020.3006100.

[25] H. Fu, G. Sun, J. Zabalza, A. Zhang, J. Ren and X. Jia, "A Novel Spectral-Spatial Singular Spectrum Analysis Technique for Near Real-Time in Situ Feature Extraction in Hyperspectral Imaging," *IEEE J. Sel. Top. Appl. Earth Obs. Remote Sens.*, vol. 13, pp. 2214–2225, 2020, doi:10.1109/JSTARS.2020.2992230.

[26] M. A. Alzubaidi, M. Otoom and H. Jaradat, "Comprehensive and Comparative Global and Local Feature Extraction Framework for Lung Cancer Detection Using CT Scan Images," *IEEE Access*, vol. 9, pp. 158140–158154, 2021, doi:10.1109/ACCESS.2021.3129597.

[27] H. Yu, C. Wang, J. Li and Y. Sui, "Automatic Extraction of Green Tide from GF-3 SAR Images Based on Feature Selection and Deep Learning," *IEEE J. Sel. Top. Appl. Earth Obs. Remote Sens.*, vol. 14, pp. 10598–10613, 2021, doi:10.1109/JSTARS.2021.3118374.

[28] N. F. M. Radzi, A. C. Soh, A. J. Ishak and M. K. Hassan, "Feature Extraction Technique Using Weighted Histogram Analysis Method (WHAM) for Herbs Discrimination Based on Gas Chromatography Signal," *IEEE Access*, vol. 9, pp. 33336–33348, 2021, doi:10.1109/ACCESS.2021.3060822.

[29] T. A. M. Elhassan, M. S. M. Rahim, T. T. Swee, S. Z. M. Hashim and M. Aljurf, "Feature Extraction of White Blood Cells Using CMYK-Moment Localization and Deep Learning in Acute Myeloid Leukemia Blood Smear Microscopic Images," *IEEE Access*, vol. 10, pp. 16577–16591, 2022, doi:10.1109/ACCESS.2022.3149637.

[30] O. M. Alyasiri, Y. N. Cheah, A. K. Abasi and O. M. Al-Janabi, "Wrapper and Hybrid Feature Selection Methods Using Metaheuristic Algorithms for English Text Classification: A Systematic Review," *IEEE Access*, vol. 10, pp. 39833–39852, 2022, doi:10.1109/ACCESS.2022.3165814.

[31] S. Dhawan, "A Review of Image Compression and Comparison of its Algorithms," *Int. J. Electron. Commun. Technol. J. Electron. Commun. Technol.*, vol. 7109, no. 1, pp. 22–26, 2011.

[32] C. Wang, T. Li, J. Ni and Q. Huang, "A New MRF-Based Lossy Compression for Encrypted Binary Images," *IEEE Access*, vol. 8, pp. 11328–11341, 2020, doi:10.1109/ACCESS.2019.2963170.

[33] T. T. Pham, X. Van Hoang, N. T. Nguyen, D. T. Dinh and L. T. Ha, "End-to-End Image Patch Quality Assessment for Image/Video with Compression Artifacts," *IEEE Access*, vol. 8, pp. 215157–215172, 2020, doi:10.1109/ACCESS.2020.3040416.

[34] X. Li and S. Ji, "Neural Image Compression and Explanation," *IEEE Access*, vol. 8, pp. 214605–214615, 2020, doi:10.1109/ACCESS.2020.3041416.

[35] H. Fu, F. Liang and B. Lei, "An Extended Hybrid Image Compression Based on Soft-to-Hard Quantification," *IEEE Access*, vol. 8, pp. 95832–95842, 2020, doi:10.1109/ACCESS.2020.2994393.

[36] M. Nassef and M. H. Alkinani, "A Novel Multilevel Lossy Compression Algorithm for Grayscale Images Inspired by the Synthesization of Biological Protein Sequences," *IEEE Access*, vol. 9, pp. 149657–149680, 2021, doi:10.1109/ACCESS.2021.3125009.

[37] B. Xie, H. Zhang and C. Jung, "WCDGAN: Weakly Connected Dense Generative Adversarial Network for Artifact Removal of Highly Compressed Images," *IEEE Access*, vol. 10, pp. 1637–1649, 2022, doi:10.1109/ACCESS.2021.3138106.

[38] B. J. Van Rensburg, W. Puech and J. P. Pedeboy, "The First Draco 3D Object Crypto-Compression Scheme," *IEEE Access*, vol. 10, pp. 10566–10574, 2022, doi:10.1109/ACCESS.2022.3144533.

[39] N. M. Zaitoun and M. J. Aqel, "Survey on Image Segmentation Techniques," *Procedia Comput. Sci.*, vol. 65, no. Iccmit, pp. 797–806, 2015, doi:10.1016/j.procs.2015.09.027.

[40] N. Senthilkumaran and R. Rajesh, "Image Segmentation—A Survey of Soft Computing Approaches," *ARTCom 2009—Int. Conf. Adv. Recent Technol. Commun. Comput.*, no. 1, pp. 844–846, 2009, doi:10.1109/ARTCom.2009.219.

[41] D. H. B. Kekre, S. Thepade, P. Mukherjee, M. Kakaiya, S. Wadhwa and S. Singh, "Image Retrieval with Shape Features Extracted using Morphological Operators with BTC," *Int. J. Comput. Appl.*, vol. 12, no. 3, pp. 1–5, 2010, doi:10.5120/1662-2238.

[42] P. Yin, R. Yuan, Y. Cheng and Q. Wu, "Deep Guidance Network for Biomedical Image Segmentation," *IEEE Access*, vol. 8, pp. 116106–116116, 2020, doi:10.1109/ACCESS.2020.3002835.

[43] S. Kaur, "Noise Types and Various Removal Techniques," *Int. J. Adv. Res. Electron. Commun. Eng.*, vol. 4, no. 2, pp. 226–230, 2015.

[44] H. Liu, W. Chu and H. Wang, "Automatic Segmentation Algorithm of Ultrasound Heart Image Based on Convolutional Neural Network and Image Saliency," *IEEE Access*, vol. 8, pp. 104445–104457, 2020, doi:10.1109/ACCESS.2020.2989819.

[45] W. Liu, C. Zhou, Z. Li and Z. Hu, "Patch-Driven Tongue Image Segmentation Using Sparse Representation," *IEEE Access*, vol. 8, pp. 41372–41383, 2020, doi:10.1109/ACCESS.2020.2976826.

[46] A. Wang, Y. Xu, X. Wei and B. Cui, "Semantic Segmentation of Crop and Weed using an Encoder-Decoder Network and Image Enhancement Method under Uncontrolled Outdoor Illumination," *IEEE Access*, vol. 8, pp. 81724–81734, 2020, doi:10.1109/ACCESS.2020.2991354.

[47] C. Liu, W. Liu and W. Xing, "An Efficient Variational-Level-Set Model Based on Adaptive Local Fitted Image for Noisy Image Segmentation," *IEEE Access*, vol. 8, pp. 17500–17526, 2020, doi:10.1109/ACCESS.2019.2957387.

[48] M. Kaliyamoorthi, "A Review of Image Classification Approaches and Techniques," *Int. J. Recent Trends Eng. Res.*, vol. 3, no. 3, pp. 1–5, 2017, doi:10.23883/ijrter.2017.3033.xts7z.

[49] S. Tiwari, S. Kumar and K. Guleria, "Outbreak Trends of Coronavirus Disease–2019 in India: A Prediction," *Disaster Med. Public Health Prep.*, vol. 14, no. 5, pp. e33–e38, 2020.

[50] M. Sheykhmousa, M. Mahdianpari, H. Ghanbari, F. Mohammadimanesh, P. Ghamisi and S. Homayouni, "Support Vector Machine Versus Random Forest for Remote Sensing Image Classification: A Meta-Analysis and Systematic Review," *IEEE J. Sel. Top. Appl. Earth Obs. Remote Sens.*, vol. 13, pp. 6308–6325, 2020, doi:10.1109/JSTARS.2020.3026724.

[51] C. Mu, J. Liu, Y. Liu and Y. Liu, "Hyperspectral Image Classification Based on Active Learning and Spectral-Spatial Feature Fusion Using Spatial Coordinates," *IEEE Access*, vol. 8, pp. 6768–6781, 2020, doi:10.1109/ACCESS.2019.2963624.

[52] Z. A. Al-Saffar and T. Yildirim, "A Novel Approach to Improving Brain Image Classification Using Mutual Information-Accelerated Singular Value Decomposition," *IEEE Access*, vol. 8, pp. 52575–52587, 2020, doi:10.1109/ACCESS.2020.2980728.

[53] H. Gao, Y. Zhang, Z. Chen and C. Li, "A Multiscale Dual-Branch Feature Fusion and Attention Network for Hyperspectral Images Classification," *IEEE J. Sel. Top. Appl. Earth Obs. Remote Sens.*, vol. 14, pp. 8180–8192, 2021, doi:10.1109/JSTARS.2021.3103176.

[54] H. Bi, J. Deng, T. Yang, J. Wang and L. Wang, "CNN-Based Target Detection and Classification When Sparse SAR Image Dataset is Available," *IEEE J. Sel. Top. Appl. Earth Obs. Remote Sens.* vol. 14, pp. 6815–6826, 2021, doi:10.1109/JSTARS.2021.3093645.

[55] Y. Xi et al., "DRL-GAN: Dual-Stream Representation Learning GAN for Low-Resolution Image Classification in UAV Applications," *IEEE J. Sel. Top. Appl. Earth Obs. Remote Sens.*, vol. 14, pp. 1705–1716, 2021, doi:10.1109/JSTARS.2020.3043109.

[56] D. Yu, Q. Xu, H. Guo, J. U. N. Lu, Y. Lin and X. Liu, "Aggregating Features From Dual Paths for Remote Sensing Image Scene Classification," *IEEE Access*, vol. 10, pp. 16740–16755, 2022, doi:10.1109/ACCESS.2022.3147543.

[57] G. Zhang et al., "Hybrid Graph Convolutional Network for Semi-Supervised Retinal Image Classification," *IEEE Access*, vol. 9, pp. 35778–35789, 2021, doi:10.1109/ACCESS.2021.3061690.

[58] K. H. Rhee, "Generation of Novelty Ground Truth Image Using Image Classification and Semantic Segmentation for Copy-Move Forgery Detection," *IEEE Access*, vol. 10, pp. 2783–2796, 2022, doi:10.1109/ACCESS.2021.3136781.

[59] C. C. Lin, C. H. Kuo and H. Te Chiang, "CNN-Based Classification for Point Cloud Object with Bearing Angle Image," *IEEE Sens. J.*, vol. 22, no. 1, pp. 1003–1011, 2022, doi:10.1109/JSEN.2021.3130268.

[60] H. Kaur, D. Koundal and V. Kadyan, "Image Fusion Techniques: A Survey," *Arch. Comput. Method. Eng.*, vol. 28, pp. 4425–4447, 2021.

28 Optimal Solution of Fully Intuitionistic Fuzzy Transportation Problems
A Modified Approach

Gourav Gupta, Deepika Rani,
Mohit Kumar Kakkar

28.1 INTRODUCTION

In today's world, there is high competition in the market, so organizations have sturdy pressure to get or discover the better way to serve their ambitious customers. It is become more challenging to transport the amount of goods to the customers along with their cost-effective manner. Transportation models give impressive structure to deal with these challenges. Hitchcock [12] initially developed the basic transportation problem. The transportation problems can also be expressed as standard form of linear programming problem and can be solve by using simplex method. The linear programming problem is one kind of mathematical programming problem that comes to use commonly in several practical applications. In real life, uncertainty exists everywhere, nothing is precise, so is true with the transportation problems also. Impreciseness may exist either/and in cost to transport unit product or stock available at the sources or requirement of the destinations, these data may not be exact always due to multiple uncontrollable factors or conditions, deficiency in information, complexity of the situations, and so on. To handle the uncertainty in transportation problems, imprecise data are expressed as fuzzy numbers. Zadeh [30] introduced the fuzzy set theory. It is prompted in the field of optimization by Bellman and Zadeh [4].

The fuzzy set theory has been used by many researchers [1,5,6,8,15] in real-life optimization problems, such as routing, scheduling, and manufacturing. Uncertainty also arises in transportation problems. We do not always have exact details of transportation problems; some impreciseness will always be there due to several reasons. To deal with transportation problems with imprecise data, many researchers [3,11,13,14,16,21–23,25] covered several types of fuzzy transportation problems.

To transport the products, data may be uncertain in nature with hesitation due to changes in fuel price, weather, errors in data, less knowledge of market and many more. In such situations, the decision-maker may have doubts and cannot

DOI: 10.1201/9781003367161-28

judge accurately. To deal with hesitation in uncertain data, a generalized form of fuzzy set theory can be used, named intuitionistic fuzzy set theory, which is distinct the grade of membership (support surface) and grade of non-membership (oppose surface) of a number in a set. Intuitionistic fuzzy set theory introduced by Atanassov [2].

In this chapter, flaws in existing methods pointed out and a modified alternative method is proposed to overcome the flaws and give the optimal solution of fully intuitionistic fuzzy transportation problems. The proposed methodology is illustrated by an existing problem [24].

28.2 PRELIMINARIES

28.2.1 Intuitionistic Fuzzy Set [26]

A set of ordered triples $\tilde{A}^I = \left\{ \left\langle \alpha, \mu_{\tilde{A}^I}(\alpha), \eta_{\tilde{A}^I}(\alpha) \right\rangle : \alpha \in U \right\}$ in U is known as intuitionistic fuzzy set, where U is universe of discourse and $\mu_{\tilde{A}^I}(\alpha), \eta_{\tilde{A}^I}(\alpha) : U \to [0,1]$ are functions such that $0 \le \mu_{\tilde{A}^I}(\alpha), \eta_{\tilde{A}^I}(\alpha) \le 1, \ \forall \alpha \in U$. $\mu_{\tilde{A}^I}(\alpha)$ and $\eta_{\tilde{A}^I}(\alpha)$ represent the grade of membership and grade of non-membership of the element $\alpha \in U$ being in $\tilde{A}^I$, respectively. The grade of hesitation for the element $\alpha \in U$ being in $\tilde{A}^I$ is given by $h(\alpha) = 1 - \mu_{\tilde{A}^I}(\alpha) - \eta_{\tilde{A}^I}(\alpha) \le 1, \forall \alpha \in U$.

28.2.2 Intuitionistic Fuzzy Number (IFN) [26]

A subset of an intuitionistic fuzzy set, $\tilde{A}^I = \left\{ \left\langle \alpha, \mu_{\tilde{A}^I}(\alpha), \eta_{\tilde{A}^I}(\alpha) \right\rangle : \alpha \in U \right\}$, of the real line R is called an IFN if

1. there exist $a \in R, \mu_{\tilde{A}^I}(a) = 1$ and $\eta_{\tilde{A}^I}(a) = 0$ (a is called the mean value of $\tilde{A}^I$) or
2. $\mu_{\tilde{A}^I}(\alpha), \eta_{\tilde{A}^I}(\alpha) : R \to [0,1]$ are piecewise continuous functions and the relation $0 \le \mu_{\tilde{A}^I}(a), \eta_{\tilde{A}^I}(a) \le 1, \forall a \in R$ holds.

28.2.3 Membership and Non-membership Function

Let $\mu_{\tilde{A}^I}(z)$ and $\eta_{\tilde{A}^I}(z)$ be the membership and non-membership function of $\tilde{A}^I$, defined as

$$\mu_{\tilde{A}^I}(z) = \begin{cases} 0 & ; \ -\infty < z \le a - \gamma \\ f_1(z) & ; \ z \in (a-\gamma, a] \\ 1 & ; \ z = a \\ h_1(z) & ; \ z \in [a, a+\delta) \\ 0 & ; \ a+\delta \le z < \infty \end{cases},$$

where the function $f_1(z)$ is increasing strictly in $(a-\gamma,a]$ and the function $h_1(z)$ is decreasing strictly in $[a,a+\delta)$, respectively. In

$$\eta_{\tilde{A}^I}(z) = \begin{cases} 1 & ; -\infty < z \le a-\gamma\,' \\ f_2(z) & ; z \in (a-\gamma\,',a]; 0 \le f_1(z)+f_2(z) \le 1 \\ 0 & ; z = a \\ h_2(z) & ; z \in [a,a+\delta\,'); 0 \le h_1(z)+h_2(z) \le 1 \\ 0 & ; a+\delta\,' \le z < \infty \end{cases}$$

the function $f_2(z)$ is increasing strictly in $(a-\gamma\,',a]$ and the function $h_2(z)$ is decreasing strictly in $[a,a+\delta\,')$, respectively. Here γ and δ are left and right expands of membership function $\mu_{\tilde{A}^I}(z)$, respectively. $\gamma\,'$ and $\delta\,'$ are called left and right expands of non-membership function $\eta_{\tilde{A}^I}(z)$, respectively. The IFN $\tilde{A}^I$ is represented by $\tilde{A}^I = (a;\gamma,\delta;\gamma\,',\delta\,')$.

28.2.4 Trapezoidal Intuitionistic Fuzzy Number (TIFN) [26]

Let $\tilde{A}^I = (\alpha_1,\alpha_2,\alpha_3,\alpha_4;\alpha_1',\alpha_2,\alpha_3,\alpha_4')$ be an intuitionistic fuzzy number. $\tilde{A}^I$ is said to be TIFN if its membership function $\mu_{\tilde{A}^I}(z)$ and non-membership function $\eta_{\tilde{A}^I}(z)$ is given by

$$\mu_{\tilde{A}^I}(z) = \begin{cases} \dfrac{z-\alpha_1}{\alpha_2-\alpha_1} & ; \alpha_1 \le z < \alpha_2 \\ 1 & ; \alpha_2 \le z \le \alpha_3 \\ \dfrac{\alpha_4-z}{\alpha_4-\alpha_3} & ; \alpha_3 < z \le \alpha_4 \\ 0 & ; \text{otherwise} \end{cases} \quad \text{and} \quad \eta_{\tilde{A}^I}(z) = \begin{cases} \dfrac{\alpha_2-z}{\alpha_2-\alpha_1'} & ; \alpha_1' \le z < \alpha_2 \\ 0 & ; \alpha_2 \le z \le \alpha_3 \\ \dfrac{\alpha_3-z}{\alpha_3-\alpha_4'} & ; \alpha_3 < z \le \alpha_4' \\ 1 & ; \text{otherwise} . \end{cases}$$

28.2.5 Arithmetic Operations on TIFN [27]

In this section, addition, subtraction, and multiplication of two TIFN is presented.

Let $\tilde{A}^I = (\alpha_1,\alpha_2,\alpha_3,\alpha_4;\alpha_1',\alpha_2,\alpha_3,\alpha_4')$ and $\tilde{B}^I = (\beta_1,\beta_2,\beta_3,\beta_4;\beta_1',\beta_2,\beta_3,\beta_4')$ be two TIFNs. Then,

1. $\tilde{A}^I \oplus \tilde{B}^I = (\alpha_1+\beta_1,\alpha_2+\beta_2,\alpha_3+\beta_3,\alpha_4+\beta_4;\alpha_1'+\beta_1',\alpha_2+\beta_2,\alpha_3+\beta_3,\alpha_4'+\beta_4')$;

2. $\tilde{A}^I \otimes \tilde{B}^I = (\alpha_1-\beta_4,\alpha_2-\beta_3,\alpha_3-\beta_2,\alpha_4-\beta_1;\alpha_1'-\beta_4',\alpha_2-\beta_3,\alpha_3-\beta_2,\alpha_4'-\beta_1')$;

3. $\tilde{A}^I \otimes \tilde{B}^I = (m_1,m_2,m_3,m_4;m_1',m_2,m_3,m_4')$,

where $m_1 = \min\{\alpha_1\beta_1, \alpha_1\beta_4, \alpha_4\beta_1, \alpha_4\beta_4\}$, $m_2 = \min\{\alpha_2\beta_2, \alpha_2\beta_3, \alpha_3\beta_2, \alpha_3\beta_3\}$, $m_3 = \max\{\alpha_2\beta_2, \alpha_2\beta_3, \alpha_3\beta_2, \alpha_3\beta_3\}$, $m_4 = \max\{\alpha_1\beta_1, \alpha_1\beta_4, \alpha_4\beta_1, \alpha_4\beta_4\}$, $m_1' = \min\{\alpha_1'\beta_1', \alpha_1'\beta_4', \alpha_4'\beta_1', \alpha_4'\beta_4'\}$, $m_4' = \max\{\alpha_1'\beta_1', \alpha_1'\beta_4', \alpha_4'\beta_1', \alpha_4'\beta_4'\}$;

4. $\lambda\tilde{A}^I = \begin{cases} (\lambda\alpha_1, \lambda\alpha_2, \lambda\alpha_3, \lambda\alpha_4; \lambda\alpha_1', \lambda\alpha_2, \lambda\alpha_3, \lambda\alpha_4'); & \lambda \geq 0, \\ (\lambda\alpha_4, \lambda\alpha_3, \lambda\alpha_2, \lambda\alpha_1; \lambda\alpha_4', \lambda\alpha_3, \lambda\alpha_2, \lambda\alpha_1'); & \lambda < 0. \end{cases}$

28.2.6 ORDERING OF TIFN [27]

Let $\tilde{A}^I = (\alpha_1, \alpha_2, \alpha_3, \alpha_4; \alpha_1', \alpha_2, \alpha_3, \alpha_4')$ and $\tilde{B}^I = (\beta_1, \beta_2, \beta_3, \beta_4; \beta_1', \beta_2, \beta_3, \beta_4')$ be two TIFNs.

Then,

1. $\tilde{A}^I \succeq \tilde{B}^I$ if $R(\tilde{A}^I) \geq R(\tilde{B}^I)$ and
2. $\tilde{A}^I \approx \tilde{B}^I$ if $R(\tilde{A}^I) = R(\tilde{B}^I)$,

    where $R(\tilde{A}^I) = \left(\dfrac{\alpha_1 + \alpha_2 + \alpha_3 + \alpha_4 + \alpha_1' + \alpha_2 + \alpha_3 + \alpha_4'}{8} \right)$

    and $R(\tilde{B}^I) = \left(\dfrac{\beta_1 + \beta_2 + \beta_3 + \beta_4 + \beta_1' + \beta_2 + \beta_3 + \beta_4'}{8} \right)$.

28.2.7 FLAWS OF THE EXISTING METHODS

Let $\tilde{A}^I = (\alpha_1, \alpha_2, \alpha_3, \alpha_4; \alpha_1', \alpha_2, \alpha_3, \alpha_4')$ and $\tilde{B}^I = (\beta_1, \beta_2, \beta_3, \beta_4; \beta_1', \beta_2, \beta_3, \beta_4')$ be two TIFNs and $\tilde{A}^I \otimes \tilde{B}^I$ is a TIFNs obtained by using special multiplication, proposed by Kumar and Hussain [19] then for the ranking function R, used by Singh and Yadav [26], the property $R(\tilde{A}^I \otimes \tilde{B}^I) = R(\tilde{A}^I) \times R(\tilde{B}^I)$ will always be satisfied, and hence, the problem (P3) [19, Section 4.10.1, Problem (P*)] can be transformed into problem (P4). However, if instead of special multiplication, the generally used multiplication of intuitionistic fuzzy numbers is used, then the property $R(\tilde{A}^I \otimes \tilde{B}^I) = R(\tilde{A}^I) \times R(\tilde{B}^I)$ will not be necessarily satisfied. For example, if $\tilde{A}^I = (3,7,10,14;2,5,12,15)$ and $\tilde{B}^I = (3,6,9,11;1,5,10,13)$ are two trapezoidal intuitionistic fuzzy numbers, then $\tilde{A}^I \otimes \tilde{B}^I = (9,42,90,154;2,25,120,195)$, and hence, $R(\tilde{A}^I \otimes \tilde{B}^I) = 79.625$. While $R(\tilde{A}^I) \otimes R(\tilde{B}^I) = 61.625$, it is obvious that $R(\tilde{A}^I \otimes \tilde{B}^I) \neq R(\tilde{A}^I) \times R(\tilde{B}^I)$. Therefore, the problem (P4) cannot be generated from the problem (P3), and it can be easily verified that existing methods [7,9,10,17,18,20,24,27–29], used the mathematical incorrect property $R(\tilde{A}^I \otimes \tilde{B}^I) = R(\tilde{A}^I) \times R(\tilde{B}^I)$. Hence, it is not genuine to use these existing methods for finding the solution of such fully intuitionistic fuzzy transportation problems.

28.3 PROPOSED METHOD

It is well known that problem (P1) represents the linear programming problem of such balanced transportation problems (total stock available = total requirement), having m sources/origins and n destinations for which the precise information about the unit transportation cost (c_{ij}) for i^{th} source to j^{th} destination, stock available (s_i) at i^{th} source and requirement (r_j) at j^{th} destination.

$$(P1)$$

$$\text{Minimize } \sum_{i=1}^{m}\sum_{j=1}^{n} c_{ij} z_{ij}$$

Subject to

$$\sum_{j=1}^{n} z_{ij} = s_i; i = 1,2,...,m,$$

$$\sum_{i=1}^{m} z_{ij} = r_j; j = 1,2,...,n,$$

$$z_{ij} \geq 0; i = 1,2,...,m; j = 1,2,...,n.$$

The step-by-step formulation of proposed method is as follows:

Step 1: By replacing c_{ij}, z_{ij}, s_i and r_j of the linear programming problem (P1) of balanced transportation problem with the TIFN $\big(c_{ij1}, c_{ij2}, c_{ij3}, c_{ij4}; c'_{ij1}, c'_{ij2}, c'_{ij3}, c'_{ij4}\big)$, $\big(z_{ij1}, z_{ij2}, z_{ij3}, z_{ij4}; z'_{ij1}, z'_{ij2}, z'_{ij3}, z'_{ij4}\big)$, $\big(s_{i1}, s_{i2}, s_{i3}, s_{i4}; s'_{i1}, s'_{i2}, s'_{i3}, s'_{i4}\big)$, and $\big(r_{j1}, r_{j2}, r_{j3}, r_{j4}; r'_{j1}, r'_{j2}, r'_{j3}, r'_{j4}\big)$, respectively, the problem (P1) is transformed into problem (P2).

$$(P2)$$

$$\text{Minimize } \sum_{i=1}^{m}\sum_{j=1}^{n} \big(c_{ij1}, c_{ij2}, c_{ij3}, c_{ij4}; c'_{ij1}, c'_{ij2}, c'_{ij3}, c'_{ij4}\big) \otimes \big(z_{ij1}, z_{ij2}, z_{ij3}, z_{ij4}; z'_{ij1}, z'_{ij2}, z'_{ij3}, z'_{ij4}\big)$$

Subject to

$$\sum_{j=1}^{n} \big(z_{ij1}, z_{ij2}, z_{ij3}, z_{ij4}; z'_{ij1}, z'_{ij2}, z'_{ij3}, z'_{ij4}\big) \approx \big(s_{i1}, s_{i2}, s_{i3}, s_{i4}; s'_{i1}, s'_{i2}, s'_{i3}, s'_{i4}\big); i = 1,2,...,m,$$

$$\sum_{i=1}^{m} \big(z_{ij1}, z_{ij2}, z_{ij3}, z_{ij4}; z'_{ij1}, z'_{ij2}, z'_{ij3}, z'_{ij4}\big) \approx \big(r_{j1}, r_{j2}, r_{j3}, r_{j4}; r'_{j1}, r'_{j2}, r'_{j3}, r'_{j4}\big); j = 1,2,...,n,$$

$$\big(z_{ij1}, z_{ij2}, z_{ij3}, z_{ij4}; z'_{ij1}, z'_{ij2}, z'_{ij3}, z'_{ij4}\big) \succeq (0,0,0,0,0,0,0,0); i = 1,2,...,m; j = 1,2,...,n,$$

$$\big(z_{ij1}, z_{ij2}, z_{ij3}, z_{ij4}; z'_{ij1}, z'_{ij2}, z'_{ij3}, z'_{ij4}\big) \text{ is a non-negative TIFN.}$$

Step 2: Using the arithmetic operations of TIFNs, the problem (P2) can be transformed into the problem (P3).

(P3)

Minimize $\displaystyle\sum_{i=1}^{m}\sum_{j=1}^{n}\left(c_{ij1}z_{ij1},c_{ij2}z_{ij2},c_{ij3}z_{ij3},c_{ij4}z_{ij4};c'_{ij1}z'_{ij1},c'_{ij2}z'_{ij2},c'_{ij3}z'_{ij3},c'_{ij4}z'_{ij4}\right)$

Subject to

$$\sum_{j=1}^{n}\left(z_{ij1},z_{ij2},z_{ij3},z_{ij4};z'_{ij1},z'_{ij2},z'_{ij3},z'_{ij4}\right)\approx\left(s_{i1},s_{i2},s_{i3},s_{i4};s'_{i1},s'_{i2},s'_{i3},s'_{i4}\right);i=1,2,...,m,$$

$$\sum_{i=1}^{m}\left(z_{ij1},z_{ij2},z_{ij3},z_{ij4};z'_{ij1},z'_{ij2},z'_{ij3},z'_{ij4}\right)\approx\left(r_{j1},r_{j2},r_{j3},r_{j4};r'_{j1},r'_{j2},r'_{j3},r'_{j4}\right);j=1,2,...,n,$$

$$\left(z_{ij1},z_{ij2},z_{ij3},z_{ij4};z'_{ij1},z'_{ij2},z'_{ij3},z'_{ij4}\right)\succeq(0,0,0,0;0,0,0,0);i=1,2,...,m;j=1,2,...,n,$$

$\left(z_{ij1},z_{ij2},z_{ij3},z_{ij4};z'_{ij1},z'_{ij2},z'_{ij3},z'_{ij4}\right)$ is a non-negative TIFN.

Step 3: Using the Section 2.2, the problem (P3) can be transformed into the problem (P4).

(P4)

Minimize $\displaystyle R\left(\sum_{i=1}^{m}\sum_{j=1}^{n}\left(c_{ij1}z_{ij1},c_{ij2}z_{ij2},c_{ij3}z_{ij3},c_{ij4}z_{ij4};c'_{ij1}z'_{ij1},c'_{ij2}z'_{ij2},c'_{ij3}z'_{ij3},c'_{ij4}z'_{ij4}\right)\right)$

Subject to

$$R\left(\sum_{j=1}^{n}\left(z_{ij1},z_{ij2},z_{ij3},z_{ij4};z'_{ij1},z'_{ij2},z'_{ij3},z'_{ij4}\right)\right)=R\left(s_{i1},s_{i2},s_{i3},s_{i4};s'_{i1},s'_{i2},s'_{i3},s'_{i4}\right);i=1,2,...,m,$$

$$R\left(\sum_{i=1}^{m}\left(z_{ij1},z_{ij2},z_{ij3},z_{ij4};z'_{ij1},z'_{ij2},z'_{ij3},z'_{ij4}\right)\right)=R\left(r_{j1},r_{j2},r_{j3},r_{j4};r'_{j1},r'_{j2},r'_{j3},r'_{j4}\right);j=1,2,...,n,$$

$$R\left(z_{ij1},z_{ij2},z_{ij3},z_{ij4};z'_{ij1},z'_{ij2},z'_{ij3},z'_{ij4}\right)\geq R(0,0,0,0;0,0,0,0);z'_{ij1}\geq0;i=1,2,...,m;$$

$$j=1,2,...,n,$$

$$z_{ijk},z'_{ijk}\geq0;i=1,2,...,m;j=1,2,...,n;k=1,2,3,4.$$

Step 4: Using the relation, $R\left(\displaystyle\sum_{i=1}^{m}\sum_{j=1}^{n}\left(\alpha_{ij},\beta_{ij},\gamma_{ij},\delta_{ij};\alpha'_{ij},\beta'_{ij},\gamma'_{ij},\delta'_{ij}\right)\right)=$ $\displaystyle\sum_{i=1}^{m}\sum_{j=1}^{n}R(\alpha_{ij},\beta_{ij},\gamma_{ij},\delta_{ij};_{ij}\alpha'_{ij},\beta'_{ij},\gamma',\delta'_{ij})$, the transformation of problem (P4) is shown by problem (P5).

(P5)

Minimize $\displaystyle\sum_{i=1}^{m}\sum_{j=1}^{n}R\left(c_{ij1}z_{ij1},c_{ij2}z_{ij2},c_{ij3}z_{ij3},c_{ij4}z_{ij4};c'_{ij1}z'_{ij1},c'_{ij2}z'_{ij2},c'_{ij3}z'_{ij3},c'_{ij4}z'_{ij4}\right)$

Subject to

$$\sum_{j=1}^{n}R\left(z_{ij1},z_{ij2},z_{ij3},z_{ij4};z'_{ij1},z'_{ij2},z'_{ij3},z'_{ij4}\right)=R\left(s_{i1},s_{i2},s_{i3},s_{i4};s'_{i1},s'_{i2},s'_{i3},s'_{i4}\right);i=1,2,...,m,$$

$$\sum_{i=1}^{m} R\left(z_{ij1}, z_{ij2}, z_{ij3}, z_{ij4}; z_{ij1}', z_{ij2}', z_{ij3}', z_{ij4}'\right) = R\left(r_{j1}, r_{j2}, r_{j3}, r_{j4}; r_{j1}', r_{j2}', r_{j3}', r_{j4}'\right); j = 1, 2, \ldots, n,$$

$$R\left(z_{ij1}, z_{ij2}, z_{ij3}, z_{ij4}; z_{ij1}', z_{ij2}', z_{ij3}', z_{ij4}'\right) \geq R(0,0,0,0; 0,0,0,0); z_{ij1}' \geq 0; i = 1, 2, \ldots, m;$$

$$j = 1, 2, \ldots, n,$$

$$z_{ijk}, z_{ijk}' \geq 0; i = 1, 2, \ldots, m; j = 1, 2, \ldots, n; k = 1, 2, 3, 4.$$

Step 5: Using the ordering of TIFNs $R\left(\alpha_{ij}, \beta_{ij}, \gamma_{ij}, \delta_{ij}; \alpha_{ij}', \beta_{ij}', \gamma_{ij}', \delta_{ij}'\right) = \dfrac{\alpha_{ij} + \beta_{ij} + \gamma_{ij} + \delta_{ij} + \alpha_{ij}' + \beta_{ij}' + \gamma_{ij}' + \delta_{ij}'}{8}$, the transformation of problem (P5) is problem (P6).

(P6)

Minimize $\displaystyle\sum_{i=1}^{m}\sum_{j=1}^{n}\left(\dfrac{c_{ij1}z_{ij1} + c_{ij2}z_{ij2} + c_{ij3}z_{ij3} + c_{ij4}z_{ij4} + c_{ij1}'z_{ij1}' + c_{ij2}'z_{ij2}' + c_{ij3}'z_{ij3}' + c_{ij4}'z_{ij4}'}{8}\right)$

Subject to

$$\sum_{j=1}^{n}\left(\dfrac{z_{ij1} + z_{ij2} + z_{ij3} + z_{ij4} + z_{ij1}' + z_{ij2}' + z_{ij3}' + z_{ij4}'}{8}\right) = \dfrac{s_{i1} + s_{i2} + s_{i3} + s_{i4} + s_{i1}' + s_{i2}' + s_{i3}' + s_{i4}'}{8};$$

$$i = 1, 2, \ldots, m,$$

$$\sum_{i=1}^{m}\left(\dfrac{z_{ij1} + z_{ij2} + z_{ij3} + z_{ij4} + z_{ij1}' + z_{ij2}' + z_{ij3}' + z_{ij4}'}{8}\right) = \dfrac{r_{j1} + r_{j2} + r_{j3} + r_{j4} + r_{j1}' + r_{j2}' + r_{j3}' + r_{j4}'}{8};$$

$$j = 1, 2, \ldots, n,$$

$$z_{ijk}, z_{ijk}' \geq 0; i = 1, 2, \ldots, m; j = 1, 2, \ldots, n; k = 1, 2, 3, 4.$$

Step 6: Find the optimal solution of crisp linear programming problem (P6).

Step 7: Using the optimal solution, obtained in step 6, the intuitionistic fuzzy optimal solution of problem (P2) is $\left\{\left(z_{ij1}, z_{ij2}, z_{ij3}, z_{ij4}; z_{ij1}', z_{ij2}', z_{ij3}', z_{ij4}'\right)\right\}$.

Step 8: Using the optimal solution, obtained in Step 7, the minimum intuitionistic fuzzy transportation cost of fully intuitionistic fuzzy transportation problem (P2) is $\displaystyle\sum_{i=1}^{m}\sum_{j=1}^{n}\left(c_{ij1}, c_{ij2}, c_{ij3}, c_{ij4}; c_{ij1}', c_{ij2}', c_{ij3}', c_{ij4}'\right) \otimes \left(z_{ij1}, z_{ij2}, z_{ij3}, z_{ij4}; z_{ij1}', z_{ij2}', z_{ij3}', z_{ij4}'\right)$.

28.4   ILLUSTRATIVE EXAMPLE

Table 28.1 representing the fully intuitionistic fuzzy transportation problem solved by Roseline and Amirtharaj [24]. In this section, the proposed method is illustrated by finding the intuitionistic fuzzy optimal solution of the same problem.

TABLE 28.1

Fully Intuitionistic Fuzzy Transportation Problem

Sources	Destinations				Stock Available ($\tilde{s}_i'$)
	D_1	D_2	D_3	D_4	
S_1	(2,5,7,10; 1,4,8,12)	(3,6,9,12; 2,5,10,14)	(3,6,9,11; 1,5,10,13)	(4,6,8,11; 3,5,9,12)	(4,6,10,14; 2,5,11,15)
S_2	(4,7,9,13; 3,6,11,15)	(2,6,10,14; 1,4,12,16)	(3,7,10,14; 2,5,12,15)	(4,7,10,13; 1,5,12,15)	(5,9,12,16; 4,8,14,18)
S_3	(4,8,10,12; 2,6,11,14)	(3,5,8,10; 2,4,9,11)	(4,7,11,13; 3,5,12,14)	(3,6,9,11; 1,4,10,14)	(8,15,20,27; 6,11,23,33)
Requi. ($\tilde{r}_j'$)	(5,8,11,15; 4,7,13,16)	(4,8,12,14; 3,6,13,18)	(3,6,9,14; 2,5,10,16)	(5,8,10,14; 3,6,12,16)	

Applying the steps of proposed method to the intuitionistic fuzzy transportation problem, presented by Table 28.1, the intuitionistic fuzzy optimal solution of fully can be obtained as follows:

Step 1: The intuitionistic linear programming problem of fully intuitionistic fuzzy transportation problem, presented by Table 28.1, can be transformed into the problem (P7).

(P7)

$$\text{Minimize}\Big((2,5,7,10;1,4,8,12)\otimes(z_{111},z_{112},z_{113},z_{114};z_{111}',z_{112}',z_{113}',z_{114}')\oplus(3,6,9,12;$$

$$2,5,10,14\otimes(z_{121},z_{122},z_{123},z_{124};z_{121}',z_{122}',z_{123}',z_{124}')\oplus(3,6,9,11;1,5,10,13)$$

$$\otimes(z_{131},z_{132},z_{133},z_{134};z_{131}',z_{132}',z_{133}',z_{134}')\oplus(4,6,8,11;3,5,9,12)$$

$$\otimes(z_{141},z_{142},z_{143},z_{144};z_{141}',z_{142}',z_{143}',z_{144}')\oplus(4,7,9,13;3,6,11,15)$$

$$\otimes(z_{211},z_{212},z_{213},z_{214};z_{211}',z_{212}',z_{213}',z_{214}')\oplus(2,6,10,14;1,4,12,16)$$

$$\otimes(z_{221},z_{222},z_{223},z_{224};z_{221}',z_{222}',z_{223}',z_{224}')\oplus(3,7,10,14;2,5,12,15)$$

$$\otimes(z_{231},z_{232},z_{233},z_{234};z_{231}',z_{232}',z_{233}',z_{234}')\oplus(4,7,10,13;1,5,12,15)$$

$$\otimes(z_{241},z_{242},z_{243},z_{244};z_{241}',z_{242}',z_{243}',z_{244}')\oplus(4,8,10,12;2,6,11,14)$$

$$\otimes(z_{311},z_{312},z_{313},z_{314};z_{311}',z_{312}',z_{313}',z_{314}')\oplus(3,5,8,10;2,4,9,11)$$

$$\otimes(z_{321},z_{322},z_{323},z_{324};z_{321}',z_{322}',z_{323}',z_{324}')\oplus(4,7,11,13;3,5,12,14)$$

$$\otimes(z_{331},z_{332},z_{333},z_{334};z_{331}',z_{332}',z_{333}',z_{334}')\oplus(3,6,9,11;1,4,10,14)$$

$$\otimes(z_{341},z_{342},z_{343},z_{344};z_{341}',z_{342}',z_{343}',z_{344}'))$$

Subject to

$$\left(z_{111},z_{112},z_{113},z_{114};z'_{111},z'_{112},z'_{113},z'_{114}\right)\oplus\left(z_{121},z_{122},z_{123},z_{124};z'_{121},z'_{122},z'_{123},z'_{124}\right)$$

$$\oplus\left(z_{131},z_{132},z_{133},z_{134};z'_{131},z'_{132},z'_{133},z'_{134}\right)\oplus\left(z_{141},z_{142},z_{143},z_{144};z'_{141},z'_{142},z'_{143},z'_{144}\right)$$

$$\approx\left(4,6,10,14;2,5,11,15\right),\left(z_{211},z_{212},z_{213},z_{214};z'_{211},z'_{212},z'_{213},z'_{214}\right)$$

$$\oplus\left(z_{221},z_{222},z_{223},z_{224};z'_{221},z'_{222},z'_{223},z'_{224}\right)\oplus\left(z_{231},z_{232},z_{233},z_{234};z'_{231},z'_{232},z'_{233},z'_{234}\right)$$

$$\oplus\left(z_{241},z_{242},z_{243},z_{244};z'_{241},z'_{242},z'_{243},z'_{244}\right)\approx\left(5,9,12,16;4,8,14,18\right),$$

$$\left(z_{311},z_{312},z_{313},z_{314};z'_{311},z'_{312},z'_{313},z'_{314}\right)\oplus\left(z_{321},z_{322},z_{323},z_{324};z'_{321},z'_{322},z'_{323},z'_{324}\right)$$

$$\oplus\left(z_{331},z_{332},z_{333},z_{334};z'_{331},z'_{332},z'_{333},z'_{334}\right)\oplus\left(z_{341},z_{342},z_{343},z_{344};z'_{341},z'_{342},z'_{343},z'_{344}\right)$$

$$\approx\left(8,15,20,27;6,11,23,33\right),\left(z_{111},z_{112},z_{113},z_{114};z'_{111},z'_{112},z'_{113},z'_{114}\right)$$

$$\oplus\left(z_{211},z_{212},z_{213},z_{214};z'_{211},z'_{212},z'_{213},z'_{214}\right)\oplus\left(z_{311},z_{312},z_{313},z_{314};z'_{311},z'_{312},z'_{313},z'_{314}\right)$$

$$\approx\left(5,8,11,15;4,7,13,16\right),\left(z_{121},z_{122},z_{123},z_{124};z'_{121},z'_{122},z'_{123},z'_{124}\right)$$

$$\oplus\left(z_{221},z_{222},z_{223},z_{224};z'_{221},z'_{222},z'_{223},z'_{224}\right)\oplus\left(z_{321},z_{322},z_{323},z_{324};z'_{321},z'_{322},z'_{323},z'_{324}\right)$$

$$\approx\left(4,8,12,14;\ 3,6,13,18\right),$$

$$\left(z_{131},z_{132},z_{133},z_{134};z'_{131},z'_{132},z'_{133},z'_{134}\right)\oplus\left(z_{231},z_{232},z_{233},z_{234};z'_{231},z'_{232},z'_{233},z'_{234}\right)$$

$$\oplus\left(z_{331},z_{332},z_{333},z_{334};z'_{331},z'_{332},z'_{333},z'_{334}\right)\approx\left(3,6,9,14;2,5,10,16\right),$$

$$\left(z_{141},z_{142},z_{143},z_{144};z'_{141},z'_{142},z'_{143},z'_{144}\right)\oplus\left(z_{241},z_{242},z_{243},z_{244};z'_{241},z'_{242},z'_{243},z'_{244}\right)$$

$$\oplus\left(z_{341},z_{342},z_{343},z_{344};z'_{341},z'_{342},z'_{343},z'_{344}\right)\approx\left(5,8,10,14;3,6,12,16\right)$$

$$\left(z_{ij1},z_{ij2},z_{ij3},z_{ij4};z'_{ij1},z'_{ij2},z'_{ij3},z'_{ij4}\right)\succeq\left(0,0,0,0;0,0,0,0\right);i=1,2,3;j=1,2,3,4.$$

$$\left(z_{ij1},z_{ij2},z_{ij3},z_{ij4};z'_{ij1},z'_{ij2},z'_{ij3},z'_{ij4}\right)\text{ is a non-negative TIFN.}$$

Step 2: Using the arithmetic operations of TIFNs, the problem (P7) can be transformed into the problem (P8).

(P8)

$$\text{Minimize}\left(2z_{111}+3z_{121}+3z_{131}+4z_{141}+4z_{211}+2z_{221}+3z_{231}+4z_{241}+4z_{311}+3z_{321}+4z_{331}\right.$$

$$+3z_{341},5z_{112}+6z_{122}+6z_{132}+6z_{142}+7z_{212}+6z_{222}+7z_{232}+7z_{242}+8z_{312}+5z_{322}$$

$$+7z_{332}+6z_{342},7z_{113}+9z_{123}+9z_{133}+8z_{143}+9z_{213}+10z_{223}+10z_{233}+10z_{243}$$

$$+10z_{313}+8z_{323}+11z_{333}+9z_{343},10z_{114}+12z_{124}+11z_{134}+11z_{144}+13z_{214}$$

$$+14z_{224}+14z_{234}+13z_{244}+12z_{314}+10z_{324}+13z_{334}+11z_{344};z'_{111}+2z'_{121}+z'_{131}$$

$$+3z'_{141}+3z'_{211}+z'_{221}+2z'_{231}+z'_{241}+2z'_{311}+2z'_{321}+3z'_{331}+z'_{341},4z'_{112}+5z'_{122}$$

$$+5z'_{132}+5z'_{142}+6z'_{212}+4z'_{222}+5z'_{232}+5z'_{242}+6z'_{312}+4z'_{322}+5z'_{332}+4z'_{342},$$

$$8z'_{113}+10z'_{123}+10z'_{133}+9z'_{143}+11z'_{213}+12z'_{223}+12z'_{233}+12z'_{243}+11z'_{313}+9z'_{323}$$

$$+12z'_{333}+10z'_{343},12z'_{114}+14z'_{124}+13z'_{134}+12z'_{144}+15z'_{214}+16z'_{224}+15z'_{234}$$

$$\left.+15z'_{244}+14z'_{314}+11z'_{324}+14z'_{334}+14z'_{344}\right)$$

Subject to

$$\left(z_{111} + z_{121} + z_{131} + z_{141}, z_{112} + z_{122} + z_{132} + z_{142}, z_{113} + z_{123} + z_{133} + z_{143}, z_{114}\right.$$
$$+ z_{124} + z_{134} + z_{144}; z'_{111} + z'_{121} + z'_{131} + z'_{141}, z'_{112} + z'_{122} + z'_{132} + z'_{142}, z'_{113} + z'_{123}$$
$$\left.+ z'_{133} + z'_{143}, z'_{114} + z'_{124} + z'_{134} + z'_{144}\right)$$
$$\approx \left(4, 6, 10, 14; 2, 5, 11, 15\right),$$

$$\left(z_{211} + z_{221} + z_{231} + z_{241}, z_{212} + z_{222} + z_{232} + z_{242}, z_{213} + z_{223} + z_{233} + z_{243}, z_{214}\right.$$
$$+ z_{224} + z_{234} + z_{244}, z'_{211} + z'_{221} + z'_{231} + z'_{241}, z'_{212} + z'_{222} + z'_{232} + z'_{242}, z'_{213} + z'_{223}$$
$$\left.+ z'_{233} + z'_{243}, z'_{214} + z'_{224} + z'_{234} + z'_{244}\right)$$
$$\approx \left(5, 9, 12, 16; 4, 8, 14, 18\right),$$

$$\left(z_{311} + z_{321} + z_{331} + z_{341}, z_{312} + z_{322} + z_{332} + z_{342}, z_{313} + z_{323} + z_{333} + z_{343}, z_{314}\right.$$
$$+ z_{324} + z_{334} + z_{344}; z'_{311} + z'_{321} + z'_{331} + z'_{341}, z'_{312} + z'_{322} + z'_{332} + z'_{342}, z'_{313} + z'_{323}$$
$$\left.+ z'_{333} + z'_{343}, z'_{314} + z'_{324} + z'_{334} + z'_{344}\right)$$
$$\approx \left(8, 15, 20, 27; 6, 11, 23, 33\right),$$

$$\left(z_{111} + z_{211} + z_{311}, z_{112} + z_{212} + z_{312}, z_{113} + z_{213} + z_{313}, z_{114} + z_{214} + z_{314}; z'_{111}\right.$$
$$\left.+ z'_{211} + z'_{311}, z'_{112} + z'_{212} + z'_{312}, z'_{113} + z'_{213} + z'_{313}, z'_{114} + z'_{214} + z'_{314}\right)$$
$$\approx \left(5, 8, 11, 15; 4, 7, 13, 16\right),$$

$$\left(z_{121} + z_{221} + z_{321}, z_{122} + z_{222} + z_{322}, z_{123} + z_{223} + z_{323}, z_{124} + z_{224} + z_{324};\right.$$
$$\left. z'_{121} + z'_{221} + z'_{321}, z'_{122} + z'_{222} + z'_{322}, z'_{123} + z'_{223} + z'_{323}, z'_{124} + z'_{224} + z'_{324}\right)$$
$$\approx \left(4, 8, 12, 14; 3, 6, 13, 18\right),$$

$$\left(z_{131} + z_{231} + z_{331}, z_{132} + z_{232} + z_{332}, z_{133} + z_{233} + z_{333}, z_{134} + z_{234} + z_{334};\right.$$
$$\left. z'_{131} + z'_{231} + z'_{331}, z'_{132} + z'_{232} + z'_{332}, z'_{133} + z'_{233} + z'_{333}, z'_{134} + z'_{234} + z'_{334}\right)$$
$$\approx \left(3, 6, 9, 14; 2, 5, 10, 16\right),$$

$$\left(z_{141} + z_{241} + z_{341}, z_{142} + z_{242} + z_{342}, z_{143} + z_{243} + z_{343}, z_{144} + z_{244} + z_{344};\right.$$
$$\left. z'_{141} + z'_{241} + z'_{341}, z'_{142} + z'_{242} + z'_{342}, z'_{143} + z'_{243} + z'_{343}, z'_{144} + z'_{244} + z'_{344}\right)$$
$$\approx \left(5, 8, 10, 14; 3, 6, 12, 16\right)$$

$$\left(z_{ij1}, z_{ij2}, z_{ij3}, z_{ij4}; z'_{ij1}, z'_{ij2}, z'_{ij3}, z'_{ij4}\right) \succeq (0,0,0,0; 0,0,0,0); i = 1,2,3; j = 1,2,3,4,$$
$$\left(z_{ij1}, z_{ij2}, z_{ij3}, z_{ij4}; z'_{ij1}, z'_{ij2}, z'_{ij3}, z'_{ij4}\right) \text{ is a non-negative TIFN.}$$

Step 3: Using Section 2.2, the problem (P8) can be transformed into the problem (P9).

(P9)

Minimize $R\left(2z_{111} + 3z_{121} + 3z_{131} + 4z_{141} + 4z_{211} + 2z_{221} + 3z_{231} + 4z_{241} + 4z_{311} + 3z_{321}\right.$

$$+ 4z_{331} + 3z_{341}, 5z_{112} + 6z_{122} + 6z_{132} + 6z_{142} + 7z_{212} + 6z_{222} + 7z_{232} + 7z_{242}$$

$+8z_{312}+5z_{322}+7z_{332}+6z_{342},7z_{113}+9z_{123}+9z_{133}+8z_{143}+9z_{213}+10z_{223}$

$+10z_{233}+10z_{243}+10z_{313}+8z_{323}+11z_{333}+9z_{343},10z_{114}+12z_{124}+11z_{134}$

$+11z_{144}+13z_{214}+14z_{224}+14z_{234}+13z_{244}+12z_{314}+10z_{324}+13z_{334}$

$+11z_{344};z'_{111}+2z'_{121}+z'_{131}+3z'_{141}+3z'_{211}+z'_{221}+2z'_{231}+z'_{241}+2z'_{311}$

$+2z'_{321}+3z'_{331}+z'_{341},4z'_{112}+5z'_{122}+5z'_{132}+5z'_{142}+6z'_{212}+4z'_{222}+5z'_{232}$

$+5z'_{242}+6z'_{312}+4z'_{322}+5z'_{332}+4z'_{342},8z'_{113}+10z'_{123}+10z'_{133}+9z'_{143}$

$+11z'_{213}+12z'_{223}+12z'_{233}+12z'_{243}+11z'_{313}+9z'_{323}+12z'_{333}+10z'_{343},$

$12z'_{114}+14z'_{124}+13z'_{134}+12z'_{144}+15z'_{214}+16z'_{224}+15z'_{234}+15z'_{244}$

$+14z'_{314}+11z'_{324}+14'z_{334}+14z'_{344})$

Subject to

$R(z_{111}+z_{121}+z_{131}+z_{141},z_{112}+z_{122}+z_{132}+z_{142},z_{113}+z_{123}+z_{133}+z_{143},z_{114}$

$+z_{124}+z_{134}+z_{144};z'_{111}+z'_{121}+z'_{131}+z'_{141},z'_{112}+z'_{122}+z'_{132}+z'_{142},z'_{113}+z'_{123}$

$+z'_{133},z'_{143},z'_{114}+z'_{124}+z'_{134}+z'_{144})$

$$= R(4,6,10,14;2,5,11,15)$$

$(z_{211}+z_{221}+z_{231}+z_{241},z_{212}+z_{222}+z_{232}+z_{242},z_{213}+z_{223}+z_{233}+z_{243},z_{214}$

$+z_{224}+z_{234}+z_{244},z'_{211}+z'_{221}+z'_{231}+z'_{241},z'_{212}+z'_{222}+z'_{232}+z'_{242},z'_{213}+z'_{223}$

$+z'_{233}+z'_{243},z'_{214}+z'_{224}+z'_{234}+z'_{244})$

$$= R(5,9,12,16;4,8,14,18)$$

$R(z_{311}+z_{321}+z_{331}+z_{341},z_{312}+z_{322}+z_{332}+z_{342},z_{313}+z_{323}+z_{333}+z_{343},z_{314}$

$+z_{324}+z_{334}+z_{344},z'_{311}+z'_{321}+z'_{331}+z'_{341},z'_{312}+z'_{322}+z'_{332}+z'_{342},z'_{313}+z'_{323}$

$+z'_{333}+z'_{343},z'_{314}+z'_{324}+z'_{334}+z'_{344})$

$$= R(8,15,20,27;6,11,23,33)$$

$R(z_{111}+z_{211}+z_{311},z_{112}+z_{212}+z_{312},z_{113}+z_{213}+z_{313},z_{114}+z_{214}+z_{314};z'_{111}$

$+z_{211}+z_{311},z_{112}+z_{212}+z_{312},z_{113}+z_{213}+z_{313},z_{114}+z_{214}+z'_{314})$

$$= R(5,8,11,15;4,7,13,16)$$

$R(z_{121}+z_{221}+z_{321},z_{122}+z_{222}+z_{322},z_{123}+z_{223}+z_{323},z_{124}+z_{224}+z_{324};$

$z'_{121}+z'_{221}+z'_{321},z'_{122}+z'_{222}+z'_{322},z'_{123}+z'_{223}+z'_{323},z'_{124}+z'_{224}+z'_{324})$

$$= R(4,8,12,14;3,6,13,18)$$

$R(z_{131}+z_{231}+z_{331},z_{132}+z_{232}+z_{332},z_{133}+z_{233}+z_{333},z_{134}+z_{234}+z_{334};$

$z'_{131}+z'_{231}+z'_{331},z'_{132}+z'_{232}+z'_{332},z'_{133}+z'_{233}+z'_{333},z'_{134}+z'_{234}+z'_{334})$

$$= R(3,6,9,14;2,5,10,16)$$

$R(z_{141}+z_{241}+z_{341},z_{142}+z_{242}+z_{342},z_{143}+z_{243}+z_{343},z_{144}+z_{244}+z_{344};$

$z'_{141}+z'_{241}+z'_{341},z'_{142}+z'_{242}+z'_{342},z'_{143}+z'_{243}+z'_{343},z'_{144}+z'_{244}+z'_{344})$

$$= R(5,8,10,14;3,6,12,16)$$

$$R\left(z_{ij1}, z_{ij2}, z_{ij3}, z_{ij4}; z'_{ij1}, z'_{ij2}, z'_{ij3}, z'_{ij4}\right) \geq R(0,0,0,0;0,0,0,0); i = 1,2,3;$$
$$j = 1,2,3,4, z_{ijk}, z'_{ijk} \geq 0; i = 1,2,3; j = 1,2,3,4; k = 1,2,3,4.$$

Step 4: Using the expression $R(a,b,c,d;a',b',c',d') = \dfrac{a+b+c+d+a'+b'+c'+d'}{8}$, the problem (P9) can be transformed into the problem (P10).

(P10)

$$\text{Minimize} \left[\frac{1}{8} \left(2z_{111} + 3z_{121} + 3z_{131} + 4z_{141} + 4z_{211} + 2z_{221} + 3z_{231} + 4z_{241} + 4z_{311} + 3z_{321} \right.\right.$$

$$+4z_{331} + 3z_{341} + 5z_{112} + 6z_{122} + 6z_{132} + 6z_{142} + 7z_{212} + 6z_{222} + 7z_{232} + 7z_{242}$$

$$+8z_{312} + 5z_{322} + 7z_{332} + 6z_{342} + 7z_{113} + 9z_{123} + 9z_{133} + 8z_{143} + 9z_{213} + 10z_{223}$$

$$+10z_{233} + 10z_{243} + 10z_{313} + 8z_{323} + 11z_{333} + 9z_{343} + 10z_{114} + 12z_{124} + 11z_{134}$$

$$+11z_{144} + 13z_{214} + 14z_{224} + 14z_{234} + 13z_{244} + 12z_{314} + 10z_{324} + 13z_{334} + 11z_{344}$$

$$+z'_{111} + 2z'_{121} + z'_{131} + 3z'_{141} + 3z'_{211} + z'_{221} + 2z'_{231} + z'_{241} + 2z'_{311} + 2z'_{321} + 3z'_{331}$$

$$+z'_{341} + 4z'_{112} + 5z'_{122} + 5z'_{132} + 5z'_{142} + 6z'_{212} + 4z'_{222} + 5z'_{232} + 5z'_{242} + 6z'_{312}$$

$$+4z'_{322} + 5z'_{332} + 4z'_{342} + 8z'_{113} + 10z'_{123} + 10z'_{133} + 9z'_{143} + 11z'_{213} + 12z'_{223}$$

$$+12z'_{233} + 12z'_{243} + 11z'_{313} + 9z'_{323} + 12z'_{333} + 10z'_{343} + 12z'_{114} + 14z'_{124} + 13z'_{134}$$

$$\left.\left.+12z'_{144} + 15z'_{214} + 16z'_{224} + 15z'_{234} + 15z'_{244} + 14z'_{314} + 11z'_{324} + 14z'_{334} + 14z'_{344} \right) \right]$$

Subject to

$$z_{111} + z_{121} + z_{131} + z_{141} + z_{112} + z_{122} + z_{132} + z_{142} + z_{113} + z_{123} + z_{133} + z_{143} + z_{114} + z_{124}$$

$$+ z_{134} + z_{144} + z'_{111} + z'_{121} + z'_{131} + z'_{141} + z'_{112} + z'_{122} + z'_{132} + z'_{142} + z'_{113} + z'_{123} + z'_{133} + z'_{143}$$

$$+ z'_{114} + z'_{124} + z'_{134} + z'_{144} = 67, z_{211} + z_{221} + z_{231} + z_{241} + z_{212} + z_{222} + z_{232} + z_{242} + z_{213}$$

$$+ z_{223} + z_{233} + z_{243} + z_{214} + z_{224} + z_{234} + z_{244} + z'_{211} + z'_{221} + z'_{231} + z'_{241} + z'_{212} + z'_{222}$$

$$+ z'_{232} + z'_{242} + z'_{213} + z'_{223} + z'_{233} + z'_{243} + z'_{214} + z'_{224} + z'_{234} + z'_{244} = 86, z_{311} + z_{321} + z_{331}$$

$$+ z_{341} + z_{312} + z_{322} + z_{332} + z_{342} + z_{313} + z_{323} + z_{333} + z_{343} + z_{314} + z_{324} + z_{334} + z_{344}$$

$$+ z'_{311} + z'_{321} + z'_{331} + z'_{341} + z'_{312} + z'_{322} + z'_{332} + z'_{342} + z'_{313} + z'_{323} + z'_{333} + z'_{343} + z'_{314}$$

$$+ z'_{324} + z'_{334} + z'_{344} = 143, z_{111} + z_{211} + z_{311} + z_{112} + z_{212} + z_{312} + z_{113} + z_{213} + z_{313}$$

$$+ z_{114} + z_{214} + z_{314} + z'_{111} + z'_{211} + z'_{311} + z'_{112} + z'_{212} + z'_{312} + z'_{113} + z'_{213} + z'_{313} + z'_{114}$$

$$+ z'_{214} + z'_{314} = 79, z_{121} + z_{221} + z_{321} + z_{122} + z_{222} + z_{322} + z_{123} + z_{223} + z_{323} + z_{124} + z_{224}$$

$$+ z_{324} + z'_{121} + z'_{221} + z'_{321} + z'_{122} + z'_{222} + z'_{322} + z'_{123} + z'_{223} + z'_{323} + z'_{124} + z'_{224} + z'_{324} = 78,$$

$$z_{131} + z_{231} + z_{331} + z_{132} + z_{232} + z_{332} + z_{133} + z_{233} + z_{333} + z_{134} + z_{234} + z_{334} + z'_{131} + z'_{231}$$

$$+ z'_{331} + z'_{132} + z'_{232} + z'_{332} + z'_{133} + z'_{233} + z'_{333} + z'_{134} + z'_{234} + z'_{334} = 65,$$

$$z_{141} + z_{241} + z_{341} + z_{142} + z_{242} + z_{342} + z_{143} + z_{243} + z_{343} + z_{144} + z_{244} + z_{344} + z'_{141} + z'_{241}$$

$$+ z'_{341} + z'_{142} + z'_{242} + z'_{342} + z'_{143} + z'_{243} + z'_{343} + z'_{144} + z'_{244} + z'_{344} = 74,$$

$$z_{ijk}, z'_{ijk} \geq 0; i = 1,2,3; j = 1,2,3,4; k = 1,2,3,4.$$

Step 5: The optimal solution of the linear programming problem (P10) is

$$z_{111} = \frac{67}{8}, \; z_{112} = \frac{67}{8}, \; z_{113} = \frac{67}{8}, \; z_{114} = \frac{67}{8}, \; z'_{111} = \frac{67}{8}, \; z'_{112} = \frac{67}{8}, \; z'_{113} = \frac{67}{8},$$

$$z'_{114} = \frac{67}{8}, z_{211} = \frac{3}{2}, z_{212} = \frac{3}{2}, z_{213} = \frac{3}{2}, z_{214} = \frac{3}{2}, z'_{211} = \frac{3}{2}, z'_{212} = \frac{3}{2}, z'_{213} = \frac{3}{2},$$

$$z'_{214} = \frac{3}{2}, z_{231} = \frac{65}{8}, z_{232} = \frac{65}{8}, z_{233} = \frac{65}{8}, z_{234} = \frac{65}{8}, z'_{231} = \frac{65}{8}, z'_{232} = \frac{65}{8},$$

$$z'_{233} = \frac{65}{8}, \; z'_{234} = \frac{65}{8}, z'_{241} = \frac{9}{4}, z'_{242} = \frac{9}{4}, z'_{243} = \frac{9}{4}, z'_{244} = \frac{9}{4}, z_{321} = \frac{39}{4}, z_{322} = \frac{39}{4},$$

$$z_{323} = \frac{39}{4}, \; z_{324} = \frac{39}{4}, z'_{321} = \frac{39}{4}, \; z'_{322} = \frac{39}{4}, \; z'_{323} = \frac{39}{4}, \; z'_{324} = \frac{39}{4}, z_{341} = \frac{65}{8},$$

$$z_{342} = \frac{65}{8}, z_{343} = \frac{65}{8}, z_{344} = \frac{65}{8}, z'_{341} = \frac{65}{8}, z'_{342} = \frac{65}{8}, z'_{343} = \frac{65}{8}, z'_{344} = \frac{65}{8}.$$

Step 5: The intuitionistic fuzzy optimal solution by using the results, obtained in Step 4, is

$$\tilde{z}^I_{11} = \left(\frac{67}{8}, \frac{67}{8}, \frac{67}{8}, \frac{67}{8}; \frac{67}{8}, \frac{67}{8}, \frac{67}{8}, \frac{67}{8} \right) \quad \tilde{z}^I_{21} = \left(\frac{3}{2}, \frac{3}{2}, \frac{3}{2}, \frac{3}{2}; \frac{3}{2}, \frac{3}{2}, \frac{3}{2}, \frac{3}{2} \right),$$

$$\tilde{z}^I_{23} = \left(\frac{65}{8}, \frac{65}{8}, \frac{65}{8}, \frac{65}{8}; \frac{65}{8}, \frac{65}{8}, \frac{65}{8}, \frac{65}{8} \right) \quad \tilde{z}^I_{24} = \left(0,0,0,0; \frac{9}{4}, \frac{9}{4}, \frac{9}{4}, \frac{9}{4} \right),$$

$$\tilde{z}^I_{32} = \left(\frac{39}{4}, \frac{39}{4}, \frac{39}{4}, \frac{39}{4}; \frac{39}{4}, \frac{39}{4}, \frac{39}{4}, \frac{39}{4} \right) \text{ and } \tilde{z}^I_{34} = \left(\frac{65}{8}, \frac{65}{8}, \frac{65}{8}, \frac{65}{8}; \frac{65}{8}, \frac{65}{8}, \frac{65}{8}, \frac{65}{8} \right).$$

Step 6: Using the intuitionistic fuzzy optimal solution, obtained in step 5, the optimal intuitionistic fuzzy transportation cost of the problem, presented by Table 28.1, is

$$(2,5,7,10;1,4,8,12) \otimes \left(\frac{67}{8}, \frac{67}{8}, \frac{67}{8}, \frac{67}{8}; \frac{67}{8}, \frac{67}{8}, \frac{67}{8}, \frac{67}{8} \right) \oplus (4,7,9,13;3,6,11,15)$$

$$\otimes \left(\frac{3}{2}, \frac{3}{2}, \frac{3}{2}, \frac{3}{2}; \frac{3}{2}, \frac{3}{2}, \frac{3}{2}, \frac{3}{2} \right) \oplus (3,7,10,14;2,5,12,15) \otimes \left(\frac{65}{8}, \frac{65}{8}, \frac{65}{8}, \frac{65}{8}; \frac{65}{8}, \frac{65}{8}, \frac{65}{8}, \frac{65}{8}; \right)$$

$$\oplus (4,7,10,13;1,5,12,15) \otimes \left(0,0,0,0; \frac{9}{4}, \frac{9}{4}, \frac{9}{4}, \frac{9}{4} \right) \oplus (3,5,8,10;2,4,9,11)$$

$$\otimes \left(\frac{39}{4}, \frac{39}{4}, \frac{39}{4}, \frac{39}{4}; \frac{39}{4}, \frac{39}{4}, \frac{39}{4}, \frac{39}{4} \right) \oplus (3,6,9,11;1,4,10,14)$$

$$\otimes \left(\frac{65}{8}, \frac{65}{8}, \frac{65}{8}, \frac{65}{8}; \frac{65}{8}, \frac{65}{8}, \frac{65}{8}, \frac{65}{8} \right) = \left(\frac{403}{4}, \frac{1855}{8}, \frac{609}{2}, \frac{3231}{8}; 59, \frac{1327}{8}, 377, \frac{3997}{8} \right).$$

28.5 CONCLUSION

A special multiplication, proposed by Kumar and Hussain [19] satisfied the property $R\left(\tilde{A}^I \otimes \tilde{B}^I\right) = R\left(\tilde{A}^I\right) \times R\left(\tilde{B}^I\right)$ for the ranking function R, used by Singh and Yadav [26], which is not genuine to use for solving fully intuitionistic fuzzy transportation problems. This chapter gives a modified alternative method to overcome this flaw and give the solution for fully intuitionistic fuzzy transportation problems. In future study, this problem can be expanded for unbalanced fully intuitionistic fuzzy transportation problems also may take the parameters of transportation problems as generalized intuitionistic fuzzy numbers or exponential fuzzy numbers.

REFERENCES

[1] Alemany M. M. E., Grillo H., Ortiz A. and Fuertes-Miquel V. S. 2015. A fuzzy model for shortage planning under uncertainty due to lack of homogeneity in planned production lots. *Applied Mathematical Modelling*, 39, 4463–4481.

[2] Atanassov K. T. 1986. Intitionistic fuzzy sets. *Fuzzy Sets and Systems*, 20, 87–96.

[3] Basirzadeh H. 2011. An approach for solving fuzzy transportation problem. *Applied Mathematical Science*, 5, 1549–1566.

[4] Bellman R. and Zadeh L. A. 1970. Decision making in fuzzy environment. *Management Science*, 17, 141–164.

[5] Cascetta E., Gallo M. and Montella B. 2006. Models and algorithms for the optimization of signal settings on urban networks with stochastic assignments models. *Annals of Operations Research*, 144, 301–328.

[6] Chinnadurai V. and Muthukumar S. 2016. Solving the linear fractional programming problem in a fuzzy environment: Numerical approach. *Applied Mathematical Modelling*, 40, 6148–6164.

[7] Dinagar D. S. and Thiripurasundari K. 2014. A navel method for solving fuzzy transportation problem involving intuitionistic trapezoidal fuzzy numbers. *International Journal of Current Research*, 6, 7038–7041.

[8] Ezzati R., Khorram E. and Enayati R. 2015. A new algorithm to solve fully fuzzy linear programming problems using the MOLP problem. *Applied Mathematical Modelling*, 39, 3183–3193.

[9] Gani A. N. and Abbas S. A. 2012. Solving intuitionistic fuzzy transportation problem using zero suffix algorithm. *International Journal of Mathematical Sciences and Engineering Applications*, 6, 73–82.

[10] Gani A. N. and Abbas S. 2014. Revised distribution method for intuitionistic fuzzy transportation problem. *International Journal of Fuzzy Mathematical Archive*, 4, 96–103.

[11] Gao Y., Yang L. and Li S. 2016. Uncertain models on railway transportation planning problem. *Applied Mathematical Modelling*, 40, 4921–4934.

[12] Hitchcock F. L. 1941. The distribution of a product from several sources to numerous localities. *Journal of Mathematical Physics*, 20, 224–230.

[13] Kaur A. and Kumar A. 2011. A new method for solving fuzzy transportation problems using ranking function. *Applied Mathematical Modelling*, 35, 5652–5661.

[14] Kaur A. and Kumar A. 2012. A new approach for solving fuzzy transportation problem using generalized trapezoidal fuzzy number. *Applied Soft Computing*, 12, 1201–1213.

[15] Kaur K. and Kumar A. 2013. A new method to find the unique fuzzy optimal value of fuzzy linear programming problems. *Journal of Optimization Theory and Applications*, 156, 529–534.

[16] Kocken H. G. and Sivri M. 2016. A simple parametric method to generate all optimal solutions of fuzzy solid transportation problem. *Applied Mathematical Modelling*, 40, 4612–4624.

[17] Kumar P. S. and Hussain R. J. 2014. A systematic approach for solving mixed intuitionistic fuzzy transportation problems. *International Journal of Pure and Applied Mathematics*, 92, 181–190.

[18] Kumar P. S. and Hussain R. J. 2015. A method for solving unbalanced intuitionistic fuzzy transportation problems. *Notes on Intuitionistic Fuzzy Sets*, 2, 54–65.

[19] Kumar P. S. and Hussain R. J. 2015. Computationally simple approach for solving fully intuitionistic fuzzy real life transportation problems. *International Journal of System Assurance Engineering and Management*, 6, 1–12.

[20] Pandian P. 2014. Realistic method for solving fully intuitionistic fuzzy transportation problems. *Applied Mathematical Sciences*, 8, 5633–5639.

[21] Rani D. and Gulati T. R. 2014. A method for unbalanced transportation problems in fuzzy environment. *Sadhana*, 39, 573–581.

[22] Rani D. and Gulati T. R. 2014. A new approach to solve unbalanced transportation problems in imprecise environment. *Journal of Transportation Security*, 7, 277–287.

[23] Rani D. and Gulati T. R. 2016. Time optimization in totally uncertain transportation problems. *International Journal of Fuzzy Systems*, 18, 1–12.

[24] Roseline S. and Amirtharaj H. 2015. New approaches to find the solution for the intuitionistic fuzzy transportation problem with ranking of intuitionistic fuzzy numbers. *International Journal of Innovative Research in Science, Engineering and Technology*, 4, 10222–10230.

[25] Singh S. and Gupta G. 2014. A new approach for solving cost minimization balanced transportation problem under uncertainty. *Journal of Transportation Security*, 7, 339–345.

[26] Singh, S. K. and Yadav, S. P. 2014. Efficient approach for solving type-1 intuitionistic fuzzy transportation problem. *International Journal of System Assurance Engineering and Management*, 6, 259–267.

[27] Singh, S. K. and Yadav, S. P. 2016. Intuitionistic fuzzy transportation problem with various kinds of uncertainties in parameters and variables. *International Journal of System Assurance Engineering and Management*, 7, 1–11.

[28] Srinivas B. and Ganesan G. 2015. Optimal solution for intuitionistic fuzzy transportation problem via revised distribution method. *International Journal of Mathematics Trends and Technology*, 19, 150–161.

[29] Thamaraiselvi A. and Santhi R. 2015. On intuitionistic fuzzy transportation problem using hexagonal intuitionistic fuzzy numbers. *International Journal of Fuzzy Logic Systems*, 5, 15–28.

[30] Zadeh L. A. 1965. Fuzzy sets. *Information and Control*, 8, 338–353.

29 Reliability Optimization of an Industrial Model Using the Chaotic Grey Wolf Optimization Algorithm

Mohit Kumar Kakkar, Jasdev Bhatti, Gourav Gupta

29.1 INTRODUCTION

In the modern industrial world, the concept of reliability is extremely important. Reliability is the ability of an item to perform a required function, under established conditions, for a specified period of time. Reliability is defined as the probability that a unit or equipment in an industry will perform a desired function for a given time window. Because reliability is defined in terms of probability, so value of reliability is always lie between 1 and 0.

In the present world, the modern technology is changing very rapidly and leading to the development of many complex systems and attract the attention of researchers towards their reliability and cost optimization. Reliability-optimization-related problems comes under the category of nonlinear optimization with number of constraints, and due to its difficulty level, it belongs to the NP-hard problem. According to the literature review, a number of metaheuristic algorithms were applied by different researchers for reliability optimization problems. Ravi et al. (1997) presented the NESA (non-equilibrium simulate annealing) for finding out the optimized value of system cost of complex reliability system. Konak et al. (2006) and Painton et al. (1995) applied genetic algorithms for optimized solution of reliability of complex industrial system. Wang et al. (2012) used the HGA (hybrid genetic algorithm) for reliability optimization of complex system. Yeh et al. (2010) also described the another nature-inspired optimization approach, which was based on Monte Carlo simulation (MCS) for optimizing the reliability of complex systems. Suman (2003) presented the simulated annealing based multi-objective algorithm for reliability optimization problems. Bhatti et al. (2020, 2021, 2021a,b) discussed the reliability of an industrial model using geometric distribution under active redundancy. Kakkar et al. (2019) presented an analysis of an industrial system using an evolutionary algorithm, i.e. genetic algorithm. Kakkar et al. (2016) analysed the two parallel unit repairable system using stochastic modelling and Markov chain concept. Bhatti et al. (2016) analysed the standby system with discrete failure. Kakkar et al. (2015) also discussed the reliability analysis of industrial system using stochastic modelling.

Al-Tashi et al. (2019) explained the application of hybrid grey wolf algorithm for feature selection. Gao & Zhao (2019) presented a modified grey wolf optimization algorithm (GWOA) with variable weights. Sheta et al. (2016) optimized the parameters used for software reliability models using GWOA. Kumar et al. (2021) examines the application of genetic algorithm for reliability analysis of settlement of pile group.

29.2 METHODOLOGY

GWOA is an optimizer algorithm which is a metaheuristic algorithm, based on the behaviour of the grey wolf in the nature. This chapter used the chaotic grey wolf optimizer algorithm (CGWOA) for solving nonlinear reliability optimization problem with constraints for complex industrial system.

29.2.1 ALGORITHM BASED ON GREY WOLF BEHAVIOUR (GREY WOLF OPTIMIZER)

It is a population-based metaheuristics intelligence algorithm for obtaining the global solution. GWO is an intelligent nature-inspired algorithm that simulates the prey mechanism of wolves in the forest proposed by Mirjalili et al. (2014). The GWOA models the behaviour of grey wolves by social hierarchy. These wolves live in a set of groups and every group having 6–12 wolves. In this group, as depicted in the Figure 29.1, the

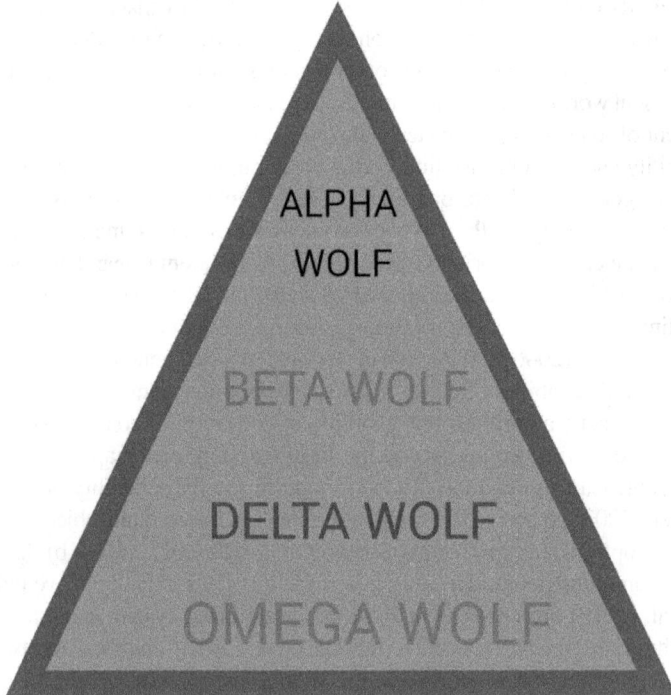

FIGURE 29.1 Group classification of grey wolves.

α-wolf (alpha wolf), having best position in the group, is responsible for making decisions about hunting, and the rest of the members must follow its orders. The α-wolf of
a group is the best in group as far as leadership skill is concerned, and the second-best-
positioned wolf is the β-wolf (beta wolf), and the third-best-positioned wolf is the δ-wolf
(delta wolf). The rest of the other wolves are ω-wolves (omega wolves).

29.3 MATHEMATICAL MODEL AND ALGORITHM

A very important concept in GWOA is "hunting the prey"; mathematically we can
describe this term as finding the optimal global solution of a function. Therefore, the
wolves move around in the area of the function and by nature wolves are leaded by
the dominant wolves and the Υ-wolves (not dominant) move towards the dominant
wolves (α, β, δ). These wolves encircle prey when they go for hunting as depicted
graphically in the Figure 29.2.

D_1 = distance from alpha wolf
D_2 = distance from beta wolf
D_3 = distance from delta wolf
a and C are parameter vectors

In GWO, each wolf (X) updates its position based on alpha (Xα), beta (Xβ), and
delta (Xδ).

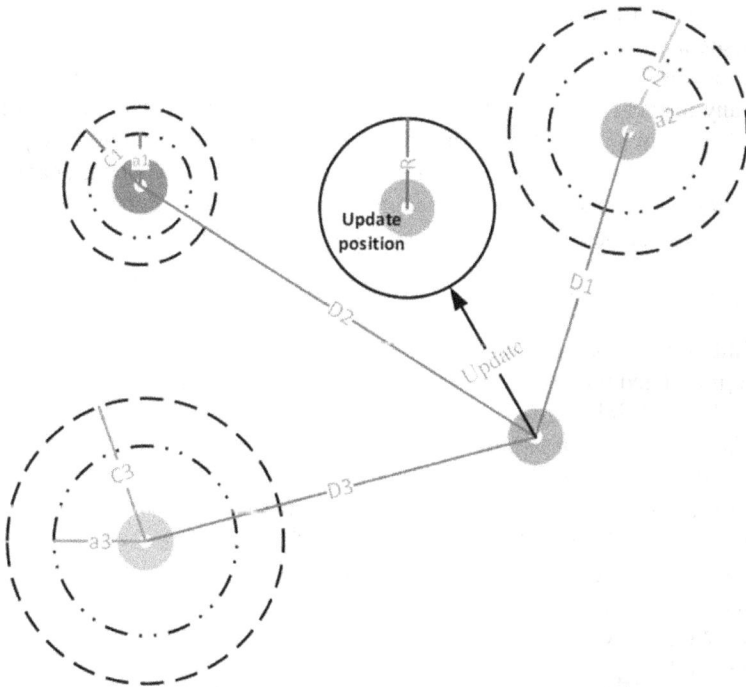

FIGURE 29.2 Encircling behaviour of grey wolves.

Position-Updated Equation

Velocity gained from α wolf, $\mathbf{D}_\alpha = |\mathbf{C}_1.\mathbf{X}_\alpha - \mathbf{X}|$

Velocity gained from β wolf $\mathbf{D}_\beta = |\mathbf{C}_2.\mathbf{X}_\beta - \mathbf{X}|$

Velocity gained from δ wolf $\mathbf{D}_\delta = |\mathbf{C}_3.\mathbf{X}_\delta - \mathbf{X}|$

Position update w.r.t α wolf, $\mathbf{X}_1 = \mathbf{X}_\alpha - \mathbf{A}_1.\,(\mathbf{D}_\alpha)$

Position update w.r.t β wolf, $\mathbf{X}_2 = \mathbf{X}_\beta - \mathbf{A}_2.\,(\mathbf{D}_\beta)$

Position update w.r.t δ wolf, $\mathbf{X}_3 = \mathbf{X}_\delta - \mathbf{A}_3.\,(\mathbf{D}_\delta)$

$\mathbf{X}_{(t+1)} = (\mathbf{X}_\alpha - \mathbf{A}_1.\,(\mathbf{D}_\alpha) + \mathbf{X}_\beta - \mathbf{A}_2.(\mathbf{D}_\beta) + \mathbf{X}_\delta - \mathbf{A}_3.(\mathbf{D}_\delta))/3$

i.e. $\mathbf{X}_{(t+1)} = (\mathbf{X}_\alpha + \mathbf{X}_\beta + \mathbf{X}_\delta - \mathbf{A}_1.\,(\mathbf{D}_\alpha) - \mathbf{A}_2.(\mathbf{D}_\beta) - \mathbf{A}_3.(\mathbf{D}_\delta))/3$

Convergence rate of GWOA is good, but as compared to the other nature-inspired algorithms like PSO, GA, and ACO, there is some scope to improve its efficiency. CGWOA is generated by presenting chaos in GWOA itself. Basically, chaos is a deterministic, random-like concept, and these chaotic systems are also described as sources of randomness.

The applicability of chaos in optimization strategies depends on the different chaotic maps, as these are helpful to increase the convergence rate of optimization algorithms, and it helps such kind of algorithms in exploring search domain globally.

Different kind of chaotic maps generated by mathematicians and scholars are existing in the literature (Lu et al., 2014; Demidova et al., 2020). Most of them have already been applied to optimization algorithms. In this study, we have used logistic map, introduced by Robert May (2004) for CGWOA:

Logistic map $X_{r+1} = ax_r\,(1\text{-}x_r)$

The logistic function uses a differential equation, where x represents the population at any instant r, with growth rate or control parameter a, or we can say that the population at any instant, (x_{r+1}) is a function of the growth rate (a) and the previous population (x_r). That is why we can say that chaos explains certain nonlinear dynamical systems that depends on initial populations or conditions. Initial value for these types of maps can be chosen from the closed interval [0, 1].

Pseudo Code for CGWOA

- Initialization of t and population X_i where (i=1, 2, . . ., n)
- x_0 initialized randomly
- a, A, and C initialized
- Fitness calculation of each type of wolf $X_\alpha, X_\beta, X_\gamma$
- While (t < iterations(max)) (loop starts)
- Sorting of the population w.r.t the wolve's fitness
- Change the chaotic number with the help of map equation
- Change the position of current wolf and values of a, A, and C as well
- $_c$hange the fitness of each wolf $X_\alpha, X_\beta, X_\gamma$
- Replace the wolf (worst with the best).
- $t = t + 1$
- End while (loop)
- Return Xα

29.4   PROBLEM DEFINITION

As shown in Figure 29.3, in this study a complex industrial system is considered, which may have different kind of components of different cost and reliability.

It is a very challenging job for the industries to optimize the reliability and cost of this kind of complex systems. Here we will reduce the cost factor under some constraints like reliability of the components and systems using CGWOA.

Reliability expression of the system shown in Figure 29.3 is

$$R_s = 1 - R_3((1-R_1)(1-R_4))^2 - (1-R_3)[1-R_2(1-(1-R_1)(1-R_4))]^2 \quad (29.1)$$

As we can see, this is an nonlinear optimization problem and our purpose is to find the minimum total cost of the system per the equation (29.2) provided reliability constraints (29.3):

$$\text{Min } C_s = \sum_{i=1}^{4} C_i = \sum \alpha_i \beta_i R_i^{\theta_i} \qquad (29.2)$$

$$\text{Subject to } X \le R_i \le 1, \; \theta_i = 0.6, \; i = 1,2,3,4 \qquad (29.3a)$$

$$Y \le R_s \le 1 \qquad (29.3b)$$

where C_s and R_s are the cost of the system and reliability of the system, respectively

α_i = number of redundancies related to unit i
β_i = multiple factor
θ_i = power exponent of R_i = 0.6 for all i

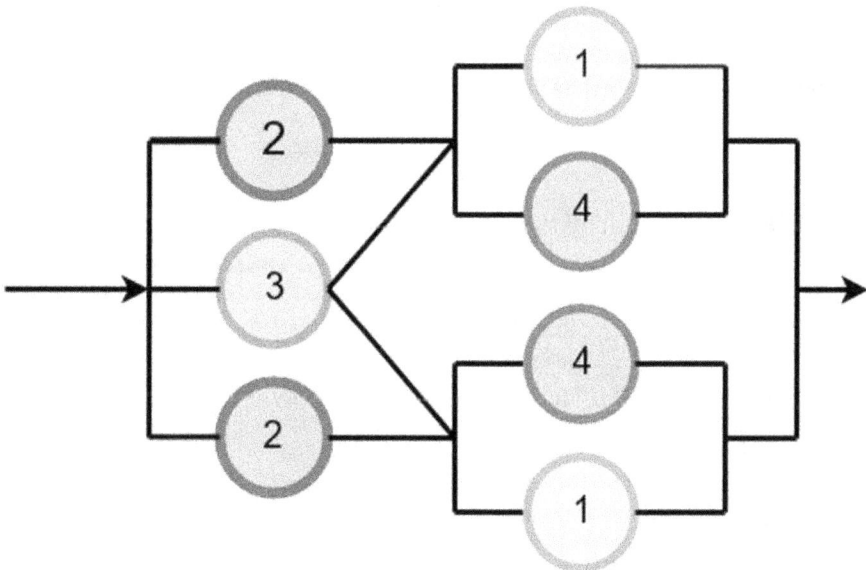

FIGURE 29.3   Complex industrial system

29.4.1 FITNESS FUNCTION

As in case of complex industrial systems, we can consider that a specific component performs better if the cost should be as minimum as possible and less violation of the constraints on the system and reliabilities of the components so the fitness function can be expressed as

$$\text{Fit}(R_i) = C(1 + P)$$

where a is a unity and P is a penalty which is used for measurement of degree of violation of the restrictions in 3a and 3b. It can be evaluated by utilizing the pseudo code-1 Wang et al. (2012).

29.5 RESULT AND CONCLUSION

In this chapter, we have find the optimized solution of reliability and cost of complex network for industrial system using nature-inspired GWOA. This is basically an NP-hard problem; that is why this problem is comparatively hard to solve than the other nonlinear programming problems. First of all, we have formulated the reliability optimization problem for complex industrial system and then solved it by using the CGWOA; here, we have used logistic map for CGWOA, and all results depict that CGWOA outperforms other algorithms GA and SA.

This section describes how GWO's performance was tested. we have implemented CGWOA on Python language under Windows 8.1 Enterprise 64-bit operating system, on a computer: Intel Core i-3. 2.10 GHz, 4 GB RAM, with the following parameters as mentioned in equations (29.2) and (29.3).

$$X = 0.5, Y = 0.9, \beta_1 = \beta_2 = 100, \beta_3 = 200, \beta_4 = 150, \alpha_1 = \alpha_2 = \alpha_4 = 2, \alpha_3 = 1$$

The results regarding CGWOA are depicted in the second column of Table 29.1. In column 3, GA-based Wang et al. (2012) results are placed for comparison. As

TABLE 29.1
Global Optimal Solution of the Problem

	CGWOA	Wang et al. (2012) GA-Based	Yeh et al. (2010)	Ravi et al. (1997) SA-Based
R_1	0.50007	0.50006	0.50001	0.50095
R_2	0.83892	0.83886	0.84062	0.83775
R_3	0.50001	0.50001	0.5000	0.50025
R_4	0.50001	0.50002	0.5000	0.50025
C_1	132.988	131.960	131.952	132.100
C_2	178.961	179.988	180.214	179.845
C_3	131.915	131.952	131.951	131.990
C_4	197.958	197.931	197.926	197.968
R_s	0.900003	0.90001	0.90050	0.900003
C_s	641.82371	641.8317	642.0400	641.90300

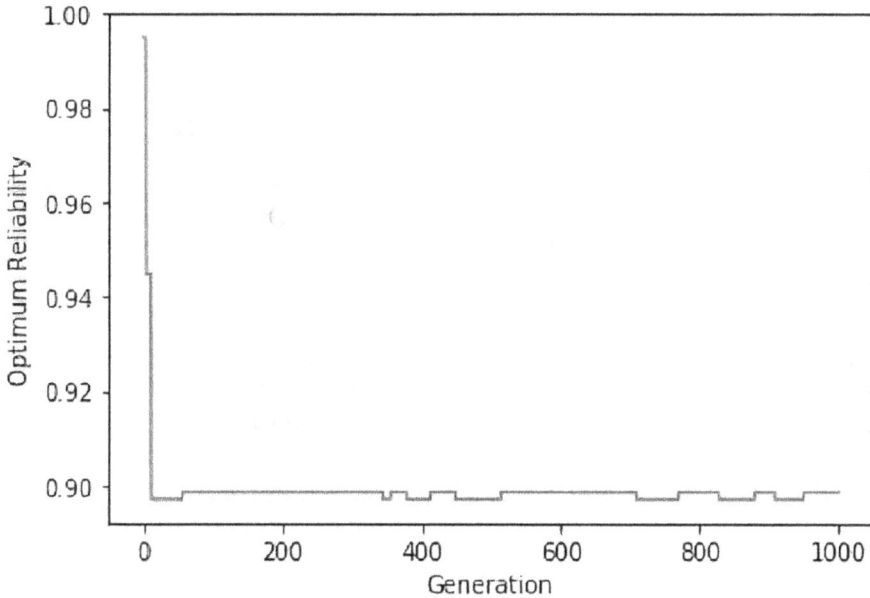

FIGURE 29.4   Performance of CGWOA.

we can see that our result is improved as compared to the other listed in columns 3 and 4.

For six out of the 10 random runs, the optimal cost found is 641. 63057, and the other results are also better than the cost listed in the column 3 and 4 of Table 29.1. Under constraints, solutions provided by the CGWOA proves its efficiency over the others for the optimization problem of complex industrial system.

In the Figure 29.4 we can see that reliability of the system is converging towards its optimal value after some iterations.

REFERENCES

Al-Tashi, Q., Kadir, S. J. A., Rais, H. M., Mirjalili, S., & Alhussian, H. (2019) Binary optimization using hybrid grey wolf optimization for feature selection. *IEEE Access* 7: 39496–39508.

Bhatti, J., Chitkara, A. K., & Kakkar, M. K. (2016) Stochastic analysis of dis-similar standby system with discrete failure, inspection and replacement policy. *Demonstratio Mathematica* 49(2): 224–235.

Bhatti, J., & Kakkar, M. K. (2021) Reliability analysis of cold standby parallel system possessing failure and repair rate under geometric distribution. *Recent Advances in Computer Science and Communications* 14(3): 968–974.

Bhatti, J., Kakkar, M. K., Bhardwaj, N., & Kaur, M. (2021a) Stochastic analysis to an power supply system through reliability modelling. *Transactions Issue Mathematics, Azerbaijan National Academy of Sciences* 41(1): 33–41.

Bhatti, J., Kakkar, M. K., Bhardwaj, N., Kaur, M., & Deepika, G. (2020) Reliability analysis to industrial active standby redundant system. *Malaysian Journal of Science* 39(3): 74–84.

Bhatti, J., Kakkar, M. K., Kaur, M., Goyal, D., & Khanna, P. (2021b) Stochastic analysis to mechanical system to its reliability with varying repairing services. *Chebyshevskii Sbornik* 22(1): 92–104.

Demidova, L. A., & Gorchakov, A. V. (2020) A study of chaotic maps producing symmetric distributions in the fish school search optimization algorithm with exponential step decay. *Symmetry* 12(5): 784–802.

Gao, Z. M., & Zhao, J. (2019) An improved grey wolf optimization algorithm with variable weights. *Computational Intelligence and Neuroscience* 2019: 1–13.

Kakkar, M., Bhatti, J., Malhotra, R., Kaur, M., & Goyal, D. (2019) Availability analysis of an industrial system under the provision of replacement of a unit using genetic algorithm. *International Journal of Innovative Technology and Exploring Engineering (IJITEE)* 9: 1236–1241.

Kakkar, M., Chitkara, A., & Bhatti, J. (2015) Reliability analysis of two unit parallel repairable industrial system. *Decision Science Letters* 4(4): 525–536.

Kakkar, M., Chitkara, A., & Bhatti, J. (2016) Reliability analysis of two dissimilar parallel unit repairable system with failure during preventive maintenance. *Management Science Letters* 6(4): 285–296.

Konak, A., Coit, D. W., & Smith, A. E. (2006) Multi-objective optimization using genetic algorithms: A tutorial *Reliability Engineering & System Safety* 91(9): 992–1007.

Kumar, M., Samui, P., Kumar, D., & Zhang, W. (2021) Reliability analysis of settlement of pile group. *Innovative Infrastructure Solutions* 6(1): 1–17.

Lu, H., Wang, X., Fei, Z., & Qiu, M. (2014) The effects of using chaotic map on improving the performance of multiobjective evolutionary algorithms. *Mathematical Problems in Engineering* 2014: 1–17.

May, R. M. (2004) Simple mathematical models with very complicated dynamics. *The Theory of Chaotic Attractors*: 85–93.

Mirjalili, S., Mirjalili, S. M., & Lewis, A. (2014) Grey wolf optimizer. *Advances in Engineering Software* 69: 46–61.

Painton, L., & Campbell, J. (1995) Genetic algorithms in optimization of system reliability. *IEEE Transactions on Reliability* 44(2): 172–178.

Ravi, V., Murty, B. S. N., & Reddy, J. (1997) Nonequilibrium simulated-annealing algorithm applied to reliability optimization of complex systems. *IEEE Transactions on Reliability* 46(2): 233–239.

Sheta, A. F., & Abdel-Raouf, A. (2016) Estimating the parameters of software reliability growth models using the Grey Wolf optimization algorithm. *International Journal of Advanced Computer Science and Applications* 7(4): 1–7.

Suman, B. (2003) Simulated annealing-based multiobjective algorithms and their application for system reliability. *Engineering Optimization* 35(4): 391–416.

Wang, J. C., Qiu, H., Chen, J. M., & Ji, G. D. (2012) Complex mechatronic system reliability optimization using hybrid genetic algorithm. *Applied Mechanics and Materials, Trans Tech Publications Ltd.* 138: 1296–1301.

Yeh, W. C., Lin, Y. C., Chung, Y. Y., & Chih, M. (2010) A particle swarm optimization approach based on Monte Carlo simulation for solving the complex network reliability problem. *IEEE Transactions on Reliability* 59(1): 212–221.

30 CT-Scan-Based Identification and Screening of Contagiously Spreading Disease Using 64-Slice CT Scanner

Jaspreet Kaur, Manjot Kaur,
Manpreet Singh Manna

30.1 INTRODUCTION

Since December 2019, the fast spread of coronavirus disease (COVID-19) has generated alarm around the world. The rapid spread of the COVID-19 pandemic, which has resulted in thousands of deaths and huge volume of infected population in different places every day, is posing major encounter in terms of controlling the virus [1, 2]. Globally, there are more than 560,280,550 COVID-19 cases and 6,366,208 deaths as of July 2022[3]. Since no effective means of vaccine or immunizing agent has been developed to date, the challenge is to find out how to better combat the coronavirus and prevent it from spreading. Many countries have imposed lockdowns in region of severe attack to control the spread of viruses and minimize casualties [4]. COVID-19 is highly frequent in less resistance people, aged people, and unhealthy people, especially those suffering from lungs problem. Symptom of COVID-19 are cold, Cough, breathing problems and high fever. COVID-19 preventive strategies include hand wash regularly, avoid touching the nose, mouth, eyes, and face, and maintain significant social distance from other persons [5].

Since there is no cure for infectious COVID-19, the infection spread at high pace. The diagnostic techniques used to determine COVID-19 infection demand that results to be examined, as the symptoms for positive COVID-19 cases increases continuously from the discovery of the virus on the earth. As the pandemic began, the rise of cases became exponential and the conventional diagnostic techniques became critical while screening the infected population.

The initially approved standard diagnostic tool is real-time polymerase chain reaction (RT-PCR), which detects viral nucleic acid [6]. Many hyper-endemic regions or

DOI: 10.1201/9781003367161-30

nations, on the other hand, are unable to provide appropriate RT-PCR monitoring for tens of thousands of suspected patients. Further, RT-PCR tests take longer time in detection of infection [6]. To compensate for the scarcity of reagents, researchers have attempted to diagnose COVID-19 using tomographic techniques. This initial determination using X-rays and CT scans controlled the disease from spreading rapidly to others. The X-rays are an appropriate and extensively accessible imaging tool, which has a significant position in scientific and epidemiological research [7,8]. Furthermore, real-time X-ray imaging would significantly speed up disease screening. The CT manual examination is a categorizing factor; if the image is standard, the patient will return home and await the laboratory test results. As a result, simple and well-designed AI models for disease detection are also helpful in overcoming this drawback.

Coronavirus detection with required sensitivity with chest X-ray is hard to obtain, which is not only because of the soft tissues covering the ribs with short differentiation but also because of the scarcity of data annotation [9]. This is particularly true in deep learning approaches, which are notorious for being data-hungry [9].

Important research has recently gained quantity and consistency in the application of machine learning for automatic disease detection. Deep learning methods are commonly used in medical problems, such as detecting cancer, classifying cancer, and detecting respiratory conditions from CT images. Convolutional neural networks (CNN) are widely used in deep learning to automatically learn features and then use the information for classification [10,16].

In a study of 51 patients [11], chest CT had a high sensitivity (up to 98%) for COVID-19 differentiate. Ophir et al. [12] used deep learning technology to diagnose COVID-19 used in CT images to speed up the screening process. Shi et al. [13] gathered an enormous scale infected dataset of CT scans and prepared a COVID-19 screening system based on machine learning.

In light of these benefits, we have made an effort to build a model based on deep learning for the detection of COVID-19 using CT images of the chest with adequate sensitivity, allowing for fast and accurate segregation. Suspected patients of COVID-19 have been identified for further clinical viral nucleic acid determination.

To address this issue, we used the public chest CT scan dataset [14] to collect images of non-infected cases. The model proposed in this chapter based on CT scan screening of COVID-19 patients encouraged by [15], to encourage the disproportion binary categorization assignment throughout the invariance recognition task, in order to confirm the data imbalance issues that exists between infected and non-infected cases. This chapter presents a deep learning modelling focused on transfer learning for rapid and programmed recognition of COVID-19. The model is initially trained with a number of CT images. The size of the training sample determines the accuracy of the results obtained. Since the infectious samples are small in size, the transfer learning becomes optimal for training the model.

30.2 CONVOLUTIONAL NEURAL NETWORK

Convolution is a linear operation that is different from mathematical operation. Neural networks having minimum one convolutional layer are convolutional neural networks. The convolution method in deep learning [16,26], on the other hand, is not the same as the convolution techniques in ordinary or simple engineering mathematics. The layers in CNN include convolution, ReLU, pooling, normalization, completely linked, and softmax [16]. The categorization process in convolutional neural networks takes place in completely linked layers and softmax layers.

The input function (*I*) and kernel function (K) in the machine learning applications are multidimensional arrays of a number of parameters, for example, for input of two-dimensional images, kernel becomes a two-dimensional matrix, and output matrix is given by the equation (30.1) [16]:

$$S_{(i,j)} = (I * K)_{(i,j)} = \sum_m \sum_n I_{(i-m,j-n)} K_{(m,n)} \qquad (30.1)$$

This equation implies that the kernel is moved depending on the data. This increases convolution invariance [17].

A matrix product is obtained to describe discrete convolution. Particular convolution process is shown in the Figure 30.1. To effectively deal with huge inputs, traditional convolution neural networks benefit from additional expertise.

Three essential considerations—parameter sharing, infrequent interactions, and covariant representations—are provided by the convolution process to improve the machine learning system. The beauty of the CNN is to perform effectively with inputs of different sizes. The connection of every output to every input is confirmed due to the use of matrix parameters, which provides the connection of each input unit and each output unit, as shown in Figure 30.2(a) and 30.2(b). Moreover, CNN have sparse links for this reason kennel made smaller than entrance. Furthermore, after each convolution, the number of pixels decreases.

The input image for image processing may have millions of pixels; however, tiny and significant characteristics like kernel edges with only tens or hundreds of pixels can be detected, resulting into saving of fewer parameters, which will reduce the

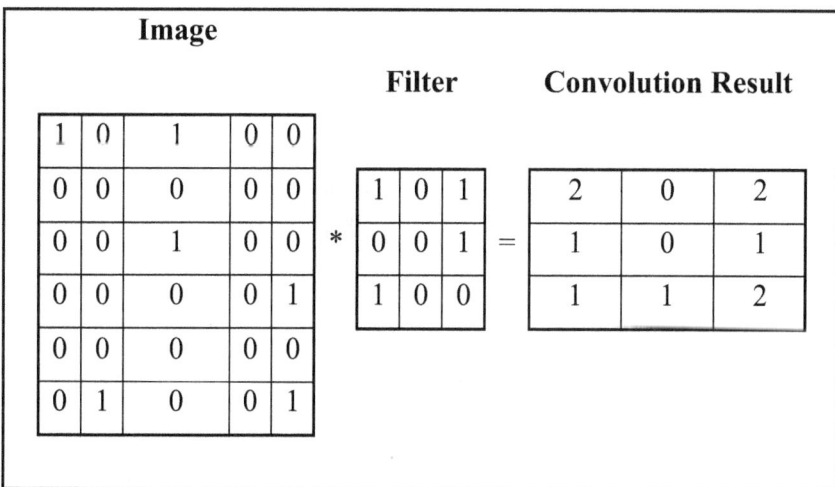

FIGURE 30.1 Convolution process.

2*2 Pooling Size
2 Stride

1	0	1	1
0	2	4	1
2	1	1	3
0	1	0	0

2	4
2	3

2*2 Pooling Size
2 Stride

1	0	1	3
1	2	4	0
2	1	3	3
3	2	5	1

1	2
2	3

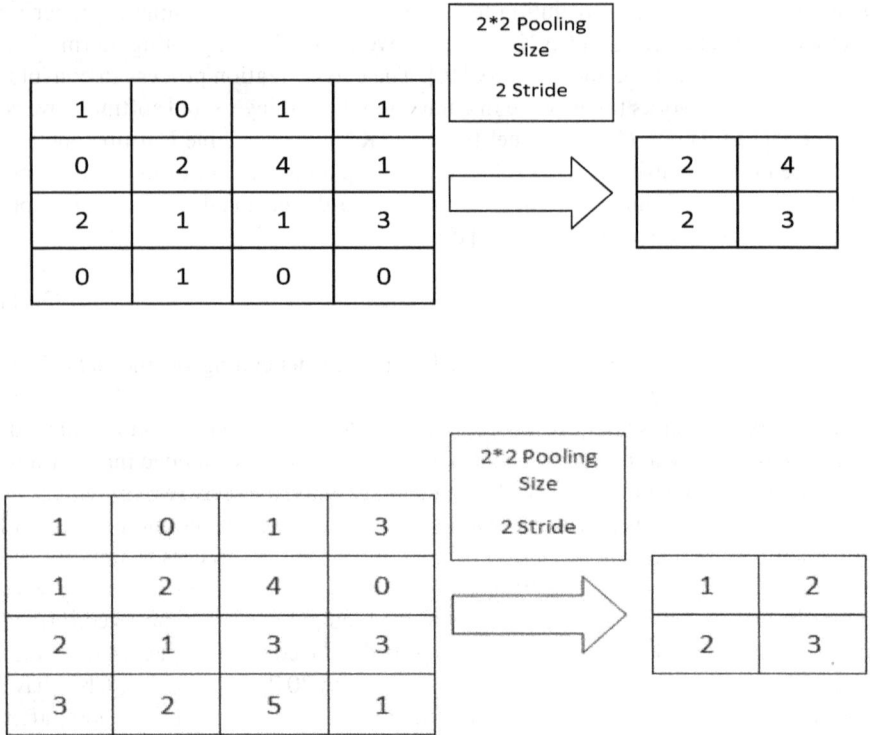

FIGURE 30.2 (a) Maximum pooling. (b) Mean pooling.

CNN model's memory requirements while also increasing its performance. It also ensures that measuring performance takes less time and effort. These increases in productivity are usually important.

The use of the same parameter for different functions in a model is referred to as parameter sharing. The performance of a layer in traditional neural network is measured by weighted matrix for each feature. This is multiplied by an entry factor and will not be looked at again. In CNN for every inserted position, all members of the core are used. Since convolution method uses parameter sharing which leads for learning a different set of constraints for one issue, only one set is learned.

Each layer uses differentiable function in ConvNet, which translates one volume activation to another. ConvNet architectures are made up of three distinct layers: convolutional layer, pool layer, and completely connected layer. A full ConvNet architecture is made up of these layers.

In artificial neural networks, the size of the data is important. When the amount of data increases, so does the amount of memory it takes up, decreasing the artificial neural network's performance and slowing down its processing speed. The entire dataset compressed to 0–1 value for simplified. The method is derived using the sum of all the datasets, which results in ranging data in the form 0–1. Standardization allows features to be rescaled for a standard normal distribution.

The feature's standard deviation is based between 1 and 0. It is also significant for the purpose of training of a number of machine learning algorithms [16]. Assembling feature switches the network's output at an explicit position with a summary of adjoining output's data. Max-pooling, for illustrations, yields the biggest in quadrilateral space as an output. Similarly, average and minimum pooling functions are two more general pooling functions.

When the parameters in next layer is achieved with the volume of feature map or the input image, any decrease in input volume increases statistical efficiency and shrinks the parameter storage requirements.

Rectified linear unit (ReLU) used as activation function can be simply applied by setting the activation matrix to 0. However, during the training process, ReLU units may become responsive and can destroy units irreversibly during training, particularly when learning rate is very high; 40% of the model will perish. With the necessary modification of the learning rate, this is a less common phenomenon. As compared to sigmoid/tanh functions, stochastic gradient descent greatly accelerates convergence.

Dropout of nodes below a certain threshold value improved performance in completely connected layers. In most cases, the dropout value is 0.5. It differs depending upon the problem and the data collection. For the dropout, the random exclusion approach may also be used. When used as the threshold value, the dropout figure is well defined as a value ranging in 0–1. Various dilution values can be used instead of using the similar dropout value in all layers.

Logistic regression multi-class termed as softmax function, which is a classifier of the classifier. The sum of normalized distribution is equal to 1. As a result, the possibility of the group to which it belongs is defined. In the given data input test the activation function gives the possibility of each value. The tag of the class will have one of the dissimilar possible values is required [26]. The result of k-dimensional vector produced by the activation function gives extrapolative possibilities.

Softmax loss function calculates the amount of error of softmax function. The function f (xi; W) = Wxi match leftovers unchanged in the softmax classifier. However, interpret the gain as standardized log probabilities for every group and apply the given cross entropy loss [16] shown by equation (30.2). For the given test, input x and activation function $j(i \ldots \ldots k)$ and forecast likelihood of p(y=j|x) [16,26].

$$L_i = -f_{yi} + \log(\sum_j e^{f_j}) \tag{30.2}$$

30.3   GOOGLENET

GoogLeNet model is one of the CNN architectures used for classification of image. InceptionV1 is another name for it. Google's study group suggested GoogLeNet for the first time in 2014 [16,18]. They showed how deeper convolution networks can be useful. It had a 6.67% top-five error rate. Compared to AlexNet (2012), ZF-Net (2013), and VGG (2014), it has a lower error rate. The results obtained are very close to expert's performance. That is why we chose GoogLeNet for fast COVID-19 screening. This model has designed with a deep CNN having 22 layers and about a dozen times fewer parameters [19].

30.4   DESIGN PRINCIPLE OF THE MODEL

The design principle of the GoogLeNet network varies from that of the ZF-Net and AlexNet networks. It employs alternative approaches, such as 1D convolution with average pooling layers, resulting in a more complex model. The following are some of the layers that were used in the design [19]:

30.4.1   1 × 1 CONVOLUTION

In the design of the first sheet, 1 × 1 convolutions are used. The number of parameters in each layer, such as weights and biases, are reduced. The network's depth can be expanded by lowering parameters. It replaces a small perceptron layer with convolutions with a bigger perceptron layer. Comparison of with and without convolution is shown in Figures 30.3(a) and 30.3(b). Performing convolution with 48 filters has been used.

30.4.1.1   Without Using 1 × 1 Convolution as Intermediate

$14 \times 14 \times 480$ $14 \times 14 \times 48$

Total number of operations without 1 × 1 convolution:

$(14 \times 14 \times 48) \times (5 \times 5 \times 480) = 112.9$ M

30.4.1.2   1 × 1 Convolution

$14 \times 14 \times 48$ $14 \times 14 \times 16$ $14 \times 14 \times 48$

Operations with 1 × 1 convolution:

$(14 \times 14 \times 16) \times (1 \times 1 \times 480) + (14 \times 14 \times 48) \times (5 \times 5 \times 16) = 5.3$ M

So the number of operations with 1 × 1 is much smaller than 112.9 M.

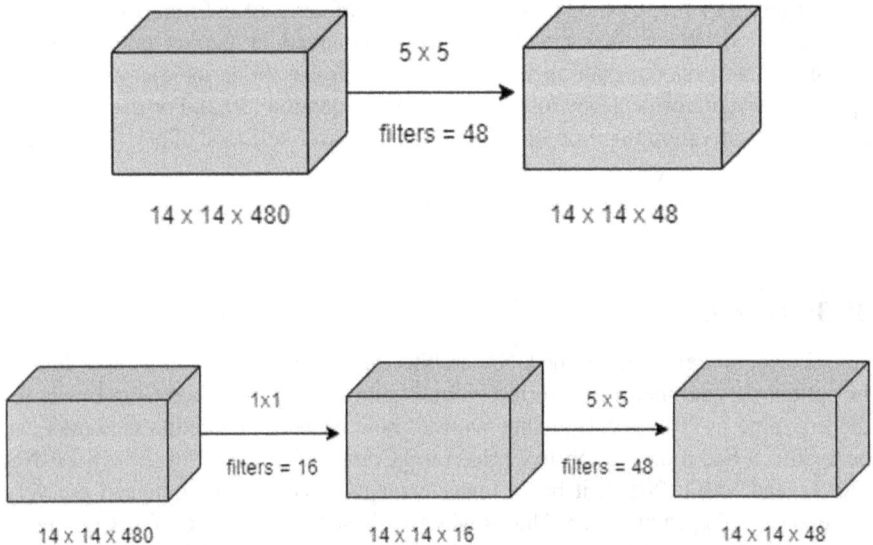

FIGURE 30.3   (a) Result without 1 × 1 convolution. (b) Result with 1 × 1 convolution.

30.4.2 INCEPTION MODEL

The large area covered with fine resolution is achieved by inception layers. These modules are proposed to boost the evaluation efficiency of the model by reducing dimensionality using stacked 1 × 1 convolutions, as mentioned earlier. Figure 30.4 (a) [19] various three-filter input of size 1 × 1, 3 × 3, and 5 × 5 along with max pooling layer in naive inception module for convolution and concatenates the outcomes. The overfitting and computational cost problem have been fixed using these modules.

Figure 30.4(b) [20] presents the updated architecture of the inception layer which reduces the computational cost in the deeper networks. Before the 3 × 3 and 5 × 5 convolution layers, a 1 × 1 convolution layer has been inserted, and the max pool layer, to reduce the input channels. Though it can seem counter intuitive to introduce an additional 1 × 1 convolution operation, it helps to minimize the data and computational cost. To generate output, the results of parallel operations 1 × 1, 3 × 3, and 5 × 5 convolution, and 3 × 3 max pooling is concatenated. Convolution filters of various sizes assist in the extraction of object features at different scales, resulting in improved performance [20].

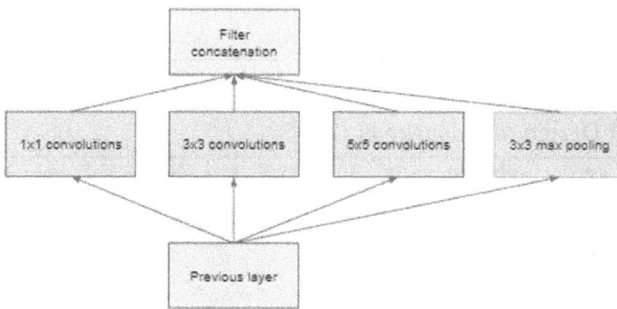

(a) Inception module, naïve version

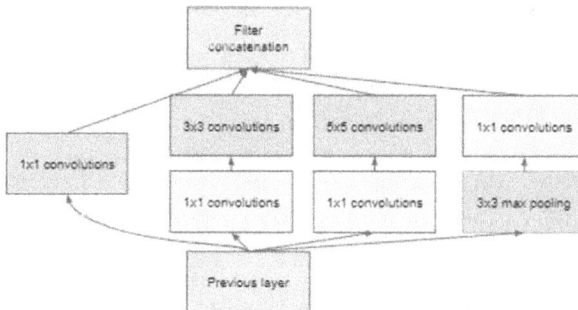

(b) Inception module with dimension reductions

FIGURE 30.4   (a) Naive version, inception module. (b) Dimension reductions inception module.

30.5 GOOGLENET MODEL ARCHITECTURE

The model was designed with 22 layers deep and 27 pooling layers. A total of 9 inception modules were stacked linearly. Global average polling layer is connected to the ends of inception modules. The average pooling layers of 5 × 5 filter and stride 3 was used. Convolutions in the model uses ReLU activation function. The auxiliary classifiers are having fully connected layer with 1,025 outputs and softmax classifier has 1,000 classes. The model can accept an image with 224 × 224 size with RGB channels. The detailed model is shown in Figure 30.5. Layer-wise detailed GoogLeNet used shown in the Table 30.1[21,22].

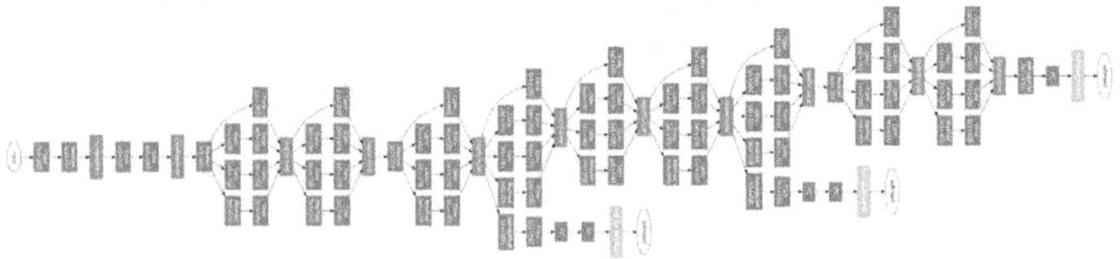

FIGURE 30.5 Architecture details of GoogLeNet.

TABLE 30.1
Architectural Design of GoogLeNet [23]

Type	patch size/stride	output size	depth	#1x1	#3x3 reduce	#3x3	#5x5 reduce	#5x5	pool proj	params	ops
convolution	7x7/2	112x112x64	1							2.7K	34M
max pool	3x3/2	56x56x64	0								
convolution	3x3/1	56x56x192	2		64	192				112K	360M
max pool	3x3/2	28x28x192	0								
inception (3a)		28x28x256	2	64	96	128	16	32	32	159K	128M
inception (3b)		28x28x480	2	128	128	192	32	96	64	380K	304M
max pool	3x3/2	14x14x480	0								
inception (4a)		14x14x512	2	192	96	208	16	48	64	364K	73M
inception (4b)		14x14x512	2	160	112	224	24	64	64	437K	88M
inception (4c)		14x14x512	2	128	128	256	24	64	64	463K	100M
inception (4d)		14x14x528	2	112	144	288	32	64	64	580K	119M
inception (4e)		14x14x832	2	256	160	320	32	128	128	840K	170M
max pool	3x3/2	7x7x832	0								
inception (Sa)		7x7x832	2	256	160	320	32	128	128	1072K	54M
inception (5b)		7x7x1024	2	384	192	384	48	128	128	1388K	71M
avg pool	7x7/1	1x1x1024	0								
dropout (40%)		1x1x1024	0								
linear		1x1x1000	1							1000K	1M
softmax		1x1x1000	0								

The GoogLeNet is used to predict COVID-19, whether a person has been infected with the coronavirus or not. It begins with a chest CT images and passed through number of convolution layers with different filters as well as with pooling layers, and completely linked layers (FC). For performance evaluation, the softmax function is used in the output layer as binary classification and binary cross entropy.

30.6 MATERIALS AND DATASETS

Fifty-three chest CT images are in the dataset used for the work in this study, which acquired 70 subjects with COVID-19 [23]. The COVID-19 cases are available at the GitHub repository, and then COVID-19 data are available at the Chest14 dataset [24].

Two datasets were prepared as Dataset-1 and Dataset-2. Each dataset includes 375 patches CT images of COVID-19-infected patients from infected portions and 375 patches of CT images of same regions of non-COVID-19 persons, as shown in Figures 30.6(a) and 30.6(b), respectively. The patch dimensions in dataset −1 is taken 16×16 and in dataset-2 the size is taken 32×32.

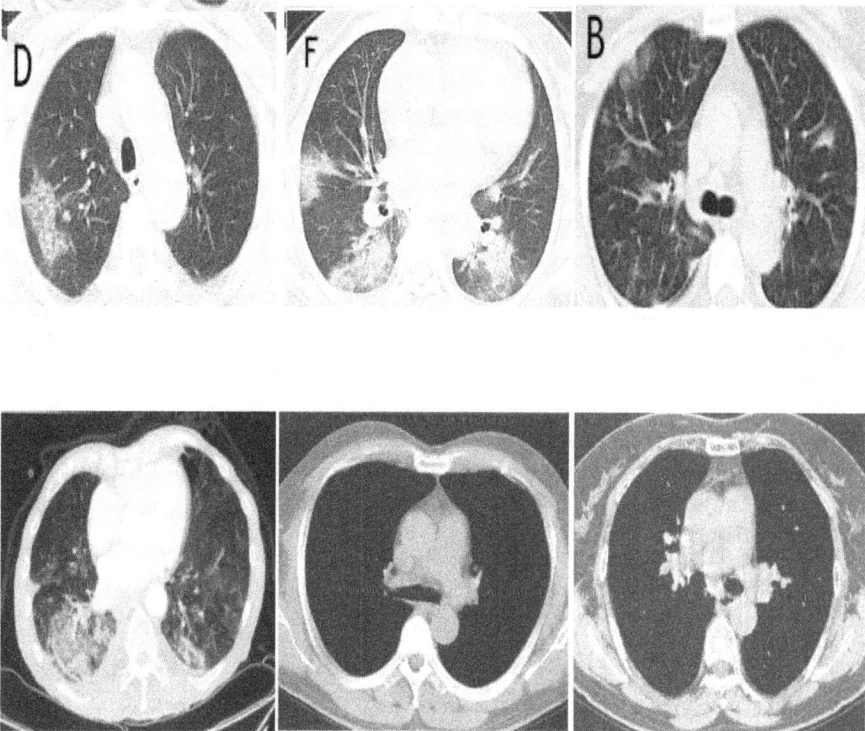

FIGURE 30.6 (a) COVID-19 chest CT scan images. (b) Non-COVID-19 chest CT scan images.

30.7 PERFORMANCE METRICS

The experimental study was carried out with texture of two datasets of different sizes. Initially the features have been obtained through fusion ranking of CNN process and categorized. The dataset is classified into four classes. The confusion matrix presented TRUE POSITIVE, TRUE NEGATIVE, FALSE POSITIVE, and FALSE NEGATIVE for both pre-training performance and test performance. From the output parameter matrix, six performance metrics have been evaluated for the proposed CNN and compared with VGG16 [25].

1. Sensitivity (S) $= \dfrac{TP}{TP + FN}$

2. Specificity (SF) $= \dfrac{TN}{TN + FP}$

3. Accuracy (ACC) $= \dfrac{TP + TN}{TP + TN + FN + FP}$

4. Precision (P) $= \dfrac{TP}{TP + FP}$

5. F-Score $= \dfrac{2XTP}{2XTP + FP + FN}$

6. Mathews Correlation Coefficient (MCC) $= \dfrac{TPXTN - FPXFN}{\sqrt{(TP + FP))(TP + FN)(TN + FP)(TN + FN)}}$

30.8 EXPERIMENTAL RESULTS

Dataset-1 comprised 1,500 pieces of CT patches of size 16 × 16. Data is equally distributed between the classes. The images used for training and testing in the proposed CNN model is in the ration of 3:1. The results obtained are presented in the Table 30.2, and the confusion matrix for dataset-1 is shown in Figure 30.7.

Similarly in the dataset-2, there are taken 1,500 patches of size 32 × 32, 750 of COVID-19-infected patients and 750 of non-COVID-19 persons. With distribution of data and same data ratio of training and testing, the results are obtained presented in the Table 30.3 and the confusion matrix results are shown in Figure 30.8.

TABLE 30.2
Classification Evaluation Results: Dataset-1

Method	TN	TP	FN	FP	Assessment Metrics (%)					
					Sensitivity	Specificity	Accuracy	Precision	F1-Score	MCC
VGG-16	327	358	17	48	95.5	87.2	91.3	88.1	91.6	82.95
GoogLeNet	371	358	17	4	95.5	98.9	97.2	98.9	97.1	94.5

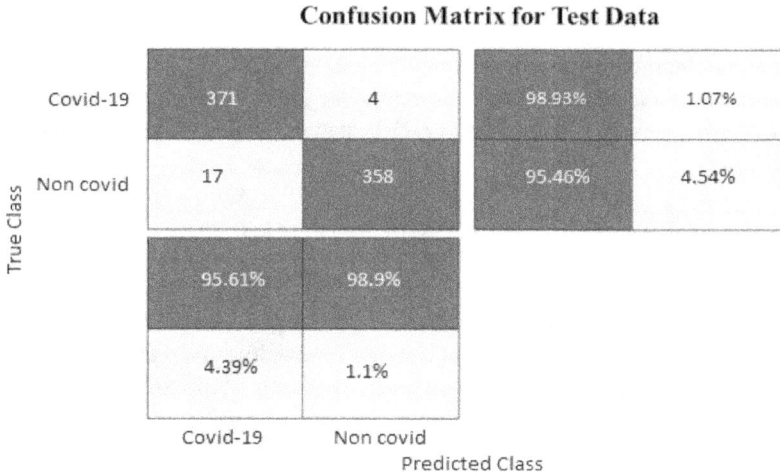

FIGURE 30.7   Dataset-1 Confusion Matrix.

TABLE 30.3
Classification Assessment Results: Dataset-2

Method	TN	TP	FN	FP	Assessment Metrics (%)					
					Sensitivity	Specificity	Accuracy	Precision	F1-Score	MCC
VGG-16	355	372	3	20	99.2	94.67	96.93	94.9	97	93.96
GoogLeNet	370	372	3	5	99.2	98.7	98.9	98.6	98.9	97.87

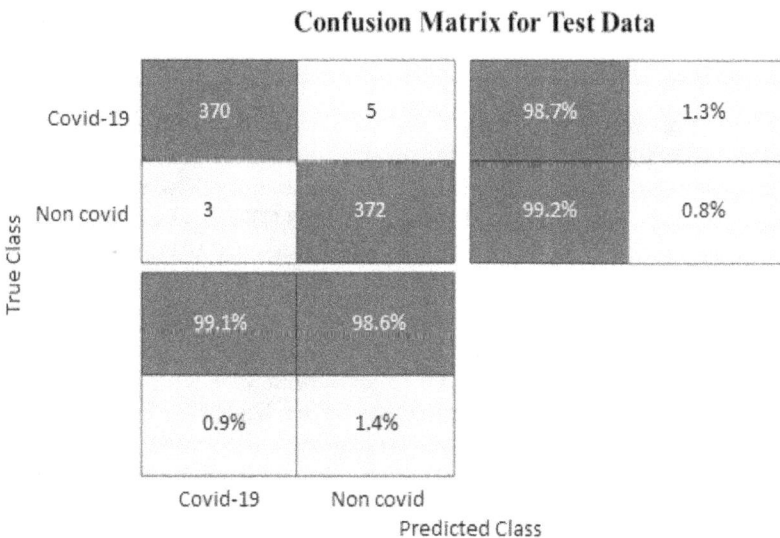

FIGURE 30.8   Dataset-2 Confusion Matrix.

30.9 PERFORMANCE ANALYSIS

The results obtained from proposed CNN model GoogLeNet for both datasets are compared with the existing VGG16 technique. Each dataset comprised 50% of COVID-19 patches and 50% non-COVID-19 patches with dimensions 16 × 16 and 32 × 32, respectively.

The accuracy level obtained for dataset-1 is 97.2%, whereas it was enhanced to 98.9% in dataset-2. On the other hand, the accuracy presented by VGG16 is lower side with 91.3% and 96.93%, respectively. The proposed model presented significant improvement in specificity and precision. Dataset-1, with smaller patch size specificity obtained with VGG16, is 87.2% and that with GoogLeNet is 98.9%. The results of the 16 × 16 patch size had shown slightly better specificity and precision as compared to the 32 × 32 size due to intensity homogeneity reasons. Similarly, the precision in classification of data acquired with proposed model is 98.9% and is 10.8% more than the existing VGG16. However, the patch size has not shown noticeable effect on the precision of measurement. The fraction of true class predicted by the proposed CNN was found to be improved from 95.5% to 99.20% with increase in dimensions of the input data and the proportion of improvement is recorded same in both the models VGG16 and GoogLeNet. The metrics had played a vital role to prevent the spread of the disease as with higher sensitivity, the predicted class is more closely correlated to the true class in the data. F1 score, the performance metrics that accounted harmonic mean of precision and recall together, displayed weighty enhancement in the classification. It is recorded 97.1% in dataset-1 and 98.9% in dataset-2. The Matthews correlation coefficient (MCC) that accounted all the four values of true and false positive and negative classes provided the better insights and arguably more elegant solution in terms of correlation between true class and predicted class and thereby resulting into better prediction. The MCC evaluated for both the models indicated enhancement of correlation between the true class and predicted class from 94.50% to 97.87% with regards to the size of input CT patches.

30.10 CONCLUSIONS

This chapter presented the application of pre-trained CNN for screening of COVID-19 and non-COVID-19 classes from CT image using the GoogLeNet model designed with 27 pooling layers and 22 layers deep. Inception modules were stacked linearly. The performance metrics from the two datasets of 53 CT images was developed. The out of 1,500 patches, 50% of COVID-19 and 50% non-COVID-19 with patch size 16 × 16 and 32 × 32 have used for diagnosis and classification. All the six performance metrics presented substantial enhancement with large patch size. The results reflected very clear picture of classification from F1 score and MCC. In comparison to VGG16, MCC obtained with GoogLeNet with V1 is very high, particularly in the dataset-2, limiting approached unity, which indicated the healthy correlation between true and predicted classes. Up gradation in the sensitivity played a vital role in the screening of two classes and highly fruitful in preventing the fast spread of the disease. Finally, it can be concluded the implementation of deep learning in the diagnosis sped up the isolation of COVID-19-infected patients from the healthy population. Work can be further explored with more CT image features and using higher inception modules.

REFERENCES

1. Yi, P. H., Kim, T. K., & Lin, C. T. (2020). Generalizability of deep learning tuberculosis classifier to COVID-19 chest radiographs. *Journal of Thoracic Imaging*, *35*, W102–W104.
2. Yang, W., Sirajuddin, A., Zhang, X., Liu, G., Teng, Z., Zhao, S., & Lu, M. (2020). The role of imaging in 2019 novel coronavirus pneumonia (COVID-19). *EurRadiol*, *15*, 1–9.
3. www.who.int/emergencies/diseases/novel-coronavirus-2019/situation-reports.
4. Hilmizen, N., Bustamam, A., & Sarwinda, D. (2020, December). The multimodal deep learning for diagnosing COVID-19 pneumonia from chest CT-scan and X-ray images. In *2020 3rd International Seminar on Research of Information Technology and Intelligent Systems (ISRITI)* (pp. 26–31). IEEE Xplore.
5. Lin, Q., Zhao, S., Gao, D., Lou, Y., Yang, S., Musa, S. S., . . . & He, D. (2020). A conceptual model for the coronavirus disease 2019 (COVID-19) outbreak in Wuhan, China with individual reaction and governmental action. *International Journal of Infectious Diseases*, *93*, 211–216.
6. Tabik, S., Gómez-Ríos, A., Martín-Rodríguez, J. L., Sevillano-García, I., Rey-Area, M., Charte, D., . . . & Herrera, F. (2020). COVIDGR dataset and COVID-SDNet methodology for predicting COVID-19 based on chest X-ray images. *IEEE Journal of Biomedical and Health Informatics*, *24*(12), 3595–3605.
7. J. P. Cohen et al. (2020). COVID-19 image data collection. *arXiv 2003.11597*.
8. Fang, Y., Zhang, H., Xie, J., Lin, M., Ying, L., Pang, P., & Ji, W. (2020). Sensitivity of chest CT for COVID-19: Comparison to RT-PCR. *Radiology*, *296*(2), E115–E117.
9. Wang, S., Kang, B., Ma, J., Zeng, X., Xiao, M., Guo, J., . . . & Xu, B. (2020). A deep learning algorithm using CT images to screen for Corona Virus Disease (COVID-19). *medRxiv*, *14*, 5. Preprint at www.medrxiv.org/content/10.1101/2020.02.
10. Xu, X., Jiang, X., Ma, C., Du, P., Li, X., Lv, S., . . . & Li, L. (2020). A deep learning system to screen novel coronavirus disease 2019 pneumonia. *Engineering*, *6*(10), 1122–1129.
11. Fang, Y., Zhang, H., Xie, J., Lin, M., Ying, L., Pang, P., & Ji, W. (2020). Sensitivity of chest CT for COVID-19: Comparison to RT-PCR. *Radiology*, *296*(2), E115–E117.
12. Gozes, O., Frid-Adar, M., Greenspan, H., Browning, P. D., Zhang, H., Ji, W., . . . & Siegel, E. (2020). Rapid ai development cycle for the coronavirus (COVID-19) pandemic: Initial results for automated detection & patient monitoring using deep learning ct image analysis. *arXiv preprintarXiv:2003.05037*.
13. Shi, F., Xia, L., Shan, F., Song, B., Wu, D., Wei, Y., ... & Shen, D. (2021). Large-scale screening to distinguish between COVID-19 and community-acquired pneumonia using infection size-aware classification. *Physics in medicine & Biology*, *66*(6), 065031.
14. Kumar, R., Wang, W., Kumar, J., Yang, T., Khan, A., Ali, W., & Ali, I. (2021). An integration of blockchain and AI for secure data sharing and detection of CT images for the hospitals. *Computerized Medical Imaging and Graphics*, *87*, 101812.
15. P. K. Sethy, S.K. Behera, Detection of coronavirus disease (COVID-19) based on deep features, 2020, Preprints 2020, 2020030300.
16. Özkaya, U., Öztürk, Ş., & Barstugan, M. (2020). Coronavirus (COVID-19) classification using deep features fusion and ranking technique. In *Big Data Analytics and Artificial Intelligence Against COVID-19: Innovation Vision and Approach* (pp. 281–295). Springer, Cham.

17. Matassoni, M., Gretter, R., Falavigna, D., & Giuliani, D. (2018, April). Non-native children speech recognition through transfer learning. In *2018 IEEE International Conference on Acoustics, Speech and Signal Processing (ICASSP)* (pp. 6229–6233). Calgary, Alberta, Canada: IEEE. DOI: 10.1109/ICASSP34228.2018

18. Szegedy, C., Liu, W., Jia, Y., Sermanet, P., Reed, S., Anguelov, D., . . . & Rabinovich, A. (2015). Going deeper with convolutions. In *Proceedings of the IEEE Conference on Computer Vision and Pattern Recognition* (pp. 1–9). IEEE.

19. Haritha, D., Swaroop, N., & Mounika, M. (2020, October). Prediction of COVID-19 cases using CNN with X-rays. In *2020 5th International Conference on Computing, Communication and Security (ICCCS)* (pp. 1–6). IEEE.

20. Jin, X., Chi, J., Peng, S., Tian, Y., Ye, C., & Li, X. (2016, October). Deep image aesthetics classification using inception modules and fine-tuning connected layer. In *2016 8th International Conference on Wireless Communications & Signal Processing (WCSP)* (pp. 1–6). IEEE.

21. Aljuhani, A. (1989). Going deeper with convolutions, https://pdfs.semanticscholar.org/e15c/f50aa89fee8535703b9f9512fca5bfc43327.pdf.

22. Ayachi, R., Afif, M., Said, Y., & Atri, M. (2018, December). Strided convolution instead of max pooling for memory efficiency of convolutional neural networks. In *International Conference on the Sciences of Electronics, Technologies of Information and Telecommunications* (pp. 234–243). Springer, Cham.

23. Kurama, V. (2020). A review of popular deep learning architectures: Resnet, inceptionv3, and squeezenet. *Consult,* August 30.

24. Barstugan, M., Ozkaya, U., & Ozturk, S. (2020). Coronavirus (COVID-19) classification using CT images by machine learning methods. *arXivpreprintarXiv:2003.09424.*

25. Kumar, R., Khan, A. A., Kumar, J., Golilarz, N. A., Zhang, S., Ting, Y., . . . & Wang, W. (2021). Blockchain-federated-learning and deep learning models for covid-19 detection using CT imaging. *IEEE Sensors Journal, 21*(14), 16301–16314.

26. Ruuska, S., Hämäläinen, W., Kajava, S., Mughal, M., Matilainen, P., & Mononen, J. (2018). Evaluation of the confusion matrix method in the validation of an automated system for measuring feeding behaviour of cattle. *Behavioural processes, 148*, 56–62.

31 Crop Protection from Animal Intrusion in the Fields Using Emerging Technologies
A Review

Nitika Goyal, Manjot Kaur

31.1 INTRODUCTION

Agriculture has played a significant part in India's economic sector. A large percentage of the Indian population is dependent on agriculture for their livelihood. Still there are major issues in Indian agriculture which need attention. As deforestation is taking place at a rapid pace due to overpopulation, the intrusion of wild animals in residential areas has become very common nowadays, which is creating conflict between man and animal. Also entry of animals in agricultural fields has led to major loss to crops as well as human lives. Intrusion of dangerous wild animals into fields has negatively impacted our lives in numerous ways and the side effects are visible as crop loss, danger to human lives, destruction of food stores, and so on. So human life, money, and assets have to be secured from this problem. Indian farmers are already facing serious issues like loss due to natural calamities, unpredictable climate, and rising prices. In this scenario, destruction by intruding animals may make the situation even worse. The rest of the factors are beyond our control, but latest inventions in technology has made it feasible to check animal intrusion into agricultural land. Traditional methods like fencing, barbed wires, and manual guarding have not only proved to be ineffective in stopping intrusion but also lead to injuries to both humans and animals. Also it is not feasible to guard the fields 24/7. So an animal detection system is the dire need of the hour so as to save our crops from controllable factors, if not from uncontrollable ones.

Technology has proved to be a boon for agriculture. Information technology has given us numerous devices and techniques which support the agriculture sector in handling some of the major issues like soil health monitoring, crop yield analysis, and farm protection. Image- or video-based processing is one of the latest techniques that give an economical and feasible alternative to solve the agriculture-related issues, especially guarding the farms. Such techniques have no requirement of any storage space due to processing of data in a real-time environment. The main component of

DOI: 10.1201/9781003367161-31

339

such a method is the detection model, which is skilfully trained in detecting, locating, and identifying the object in the image using an effective learning method. After the detection of an animal, either a sound buzzer may be used to make the animal run away from the field, or an alert system may be placed in the model, which can alert the farmer about entry of an animal in the fields.

31.2 AGRICULTURE: PAST AND PRESENT

Agriculture has passed through four stages to reach its present situation. There have been four different stages in the history of agricultural development, depicted in Figure 31.1[1]. These four stages are as follows:

1. Stage of conventional agriculture dominated by animal power and human effort
2. Stage of mechanization identified by reverberant noise
3. Stage of automation characterized by rapid development
4. Stage of smart farming featured by it and other emerging technologies [2], [3]

Therefore, the fourth stage, smart farming, is the current one. Smart agriculture refers to increasing sustainability by utilizing ICT (information and communication technologies) in the field of agriculture. Various emerging technologies, like internet of things (IoT) and cloud computing, which make agriculture smart are elaborated here [4], [5]:

1. Internet of things: IoT can be considered as a vital part of smart agriculture. Various authors have defined IoT in their own ways as the meaning of "things" has been changing along with technological advancements [6], [7]. However, in simple terms, IoT can be defined as integrated heterogeneous components which are capable of sensing and transferring data across a network automatically. IoT is an imminent advancement in the field of information technology, which supports the communication among electronic components and sensors using the internet as the medium in order to make

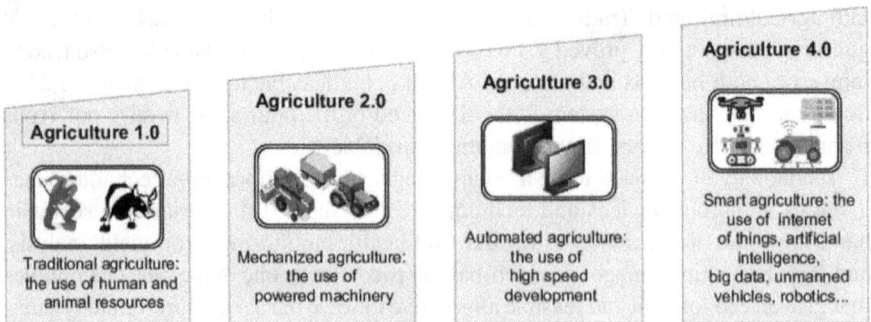

Agriculture 1.0
Traditional agriculture: the use of human and animal resources

Agriculture 2.0
Mechanized agriculture: the use of powered machinery

Agriculture 3.0
Automated agriculture: the use of high speed development

Agriculture 4.0
Smart agriculture: the use of internet of things, artificial intelligence, big data, unmanned vehicles, robotics...

FIGURE 31.1   The four agricultural revolutions.

our lives easier. IoT connects smart devices via the internet for providing an effective solution to complex situations related to different fields like commercial organizations, defence, agriculture, and so on [8]. With the invention of technologies like IoT, agricultural issues like crop security, water management, and soil management can be sorted out to a large extent.

2. Image processing: Image processing is a term used for technical analysis and manipulation of an image using some algorithms to obtain relevant information from it. To extract information from an image we need to perform some operations on it for which we give the image as input to the computer. The computer further reads it as a matrix of various colours and a different value is assigned for each colour in the image. Image processing is being utilized in different fields—defence, automobiles, medicine, and so on. Image processing has proved to be a boon to agriculture sector and is used to automate agriculture processes like crop management, weed detection, and farm protection.

3. Deep learning: Deep learning is a subset of machine learning in which multi-layer processing process is followed in order to extract features from an image by using artificial neural networks. Deep learning algorithms may find application in various fields of agriculture like plant disease identification, soil health monitoring, yield prediction, and crop protection.

31.3 RELATED WORK

31.3.1 SOUND-BASED SYSTEM

A sound-based system is the one where a loud and irritating sound is used to distract the animal and to divert it from approaching the field. Such type of systems has been successful to some extent, but some loopholes need to be plug in.

S. Giordano et al. [9] carried out research on protecting crops from any type of intrusion in the fields. He worked on developing an IoT application for the same. Wireless technologies, such as ZigBee, 6LoWPAN, and Wi-Fi, were used along with IoT gateway to monitor the fields in order to detect any intrusion. The authors developed a sound repeller by using solar technology to ensure that the developed model works efficiently even in total or partial darkness. A passive infrared sensor was used for network operation and frequency transmission, which used to transmit small frames to 50 m of distance. RIOT-OS software was used for this type of transmission. A loud sound of around 120 dB was produced on detection of an animal in the fields. The drawback of this device is that its performance use to get reduced below 90% if distance gets increased beyond 60 metres and the device will not work in case the distance goes beyond 100 metres.

Iniyaa K. K. et al. [10] utilized a convolutional neural network (CNN) and deep learning to check animal intrusion in the fields. The motive of this work was not only to guard the fields from animal entry but also to ensure that neither the animal nor the farmer gets injured due to the conflict. So this model is based on the principle of diverting the animals away from the farms. A model that combines neural network and machine learning has been developed to monitor animals approaching the farms using computer vision technology. Animal movement is detected by monitoring the fields using a camera and the algorithm uses neural network concepts and library functions to identify the animal in camera frame and a relevant sound is played to make the animal run away from the fields.

D. Kalra et al. [11] worked to develop an IoT-based model to guard the fields from small animals and insect intrusion using sensors. This model not only helps to guard fields but also helps to manage proper irrigation. The heart of this model is the Arduino Uno microcontroller. To control the irrigation dampness of the farm is measured using sensors, and water siphons are switched on or off accordingly. To guard the farm from animals and small insects, sensors are used; these sensors detect the motion of animals approaching the farm, and a signal is sent to the Arduino Uno microcontroller to calculate its distance from the field. The microcontroller produces a sound of a frequency based on the distance and type of animal coming to the field.

Nerukar et al. [12] worked on a crop protection system using image processing. Several types of work integrated with the Raspberry Pi model for motion detection. In case of detection of any movement indicated by the sensor, the Raspberry Pi checks if there is an intrusion which may damage the crops. If any intrusion is found, sound is produced to distract the animal and to alert the farmer. The main components of this system are PIR sensors, Raspberry Pi, and buzzer etc. This system works efficiently if the field area is large and we want to keep minimum people to guard the field.

Giordano et al. [9] worked on a deep-learning-based model for protection of crops from destruction by animals. It was not a camera-based system. This buzzer-based system used to generate alarms on entry of an object in the fields. Such system does not identify the object which can prove to be harmful for farmer's life if any wild animal enters the fields.

31.3.2 Alert-Based System

An alert-based system does not take any action if anything undesirable happens but alerts the farmer about the issue so that required action may be taken. In this section, two types of alert systems are discussed. The first is the buzzer-based system that generates a sound using a buzzer to alert the farmer about the event, and in the second one, an SMS is sent to the farmer to inform him/her about the alarming situation.

31.3.2.1 Buzzer Alert

Raksha and Surekha [13] worked on a model based on machine learning and IoT to protect crops from the intrusion of wild animals in the fields. The IoT devices used include GSM module, Arduino Uno microcontroller, pan-tilt-zoom camera, and sensors. A 605-image dataset was used for classification, which included images of animals like horses, lions, and elephants. For the classification of these animals, logistic regression, support vector machine (SVM) algorithm, and k-nearest neighbour (KNN) algorithm were used. It was concluded that 89.6% accuracy can be obtained using SVM algorithm, and it performed better than logistic regression and KNN algorithm in case of iterated regularization parameter of C= 100.

Goyal et al. [14] developed a model based on deep learning for protection of crops from any damage. This model not only supports protection of crops but also helps in irrigation and moisture control. Installation of sensors at different locations in the field helps us in collecting data related to moisture and temperature, and accordingly water pumps and sprayers get activated whenever required. Moisture sensor and temperature sensor are used to measure humidity level and temperature in the warehouse where crops are stored. The cooler or heater gets switched on/off per the value of these parameters. Another sensor used in this system is a motion detection sensor. When a person tries to enter the warehouse, this motion sensor comes into action and triggers the detection alarm which produces sound to alert the farmer. The drawback of this system is that only a buzzer has been used to alert the farmer, so someone has to be present in that area to handle the situation. No message is being sent to the farmer.

Prabha et al. [15] developed a model to trace flying objects and terrestrial animals in and around the agricultural land. It was a sound-based system. The drawback of this model was low range of sensors. The sensor was not able to track a large area; on the other hand, it was not feasible to place a large number of sensors all around the field.

31.3.2.2 SMS Alert

Balakrishna et al. [16] proposed an IoT- and machine-learning-based system to protect the crops from intrusion of animals. To recognize the animals in the captured video, a CNN-based algorithm was used. The accuracy of this model was more than 80%. The only loophole of this model was its inability to give accurate results in low light at night.

Agale and Gaikwad [17] designed a security system to check intrusion of animals into fields. A motion detection system was used to track an object entering the farm, and an alert system had been used to alert the farmer about intrusion. Cameras have been used for 24/7 monitoring of fields, and on detection of something suspicious, an alarm gets triggered and a message regarding intrusion is sent to the farmer through Wi-Fi or GSM. This system not only checks intrusion but also monitors soil health using different sensors, such as moisture sensor and temperature sensor. Dampness in the soil is measured by moisture sensors, and the value is sent to the microcontroller for further processing. The microcontroller controls the on/off switch of the water pump and turns the pump on or off per the requirement. Raspberry Pi and IoT have been used to implement this model.

Prabha and Ramprabha [15] developed an IoT-based model by using three components—ultrasonic sensor, GSM module, and IR sensor. Besides these devices, some other components like a microcontroller and voice module had also been used. The IR sensor detects the movement of an object around the farm and transmits a signal indicating the intrusion. This signal is received by the microcontroller (Arduino Uno) for further processing. A GSM module has been used to forward the message to the farmer. An alert message reaches the farmer indicating an intrusion by an animal, human being, or bird. To track birds, an ultrasonic sensor has been used, which is installed on the DC engine in order to rotate it to 360 degrees so as to identify flying objects approaching the farm. This system not only alerts the farmers but also generates sound using buzzers to keep the intruders away from the fields until the farmer reaches and takes appropriate action. The only limitation of this system is that there has been no use of camera or LED lights for night vision.

Balaji and Nandhnini [18] worked on developing a model for guarding fields from intrusion. This crop monitoring system was developed using IoT and deep learning. The concept of IoT—the storage and retrieval of data from anywhere, anytime—has been utilized in this model. Temperature sensor, soil moisture sensor, and humidity sensor have been used to measure their respective parameters. Arduino microcontroller ATmega328 receives the values of parameters measured by these sensors and processes it. The processed information is displayed on a webpage periodically and also sent to the farmer by a GSM module, which has been integrated with the microcontroller. The authors suggest that this work can be enhanced by using it for guarding the fields by video surveillance using deep learning to check animal intrusion into the fields.

Suchithra et al. [19] proposed a system which consists of sensors which measure the values of field parameters then the measured values get verified and further get transmitted to the Wi-Fi module. The Wi-Fi module forwards this information on the farmers mobile via cloud computing. This work also includes an algorithm with threshold values of humidity and temperature, which is used to control water inflow in the fields. Using this system, a farmer can operate the motor from anywhere, anytime. The system can be upgraded by incorporating a field protection module to check intrusion in the agriculture land by installing sensors along the farm boundary which sends a signal when some object tries to cross the boundary and integrating a camera module which takes pictures of the object and forwards it. Moreover, the system can be further automated by adding self-learning techniques to grasp the nature of sensed data and take decisions automatically.

Santhiya et al. [20] proposed a system for protection of fields from animal intrusion with the help of the Raspberry Pi. A radio frequency identification device (RFID) module and GSM have been used to develop this system. The alert message is forwarded to the farmer indicating an intrusion along with the information about identification and location of the coming object. This system overcomes the shortcomings of buzzer-based systems, which introduced a process of identifying the animal and generating a distinguished sound which irritates that particular animal, thus making it run from the field. The main drawback of this system is that an animal can get infected by the RFID vaccine, which gets injected under the skin of the animal. This system can be modified by incorporating a module to trace the exact location of the animal using GPS and RFID.

31.4 CHALLENGES

The most significant challenge faced by this type of systems is facing natural conditions like change in weather and the similarity between the object and the background and shadow in case of sunny weather [21], [22]. Another issue faced by such systems is night vision. Inexpensive cameras do not support night vision, whereas expensive products increase the cost of the system, thus making it unaffordable. Another factor is the inability to put cameras everywhere in the field. If we use a moving camera as an alternative, we may face problems like speed vibrations and damage to crops by moving objects. One of the major issues faced by farmers in developing nations is arranging infrastructure for such projects, as they may not get proper devices due to lack of guidance and financial constraints. Also, such models require a stable high-speed internet connection, which is still not available in some parts of the world. Images recorded by the camera in the form of images need to be processed by a leaning model, and the information is to be passed to the destination. If the duplicated frames keep on being processed, there will be a delay in passing information, so the destruction becomes unavoidable in such a situation.

Object monitoring is generally considered as a tedious task due to the presence of uncontrollable parameters like changing position of the object, camera motion, and different structures of the same object [23]. Another issue faced in the field of image processing is to identify multiple objects in the same video sequence. The problem gets aggravated when the objects to be identified look similar [24]. Solimon et al. [25] elaborated that the major challenge in image processing is to perform feature detection with speed, consistency, and accuracy.

31.5 CONCLUSION

Research work done by different scholars to check crop damage by intrusion of animals in the fields has been discussed thoroughly in this chapter. A detailed review of emerging technologies in the field of agriculture has been done. By extensive research and analysis, we have been able to conclude that the models developed for crop protection have their own pros and cons. Some of them are capable of giving good performance, but accuracy needs to be increased further, and also a few drawbacks which each model has need to be worked upon to give a perfectly accurate model for animal identification and tracking for welfare of farmer fraternity.

REFERENCES

[1] O. Friha, M. A. Ferrag, L. Shu, L. Maglaras, and X. Wang, "Internet of Things for the Future of Smart Agriculture: A Comprehensive Survey of Emerging Technologies," *IEEE/CAA J. Autom. Sin.*, vol. 8, no. 4, pp. 718–752, Apr. 2021, doi:10.1109/JAS.2021.1003925.

[2] Y. Liu, X. Ma, L. Shu, G. P. Hancke, and A. M. Abu-Mahfouz, "From Industry 4.0 to Agriculture 4.0: Current Status, Enabling Technologies, and Research Challenges," *IEEE Trans. Ind. Informatics*, vol. 17, no. 6, pp. 4322–4334, June 2021, doi:10.1109/TII.2020.3003910.

[3] K. Huang et al., "Photovoltaic Agricultural Internet of Things Towards Realizing the Next Generation of Smart Farming," *IEEE Access*, vol. 8, pp. 76300–76312, 2020, doi:10.1109/ACCESS.2020.2988663.

[4] G. Fortino, W. Russo, C. Savaglio, W. Shen, and M. Zhou, "Agent-Oriented Cooperative Smart Objects: From IoT System Design to Implementation," *IEEE Trans. Syst. Man. Cybern. Syst.*, vol. 48, no. 11, pp. 1949–1956, Nov. 2018, doi:10.1109/TSMC.2017.2780618.

[5] X. Yang et al., "A Survey on Smart Agriculture: Development Modes, Technologies, and Security and Privacy Challenges," *IEEE/CAA J. Autom. Sin.*, vol. 8, no. 2, pp. 273–302, Feb. 2021, doi:10.1109/JAS.2020.1003536.

[6] L. Atzori, A. Iera, and G. Morabito, "The Internet of Things: A Survey," *Comput. Networks*, vol. 54, no. 15, pp. 2787–2805, Oct. 2010, doi:10.1016/J.COMNET.2010.05.010.

[7] A. Al-Fuqaha, M. Guizani, M. Mohammadi, M. Aledhari, and M. Ayyash, "Internet of Things: A Survey on Enabling Technologies, Protocols, and Applications," *IEEE Commun. Surv. Tutor.*, vol. 17, no. 4, pp. 2347–2376, Oct. 2015, doi:10.1109/COMST.2015.2444095.

[8] A. R. Sfar, Z. Chtourou, and Y. Challal, "A Systemic and Cognitive Vision for IoT Security: A Case Study of Military Live Simulation and Security Challenges," *2017 Int. Conf. Smart, Monit. Control. Cities, SM2C 2017*, pp. 101–105, Oct. 2017, doi:10.1109/SM2C.2017.8071828.

[9] S. Giordano, I. Seitanidis, M. Ojo, D. Adami, and F. Vignoli, "IoT Solutions for Crop Protection Against Wild Animal Attacks," *2018 IEEE Int. Conf. Environ. Eng. EE 2018—Proc.*, pp. 1–5, Jun. 2018, doi:10.1109/EE1.2018.8385275.

[10] Iniyaa, K. K., Divya, J. K., Devdharshini, S., & Sangeethapriya, R. (2021). Crop Protection from Animals Using Deep Learning. *International Journal of Progressive Research in Science and Engineering*, 2(3), pp. 41–44.

[11] D. Kalra, P. Kumar, K. Singh, and A. Soni, "Sensor based Crop Protection System with IOT monitored Automatic Irrigation," *Proc.—IEEE 2020 2nd Int. Conf. Adv. Comput. Commun. Control Netw., ICACCCN 2020*, pp. 309–312, Dec. 2020, doi:10.1109/ICACCCN51052.2020.9362739.

[12] D. Karthik, D. Karthik, and R. R. Babu, "Smart Crop Protection System with Image Capture Over IoT," *Int. J. Innov. Technol. Res.*, vol. 5, no. 6, pp. 7434–7436, Nov. 2017.

[13] R. Raksha and P. Surekha, "A Cohesive Farm Monitoring and Wild Animal Warning Prototype System Using IoT and Machine Learning," *Proc. Int. Conf. Smart Technol. Comput. Electr. Electron. ICSTCEE 2020*, pp. 472–476, Oct. 2020, doi:10.1109/ICSTCEE49637.2020.9277267.

[14] S. Goyal, U. Mundra, and S. Shetty, "Smart Agriculture Using IoT," *Int. J. Comput. Sci. Mob. Comput.*, vol. 8, no. 5, 2019.

[15] M. J. Prabha, R. Ramprabha, V. V. Brindha, and C. A. Beaula, "Smart Crop Protection System from Animals," *Int. J. Eng. Adv. Technol.*, vol. 9, no. 4, pp. 2064–2067, 2020.

[16] K. Balakrishna, F. Mohammed, C. R. Ullas, C. M. Hema, and S. K. Sonakshi, "Application of IOT and Machine Learning in Crop Protection Against Animal Intrusion," *Glob. Trans. Proc.*, vol. 2, no. 2, pp. 169–174, Nov. 2021, doi:10.1016/J.GLTP.2021.08.061.

[17] R. R. Agale and D. P. Gaikwad, "Automated Irrigation and Crop Security System in Agriculture Using Internet of Things," *2017 Int. Conf. Comput. Commun. Control. Autom. ICCUBEA 2017*, Sep. 2018, doi:10.1109/ICCUBEA.2017.8463726.

[18] G. N. Balaji, V. Nandhini, S. Mithra, N. Priya, and R. Naveena, "IoT Based Smart Crop Monitoring in Farm Land," *Imp. J. Interdiscip. Res.*, vol. 4, no. 1, pp. 88–92, 2018.

[19] M. Suchitra, T. Asuwini, M. C. Charumathi, and N. L. Ritu, "Monitoring of Agricultural Crops Using Cloud and IOT with Sensor Data Validation," *Int. J. Pure Appl. Math.*, vol. 12, 2018.

[20] S. Santhiya, Y. Dhamodharan, N. E. K. Priya, C. S. Santhosh, and M. Surekha, "A Smart Farmland Using Raspberry Pi Crop Prevention and Animal Intrusion Detection System," *Int. Res. J. Eng. Technol*, vol. 5, no. 3, pp. 3829–3832, 2018.

[21] E. Hamuda, B. Mc Ginley, M. Glavin, and E. Jones, "Automatic Crop Detection Under Field Conditions Using the HSV Colour Space and Morphological Operations," *Comput. Electron. Agric.*, vol. 133, pp. 97–107, Feb. 2017, doi:10.1016/J.COMPAG.2016.11.021.

[22] D. C. Slaughter, D. K. Giles, and D. Downey, "Autonomous Robotic Weed Control Systems: A Review," *Comput. Electron. Agric.*, vol. 61, no. 1, pp. 63–78, Apr. 2008, doi:10.1016/J.COMPAG.2007.05.008.

[23] A. Yilmaz, O. Javed, and M. Shah, "Object Tracking," *ACM Comput. Surv.*, vol. 38, no. 4, p. 45, Dec. 2006, doi:10.1145/1177352.1177355.

[24] S. S. Pathan, A. Al-Hamadi, and B. Michaelis, "Intelligent Feature-Guided Multi-Object Tracking Using Kalman Filter," *2009 2nd Int. Conf. Comput. Control Commun. IC4 2009*, 2009, doi:10.1109/IC4.2009.4909260.

[25] M. M. El-Gayar, H. Soliman, and N. Meky, "A Comparative Study of Image Low Level Feature Extraction Algorithms," *Egypt. Inform. J.*, vol. 14, no. 2, pp. 175–181, Jul. 2013, doi:10.1016/J.EIJ.2013.06.003.

32 Design of Enhanced Magnitude and Phase Response FIR Fractional-Order Digital Differentiator Using Lightning Attachment Procedure Optimization Algorithm

Puneet Bansal, Sandeep Singh Gill

32.1 INTRODUCTION

A new field that has great promise for applications in signal processing, control, and biomedical instrumentation is that of fractional-order systems. Recently, several engineering, technological, and scientific fields, including automated control, image processing, VLSI design, automated regulation, fluid mechanics, electromagnetic theory, and electrical networks, have shown a lot of interest in this particular fractional derivative concept (Mohan & Rao, 2019; Das, 2008; Miller & Ross, 1993; Engheta, 1999). Fractional order operators simulate the dynamics of physical systems more effectively than integer order operators. Fractional order differentiators (FODs) based on finite impulse response (FIR) gained popularity over FODs based on the infinite-length impulse response (IIR) due to linear phase response of earlier design.

In Singh et al. (2015), a continuous fraction expansion approach is used to create another digital FOD. These FODs underwent time domain analysis, and the results were compared with one another. FODs are designed using the radial basis function interpolation (RBFI) approach (Tseng et al., 2010). The discrete-time sequence non-integer delay sample estimate is obtained using the RBFI approach.

Optimization methods like PSO (Kennedy & Eberhart, 1995), cat swarm optimization (Pradhan & Panda, 2012), CSA (X.-S. Yang & Deb, 2009), and LAPO

348

DOI: 10.1201/9781003367161-32

algorithm (Nematollahi et al., 2017) are used to find best solution. In Krishna & Rao (2012), theory of operation, application, and implementation of digital differentiator is presented. Digital differentiator based on simulated annealing is presented in (Mohit Kumar & Rohilla, 2014). Latest design of a digital differentiator is illustrated in (Aggarwal et al., 2019). Particle swarm optimization is implemented in Manjeet Kumar (2019) for the construction of fractional order FIR differentiators. Manjeet Kumar & Rawat (2015) present an optimal design for a cuckoo-search-based FIR-FOD. Digital FODs are discussed by employing inverse multiquadric radial basis function (N. Kumar & Rawat, 2012).

Here, frequency response of ideal digital FOD is written as

$$H_{id}(\omega) = (j\omega)^{\alpha}. \tag{32.1}$$

Here α is a fractional number. ω is a normalized frequency which varies between [0,1] and $j = \sqrt{-1}$. There are following steps included for designing FOD, as shown in Figure 32.1.

For the design of digital FIR-FODs, a number of hybrid optimization approaches have been employed. Digital differential operators have been optimized using a PSO and genetic algorithm hybrid, making them more suited to represent their new, superior fractional order differentiator counterparts (Gupta & Yadav, 2014). Similar techniques include hybrid firefly algorithm and particle swarm optimization (Ray

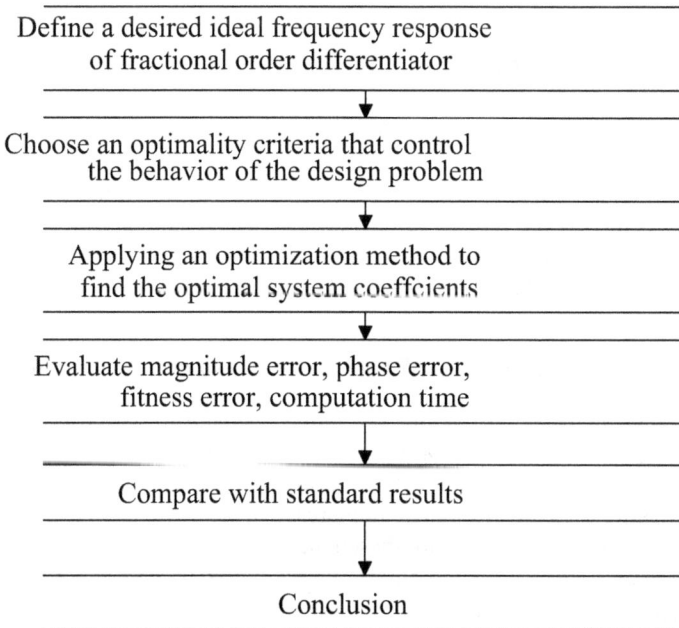

FIGURE 32.1 The design procedure of FIR-FOD.

et al., 2019) and hybrid firefly algorithm and ant colony optimization (ACO) (Xue et al., 2016).

The order of the remaining paper is as follows: In Section 32.2, the formulation of the design problem for the FIR-FOD based on the fitness response is demonstrated. Additionally, a general formulation of the Grünwald-Letnikov integral-differential operator and an explanation of fractional derivatives are given. The fundamentals of the PSO, CSA, and LAPO algorithms are covered in Section 32.3. The results of the produced simulations and the statistical "Wilcoxon rank-sum test" are explained in Section 32.4. Section 32.5 concludes the work presented in this chapter.

32.2 PROBLEM FORMULATION

Using the definition in Scherer et al. (2011), the fractional derivative is calculated. An ideal FIR-FOD response is expressed as

$$H_{ideal}(\omega) = (j\omega)^\alpha, \tag{32.2}$$

where α is a fractional number, $H_{ideal}(\omega)$ is an ideal FOD frequency response, and $0 \leq \omega \leq 1$ is a normalized frequency. For length N, Z-transform of FIR-FOD function is used as

$$H(z) = \sum_{n=0}^{N-1} h(n) z^{-n}. \tag{32.3}$$

Here, weighted least square fitness function is employed to obtain best design and is expressed by

$$J_f = \int_0^1 W_1(\omega) \left| abs \left(H_{ideal}(\omega) \right) - abs \left(H(\omega) \right) \right|^2 + W_2(\omega) \left| phase \left(H_{ideal}(\omega) \right) \right.$$

$$\left. - phase \left(H(\omega) \right) \right|^2 d\omega. \tag{32.4}$$

Here $W_1(\omega) = 0.8$ and $W_2(\omega) = 0.2$ are non-negative weighting functions. The digital FIR FOD presented has a filter order = 4 and fractional order $\alpha = 0.5$. The specific fitness function given in equation (32.4) is minimized in this study utilizing optimization methods like PSO, CSA, and LAPO algorithm.

32.2.1 LIGHTNING ATTACHMENT PROCEDURE OPTIMIZATION ALGORITHM

The physical phenomena of lightning attachment mechanism is used as the model for the LAPO algorithm, a heuristic optimization technique. The clouds store enormous quantities of electric charge, and as those charges rise, lightning is created. As can be seen in Figure 32.2, this results in an increase in electrical strength.

In contrast to previous population-based methodologies, the optimization process of the LAPO algorithm is carried out in two basic phases: downward leader movement and upward leader propagation; further information is provided in Nematollahi et al. (2017). Also, details about employed PSO and CSA can be found in Kennedy & Eberhart (1995) and X.-S. Yang & Deb (2009).

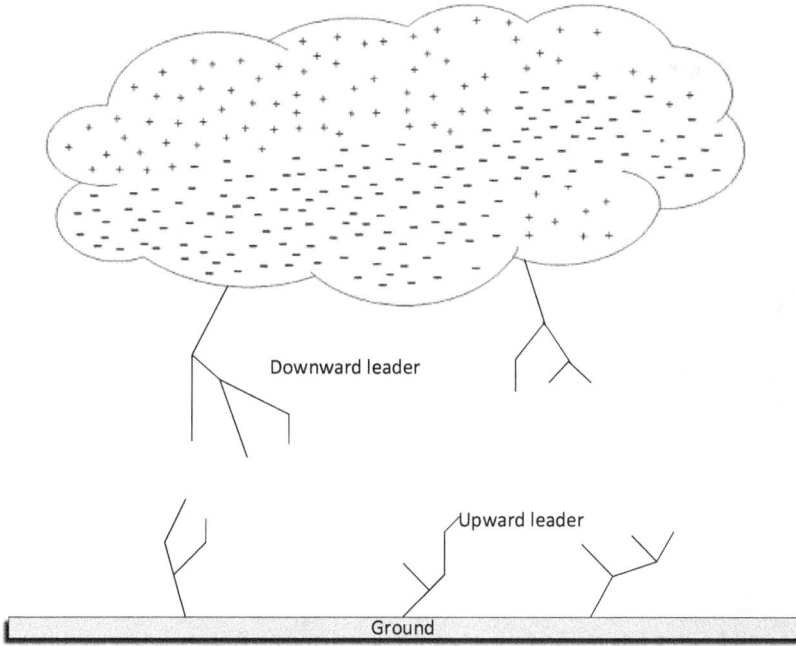

FIGURE 32.2 Charge distribution with lightning from the cloud (Nematollahi et al., 2017).

32.3 SIMULATION RESULTS AND DISCUSSIONS

These optimal outcomes were attained during 100 simulation attempts using the Intel® CoreTM i5, 1.60 GHz, 6GB RAM, and MATLAB 2015a version, with random modifications to the control settings presented in Table 32.1 with optimal coefficients as shown in Table 32.2.

Figures 32.3–32.4 provide a graphical comparison achieved by the PSO, CSA, and LAPO algorithms with the ideal FIR-FOD.

32.4 PERFORMANCE ANALYSIS

Performance analysis for LAPO-based FIR-FOD design is presented in Table 32.3 and compared with designs based on PSO and CSA. In terms of these performance parameters for fourth-order FIR-FOD designs, it is determined that LAPO-based FIR-FOD designs are better than PSO- and CSA-based designs.

The minimum, maximum, and mean values of absolute magnitude error obtained by using LAPO algorithm for fourth-order digital FIR-FOD are 0.0002, 0.02423, and 0.00813, which are 97.33%, 54.55%, 66.64%, 24.61%, 62.60%, and 12.77% better as compared to minimum, maximum, and mean values of absolute magnitude error obtained by PSO and CSA techniques, respectively, as shown in Table 32.3. Also, the standard deviation of absolute magnitude error obtained by using LAPO algorithm

TABLE 32.1

FIR-FOD Control Parameters (X. Yang, 2014)

Parameters	Symbol	PSO	CSA	LAPO
Population size	Popsize	25	25	25
Max. iteration cycle	Epoch	1,000	1,000	1,000
Inertia weight	w	0.9–0.4	-	-
Learning parameters	C_1, C_2	2,2	-	-
Particle velocity	v_{min}, v_{max}	0.01, 1	-	-
Discovering rate of alien eggs	P_a	-	0.25	-
Number of nests	n	-	25	-
No. of design variables	-	-	-	N + 1
Limits of filter coefficients		−1, +1	−1, +1	−1, +1

TABLE 32.2

Optimized Filter Coefficients

Techniques	Optimized Coefficients (h_k)
PSO	0.9859, −0.2413, −0.9678, 0.8468, −0.4493
CSA	0.9648, −0.2348, −0.9751, 0.8578, −0.4725
LAPO	0.9853, −0.2367, −0.9714, 0.8542, −0.4534

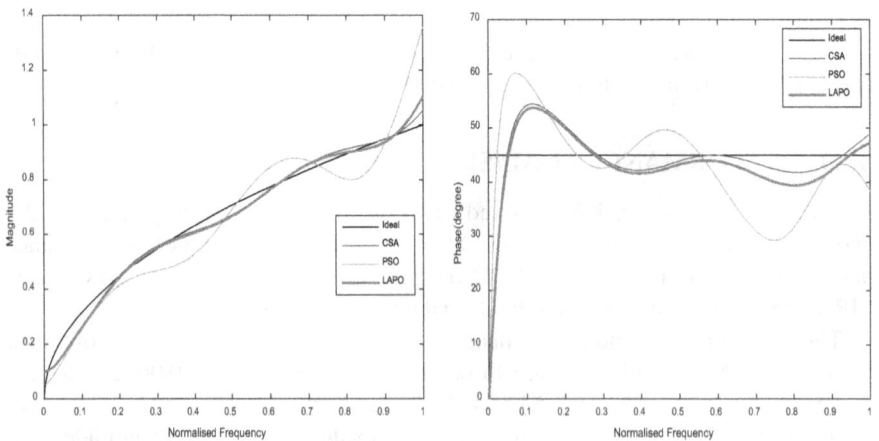

FIGURE 32.3　Fourth-order digital FIR-FOD enhanced magnitude response and phase response.

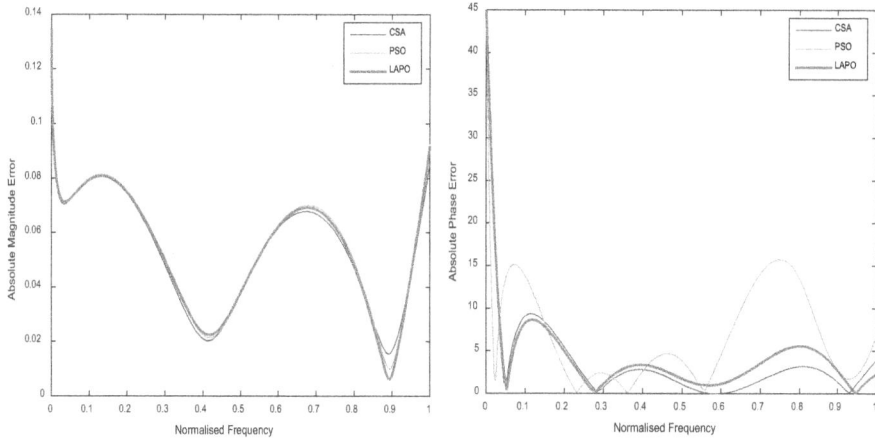

FIGURE 32.4   Fourth-order digital FIR-FOD absolute magnitude and phase errors

TABLE 32.3
Performance Analysis of Absolute Magnitude Error

Algorithms	Minimum	Maximum	Mean	Variance	Standard Deviation
PSO	0.00777	0.07264	0.02174	6.61E-05	0.0081
CSA	0.00044	0.03214	0.00932	6.31E-06	0.0025
LAPO	0.0002	0.02423	0.00813	3.36E-06	0.0018

for fourth-order digital FIR-FOD is 0.0018 which is 77.78% and 28.00% better as compared to the standard deviation of absolute magnitude error obtained by PSO and CSA techniques as shown in Table 32.3. The shows the robustness of the LAPO algorithm used for the design.

Table 32.4 accounts for the performance study of the fourth-order FIR-absolute FOD's phase inaccuracy. The suggested LAPO-based FIR-FOD design's computed minimum, maximum, mean, variance, and standard deviation for absolute phase error are provided and compared with PSO- and CSA-based designs. In terms of these performance parameters for fourth-order FIR-FOD designs, it is established that LAPO-based FIR-FOD designs are better than PSO- and CSA-based designs.

The minimum, maximum and mean values of absolute phase error obtained by using LAPO algorithm for fourth-order digital FIR-FOD are 00.0043, 0.2851, and 0.0082, which are −16.22%, 30.65%, 34.08%, 5.41%, 91.9%, and 16.33% better as compared to minimum, maximum, and mean values of absolute phase error obtained by PSO and CSA techniques, respectively, as shown in Table 32.4. Also, the standard deviation of absolute phase error obtained by using LAPO algorithm for fourth-order digital FIR-FOD is 7.95E-08, which is 35.89% and 86.75% better as

TABLE 32.4

Performance Analysis of Absolute Phase Error

Algorithms	Minimum	Maximum	Mean	Variance	Standard Deviation
PSO	0.0037	0.4325	0.1012	1.54E-14	1.24E-07
CSA	0.0062	0.3014	0.0098	3.60E-13	6.00E-07
LAPO	0.0043	0.2851	0.0082	7.00E-15	7.95E-08

compared to the standard deviation of absolute phase error produced by using PSO and CSA techniques, as shown in Table 32.4. The shows the robustness of the LAPO algorithm used for the design.

The maximum values of absolute magnitude and phase errors using PSO, CSA, and LAPO algorithms with worst, average, and best conditions are summarized in Table 32.5. "The worst condition is reached after 100 iterations, the average condition after 500 iterations, and the best condition after 1000 iterations" (Manjeet Kumar, 2019). This study comes to the conclusion that the LAPO algorithm produces less errors than PSO- and CSA-based FIR-FOD designs.

In case of worst conditions, absolute magnitude and phase errors obtained by using LAPO algorithm for fourth-order digital FIR-FOD are 0.04214 and 0.3256, which are 55.28%/43.44% and 49.34%/9.45% better as compared to absolute magnitude and phase errors obtained by PSO and CSA techniques, respectively, as shown in Table 32.5.

In case of average conditions, absolute magnitude and phase errors obtained by using LAPO algorithm for fourth-order digital FIR-FOD are 0.03127 and 0.3041, which are 62.84%/39.2% and 43.12%/5.44% better as compared to absolute magnitude and phase errors obtained by PSO and CSA techniques, respectively, as shown in Table 32.5.

In case of best conditions, absolute magnitude and phase errors obtained by using LAPO algorithm for fourth-order digital FIR-FOD are 0.02423 and 0.2851, which are 66.64%/24.61% and 34.08%/5.41% better as compared to absolute magnitude and phase errors obtained by PSO and CSA techniques, respectively, as shown in Table 32.5.

The optimized filter coefficients h_k for the fourth-order FIR-FOD design obtained by the PSO, CSA, and LAPO algorithms with order of four with worst, average, and best conditions are shown in Table 32.6.

The calculation time for a fourth-order FIR-FOD design based on LAPO is 24.01 seconds, compared to 21.65 seconds and 22.96 seconds for designs based on CSA and PSO, respectively. It is concluded that the LAPO method achieved optimum FIR-FOD designs faster than the PSO and CSA algorithms. Table 32.7 compares the performance consistency of PSO-, CSA-, and LAPO-based digital FIR-FOD using the Wilcoxon rank-sum test at 99% level of confidence. It is observed that the p-value is less than 0.01, LAPO outperforms PSO and CSA.

TABLE 32.5

Fourth-Order FIR-FOD Maximum Magnitude and Phase Errors by Employing PSO, CSA, and LAPO for the Worst, Average, and Best Conditions

Conditions	Absolute Magnitude Error			Absolute Phase Error		
	PSO	CSA	LAPO	PSO	CSA	LAPO
Worst	0.09423	0.07451	0.04214	0.6427	0.3596	0.3256
Average	0.08414	0.05143	0.03127	0.5346	0.3216	0.3041
Best	0.07264	0.03214	0.02423	0.4325	0.3014	0.2851

TABLE 32.6

Optimized FIR-FOD Coefficients

Condition	Algorithm	Optimized Coefficients (h_k), $0 \leq k \leq N$
Worst	PSO	0.9883, −0.9822, 0.8765, −0.9875, 0.2477
	CSA	0.9920, −0.3510, −0.6233, 0.4209, −0.2391
	LAPO	0.9786, −0.2413, −0.9947, 0.8468, −0.4279
Average	PSO	0.9854, −0.2425, −0.9671, 0.8477, −0.4345
	CSA	0.8947, −0.2412, −0.9872, 0.8464, −0.4293
	LAPO	0.9791, −0.2417, −0.8596, 0.8178, −0.4548
Best	PSO	0.9859, −0.2413, −0.9678, 0.8468, −0.4493
	CSA	0.9648, −0.2348, −0.9751, 0.8578, −0.4725
	LAPO	0.9853, −0.2367, −0.9714, 0.8542, −0.4534

TABLE 32.7

Statistical Analysis Using Wilcoxon Rank-Sum Test

Comparison	Absolute Magnitude Error p-value	Absolute Phase Error p-value
LAPO versus PSO	5.38302E-15	0.00025
LAPO versus CSA	0.00421	0.00134

The convergence curve of fourth-order Digital FIR-FOD using PSO, CSA and LAPO algorithms is shown in Figure 32.5.

When compared to PSO and CSA algorithms, proposed FIR fractional-order digital differentiator design employing the LAPO algorithm quickly reached convergence. Additionally, compared to designs based on PSO and CSA algorithms, LAPO-based design is better able to effectively carry out both exploitation and exploration.

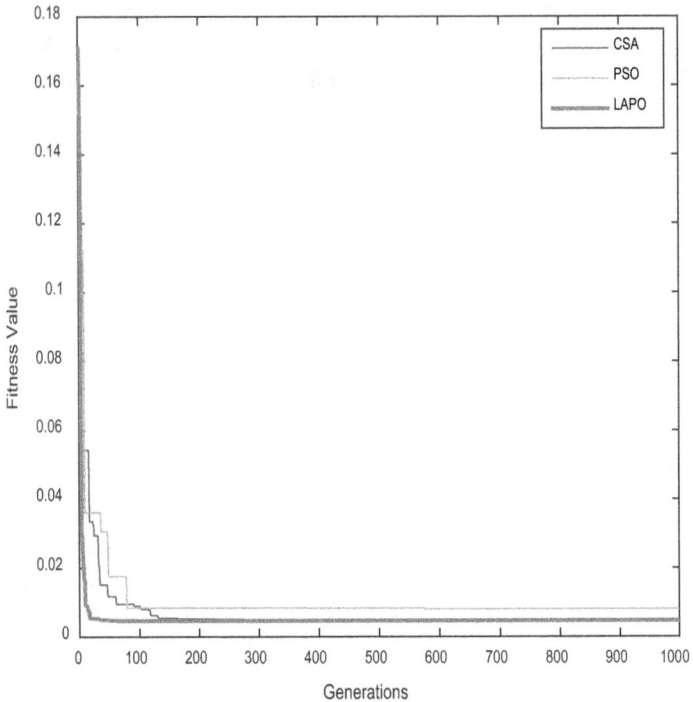

FIGURE 32.5 Convergence curve of FIR fractional-order digital differentiator

32.5 CONCLUSIONS AND FUTURE SCOPE

In this chapter, PSO, CSA, and LAPO algorithms are used to develop FIR fractional-order digital differentiators with a fourth order. The calculation of optimal coefficients yields the enhanced the magnitude and phase responses and reduce their absolute magnitude and phase errors. To demonstrate the reliability of the findings, the Wilcoxon rank-sum test is run at 99% level of confidence. In comparison to PSO and CSA approaches, the LAPO-based FOD design exhibited faster convergence and superior exploration and exploitation capabilities. In future, this work may be extended to design the optimal 2D FIR fractional-order digital differentiator using LAPO algorithm.

32.6 ACKNOWLEDGEMENTS

This work is supported by the I. K. G. Punjab Technical University, Kapurthala, Punjab-144603, India. We are very thankful to the university for all support.

REFERENCES

Aggarwal, A., Kumar, M., & Rawat, T. K. (2019). Design of digital differentiator using the L1-method and swarm intelligence-based optimization algorithms. *Arabian Journal for Science and Engineering*, *44*(3), 1917–1931. https://doi.org/10.1007/s13369-018-3188-0

Das, S. (2008). *Functional Fractional Calculus for System Identification and Controls.* Springer. https://doi.org/10.1007/978-3-540-72703-3

Engheta, N. (1999). Fractional derivatives, fractional integrals and electromagnetic theory. *1999 international conference on computational electromagnetics and its applications. ICCEA 1999—Proceedings*, 1–4. https://doi.org/10.1109/ICCEA.1999.825051

Gupta, M., & Yadav, R. (2014). New improved fractional order differentiator models based on optimized digital differentiators. *The Scientific World Journal, 2014.*

Kennedy, J., & Eberhart, R. (1995). Particle swarm optimization. *Neural Networks, 1995. Proceedings., IEEE International Conference on, 4, 1942–1948*, vol. 4. https://doi.org/10.1109/ICNN.1995.488968

Krishna, B. T., & Rao, S. S. (2012). On design and applications of digital differentiators. *4th International Conference on Advanced Computing, ICoAC 2012, 2*, 1–7. https://doi.org/10.1109/ICoAC.2012.6416802

Kumar, M. (2019). Fractional order FIR differentiator design using particle swarm optimization algorithm. *International Journal of Numerical Modelling: Electronic Networks, Devices and Fields, 32*(2), e2514. https://doi.org/https://doi.org/10.1002/jnm.2514

Kumar, M., & Rawat, T. K. (2015). Optimal design of FIR fractional order differentiator using cuckoo search algorithm. *Expert Systems with Applications, 42*(7), 3433–3449. https://doi.org/10.1016/j.eswa.2014.12.020

Kumar, M., & Rohilla, K. (2014). Optimal design of digital differentiator using simulated annealing. *International Journal of Engineering Science and Innovative Technology (IJESIT), 3*(4), 232–239.

Kumar, N., & Rawat, T. K. (2012). *Design of Fractional Order Digital Differentiator Using Inverse Multiquadric Radial Basis Function BT—Advances in Computer Science and Information Technology. Computer Science and Engineering* (N. Meghanathan, N. Chaki, & D. Nagamalai, eds., pp. 32–49). Springer, Berlin and Heidelberg.

Miller, K. S., & Ross, B. (1993). *An Introduction to the Fractional Calculus and Fractional Differential Equations* (1st ed.). New York, US: Wiley.

Mohan, G. S. S. S. S. V. K., & Rao, Y. S. (2019). An efficient design of fractional order differentiator using hybrid Shuffled frog leaping algorithm for handling noisy electrocardiograms. *International Journal of Computers and Applications*, 1–7. https://doi.org/10.1080/1206212X.2019.1573948

Nematollahi, A. F., Rahiminejad, A., & Vahidi, B. (2017). A novel physical based metaheuristic optimization method known as Lightning Attachment Procedure Optimization. *Applied Soft Computing Journal, 59*, 596–621. https://doi.org/10.1016/j.asoc.2017.06.033

Pradhan, P. M., & Panda, G. (2012). Solving multiobjective problems using cat swarm optimization. *Expert Systems with Applications, 39*(3), 2956–2964. https://doi.org/10.1016/j.eswa.2011.08.157

Ray, P. K., Paital, S. R., Mohanty, A., Foo, Y. S. E., Krishnan, A., Gooi, H. B., & Amaratunga, G. A. J. (2019). A hybrid firefly-swarm optimized fractional order interval type-2 fuzzy PID-PSS for transient stability improvement. *IEEE Transactions on Industry Applications, 55*(6), 6486–6498. https://doi.org/10.1109/TIA.2019.2938473

Scherer, R., Kalla, S. L., Tang, Y., & Huang, J. (2011). The Grünwald—Letnikov method for fractional differential equations. *Computers and Mathematics with Applications, 62*(3), 902–917. https://doi.org/10.1016/j.camwa.2011.03.054

Singh, K. J., Mehra, R., & Pal, G. P. (2015). Improved design of IIR-type digital fractional-order differentiators using continuous fraction expansion. *2015 2nd International Conference on Recent Advances in Engineering & Computational Sciences (RAECS)*, 1–5. https://doi.org/10.1109/RAECS.2015.7453368

Tseng, C., Member, S., & Lee, S. (2010). Design of fractional order digital differentiator using radial basis function. *IEEE Transactions on Circuits and Systems I: Regular Papers*, *57*(7), 1708–1718.

Xue, H., Shao, Z., Pan, J., Zhao, Q., & Ma, F. (2016). A hybrid firefly algorithm for optimizing fractional proportional-integral-derivative controller in ship steering. *Journal of Shanghai Jiaotong University (Science)*, *21*(4), 419–423. https://doi.org/10.1007/s12204-016-1741-0

Yang, X. (2014). *Nature-Inspired Metaheuristic Algorithms* (2nd ed., Issue July 2010). Frome, BA11 6TT, UK: Luniver Press, UK.

Yang, X.-S., & Deb, S. (2009). Cuckoo search via levey flights. *2009 World Congress on Nature & Biologically Inspired Computing (NaBIC 2009)*, 210–214. https://doi.org/10.1109/NABIC.2009.5393690

33 A Network Analysis of AI and ML in the Insurance Industry through Systematic Literature Review

Pankaj, Dr Cheenu Goel, Dr Payal Bassi

33.1 INTRODUCTION

The insurance industry has always been a lucrative sector of the economy despite being volatile in nature and has contributed exclusively as a prominent domain of the financial service industry. Moreover, it continues to generate an exclusive amount of data in routine; the data is generated through the transaction taking place by five main departments of this industry: claims, finance, legal, marketing, and underwriting. The outbreak of artificial intelligence (AI) has shaped up the insurance industry by reducing time and handling customers' queries in a much more effective manner. A sudden outbreak of COVID-19 during the last two years has turned into a turning point for the insurance industry as companies started making apt use of AI to build new products to serve their clients better than ever before and take data-driven decisions. Further, AI enables insurance companies to prepare machine learning (ML) models to predict and forecast the behaviour of prospective clients. The advent of deep learning has further revolutionized the way industries operated; companies now make a hybrid use of deep learning and ML to attain better results.

Since an enormous amount of data is being generated in the insurance industry only human resources are not sufficient to make full use of data. It is important to create AI-enabled predictive models which can process a large amount of data in a short span of time and thus reducing the cost and energy involved in the task. This will further enable customers to have computer-generated insurance advice, thus facilitating direct marketing, handling claims, conducting audits, and retaining customers.

33.2 REVIEW OF LITERATURE

Usage of AI has revolutionized the way each industry was performing, especially in healthcare AI and ML has changed the way diagnosis was being performed (Rathi et al. 2022). Benedek et al. (2022) conducted a systematic review in the field of

DOI: 10.1201/9781003367161-33

359

automobile insurance and detection of frauds in the era of big data. Haque and Tozal (2022) mention AI not just develops ML models but also are very instrumental in validating automated AI systems. Apps where direct-to-consumer artificial intelligence/ machine learning (DTC AI/ML) are enabled can secure the privacy of consumers and can provide customized solutions for every user (Gerke and Rezaeikhonakdar, 2022).

Thomas Davenport (2019) explored the potential of AI with special reference to healthcare and concluded that AI-driven decisions are better than human-integrated solutions. ML has emerged as an effective statistical technique to create data-driven models. Thus companies are largely implementing AI-driven approaches in their processes to handle the data generated by their customers and employees. Nudge (2018) states that the use of AI has reduced the tasks of administrative activities performed by employees as it records and stores the data in the appropriate format, thus making relevant information readily available for use.

Insurance companies are daily dealing with an enormous amount of information thus creating a decent amount of structured and unstructured data which individually might not directly or indirectly affect the industry. But collectively it can emerge as an attractive issue that must be addressed while designing strategies for any company and thus designing effective ML-enabled solutions for the business (Bassi and Kaur, 2022). According to Takura et al. (2021), insurance is largely a concern in the healthcare industry where examining patients' health, recording the progress of medications, and keeping the track of patient's recovery especially while claiming insurance. Further, a predictive model, Adherence Score for Healthcare Resource Outcome (ASHRO), was developed as a framework where the entire patient's record can be stored. Xu et al. (2019) mention that people are often stressed due to the reports that are required while claiming insurance, thus creating a burden on the healthcare system. The adoption of deep learning and ML models acts as one spot for retrieving that information as this data is stored in the cloud and can be directly linked with insurance companies. Bureau MFB (2019) states that the healthcare industry is dealing with profuse diseases, which are emerging in people of all age groups, and one of the main reasons for this is a changing lifestyle, which in turn is creating gaps while issuing claims of insurance.

Concerns about the dangers involved in growing a company's revenue were discussed by Lee and Ezekowitz (2014) and Marsh et al. (2017), particularly when it comes to the public insurers' ability to continue doing business. Chen et al. (2020), in his research, examined the operational risk faced by insurance companies, and thus, he developed a novel numerical method for resolving the ruin probability equation and created a ML and deep learning risk model for greater stability and fewer deviations. Insurers were found to be engaging in abusive and fraudulent behaviour, which Kose et al. (2015) discovered and further investigated in his study.

Misra et al. (2021) mention that effective ML models can be created utilizing clinical and administrative data to predict septic shock. It is possible to create and incorporate intelligent decision support tools into the electronic health record, which will enhance clinical results and make real-time resource optimization easier.

The following research questions were framed to find answers after carefully examining the papers chosen as a sample for the study and are the subject of the current investigation:

1. Which publications and nations in the area of AI, ML, and insurance sector have published the maximum number of articles?
2. What is the year-wise count of publications that were recorded in the chosen database?
3. Extract the graphical visualization of the top ten researchers and their countries supporting the research in the chosen topic along with their citations.
4. Which keywords are most frequently used in the documents published from various nations?

33.3 RESEARCH METHODOLOGY

The present study focuses on the application of Machine Learning and Artificial Intelligence in the insurance field. The present data has been collected from two reliable sources i.e. ProQuest and Scopus database. After applying filters of inclusion and exclusion criteria, a total of 123 papers which directly supports as a published literature in the undertaken study. In both the databases same string was applied: "Artificial Intelligence" AND "Machine Learning" AND "Insurance." The search started with 7,300 papers, and after inclusion of relevant decade, document type, language, subject areas, and publication type, the relevant sample of 123 papers was attained in the light of title, abstract, and full-text papers, as shown in Figure 33.1.

33.4 DATA ANALYSIS

33.4.1 TOTAL NUMBER OF PUBLICATIONS

Figure 33.2 represents the details of the annual publications that took place during the time period of 2015 till 2022. The actual time period was chosen as 2012 till 2022, but no paper got registered in the database of Scopus and ProQuest. The year 2021 emerged as a year with the highest publications, with a total document type of

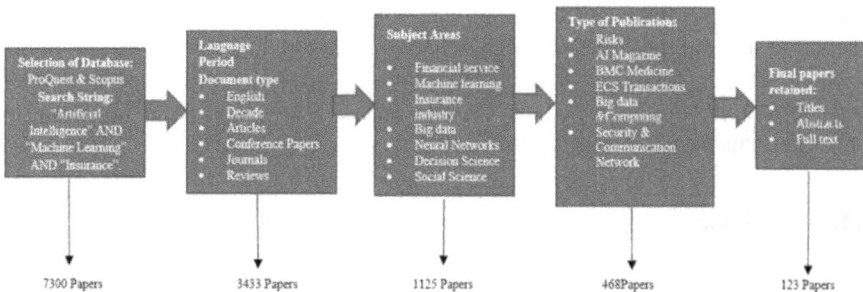

FIGURE 33.1   PRISMA model.

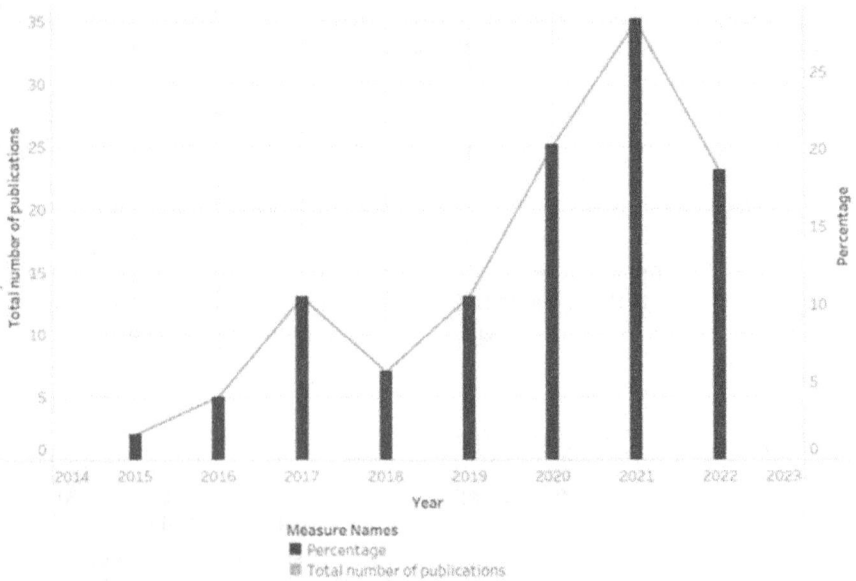

FIGURE 33.2 Year-wise publication count.

35 research publications, followed by the year 2020 and the year 2022 (till August), with the publication of 25 and 23 documents, respectively.

33.4.2 DOCUMENT TYPE

Figure 33.3 represents the total number of documents per their type of publication. It is evident that the present study has a total of 123 publications from both databases, out of which 56 documents are in the form of articles, followed by 24 conference papers, 31 journals, and 12 review papers.

33.4.3 TOP TEN JOURNALS

Figure 33.4 shows the detail of the top ten journals in the study undertaken in this chapter to achieve the research objectives. *Risks* appeared as the top journal in the selected sample, followed by the *International Journal of Advanced Computer Science & Applications*, then *Security & Communication Networks*. However, rest of the journals have two articles each.

33.4.4 WORD CLOUD

With the help of Word Art, word cloud has been derived from the author's keywords and titles, as represented in Figures 33.5 (a) and 33.5 (b). Keywords—"Machine Learning," "Insurance," Healthcare," "Artificial Intelligence," "Algorithm," "Neural

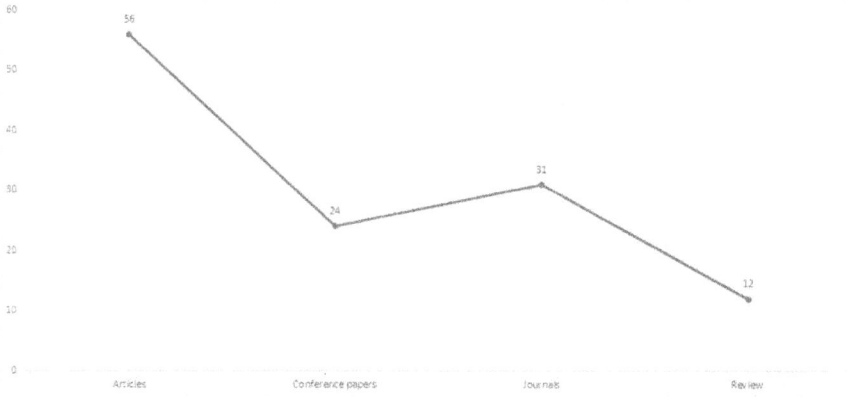

FIGURE 33.3 Count of document type.

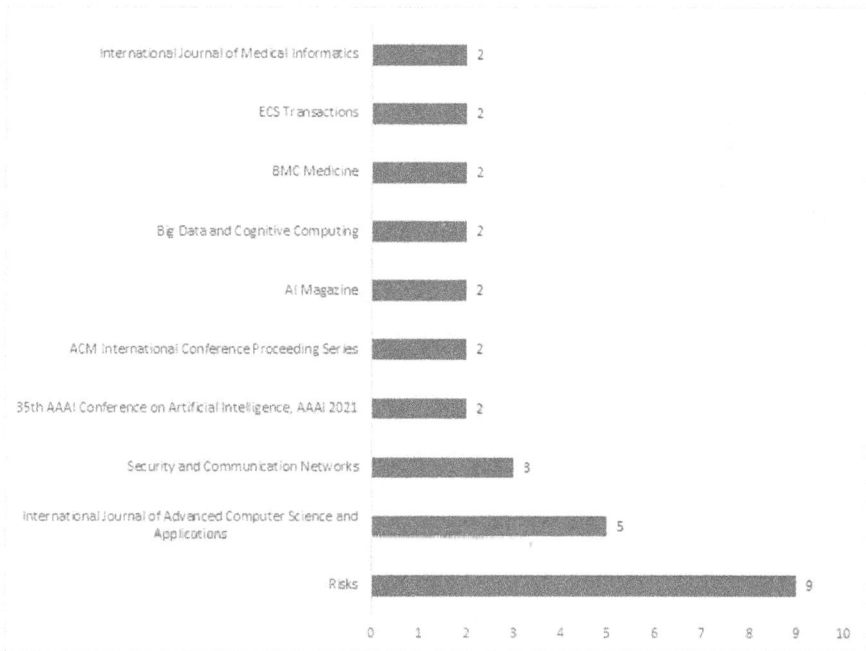

FIGURE 33.4 Top ten journals.

Networks," "Risk Assessment," "Health Insurance," and "Decision Tree"—are among the few areas of research related to the existing literature in the field of AI, ML, and insurance. Figure 33.5 (b) mentions "Artificial Intelligence," "Machine Learning," "Insurance Industry," "Fraud Detection," "Artificial Intelligence Technology," and "Big Data" as some of the words represented in the word cloud of titles.

insurance claim　　　personalized medicine

artificial neural network

comparative study　fraud detection

supervised learning　neural network　　support vector machine

risk assessment　　retrospective study

insurance company　health care delivery

health insurance　internet of things

machine learning

artificial intelligence　big data

machine learning techniques

electronic health record

algorithm　health care

random forest

health care personnel

deep learning

predictive value

insurance

human

language processing system

vector machine　decision tree　natural language processing

long term care

controlled study

data mining　machine learning model　insurance industry

risk factors　decision support system

anomalies detection

delivery of health

regression analysis　major clinical study　decision making

precision medicine

health care policy　insurance fraud

machine learning approach

recent advance　machine learning approach　actuarial science

personalized health insurance　transparency of underwriting　analysis of insurance

tree classification algorithm　machine learning performance　solutions of risks

learning performance optimisation　artificial intelligence approach

insurance claim generation　artificial intelligence technology　medicare fraud detection

big data analytics　automated payroll management　decision support system

vehicle insurance company

comparative study

artificial intelligence

systematic mapping study　mental health care

clinical decision support　automobile insurance fraud　decision tree classification

machine learning health

future care resources

decision tree algorithm　insurance industry　machine learning

national health insurance　improved fraud detection

intelligence predicts cost

world data　fraud detection　big data　life insurance industry

guided decision tree

insurance fraud detection　deep learning

continuous time model

insurance business model

file insurance risk

risk assessment　data science

health care

explainable artificial intelligence

insurance marketing planning　deep learning approach

FIGURE 33.5　(a) Word Cloud of author's keywords. (b) Word cloud of titles.

33.4.5　HIGHEST CONTRIBUTING AUTHORS AND COUNTRIES

Figure 33.6 represents the Sankey diagram for the top-ten contributors with respect to their countries and the citations of their research work in the field of AI/ML and deep learning in the insurance industry. The filters used were the names of the authors, citations received by their published work, the country to which they belong,

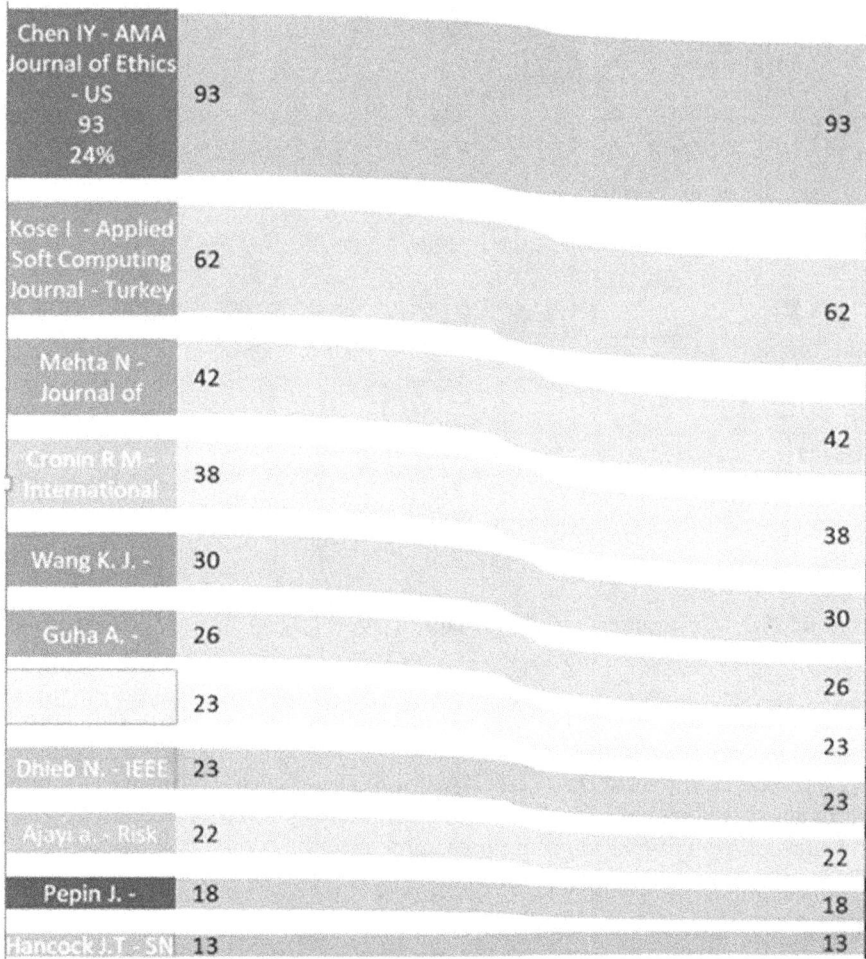

FIGURE 33.6 Sankey diagram showing top ten researchers contributing to the machine learning and deep learning in the insurance industry.

and the name of the journal where the research has been published. The size of each node represents total publication with attention from other researchers in the same field. Chen I. Y. from United States has significantly contributed to the research in the said field, with a contribution of 93 citations; followed by Kose I. from Turkey, with 62 citations; and then Mehta N. India, with 42 citations.

33.4.6 CO-OCCURRENCE OF ALL KEYWORDS

Figure 33.7 highlights the information regarding co-occurring keywords, after retrieving the network diagram from the input of all keywords, 4 clusters with 51

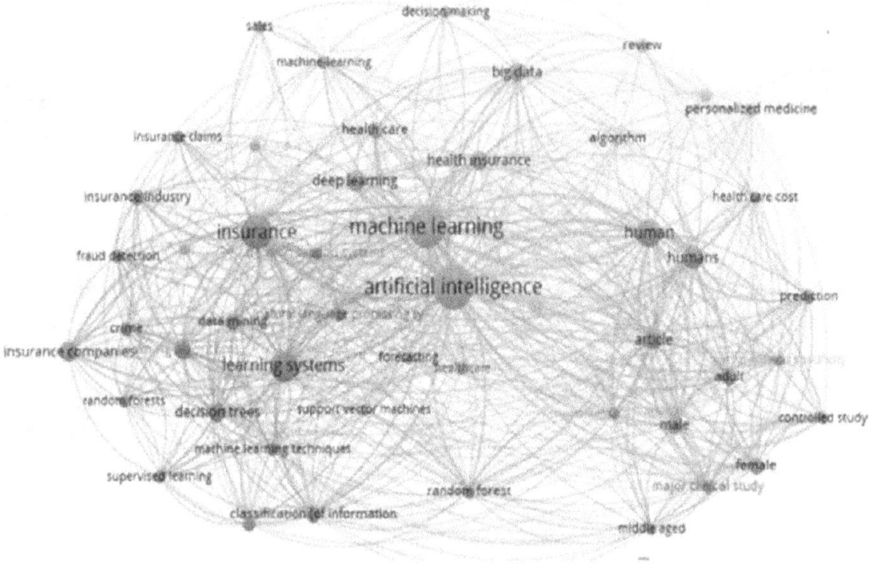

FIGURE 33.7 Co-occurrence of all keywords.

items and total link strength of 2,443 were obtained. Each colour in the network represents the keywords that are interconnected when the researchers are performing their research in their expert area. "Artificial Intelligence" appeared as the most occurring keyword with 75 occurrences, followed by "Machine Learning" and "Insurance," with 65 and 44 occurrences, respectively.

33.4.7   CO-OCCURRENCE OF INDEX WORDS

The co-occurrence of index keywords is represented in Figure 33.8. The clusters that appeared in the network diagram were retrieved with co-occurrence on the basis of full count of indexed keywords. The minimum number of occurrences is restricted to 5 keywords, thus meeting the threshold of 47 out of 1,072 keywords. The network diagram appeared with 3 clusters with 47 items, and total link strength of 2020. "Artificial Intelligence" appeared as the most occurring indexed keyword with 62 occurrences, while "Controlled Study" appeared as the least occurring indexed keyword with 6 occurrences. The size of different keywords depends upon the number of occurrences in the sample.

33.4.8   BIBLIOGRAPHIC COUPLING OF DOCUMENTS

The network diagram of bibliographic coupling of documents is represented in Figure 33.9. The threshold of 50 out of 123 documents was derived from when a minimum number of 4 cited documents. Research documents authored by Wu et al. (2019) gained 93 citations, followed by Kose et al. (2015) with 62 documents and

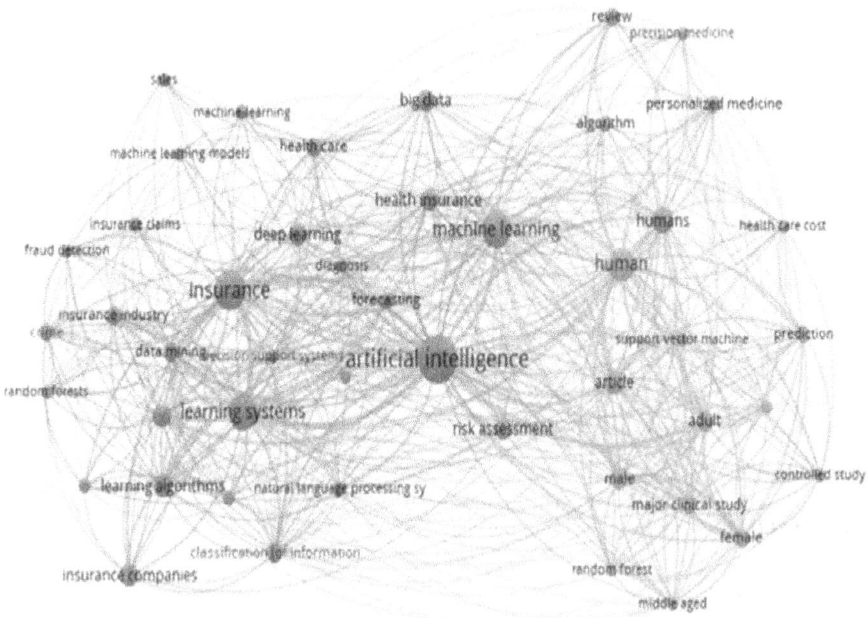

FIGURE 33.8 Co-occurrence of index keywords.

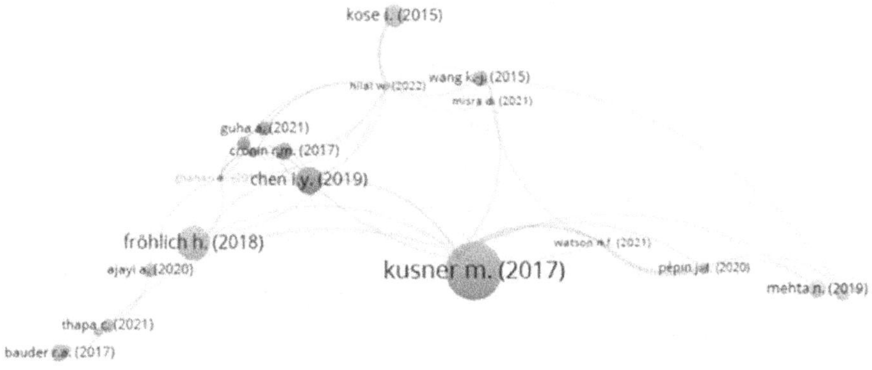

FIGURE 33.9 Bibliographic coupling of documents.

Mehta (2019) received 42 documents till date. Chen I. Y.'s research work is related to AI in the medical sector and his research publication is considered the most novel and most considered work in his research domain.

33.4.9 BIBLIOGRAPHIC COUPLING OF COUNTRIES

The network document is derived with a full counting method for the bibliographic coupling of countries and the results are reflected in Figure 33.10. The

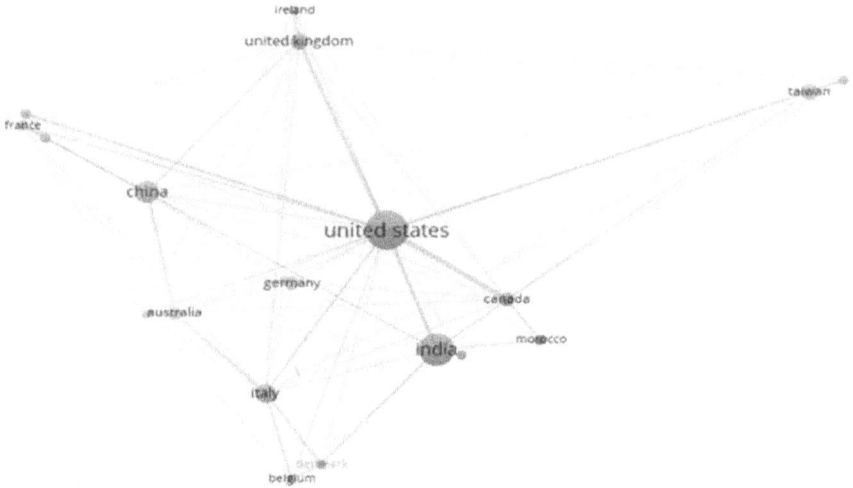

FIGURE 33.10   Bibliographic coupling of countries.

network was restricted to a minimum of 2 documents per country. United States emerged as the country with 31 published documents and citations of 1,038; followed by India with 21 documents with citations of 141 and the contribution of authors from China as the third largest country with 10 documents and 67 citations.

33.4.10   CO-CITATION OF CITED SOURCES

The co-citation of cited sources in Figure 33.11 identifies 14 clusters with 384 items with a link score of 4,785 and total link strength of 17,273. *PLOS One* is the source that has received maximum citations for the authors who have contributed with their research work in their data repository, followed by *Expert Systems with Applications* and *Advances in Neural Information Processing Systems* as the second and third most cited source of citations.

33.4.11   CO-CITATION OF CITED REFERENCES

Figure 33.12 reveals the information of the co-citation of cited references with a minimum of 2 citations in the research domain. Out of 4,184 cited references, 26 met the threshold. The data was represented in 13 items, 3 clusters with total link strength of 53. There are several authors whose research works are providing literature and thoughtful support for the researchers of the related domains.

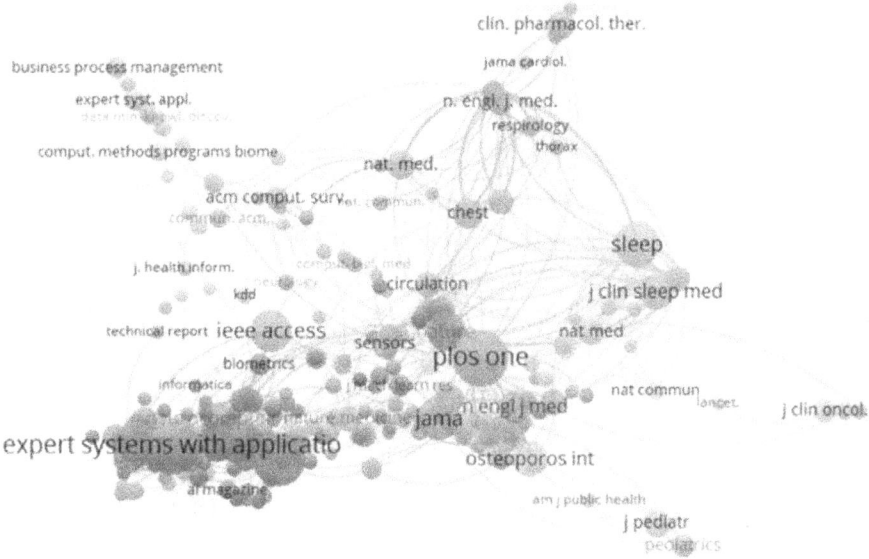

FIGURE 33.11 Co-citation of cited sources.

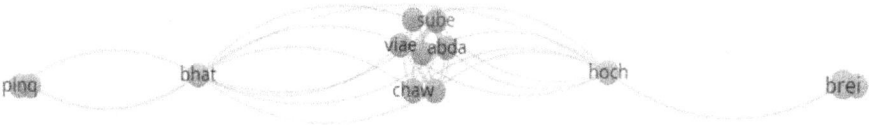

FIGURE 33.12 Co-citation of references.

33.5 CONCLUSION

The insurance industry experience lot of tumultuousness due to the emergence of technology and introduction of AI has unlocked several avenues, thus leading to better forecast the future of insurance industry, mitigating the risks involved, sensing frauds, enhancing claim management, lead management, insurance underwriting, virtual assistants, and much more. Although the concept of AI and ML is still at very nascent stage when it comes to insurance industry, but its significant contributions cannot be denied. In the present study, we delved deep into the published literature and identified that *International Journal of Advanced Computer Science & Applications* and *Security & Communication Networks* are among the top journals contributing to the role and contribution of AI, ML, and deep learning in insurance sector. Intensive research is being performed by authors in the fields of machine learning, insurance, healthcare, AI technology, algorithms, neural networks, fraud detection, big data, and related areas. Research publications authored by Wu et al. (2019), Kose et al. (2015), and Mehta et al. (2019) in the US, Turkey, and India, respectively, have gained most of the citations in the said work, as represented by Sankey diagram. The study will be helpful for the business avenues who are framing the strategies for insurance industries at much faster pace, with lower economical cost, and with better accuracy. It will further allow companies to predict needs, behaviours, and demands of customers, thus devising customized solutions for their needs.

REFERENCES

Bassi, P., & Kaur, J. (2022). Comparative predictive performance of BPNN and SVM for Indian insurance companies. In *Big Data Analytics in the Insurance Market*, Emerald Publishing Limited, Bingley, 21–30. https://doi.org/10.1108/978-1-80262-637-720221002.

Benedek, B., Ciumas, C., & Nagy, B.Z. (2022). Automobile insurance fraud detection in the age of big data—a systematic and comprehensive literature review. *Journal of Financial Regulation and Compliance*, *30*(4), 503–523. https://doi.org/10.1108/JFRC-11-2021-0102

Bureau MFB. (2019). Financial situation in our country: FY 2020 budget proposal. In *Ministry of Finance*. www.mof.go.jp/budget/budger_workflow/budget/fy2020/seifuan2019/04. pdf. Accessed 20 August, 2020.

Chen, Y., Yi, C., Xie, X., Hou, M., & Cheng, Y. (2020). Solution of ruin probability for continuous time model based on block trigonometric exponential neural network. *Symmetry*, *12*(6), 876. https://doi.org/10.3390/sym12060876

Davenport, T., & Kalakota, R. (2019). The potential for artificial intelligence in healthcare. *Future Healthcare Journal*, *6*(2), 94–98. https://doi.org/10.7861/futurehosp.6-2-94

Gerke, S., & Rezaeikhonakdar, D. (2022). Privacy aspects of direct-to-consumer artificial intelligence/machine learning health apps. *Intelligence-Based Medicine*, *6*. https://doi.org/10.1016/j.ibmed.2022.100061

Haque, Md. E., & Tozal, M.E. (2022). Negative insurance claim generation using distance pooling on positive diagnosis-procedure bipartite graphs. *Journal of Data and Information Quality*, *14*(3), 1–26. https://doi.org/10.1145/3531347

Kose, I., Gokturk, M., & Kilic, K. (2015). An interactive machine-learning-based electronic fraud and abuse detection system in healthcare insurance. *Applied Soft Computing*, *36*, 283–299. https://doi.org/10.1016/j.asoc.2015.07.018

Lee, D. S., & Ezekowitz, J. A. (2014). Risk stratification in acute heart failure. *Canadian Journal of Cardiology*, *30*(3), 312–319. https://doi.org/10.1016/j.cjca.2014.01.001

Marsh, A. M., Nguyen, A. H., Parker, T. M., & Agrawal, D. K. (2017). Clinical use of high mobility group box 1 and the receptor for advanced glycation end products in the prognosis and risk stratification of heart failure: A literature review. *Canadian Journal of Physiology and Pharmacology*, *95*(3), 253–259. https://doi.org/10.1139/cjpp-2016-0299

Mehta, N., Pandit, A., & Shukla, S. (2019). Transforming healthcare with big data analytics and artificial intelligence: A systematic mapping study. *Journal of Biomedical Informatics*, *100*, 103311. https://doi.org/10.1016/j.jbi.2019.103311

Misra, D., Avula, V., Wolk, D. M., Farag, H. A., Li, J., Mehta, Y. B., . . . & Abedi, V. (2021). Early detection of septic shock onset using interpretable machine learners. *Journal of Clinical Medicine*, *10*(2), 301. https://doi.org/10.3390/jcm10020301.

Nudge, B. S. (2018). *Theory Explored to Boost Medication Adherence*, American Medical Association, Chicago. www.ama-assn.org/delivering-care/patient-support-advocacy/nudge-theory-explored-boost-medication-adherence.

Rathi, H.K., Dawande, P., Kane, S., Gaikwad, A., & Narlawar, M.S. (2022). Artificial intelligence, machine learning and deep learning in health care. *1st International Conference on Technologies for Smart Green Connected Society, 107*(1), 15981–15987.

Takura, T., Hirano Goto, K., & Honda, A. (2021). Development of a predictive model for integrated medical and long-term care resource consumption based on health behaviour: Application of healthcare big data of patients with circulatory diseases. *BMC Medicine, 19*(15). https://doi.org/10.1186/s12916-020-01874-6

Wu, Q., Zhang, D., Zhao, Q., Liu, L., He, Z., Chen, Y., Huang, H., Hou, Y., Yang, X., & Gu, J. (2019). Effects of transitional health management on adherence and prognosis in elderly patients with acute myocardial infarction in percutaneous coronary intervention: A cluster randomized controlled trial. *PLoS ONE, 14*(5), e0217535. https://doi.org/10.1371/journal.pone.0217535

Xu, W. Y., Song, C., Li, Y., & Retchin, S. M. (2019). Cost-sharing disparities for out-of-network care for adults with behavioral health conditions. *JAMA Network Open, 2*(11). https://doi.org/10.1001/jamanetworkopen.2019.14554

34 # State of the Art of Quantum Teleportation and Quantum Key Distribution Methods
A Survey

Jatin Arora, Saravjeet Singh

34.1 INTRODUCTION

Due to high processing power, quantum computing is having the potential to transform the computation process involved in different industries like transportation, security, and drug development. Earlier quantum computing was the theoretical concept, but now with the advancement of hardware and software, many industries are providing cloud-based solutions for quantum processing. Companies like Microsoft, Amazon, Google, Pasqal, and IBM provide cloud-based access to quantum processors and work to provide hardware to users. In comparison to classical computation, quantum computing provides effective and different solutions to computational problems. According to experimental results, a solution using quantum computing is scalable and fast. According to theoretical research, the quantum computing-based technique can solve classical problems within hours that requires billion of hours to be solved using traditional approaches. These quantum computers are hypothetical machines that would exploit the principles of quantum mechanics, which govern the subatomic world, in order to solve calculations are critical problems at a much faster rate than a normal classical computer.

A quantum computer is used to perform operations and calculations based on the state of an object (0s or 1s). Quantum computers are very different from the classical ones. According to function prospective, quantum computers have gates, registers, buses, storage spaces, and so on, but according to physical implementation, these components are based on the concept of the quantum. There are a variety of components which help to process the information using quantum computing. This information is made up of bits, and in terms of quantum computing, these are called qubits. These qubits are basically those state with undefined properties of the object which obey the laws of physics like, polarization, superposition, etc. It is basically a quantum mechanical system that under some suitable circumstances can be treated as having only two quantum levels, and once you have it, you can use it to encode

DOI: 10.1201/9781003367161-34

quantum information in a similar way as it would have been done for a normal classical computer. Some of the basic terms and definitions related to quantum teleportation and quantum key are as follows:

1. **Qubits:** Quantum computing is defined as the basic use of quantum physics to perform complex computations. These quantum computers are based on qubits, in quantum superposition of the 0 and 1 states. These bits can be 0 and 1 at the same time, and at some point, in time these bits are measured to figure out the plausible answers, which can be done by amplitudes.

2. **Amplitudes:** An amplitude is just like the square root of a probability, and the same number can have more than one square root; for example, 2 and −2 both have same square root (4). Corresponding to this, we can either have a positive or negative amplitude, and a quantum computer would exploit this phenomenon of interference between positive and negative amplitudes on a massive scale. Similarly, the goal would be to choreograph so that all the different paths leading to a wrong answer would be out of phase; some would have positive amplitude, and some would have negative amplitude. Such amplitudes would cancel out each other, whereas the path leading to a right answer would be in a same state (same sign).

3. **Superposition:** Superposition is one of the fundamental principles of quantum mechanics. Quantum superposition is a system that has two different states that can define it, and it is possible for it to exist in both. For example, electrons possess quantum feature called spin; each electron, until it is measured, will have a finite chance of being in either state. If one could align them in the right way, it is possible to effectively harness the power of that superposition [6].

4. **Quantum entanglement:** One of the other counterintuitive phenomena in quantum physics is entanglement. A pair or group of quantum particles is entangled when the quantum state of each particle cannot be described independently of the quantum state of the other particles. Quantum entanglement is a quantum mechanical phenomenon in which the quantum states of two or more objects have to be described with reference to each other, even though the individual objects may be spatially separated [7].

5. **Quantum annealing:** Such type of quantum computer is considered to be the best for solving critical problems mainly industry problems like Airbus. Furthermore, quantum annealing is the least powerful amongst all.

6. **Quantum simulations:** Quantum simulations mainly explore those problems which are specific to quantum physics and are beyond the capability and potential of classical computers.

7. **Universal quantum computing:** Universal quantum computers are the most powerful quantum computers at present. Such type of quantum computers usually uses 100,000 qubits. The main logic behind the universal quantum computers is that one can direct the machine at any complex computation; such complex computations are processed at a very fast rate.

34.2 QUANTUM CRYPTOGRAPHY

Quantum cryptography is combination of quantum mechanics and cryptography and ensure the security of data while transmission using quantum physics. Elements of quantum cryptography is shown in Figure 34.1. Quantum cryptography provides security to data using the concept of both bits and qubits. According to current state of the art, Quantum cryptography is largely used for security of classical bits instead of qubits. Quantum cryptography has three components: quantum entanglement, quantum teleportation, and quantum measurement. Quantum entanglement defines the connection between two microscopic particles from same source. If two particles are entangling, then changes in one particle will make the change in another particle as well. Quantum measurement is processed to extract the information from quantum elements. Quantum teleportation is a process to transfer information between two quantum elements.

34.2.1 QUANTUM KEY DISTRIBUTION (QKD)

QKD is method to generate and share secure key for the encryption and decryption using the concept of quantum physics. Random key using quantum data is created and send to receiver as photon through quantum link, it uses quantum engine for key creation and fibre optics for the quantum link. Key is completely random and if any intruder tries to access or modify the key, photon channel will create alert. A network of trusted nodes is used to create a channel of long transmission. Figure 34.2 represents the QKD scenario of secure transmission of information between sender and receiver.

Quantum channel can be the fibre optic cable or free space link and public channel is link provides the details of the quantum but transmission. The key distribution in quantum scenario can be categorized into two parts: continuous variable quantum key distribution protocol (CV-QKD) and discreate variable quantum key distribution protocol (DV-QKD). In first DV-QKD (i.e. BB84), single photo polarization was used for the encoding. For first CV-QKD, Gaussian modulation technique with photon packets were used for the encoding [1].

Brief information of basic CV-QKD and DV-QKD protocols is shown in Table 34.1. DV-QKD uses polarization states act as base for the information transmission. Discrete variable quantum key distribution was the first QKD protocol. In DV-QKD system, rectilinear and diagonal base of single-photon polarization is used to send

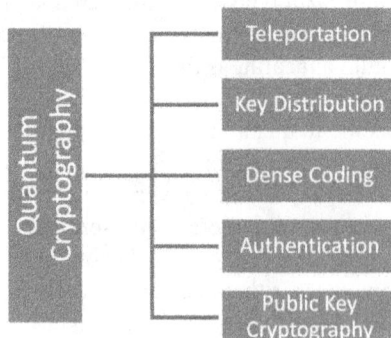

FIGURE 34.1 Elements of quantum cryptography

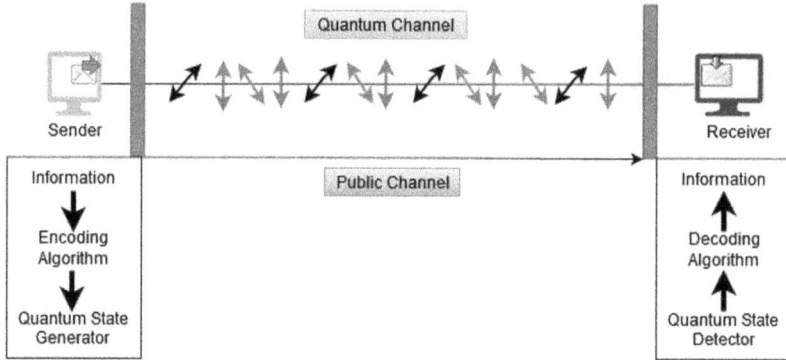

FIGURE 34.2 Principle of quantum key distribution process.

TABLE 34.1

The State of the Art of Quantum Key Distribution Studies

Reference	Base Protocol	No. of States	Remarks
[7]	Semi quantum key distribution	4	Enhanced security, robustness against collective, independent, and dependent channel attack.
[8]	Phase-matching quantum key distribution	3	Keys were generated using flip-flops. Good for long transmission.
[9]	Ekert quantum key distribution protocol	4	Semiconductor quantum dots were used for the key generation. Provides good security against eavesdrop attacks.
[10]	SARG04	6	Secret keys were generated from source emitting multi-photons. Robust against PNS attack.
[11]	BB84- and CHSH-based	2	Conditional von Neumann entropy is used for the key generation. Applicable for different device. Robust to network noise.
[12]	BB84	4	Based on Heisenberg's uncertainty. Simple approach. Provides effective use of resources. Effective to monitor and detect the eavesdropping.
[13]	DV-QKD and CV-QKD	4	Based on both DV-QKD and CV-QKD. Suitable for sharing of different phase references. Post-measurement decoding will be there. Good for high-speed key transmission with in distance of 15 km.
[14]	BB84	4	Increase key rate using decoy pulses. High security against eavesdropping for long distance. Can be used for both line-of-sight and non-line-of-sight.

the information to the receiver. Receiver uses both bases to extract the information. The non-orthogonal states provide security against eavesdropping. In CV-QKD the Gaussian modulation is used to provide the continuous states to encoded data. Homodyne is used to decode the data at receiver end. Table 34.2 provides information about basic quantum key distribution protocols.

TABLE 34.2

Basic Quantum Key Distribution Protocols

Protocol	No of States	Attack Robustness				Polarization	Security	Efficiency	Principals
		DoS	Beam-Splitter	PNS	Middle-Man				
BB84 Protocol [2]	4	N	N	N	N	Orthogonal	Good long distance	Low	Heisenberg
BB92 [2]	2	N	N	N	Y	Non-orthogonal	Average	Average	Heisenberg
SARG04 [3]	4	N	Y	N	Y	Orthogonal	Average	Average	Heisenberg
KMB09 [2]	2	N	Y	Y	Y	Arbitrary	Average	Low	Heisenberg
S09 [4]	No fixed	N	NA	N	Y	Bit-flip phase-Flip	Good small distance	Good	Public private key
S13 [5]	4	N	NA	NA	NA	2-Orthogonal	Average	Average	Heisenberg
EPR [6]	4	Y	NA	NA	Y	Rectilinear and circular	Good	Good	Heisenberg

34.2.2   Challenges of Quantum Key Distribution

1. Key Size
2. Transmission over long distance
3. Cost-effective method to share keys

34.2.3   Teleportation

Quantum teleportation is a method of quantum data exchange between two parties. In involves sharing of quantum information between sender and receiver using quantum channel and quantum technology. In quantum teleportation, qubits were instead transferred using the transmission medium. A successful experiment of quantum teleportation by demonstrated using the qubits of photons. The data was teleported using the optical fibre over 44 km long with 90% fidelity [15]. Quantum teleportation uses quantum entanglement to transfer quantum states from one place to another place using action at a distance rule. The action at a distance rule says that physical object can be affected, changed, and moved by another object with a physical touch. These affected objects are connected as entangled pair and can share information [16]. Sample scenario of quantum teleportation is shown in Figure 34.3. In this figure, three photons are connected using quantum entanglement. State |Φ> is transmitted using teleportation.

Quantum internet is considered to be a secure, high-speed information-processing platform. Quantum teleportation is predicted as technique to transfer the unknown quantum states over long distance. External qubit, internal qubit, and squeezed state can be used for the teleportation. According to a recent study, long-distance

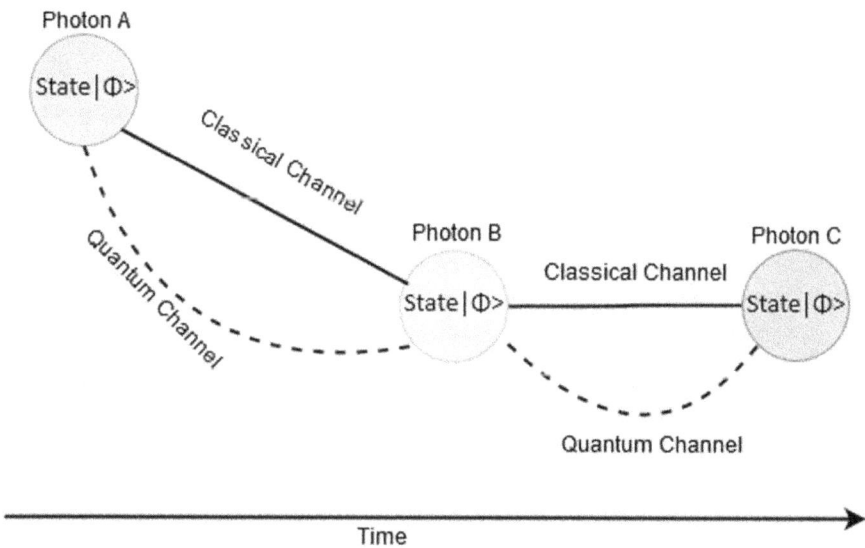

FIGURE 34.3   Sample Scenario of Quantum Teleportation

transmission is major concern, and many experiments were conducted to estimate the error rate and data fidelity over long distance. Few of the recent studies based on quantum teleportation are briefly presented in Table 34.3.

34.2.4 Challenges of Quantum Key Distribution

1. Noise while data transmission
2. Distribution of qubit keys over long distance
3. Ability to generate more photons
4. Data fidelity with less qubits and over long distance

TABLE 34.3
The State of the Art of Quantum Teleportation

Reference	Basic Component	Advantages	Disadvantages
[17]	Photons of solid-state emitter were used as teleportation element.	Overcomes the problem of input state dependency on data transmission.	Practical implementation is a challenge.
[18]	Photons with bi- and tripartite entangled channels were used for the teleportation.	Provides 83% data fidelity. Tripartite entanglement is not mandatary for teleportation.	Requirement of sufficient channel for transportation.
[19]	Path-polarization conversion technique was used for chip-to-chip quantum teleportation.	Single qubit used for the quantum teleportation. Low cost and high-speed transmission.	Exponential complexity for the calculations. Coincidence measurements from single source are difficult to detect.
[20]	Two to three qubit states and 5–7 cluster qubit states were used to control the teleportation.	One hundred percent efficiency to recover the data. Reduced cost and effective against Charlie's attacks.	Data fidelity is 40%.
[21]	Six qubit cluster states were used for the teleportation. GHz-state, bell-state, and von Neumann measurement were used for the bi-directional transmission	Sender and receiver can transmit at same time. High transmission rate and high intrinsic efficiency. High security against spoofing.	High resource requirement and high computation complexity.
[22]	Five qubit entanglements with mutually unbiased particles were used. Decoy particles concept with open teleportation was opted for the transmission.	Satisfy von Neumann, Bell measurement, and single-particle measurement. Provides signature integrity and high stability.	Low quantum efficient. Requirement of high number of bits.
[23]	Single photon detector, assisted clock synchronization, and frequency-controlled polarization-based entanglement were used.	Achieved good state fidelity. Achieved teleportation without feed-forward process. Also suitable for satellite-based teleportation	Low fidelity due to high altitude

34.3 CONCLUSION

This chapter provides brief overview of quantum cryptography. A survey of quantum key distribution protocols and their used in recent research studies is provided in this chapter. Based on the survey, benefits, state requirements, and security concerns associated with basic key distribution protocols are highlighted in this chapter. This chapter also provides the details of quantum teleportation techniques, and recent state of art of teleportation techniques. Use of single photon, cluster photon, solid-state devices, and path polarization for teleportation is recent study area for the state transmission and data entanglement. Impact of number of qubits used for the data transmission is also analysed by the different researcher. Quantum teleportation was a theoretical concept, but now many researchers are conducting different experiments to identify the use of qubit and photon to identify the data transmission distance. Based on the recent experiment, quantum teleportation is possible for secure data transmission over 143 km distance with good fidelity. Few studies also analysed the impact of number of qubits on data fidelity. In addition to the recent state of the art, this chapter also lists the challenges associated with quantum key distribution and teleportation methods.

REFERENCES

[1] A. Nurhadi, "Quantum key distribution (QKD) protocols: A survey," *ieeexplore.ieee. org*, 2018, Accessed: Jan. 18, 2022 [Online]. Available: https://ieeexplore.ieee.org/abstract/document/8527822/.

[2] P. Patil and R. Boda, "Analysis of cryptography: Classical verses quantum cryptography," *academia.edu*, 2016, Accessed: Jan. 18, 2022 [Online]. Available: www.academia.edu/download/54581978/IRJET-V3I5281.pdf.

[3] V. Scarani, A. Acín, G. Ribordy, and N. Gisin, "Quantum cryptography protocols robust against photon number splitting attacks for weak laser pulse implementations," *Phys. Rev. Lett.*, vol. 92, no. 5, p. 4, 2004, doi:10.1103/PHYSREVLETT.92.057901.

[4] E. Esteban and H. Serna, "Quantum key distribution protocol with private-public key," Aug. 2009, Accessed: Jan. 18, 2022 [Online]. Available: http://arxiv.org/abs/0908.2146.

[5] Eduin H. Serna, "Quantum key distribution from a random seed," *arxiv.org*, 2013, Accessed: Jan. 18, 2022 [Online]. Available: https://arxiv.org/abs/1311.1582.

[6] A. Kumar, P. Dadheech, V. Singh, L. Raja, and R. C. Poonia, "An enhanced quantum key distribution protocol for security authentication," *J. Discret. Math. Sci. Cryptogr.*, vol. 22, no. 4, pp. 499–507, May 2019, doi:10.1080/09720529.2019.1637154.

[7] H. Hajji and M. El Baz, "Qutrit-based semi-quantum key distribution protocol," *Quantum Inf. Process.*, vol. 20, no. 1, Jan. 2021, doi:10.1007/S11128-020-02927-8.

[8] Z. Li, D. Han, C. Liu, and F. Gao, "The phase matching quantum key distribution protocol with 3-state systems," *Quantum Inf. Process*, vol. 20, no. 1, pp. 1–9, Jan. 2021, doi:10.1007/S11128-020-02942-9.

[9] F. B. Basset et al., "Quantum key distribution with entangled photons generated on demand by a quantum dot," *Sci. Adv.*, vol. 7, no. 12, p. 6379, Mar. 2021, doi:10.1126/SCIADV.ABE6379/SUPPL_FILE/ABE6379_SM.PDF.

[10] C. Sekga and M. Mafu, "Security of quantum-key-distribution protocol by using the post-selection technique," *Phys. Open*, vol. 7, p. 100075, May 2021, doi:10.1016/J.PHYSO.2021.100075.

[11] E. Woodhead, A. Acín, and S. Pironio, "Device-independent quantum key distribution with asymmetric CHSH inequalities," *Quantum*, vol. 5, p. 443, Apr. 2021, doi:10.22331/q-2021-04-26-443.

[12] B. Wang, B. F. Zhang, F. C. Zou, and Y. Xia, "A kind of improved quantum key distribution scheme," *Optik (Stuttg).*, vol. 235, p. 166628, Jun. 2021, doi:10.1016/J.IJLEO.2021.166628.

[13] I. W. Primaatmaja, C. C. Liang, G. Zhang, J. Yan Haw, C. Wang, and C. C.-W. Lim, "Discrete-variable quantum key distribution with homodyne detection," doi:10.22331/q-2022-01-03-613.

[14] A. Biswas et al., "Quantum key distribution with multiphoton pulses: An advantage," *Opt. Contin.*, vol. 1, no. 1, pp. 68–79, Jan. 2022, doi:10.1364/OPTCON.445727.

[15] Philip Stevens, "NASA scientists achieve long-distance 'quantum teleportation'," Accessed: Jan. 19, 2022. Available: www.designboom.com/technology/nasa-long-distance-quantum-teleportation-12-22-2020/.

[16] Ahmed Banafa, "Quantum teleportation: Facts and myths | OpenMind," Accessed: Jan. 19, 2022. Available: www.bbvaopenmind.com/en/technology/digital-world/quantum-teleportation-facts-and-myths/.

[17] M. Reindl et al., "All-photonic quantum teleportation using on-demand solid-state quantum emitters," 2018, Accessed: Jan. 19, 2022 [Online]. Available: www.science.org.

[18] A. Barasiński, A. Černoch, K. Lemr, "Demonstration of controlled quantum teleportation for discrete variables on linear optical devices," *APS*, 2019, Accessed: Jan. 19, 2022 [Online]. Available: https://journals.aps.org/prl/abstract/10.1103/PhysRevLett.122.170501.

[19] D. Llewellyn et al., "Chip-to-chip quantum teleportation and multi-photon entanglement in silicon," *nature.com*, vol. 3, no. 10, 2020, Accessed: Jan. 19, 2022 [Online]. Available: www.nature.com/articles/s41567-019-0727-x.

[20] A. Kumar, S. Haddadi, M. Pourkarimi, B. K. Behera, and P. K. Panigrahi, "Experimental realization of controlled quantum teleportation of arbitrary qubit states via cluster states," *nature.com*, Accessed: Jan. 19, 2022 [Online]. Available: www.nature.com/articles/s41598-020-70446-8.

[21] R. Zhou, R. Xu, and H. Lan, "Bidirectional quantum teleportation by using six-qubit cluster state," *ieeexplore.ieee.org*, 2019, Accessed: Jan. 19, 2022 [Online]. Available: https://ieeexplore.ieee.org/abstract/document/8653915/.

[22] D. Lu, Z. Li, J. Yu, and Z. Han, "A verifiable arbitrated quantum signature scheme based on controlled quantum teleportation," *Entropy*, vol. 24, no. 1, p. 111, Jan. 2022, doi:10.3390/E24010111.

[23] X.-S. Ma et al., "Quantum teleportation over 143 kilometres using active feed-forward," *Nature*, vol. 489, 2012, doi:10.1038/nature11472.

35 An Analysis of Dimensionality Reduction Techniques to Predict the Effective Model for 5G Dataset

Sangeetha Annam, Mithillesh Kumar P.

35.1 INTRODUCTION

The newest mobile network technology, known as fifth-generation (5G), promises extended bandwidth of 10 Gbps, reduced latency rate of between 1 and 2 milliseconds, and improved reliability leading to an enhanced user experience. It is intended to expand upon and improve the existing services and application scenarios, such as internet of things (IoT), linked cars, smart matrix, and multimodal networks. The 5G mobile networks claim to enable extremely fast data rates, extremely low latency, large network capacity, and dependability. 5G will also offer these features on a worldwide scale. These built-in characteristics of the technology stimulate the development of novel business cases and specific applications with the aid of supporting technologies [1]. In most nations throughout the world, fourth-generation (4G) mobile networks are replaced by fifth-generation (5G) networks. While discussing about the current developments in wireless and networking technologies, software-defined networking and virtualization are the next generation wireless network technology, which is the ongoing process. A key advantage for the massive, linked items with respect to IoT is that 5G technologies are distinguished from 4G for faster retrieval speed of data with 10 gigabits/second and with additional capacity and extremely reduced latency. Additionally, because 5G ecosystem is internet protocol based, it is subject to every specific susceptibility. These findings make it clear that among the key elements for the successful distribution of 5G networks is its high degree of data security and data privacy [2].

In addition to being highly needed by location-based services, accurate and real-time positioning might be helpful for radio-resource in 5G networks, being developed to attain considerable performance increase over currently used cellular networks. Massive MIMO, mm waves, ultra-dense networks, and device-to-device (D2D) communications are just a few of the many new technologies being included

DOI: 10.1201/9781003367161-35

into 5G networks to help boost communication performance and location accuracy [3]. The important aspects of 5G are high speed, low latency rate, and higher bandwidth. Limitations include proximity, security, and limited coverage. This chapter's primary contributions are as follows:

- To predict throughput parameters
- To reduce the space and time complexity, using different dimensionality reduction techniques
- To apply different machine-learning (ML) and deep-learning (DL) models for estimation

The rest of this chapter is structured as follows. Section 2 showcases the previous literatures based on the surveys that are currently available for 5G cellular networks with different models. The dimensionality reduction techniques, ML and DL models, and performance metrics are elaborated in Section 3. Methodology is proposed in Section 4. Dataset attributes are presented in Section 5. The findings, results, and analysis are discussed in Section 6. At the end, the conclusion is portrayed in Section 7.

35.2   LITERATURE REVIEW

In this section, the authors describe the benefits of 5G networks, highlight the most relevant cellular positioning technologies, and explore how ML and DL technologies may be used to increase their accuracy. The application of ML in many aspects of wireless networks has been suggested in many studies in the literature [3]. System modelling is also possible with the use of ML techniques [4]. A structural overview of distributed ML is presented in this chapter [5], along with a thorough examination of the several distributed ML algorithms and their variations. Although the estimation accuracy of the various models will not change, we can anticipate a shift in favour of higher quality of estimation (QoE) ratings in the gathered ground-truth data collection that concentrate on the QoE parameter that shows that it can be performed in 5G devices with efficiency and dependability using ML [6]. We create a realistic dataset for testing ML-based techniques using our experimental 5G OpenAirInterface (OAI) prototype, along with two baseline solutions. According to simulation findings, utilizing regression models for classification and prediction, in conjunction with an adaptive-resource management based on data retrieval language (DRL), outperforms competing strategies relying on prediction accuracy, resource smoothing, system usage, and network throughput [7].

An investigational framework to forecast channel quality indicator (CQI) utilizing signal-to-noise ratio (SNR) for various environmental circumstances and, separately, use experiment (UE) properties such as speed and delay profile. The experimental results outperform existing methods in terms of prediction error and CQI to SNR mapping [8]. With virtualized 5G network, the author discussed an integrated methodology for cognitive network and slice management [9]. An automated, adaptable, and scalable network defence system is suggested by this study because of software defined security (SDS). To create a CNN with neural architecture search (NAS), SDS will take advantage of recent developments in ML. An overall defence for this 5G network may be made by integrating SDS with an intrusion detection system. This hypothesis has been put to the test by gathering both regular and unusual network traffic from a simulated environment and using a CNN to assess them. The model correctly recognized traffic with a full accuracy rate and with anomalous traffic, a little low of 96.4% detection rate, which indicates that findings from this approach are promising [10]. This study analyses and investigates the use of several ML models for intrusion detection. To model and evaluate the efficacy of several ML algorithms in identifying cyberattacks, the CSE-CIC IDS2018 dataset was employed. The dataset subset that was chosen had an even distribution of flows for both regular traffic and assaults; therefore, binary classification was used on it. The dataset was cleaned and pre-processed before feature selection was used to find 20 pertinent features [11]. Every model, according to the literature study, corresponds to a certain issue statement. The aim of this research is to choose the ML and DL models that perform the best. Dimensionality reduction techniques are used to examine the impact from the dataset's reduced number of features.

35.3   MODELS EMPLOYED AND PERFORMANCE METRICS

In this section, different dimensionality reduction (DR) techniques and ML and DL models utilized in this study are described in brief. Additionally, a brief explanation of the regression measures required for performance comparison is provided.

35.3.1   DIMENSIONALITY REDUCTION (DR) TECHNIQUES

DR strategies are helpful for extracting the meaningful features from the dataset. In simple terms, it is the removal of features from a dataset that occupy a massive space in the dataset. When the number of features is large in number, then it is difficult for analysis. It becomes more difficult to visualize the training samples in the dataset. In addition to this, it also avoids overfitting of the dataset which improves the accuracy. The various DR techniques are shown in Figure 35.1.

Table 35.1 briefly discusses the linear and non-linear methods of DR with feature extraction and feature selection.

35.3.2   MACHINE LEARNING (ML) AND DEEP LEARNING (DL) MODELS

The application of ML teaches computers to handle the data in an effective manner. At times, even after examining the dataset, it is difficult to evaluate and extrapolate some basic information. In such situations, ML models are explored. The distribution of many datasets has increased the demand for ML which is used in many sectors to retrieve pertinent data. In addition to that, ML is used to get knowledge from the data [13]. Various ML and DL algorithms are discussed in this section. Some of them include extra-trees regressor (ETR), AdaBoost regressor (ABR), support-vector regressor (SVR), linear regressor (LR), ridge regressor (RR), lasso regressor, elastic net regressor (ENR), and multi-layer perceptron (MLP), which are the different models applied to estimate and compare the performance measures.

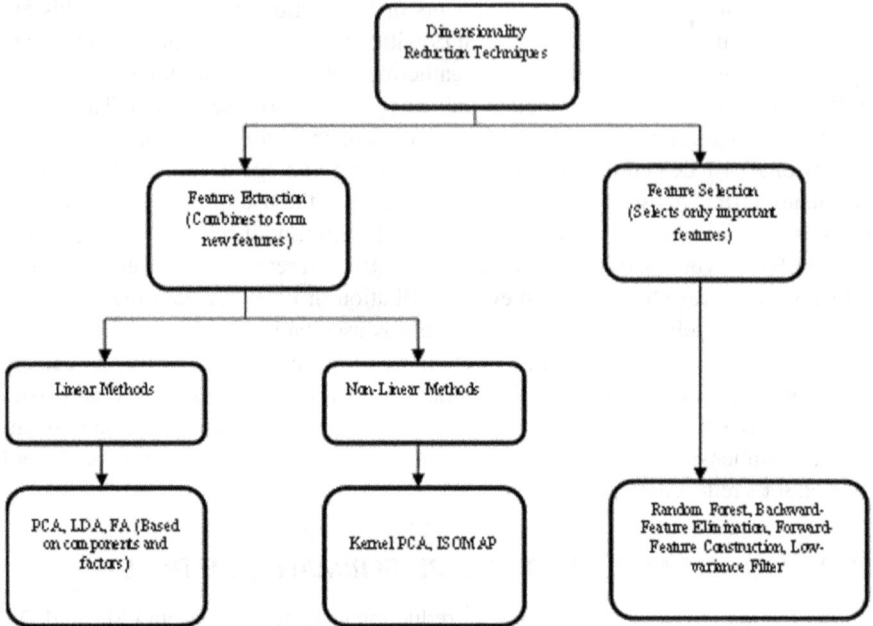

FIGURE 35.1 Dimensionality reduction techniques [12].

TABLE 35.1
Dimensionality Reduction (DR) Techniques with Its Features [12]

DR Techniques	Feature Importance
Principal component analysis (PCA)	With this, the data are directly mapped to a lower-dimensional space in a way that optimizes the variance of the data there.
Linear discriminant analysis (LDA)	With LDA, class separability is maximized in the representation of data.
Factor analysis (FA)	When analysing a dataset of variables, FA is used to determine the underlying "cause" of the factors.
Kernel PCA	Extends PCA using kernel.
ISOMAP	Creates a neighbourhood network based on spectral theory.
Random forest (RF)	Identifies the most important subset of features.
Backward-feature elimination	Determines the least number of features required to achieve that performance of the classifier with the provided ML method; chooses the maximum tolerable error rate.
Forward-feature construction	Features are added in the forward direction.
Low-variance filter	All data columns with variation below the threshold value will be removed.

35.3.3 Performance Metrics

Table 35.2 clearly shows the performance evaluation of the models applied for the dataset. The mean absolute error (MAE), mean squared error (MSE), and R-square (R^2) are calculated according to different models after undergoing different DR techniques.

35.4 METHODOLOGY USED

The methodology proposed for the current study is shown in Figure 35.2.

TABLE 35.2
Performance Evaluation Metrics

Measure	Evaluation of the measure
MSE	Difference between the actual and predicted values
MAE	Impact of the error magnitude
R^2	Representation of the coefficient of determination

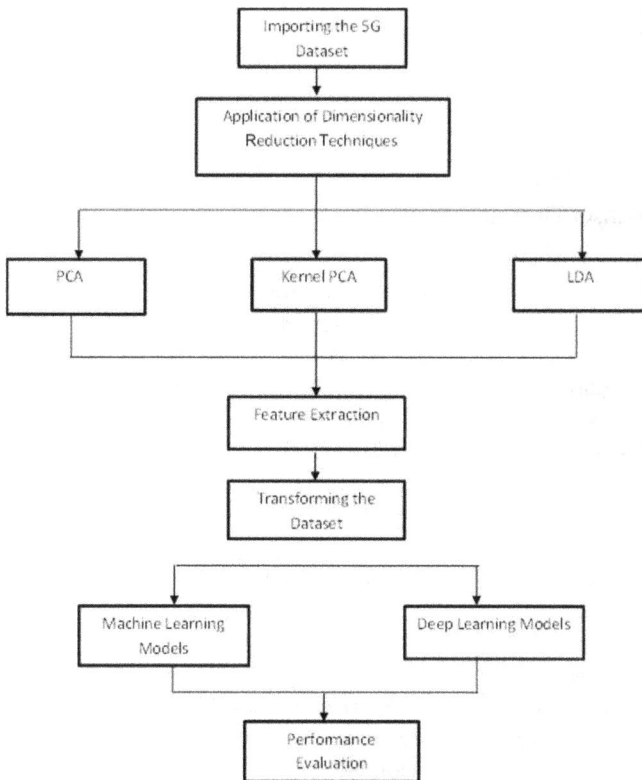

FIGURE 35.2 Proposed methodology for the study.

35.5 DATASET DESCRIPTION

Data for the 5G performance dataset was obtained from the IEEE data port website [14] reported by Android API. With the help of 300 km of walking, 130 km of driving, and 35 TB of data download, 68,118 records of data were obtained. A loop region with a 1,300 m loop length is used to collect the data. The dataset includes the following information shown in Table 35.3.

35.5.1 PRE-PROCESSING OF THE DATASET

The aim parameter in this study is throughput, which is the output rate for such a specific condition at a particular time point. Every other parameter describes the signal characteristics at a certain run and sequence number [15]. For identification, we use the run and sequence numbers. The dataset is first pre-processed to eliminate any discovered null values, which are then replaced with 0. Using pre-processing, the ML and DL algorithms will be able to fully understand the dataset and produce a more accurate model. This might thus reduce inaccuracies and misclassification in the subsequent step. The dataset is then converted to a shape with 68,118 records and converted to eight features where only the regression parameters are concerned. Regression measures such the MAE, MSE, and R^2 values are used to evaluate each model's performance.

TABLE 35.3
Dataset Description

Dataset Features	Description
AbstractSignalStr	Abstract-signal strength
Latitude	Latitude (in degrees)
Longitude	Longitude (in degrees)
movingSpeed	Moving speed of the user experiment
Compass Direction	Bearing (in degrees)
NRstatus	Connection to 5G network status
lte_rsrq	To get reference-signal-received-quality (RSRQ)
lte_rssnr	For reference signal signal-to-noise ratio (RSSNR)
lte_rsrp	For reference signal received power (RSRP) in dBm
nr_ssRsrp	Obtained by parsing the raw string representation of
nr_ssRsrq	`Signal Strength` object
nr_ssSinr	
Throughput	Throughput perceived by the user experiment
mobility_mode	Walking status or driving status
trajectory_direction	Indicates the ground-truth direction, i.e. clockwise or anticlockwise
tower_id	Indicates the tower identifier

35.6   FINDINGS AND ANALYSIS OF THE VARIOUS DIMENSIONALITY REDUCTION TECHNIQUES

The DR techniques, such as PCA [16], kernel PCA, and LDA, are performed for different ML and DL models with the default values. The findings are tabulated in Table 35.4.

The findings from the Table 35.4 are that the top-performing models' performance was unaffected with less information loss and with a higher degree of correlation. This suggests that the DR technique kernel PCA can be used to reduce the complexity of the data.

35.7   CONCLUSION AND FUTURE WORK

In this study, DR techniques coupled with multiple ML and DL models are used to simulate throughput, considered to be one of the important network performance measures. It is nothing but the measurement of the amount of information received

TABLE 35.4
Analysis of the DR Techniques with ML and DL Models

DR Techniques	Models	MAE Score	MSE Score	R-Square Score
PCA	ETR	140.7185	46350.9544	0.7835
	ABR	228.8582	100119.8216	0.5324
	SVR	391.1600	219580.3770	−0.0251
	LMR	261.4472	124199.2382	0.4084
	RR	261.8604	120868.8743	0.4248
	Lasso Regression	261.9432	121145.7204	0.4237
	ENR	260.9441	121342.8296	0.4224
	MLP	405.2318	261546.2201	−0.2451
Kernel PCA	ETR	139.2482	46350.4432	0.8131
	ABR	231.5882	118769.2853	0.4451
	SVR	390.2315	214709.9770	−0.0318
	LMR	262.2231	126123.2382	0.4183
	RR	261.4443	124198.8743	0.4283
	Lasso Regression	267.1267	122134.7412	0.4871
	ENR	260.8521	123512.7512	0.4148
	MLP	408.1015	281642.954	−0.3821
LDA	ETR	163.2403	58961.1725	0.7191
	ABR	260.8838	119955.2138	0.4286
	SVR	242.7974	118693.4531	0.4475
	LMR	262.5329	120441.8832	0.4263
	RR	262.5329	120441.8831	0.4263
	Lasso Regression	262.5261	120445.0068	0.4263
	ENR	268.3888	123411.5564	0.4122
	MLP	442.6204	111326.4384	−0.4697

at the receiver end. The study's conclusion is that the construction of optimized models with fewer space and temporal complexity is possible by using techniques for reducing dimensionality to minimize the number of associated elements involved in forecasting throughput. Models based on additional performance measures and parameters can also be developed as an extension of the current work.

REFERENCES

[1] Farooqui, M. N. I., Arshad, J., & Khan, M. M. (2021). A bibliometric approach to quantitatively assess current research trends in 5G security. *Library Hi Tech*, 39(4), 1097–1120.

[2] Ferrag, Mohamed Amine, et al. "Security for 4G and 5G cellular networks: A survey of existing authentication and privacy-preserving schemes." *Journal of Network and Computer Applications* 101 (2018): 55–82.

[3] Mogyorósi, Ferenc, et al. "Positioning in 5G and 6G networks—A survey." *Sensors* 22.13 (2022): 4757.

[4] Ruiz Sicilia, Juan Carlos, and Maria del Carmen Aguayo-Torres. "Logistic regression for BLER prediction in 5G." (2020), https://hdl.handle.net/10630/19923

[5] Nassef, Omar, et al. "A survey: Distributed machine learning for 5G and beyond." *Computer Networks* 207 (2022): 108820.

[6] Schwarzmann, S., Marquezan, C. C., Trivisonno, R., Nakajima, S., Barriac, V., & Zinner, T. (2022). Ml-based qoe estimation in 5g networks using different regression techniques. *IEEE Transactions on Network and Service Management*, 19(3), 3516-3532.

[7] Salhab, Nazih, Rami Langar, and Rana Rahim. "5G network slices resource orchestration using machine learning techniques." *Computer Networks* 188 (2021): 107829.

[8] Saija, Krunal, et al. "A machine learning approach for SNR prediction in 5G systems." *2019 IEEE International Conference on Advanced Networks and Telecommunications Systems (ANTS)*. IEEE, 2019.

[9] Vasilakos, Xenofon, et al. "Integrated methodology to cognitive network slice management in virtualized 5g networks." *arXiv preprint arXiv:2005.04830* (2020).

[10] Lam, Jordan, and Robert Abbas. "Machine learning based anomaly detection for 5g networks." *arXiv preprint arXiv:2003.03474* (2020).

[11] Kulshreshtha, Piyush, and Amit Kumar Garg. "Securing 5G networks through machine learning–a comparative analysis." *2022 IEEE 11th International Conference on Communication Systems and Network Technologies (CSNT)*. IEEE, 2022.

[12] Huang, Xuan, Lei Wu, and Yinsong Ye. "A review on dimensionality reduction techniques." *International Journal of Pattern Recognition and Artificial Intelligence* 33.10 (2019): 1950017.

[13] Mahesh, Batta. "Machine learning algorithms-a review." *International Journal of Science and Research (IJSR)* 9 (2020): 381–386.

[14] Narayanan, A., Ramadan, E., Mehta, R., Hu, X., Liu, Q., Fezeu, R. A., ... & Zhang, Z. L. (2020, October). Lumos5G: Mapping and predicting commercial mmWave 5G throughput. In *Proceedings of the ACM Internet Measurement Conference* (pp. 176–193).

[15] Mithillesh Kumar, P., & Supriya, M. (2022, September). Modelling 5g data using tree-based machine learning models. In *International Conference on Innovative Computing and Communications: Proceedings of ICICC 2022*, Volume 1 (pp. 81–90). Singapore: Springer Nature Singapore.

[16] Supriya, M. (2022, September). Throughput Analysis with Effect of Dimensionality Reduction on 5G Dataset using Machine Learning and Deep Learning Models. In *2022 International Conference on Industry 4.0 Technology* (I4Tech) (pp. 1–7). IEEE.

36 Mango Leaf Disease Detection Using Deep Convolutional Neural Networks

Isha Gupta, Amandeep Singh

36.1 INTRODUCTION

In several nations around the world, mangoes are one of the major fruit harvests and are frequently referred to as "the king of fruits." India is the world's top producer of mangoes, accounting for over 40% of global production. Over 30 to 40% of agricultural output is considered to be harmed by pests and diseases [1]. Gall infestation, Webber's attack, mango deformity, anthracnose, *Alternaria* leaf spots, and stem miner are a few of the prevalent diseases that affect mango plants. These illnesses can be brought on by pathogens such as bacteria, viruses, fungi, parasites, and more, as well as by unfavourable environmental circumstances. Photosynthesis is damaged by disease, which kills the plant. The symptoms and locations of the affected leaves help identify the illness type. In the past, knowledgeable farmers who regularly examined plants could spot plant illnesses. Local farmers enabled prompt disease detection and the implementation of preventative and control measures. However, large farms find it to be both time- and cost-intensive. Finding an automated, accurate, swift, and less expensive technique to identify plant diseases is therefore essential. The methods that are most well-liked and frequently employed for identifying and categorizing plant leaf diseases include image processing and machine learning. Deep learning, a branch of machine learning in general, employs neural networks. With a variety of applications, it has spread its wings across several industries. The development of these computer technologies can help farmers identify and treat plant diseases.

In this study, a novel paradigm for deep convolutional network-based malaria detection is proposed. Sections 36.2 and 36.3 contain details of the experimental setup and related works, respectively. System design and performance measurements for mango leaf disease detection and then the conclusions are covered in Sections 36.4 and 36.5, respectively.

36.2 RELATED WORK

Over the past few decades, there has been a long-running body of research on the detection of leaf disease. Researchers have looked into a range of machine

DOI: 10.1201/9781003367161-36

learning and pattern recognition algorithms to increase the accuracy of disease detection. Convolutional neural networks (CNNs) [2], artificial neural networks [3], backpropagation neural networks [4], support vector machines [5], and other image processing techniques [6, 7] are some of the machine learning approaches used. CNNs are used in the aforementioned methods to classify and extract features. Other feature extraction techniques include the use of the colour co-occurrence matrix [8], angle code histogram [4], zooming algorithm [9], and Canny edge detector [6] algorithms, among others. Both a single ailment in a wide range of plant species and numerous illnesses in a single plant variant have been the subjects of research. A range of crops, including cotton [12], rice [2], wheat, and maize are grown using these cutting-edge methods [10,11]. In comparison to other methods, CNN also requires little or no picture pre-processing. Recent studies on automated plant disease diagnosis have incorporated deep learning algorithms. Yang Lu et al. recommended using a deep CNN as a method for identifying and categorizing images of rice plant sick leaves. In an experimental rice field, a series of 500 naturally occurring pictures of damaged and healthy rice stems and leaves were employed. The tenfold cross-validation method is used to train CNNs to recognize ten common rice diseases, with encouraging results. When colour photos are used, the steady feature does not maintain across colour channels. The images are rescaled in [1] and then put through principal component analysis and whitening [2] to obtain training features and testing features. Alvaro Fuentes et al. [13] suggested a deep-learning-based detection method for real-time leaf disease and pest recognition in tomato plants. Nine different tomato diseases are identified using a dataset of 5,000 images and three different detectors: single-shot multibox detector (SSD), region-based fully convolutional network (R-FCN), and faster region-based CNN (faster R-CNN). Additionally, to cut down on false positives and boost accuracy, data annotation and augmentation are used. For the purpose of diagnosing leaf illness, Mohanty [14] et al. proposed a deep CNN-based method. This researcher was able to group 38 classes of 14 crop species and 26 disease variations using a batch of 54,306 images from the PlantVillage dataset. An automated method for diagnosing mango leaf disease using deep CNNs is presented in this study. To generate results for recognition rapidly and precisely, this method uses batch normalization and the ReLU activation function. This study focuses on the five different mango leaf diseases anthracnose, *Alternaria* leaf spots, leaf gall, leaf burn, and leaf webber. The CNN analyses the raw inputs while performing automated feature extraction. The selection of the features with the highest probabilities values serves as the foundation for categorization. The following sections make up the remaining text of this essay. Section 3 goes over how the CNN model was created. The application of mango leaf disease identification is covered in Section 36.4.

36.3 EXPERIMENTAL SETUP

On a PC running Windows, an Intel Core (TM) i3-7100 CPU running at 2.40 GHz, a 1 TB hard drive, RAM of 64 GB, Python 3.7, and a Kaggle database online were used for all of these tests.

36.4   SYSTEM DESIGN FOR DETECTION
OF DISEASE IN MANGO LEAF

Figure 36.1 shows the various stages of the proposed method for identifying sick mango leaves using pictures. Samples are split into training and testing groups after pre-processing the images of mango leaves. The practice data is then used to train the CNN classifier. After being trained, the CNN classifier uses testing samples to distinguish between healthy and diseased mango leaves as input. The system is then evaluated using criteria that take accuracy, sensitivity, specificity, and classification error into account.

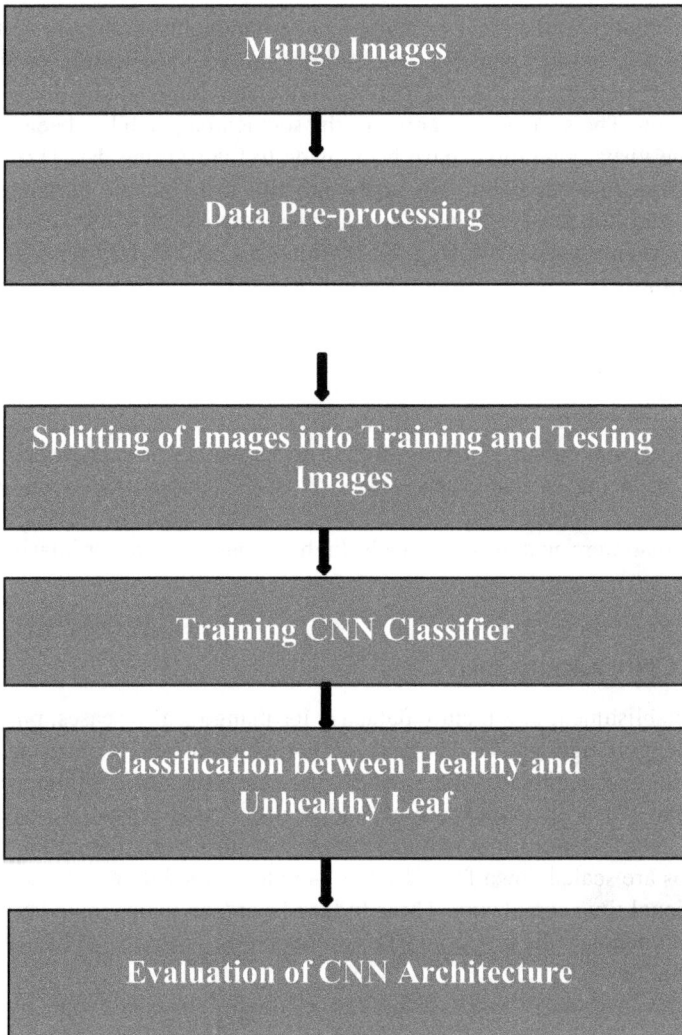

FIGURE 36.1   Block diagram for detection of mango leaf.

36.4.1   DATA PRE-PROCESSING

In this section, several pre-processing techniques have been experimented with, and the outcomes are reported here.

36.4.1.1   Rescaling (Min-Max Normalization)

The patches of images are rescaled to map the range of characteristics to 0 to 1, which speeds up convergence. The greatest and lowest pixel values for the red blood cell patches are 255 and 0, respectively, because they are RGB colour images.

36.4.1.2   Data Augmentation

Data augmentation is the process of adding new data points to the base data. It has been applied to several medical datasets to improve classification performance [15,16]. Here, performance during the hold out test is improved by the use of data augmentation. The semantic meaning of the segmented patch has been preserved while some visual alterations have been made to the training data [17]. Some of the data augmentation techniques applied in this case include flipping the data vertically and horizontally, Gaussian blurring, rotation, horizontal and vertical shifting, darkening and whitening, ZCA whitening, and feature-wise normalization [18, 19].

36.4.2   DATA SPLITTING

The dataset is split into two sets with an 80:20 ratio: a training set and a test set. Cross-validation is carried out to ensure the dependability of the network architectures that are being assessed. The value of k (in k-fold cross-validation) is chosen to be 5 because empirical studies indicated that this quantity yielded error rate estimates that were neither excessively high in bias nor extraordinarily high in variance.

36.4.3   CNN ARCHITECTURE

For the establishment of a picture database for mango leaf diseases, photographs with challenging lighting, size, posture, and orientation were taken using a digital camera with a resolution of 4608 × 3456. There are a total of 1,200 images, which include leaves that are both diseased and healthy. Anthracnose, *Alternaria* leaf spot, leaf gall, leaf webber, and mango plant leaf burn are a few of the ailments. The photos are scaled down from 4608 × 3456 to 256 × 256 in order to lower the computational time complexity. Three hidden layers, an image input layer, and an output layer make up the proposed CNN architecture. Table 36.1 presents the layer implementation.

The input layer receives the 256 × 256 × 3 leaf pictures as input. By creating fictional data, data augmentation is used to expand the dataset. The concealed

TABLE 36.1
Layer Implementation of the CNN Model

Layer	Filter Size	Output Size
Input layer		$256 \times 256 \times 3$
Convolutional layer	11	$127 \times 127 \times 32$
Max-pooling layer	5	$123 \times 123 \times 32$
Convolutional layer	7	$62 \times 62 \times 64$
Max-pooling layer	3	$60 \times 60 \times 64$
Convolutional layer	5	$31 \times 31 \times 128$
Max-pooling layer	3	$29 \times 29 \times 128$
Output		2×1

layers are subsequently traversed by the photographs. Convolutional, batch normalization, rectified linear unit, and maximum pooling layers make up each hidden layer. Convolutional and pooling layers are used for feature extraction, while the fully connected layer is used for classification. Each pooling layer and convolutional layer has a varied number and size of filters in it. According to the formula, the three convolution layers each include 32, 64, and 128 filters with sizes of 11×11, 7×7, and 5×5, respectively (3.3). Both network performance and the training process are enhanced by the batch normalization layer and the ReLU layer. The first maximum pooling layer has P=1, whereas layers two and three have P=0. The three maximum pooling layers are built up of 5×5, 3×3, and 3×3 filters with stride 1 and padding. The least-learned features are then deactivated using the 50% dropout technique. The convolutional and pooling layers' discovered features are categorized using two fully connected layers of size 64 and 6, respectively. The number of classes is equal to the thickness of the second layer that is totally connected. Each class's probability distribution is listed. Figure 36.2 illustrates how the suggested CNN model was trained using the steepest gradient descent approach. The labels are taught using images because CNN is a supervised learning network. The training progress plot depicts how the number of iterations in the training and validation methods increases training accuracy while decreasing loss. The loss is the sum of all errors in the training and validation sets for each sample. Lowering the loss produces better model and recognition performance.

The suggested CNN model [20] was trained using 1,229 total images across all classes. The model was tested with the final 307 photos. The testing accuracy score was 96.67%. In Table 36.2, the confusion matrix is displayed.

The performance metrics for the CNN architecture to categorize the photos as infected or uninfected mango leaf are shown in Figure 36.3. The best accuracy, 97.9%, is noted to be attained.

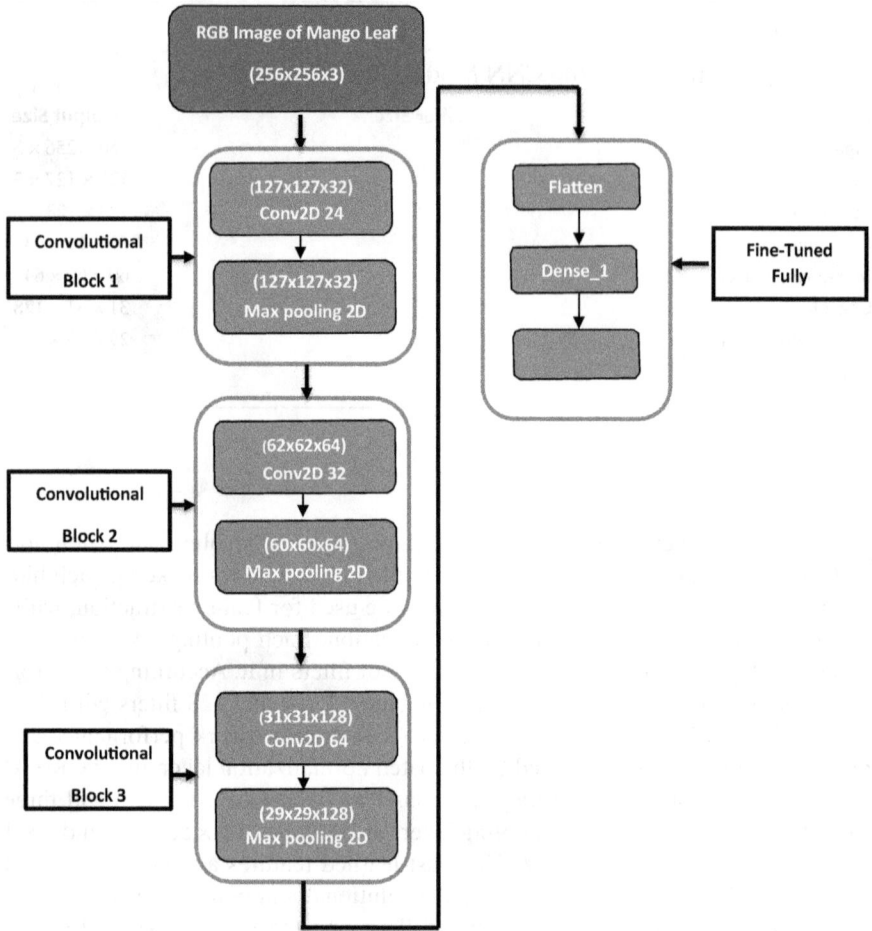

FIGURE 36.2 Nine-layer convolutional neural networks.

TABLE 36.2
Analysis of Parameters for Mango Leaf Detection

Parameter	Values
Accuracy	97.9
Sensitivity	96.2
Specificity	95.1
Precision	94.8

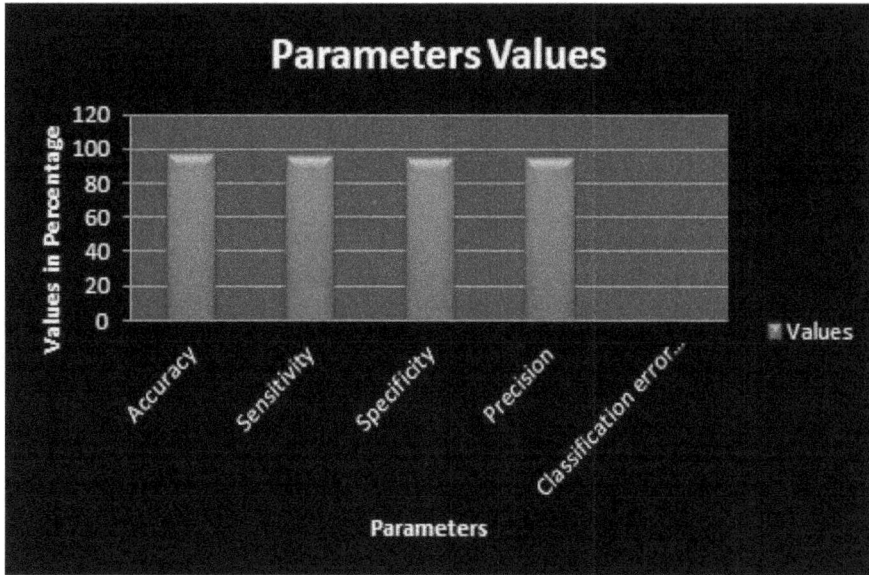

FIGURE 36.3 Performance Parameters for Classifying Images as Infected or Uninfected Mango Leaf

36.5 CONCLUSION

Different mango leaves and healthy leaves can be distinguished using the proposed CNN-based algorithm for recognizing leaf illnesses. CNN is preferred over conventional techniques in many applications due to its quick convergence rate, improved training performance, and lack of time-consuming pre-processing of input images. The accuracy of classification can be achieved even further only by adding more photographs to the image dataset and modifying the CNN model. It remains a scientific problem to determine the ideal CNN model parameters.

REFERENCES

[1] www.sciencedirect.com/topics/agricultural-andbiological-sciences/plant-diseases.
[2] Lu, Y., Yi, S., Zeng, N., Liu, Y., & Zhang, Y. (2017). Identification of rice diseases using deep convolutional neural networks. *Neurocomputing, 267*, 378–384.
[3] Omrani, E., Khoshnevisan, B., Shamshirband, S., Saboohi, H., Anuar, N. B., & Nasir, M. H. N. M. (2014). Potential of radial basis function-based support vector regression for apple disease detection. *Measurement, 55*, 512–519.
[4] VigneshJanarthanan, V. R. (2017). Hybrid multi-core algorithm based image segmentation for plant disease identification using mobile application. *Journal of Excellence in Computer Science and Engineering, 3*(2).

[5] Daniya, T., & Vigneshwari, S. (2019). A review on machine learning techniques for rice plant disease detection in agricultural research. *System, 28*(13), 49–62.

[6] Preethi, R., Priyanka, S., Priyanka, U., & Sheela, A. (2015). Efficient knowledge based system for leaf disease detection and classification. *International Journal of Advanced Research in Science, Engineering, 4*, 1134–1143.

[7] Revathi, P., & Hemalatha, M. (2012, December). Classification of cotton leaf spot diseases using image processing edge detection techniques. In *2012 International Conference on Emerging Trends in Science, Engineering and Technology (INCOSET)* (pp. 169–173). IEEE.

[8] Arivazhagan, S., Shebiah, R. N., Ananthi, S., & Varthini, S. V. (2013). Detection of unhealthy region of plant leaves and classification of plant leaf diseases using texture features. *Agricultural Engineering International: CIGR Journal, 15*(1), 211–217.

[9] Phadikar, S., & Sil, J. (2008, December). Rice disease identification using pattern recognition techniques. In *2008 11th International Conference on Computer and Information Technology* (pp. 420–423). IEEE.

[10] Kumar, A., Malhotra, S., Kaur, D. P., & Gupta, L. (2022, November 9). Weather monitoring and air quality prediction using machine learning. In *2022 1st International Conference on Computational Science and Technology (ICCST)* (pp. 364–368). IEEE.

[11] Khairnar, K., & Dagade, R. (2014). Disease detection and diagnosis on plant using image processing—a review. *International Journal of Computer Applications, 108*(13), 36–38.

[12] Zhang, L. N., & Yang, B. (2014). Research on recognition of maize disease based on mobile internet and support vector machine technique. In *Advanced Materials Research* (Vol. 905, pp. 659–662). Trans Tech Publications Ltd.

[13] Shicha, Z., Hanping, M., Bo, H., & Yancheng, Z. (2007). Morphological feature extraction for cotton disease recognition by machine vision. *Microcomputer Information, 23*(4), 290–292.

[14] Fuentes, A., Yoon, S., Kim, S. C., & Park, D. S. (2017). A robust deep-learning-based detector for real-time tomato plant diseases and pests recognition. *Sensors, 17*(9), 2022.

[15] Mohanty, S. P., Hughes, D. P., & Salathé, M. (2016). Using deep learning for image-based plant disease detection. *Frontiers in plant science, 7*, 1419.

[16] Hung, J., & Carpenter, A. (2017). Applying faster R-CNN for object detection on malaria images. In *Proceedings of the IEEE Conference on Computer Vision and Pattern Recognition Workshops* (pp. 56–61). IEEE Xplore.

[17] Pan, W. D., Dong, Y., & Wu, D. (2018). Classification of malaria-infected cells using deep convolutional neural networks. *Machine Learning: Advanced Techniques and Emerging Applications, 159*.

[18] Bibin, D., Nair, M. S., & Punitha, P. (2017). Malaria parasite detection from peripheral blood smear images using deep belief networks. *IEEE Access, 5*, 9099–9108.

[19] Krishnadas, P., Chadaga, K., Sampathila, N., Rao, S., & Prabhu, S. (2022, September). Classification of malaria using object detection models. In *Informatics* (Vol. 9, No. 4, p. 76). MDPI.

[20] Aggarwal, S. (2022, April 24). Prediction of proteins subcellular location in microscopic images using fine-tuned deep learning architectures. *ECS Transactions, 107*(1), 6309.

37 State of the Art of Hardware Security Attacks
Threats and Challenges

Shaminder Kaur, Geetanjali, Poonam Jindal

37.1 INTRODUCTION

Embedded devices involve the use of third-party vendors in building and designing phase. Their security gets compromised since they undergo to different vendors each with varying establishment level of trust and security. It arises a critical situation where hardware gets compromised in terms of its security. Adversary gets an access to hardware as back door entry or remote access without actually requiring physical access to target device. As a result hardware device gets exploited by attacker and become victim against security threats. Recently it has been observed that US government spent 7.5 billion on cybersecurity in 2007, and it kept on increasing to 28 billion in 2017 as depicted in Figure 37.1 [1–4]. It is alarming situation for electronics industry where almost every embedded device gets affected by one or another attack. Hardware threats can result in any stage ranging from manufacturing till fabrication process. They are vulnerable at each and every stage, and it becomes a big challenge for industry experts to secure embedded devices at each and every stage against various attacks. Hardware attacks are of many

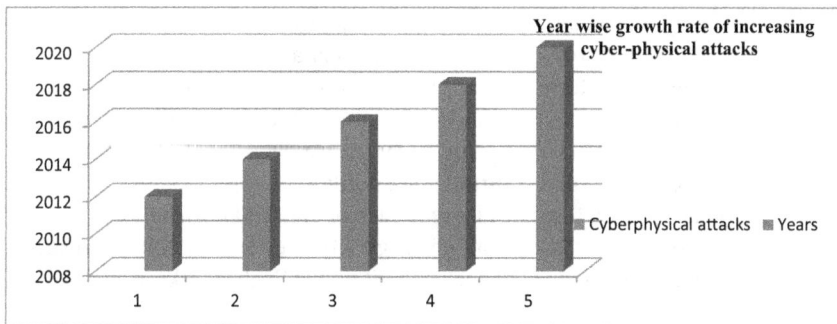

FIGURE 37.1 Cyber-physical attacks growing every year [2].

DOI: 10.1201/9781003367161-37

397

types: invasive attacks, non-invasive attacks, and semi-invasive attacks. Hardware attacks also known as physical attacks can result due to various reasons depending upon the attacking technique used by adversary. Attacks can be injected through various methods: hardware based, software based, simulation based, and emulation based. It is unto to the attacker to follow any of the technique and get access to the secret information easily [5–8]. Lots of research have been carried out so far on attacking techniques, but recent attacks still need to be explored. In this chapter, we have presented the state of the art of hardware attacks, which needs to be explored so that engineers can work in direction to build suitable mitigations/ countermeasures against such attacks.

37.2 HARDWARE MICRO-ARCHITECTURAL ATTACKS BASED ON OUT-OF-ORDER TRANSIENT EXECUTION

In this section description of micro-architectural attacks is presented. We present recent attacks performed on CPU processors:

- Malware attacks
- Spectre attacks
- Meltdown attacks
- Foreshadow attacks
- Address translation attacks
- X85/85 backdoor entries attacks
    - 1. **Malware attacks**: It has been observed that with time, vulnerabilities against both software and hardware have increased. It becomes very difficult to identify attacks against hardware due to their complex architecture designs. Earlier software architectures were an easy target but nowadays even hardware architecture can easily get attacked. It can also be exploited to gain the unintended information by inserting malware attacks. The big issue and concern here is on the impact of system security wherein data extraction is easily performed by exploiting hardware architecture [9, 10].
    - 2. **Spectre attacks**: Microprocessor is one of the very important hardware parts of any computer. Spectre and meltdown attacks deals with exploiting the faults associated with such attacks. Microprocessor x86 executes instructions only with clear format and valid instructions only. Somehow if the format or op-code of any instructions is not valid or incorrectly written then that instruction is not executed. At the same time, it is notified to upper layers by raising flags condition, an exception is generated, and it furthers get notified to upper operating system (OS) layers. Such activities leave a trace of data and unintentionally give access to secret information which gets further exploited by attackers. Spectre attacks deals with such exploitations. Such attacks are very difficult to design and identify due to increasing complexity with architecture designs [11, 12].

37.2.1 TARGET AREAS

The target areas are intel micro-architecture, speculative execution, branch prediction.

37.3 MELTDOWN

Meltdown and spectre exploit the hardware vulnerabilities, and they work on similar grounds i.e. exploiting secret data while system is still working. They can steal secret information stored in memory such as user's passwords, critical documents, emails, and so on. They work between layers of user application and operating system. They basically target the memory which is not supposed to be accessed by user. These memories are generally kept secretive and not easily accessible. Meltdown exploits the fact that sometimes processor works abnormally and in out-of-order condition. It further leads to faulty output and undesired results. It is this fact which is exploited by meltdown attack. Moreover this attack does not depend on the operating system. Meltdown attacks break memory isolation, which is one of the most important features in modern processor to save contents of memory and steals the hidden data. Memory isolation does not allow other parallel applications running to access each other's data [13]. Meltdown reads all kernel memory including physical memory which is mapped in that particular kernel region. Figure 37.2 represents the fact that these attacks exploit the fact of memory isolation between two domains. Their target areas are generally Intel CPU processors or micro-architectures [14].

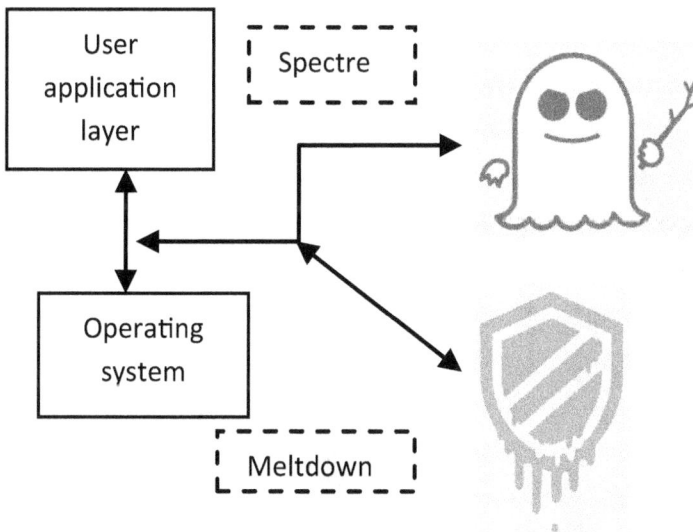

FIGURE 37.2   Spectre and meltdown attacks.

37.3.1 TARGET AREAS

The target areas are out-of-order execution and intel micro-architecture.

37.4   FORESHADOW-NEXT GENERATION (NG)

Foreshadow attacks are also referred to as L1 terminal fault attacks by Intel. It steals memory contents from L1 memory cache. CPU memory is divided into L1, L2, and L3 cache levels. L1 is quite faster as compared to the L2 level. Cache memory can be easily accessed as compared to main memory for speeding up the execution of processor. These attacks pose a serious threat compared to meltdown attacks. As compared to meltdown attacks, these attacks completely bypass virtual memory abstraction, which is not the case with former attack [15].

Foreshadow-NG attacks prove to be more dangerous as compared to other attacks in a way that they successfully completely detour the virtual memory abstraction. Foreshadow attacks manifest cached physical memory data to unintended applications and temporary/guest virtual machines. They come under category of micro-architectural attacks. It is requirement to make memory isolation among various applications/domains. Foreshadow breaks this isolation and is able to steal secret data among themselves [16]. Engineers face lot of difficulties in mitigating this attack due to its high cost. One of the mitigation against foreshadow attacks is to hide physical address of processor. Figure 37.3 describes the operation of foreshadow attack which demonstrates breaking of memory isolation between two domains. There are three variants of foreshadow attacks, as shown in Figure 37.4: foreshadow OS, foreshadow SGX, foreshadow; VMM.

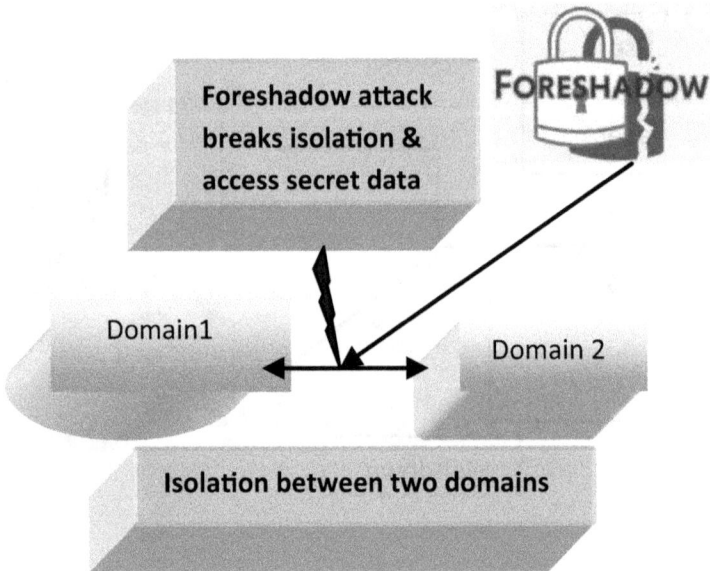

FIGURE 37.3   Foreshadow attack exploiting memory isolation between two domains.

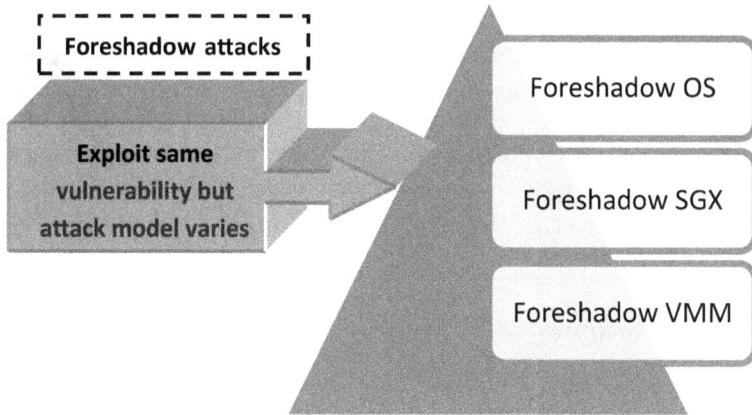

FIGURE 37.4 Foreshadow attack variants.

37.4.1 Target Areas

The target areas are Intel SGX technology and out-of-order execution.

37.5 ADDRESS TRANSLATION ATTACK

It is also known as address translation redirection attack (ATRA). ATRA attacks page table data entries of the CPU, redirecting them to wrong place for extracting the secret data. The adversary intent to unsettle the OS under the presence of hardware-based external monitor. It is assumed that the attacker gains access to the host's kernel root kit, which manipulates the entire operating system and further aid the adversary to malfunction the paging data structure. The hardware-based external monitor does not share its memory with host memory, so any changes made by the attacker, such as changes in register value of the CPU, are not visible to it. Address translation attack exploits the fact that there is gap between external host monitor and host machine. Also, the current external host machine system monitors the host memory registers only [17]. They are still lacking in monitoring the CPU states: flags and registers. These facts are exploited are attacker that give rise to ATRA attacks. They are categorized as follows:

- Memory-bound ATRA
- Register-bound ATRA

Memory-bound ATRA and register-bound ATRA attacks are described here. Memory-bound ATRA modifies the data structures related to the page table, while register-bound ATRA achieves the translation redirection by directly or indirectly compromising the values related to the CPU registers, such as CR3.

37.6 X86/85 PROCESSOR BACKDOOR ENTRY ATTACKS

Backdoor entry attacks exist because it is a known fact that every processor has evolved from its predecessor processor which carried unresolved vulnerabilities along with them. They were left unattended due to the complexity associated with hardware designs. It kept on carried forward to advanced processor creating a loophole and opportunity for attackers to exploit the secret information. These types of vulnerabilities can easily bypass security check points, which will become a big concern if left unattended.

37.7 CONCLUSION

We presented a brief survey on recent state of the art of the attacks which are less researched upon. Attacks are described in detail along with their execution methods. We have conducted this study since the current research focus more on mitigation process rather than providing deep perception into the attacks themselves and their reverberation. The aim of this chapter is to take the edge of the current situation by exhaustively scrutinizing the attacks and their effects on embedded devices.

REFERENCES

[1] K. F. Li and N. Attarmoghaddam, "Challenges and methodologies of hardware security," *2018 IEEE 32nd International Conference on Advanced Information Networking and Applications (AINA)*, 2018, pp. 928–933, doi:10.1109/AINA.2018.00136.

[2] S. Walker-Roberts, M. Hammoudeh, O. Aldabbas, M. Aydin, and A. Dehghantanha, "Threats on the horizon: Understanding security threats in the era of cyber-physical systems," *The Journal of Supercomputing*, vol. 76, no. 4, pp. 2643–2664, 2020.

[3] Y. Lyu and P. Mishra, "A survey of side-channel attacks on caches and countermeasures," *Journal of Hardware and Systems Security*, vol. 2, no. 1, pp. 33–50, 2018.

[4] S. E. Quadir, J. Chen, D. Forte, N. Asadizanjani, S. Shahbazmohamadi, L. Wang, J. Chandy, and M. Tehranipoor, "A survey on chip to system reverse engineering," *The Journal of Emerging Technologies in Computing Systems*, vol. 13, no. 1, 2016.

[5] B. Yuce, P. Schaumont, and M. Witteman, "Fault attacks on secure embedded software: Threats, design, and evaluation," *Journal of Hardware and Systems Security*, vol. 2, pp. 111–130, 2018.

[6] M. Alioto, "Trends in hardware security: From basics to ASICs," *IEEE Solid-State Circuits Magazine*, vol. 11, no. 3, pp. 56–74, 2019, doi:10.1109/MSSC.2019.2923503.

[7] A. Barnghi, L. Breveglieri, I. Korean, and D. Naccache, "Fault injection attacks on cryptographic devices: Theory, practice, and countermeasures," *IEEE Proceedings*, vol. 100, no. 11, pp. 3056–3076, 2012.

[8] E. Valea, M. Da Silva, G. Di Natale, M. Flottes, and B. Rouzeyre, "A survey on security threats and countermeasures in 'IEEE Design & Test'," *IEEE Design & Test*, vol. 36, no. 3, pp. 95–116, 2019.

[9] A. Qamar, A. Karim, and V. Chang, "Mobile malware attacks: Review, taxonomy & future directions," *Future Generation Computer Systems*, vol. 97, pp. 887–909, 2019.

[10] S. Selvaganapathy, S. Sadasivam, and V. Ravi, "A review on android malware: Attacks, countermeasures and challenges ahead," *Journal of Cyber Security and Mobility*, pp. 177–230, 2021.

[11] P. Kocher, J. Horn, A. Fogh, D. Genkin, D. Gruss, W. Haas, M. Hamburg, M. Lipp, S. Mangard, T. Prescher, and M. Schwarz, "Spectre attacks: Exploiting speculative execution," *Communications of the ACM*, vol. 63, no. 7, pp. 93–101, 2020.

[12] J. Fustos, F. Farshchi, and H. Yun, "Spectreguard: An efficient data-centric defense mechanism against spectre attacks," in *Proceedings of the 56th Annual Design Automation Conference 2019*, 2019, June, pp. 1–6, ACM Digital Library.

[13] M. Lipp, M. Schwarz, D. Gruss, T. Prescher, W. Haas, S. Mangard, P. Kocher, D. Genkin, Y. Yarom, and M. Hamburg, "Meltdown," *ArXiv e-prints*, 2018.

[14] B. A. Ahmad, "Real time detection of spectre and meltdown attacks using machine learning," *arXiv preprint arXiv:2006.01442*, 2020.

[15] O. Weisse, J. Van Bulck, M. Minkin, D. Genkin, B. Kasikci, F. Piessens, M. Silberstein, R. Strackx, T. F. Wenisch, and Y. Yarom, "Foreshadow-NG: Breaking the virtual memory abstraction with transient out-of-order execution," 2018, jovanbulck.github.io.

[16] A. Johnson and R. Davies, "Speculative execution attack methodologies (SEAM): An overview and component modelling of Spectre, Meltdown and Foreshadow attack methods," in *2019 7th International Symposium on Digital Forensics and Security (ISDFS)*, 2019, June, pp. 1–6. IEEE.

[17] S. Akashi and Y. Tong, "A vulnerability of dynamic network address translation to denial-of-service attacks," in *2021 4th International Conference on Data Science and Information Technology*, 2021, July, pp. 226–230.

38 Transition in E-Learning in Higher Education during the Pandemic

Rubina Dutta, Dimple Nagpal, Gursleen Kaur

38.1 INTRODUCTION

Information and communication technologies (ICT) have its own impact on almost all the aspects of modern society, from science, entertainment, e-commerce, medicine, and lifestyle. From the beginning of technological growth till date, it is emerging in various forms, proving out to be helpful. It has been guiding a supportable and comfortable life for human beings. Student–teacher interaction is crucial to create a healthy learning environment [1]. India, along with the whole world is struck by the novel coronavirus disease (COVID-19). As the primary measure to stop the coronavirus was to stay at home and not go out for any purpose and not come in contact with other people, On 25 March 2020, the Punjab region's administration made the lockdown a formal announcement. Schools and other educational institutions were shuttered in accordance with the government's rules, making it impossible to continue with the customary classroom teaching method. Thus, online instruction was the only method of T&L. The necessity of online teaching was felt so as to maintain social distancing and so that the spread of the virus could be reduced. While medical experts led the fight against it, instructors had to step up and take ownership to persuade their students to provide a helping hand during COVID-19. Conventional teaching methods are often teacher-centred, with pupils being guided in such a way that they must sit and listen. It is sometimes argued that this kind of T&L may not be beneficial to students since it does not teach significant learning skills; instead, non-traditional ways to T&L may supply students with such skills more effectively [2].

Each one has a unique psychological approach to learning; conventional learning frequently fails to address these differences. Students are either hesitant to address their concerns in front of a group of peers, or they may not grasp lectures due to a lack of concentration [3]. To promote successful learning, an outcome-focused approach is essential, as well as student-centred learning, which stimulates cognitive engagement with the concepts of the subject being taught. Students are also driven to learn in a creative T&L atmosphere [1], [3]. Online education provides lower-quality classroom learning than regular classrooms since learning is asynchronous and non-interactive. Many effective technologies have had an impact on engineering education and have dramatically altered how students study [4]. For instance, in the present climate of a

                    DOI: 10.1201/9781003367161-38

worldwide epidemic, tools like Zoom, Microsoft Team, and Google Classroom have significantly altered the way that people learn. Online learning environments include both working and learning environments. There are several references and bibliographic sources where experts discovered that the traditional classroom-based learning model was being disrupted by online learning platforms [5]. E-learning is a method of education based on traditional teaching but utilizing electronic resources. E-learning, as its name implies, combines two of the most important fields: learning and technology [6]. Earlier, learning principles were based on classroom instruction, but with the advancement of technology, T&L processes may now be used both in-person in the classroom and remotely through the use of computers and the internet. In other words, e-learning is described as an internet-directed method of imparting knowledge and skills that may reach many students at once or over the course of several days. Many different tools are used in e-learning, including writing technologies, communication technologies, visualising, and storing. The scope of e-learning is already quite broad and expanding quickly. Different learning styles may be used to distinguish between the many methods that a learner tries to comprehend the knowledge [7]. However, since technology develops quickly and with the development of T&L systems, more people are using it. There are several benefits to e-learning, including the ability to teach and study at any time and from any location. E-learning allows for one-on-one interaction between students and instructors, as well as the ability for students to take online tests and receive immediate feedback [8]. There is no longer a delay between the evaluation and the result since the process has become so quick.

The finest way of T&L in the business sector has been determined to be e-learning, particularly when multinational corporations are organizing training sessions for professionals all over the world. In general, E-learning-using schools are regarded as being superior to those that continue to employ T&L techniques. Since they do not feel comfortable asking questions in front of everyone, students typically hesitate to ask questions when studying in class, as is well known. Additionally, the teachers' short attention spans contribute to the fact that most questions go unanswered. So the student may clear up any issues they have by speaking with the teacher immediately and without any reluctance thanks to e-learning [9].

The teaching schedule for all departments was suspended beginning on 19 March in accordance with government regulations. Our college was likewise shuttered during this time. The need for online instruction for the pupils became apparent as the days went by. In order to determine whether the pupils could adjust to the situation because it was first challenging, a poll was done. In order to do this, we tried a number of platforms, including Zoom, GoTo Meeting, GoTo Webinar, and Google Meet. We solicited input from the students to determine which platform they like the best. They gave solid responses, and as things grew more adaptable, the process of online instruction proceeded well.

Since it saves time and money on travel and is less expensive than attending traditional universities, online learning has emerged as a highly practical choice. But there are a few misconceptions regarding online education. Online education is frequently perceived as being more expensive than traditional education. Universities with online programmes are more affordable than those with traditional classrooms. In actuality, institutions with purely distant learning programmes have lower operating costs. Understanding current research trends in learning technologies is crucial for education technology researchers. They also need to take into account what teachers and students think of these technologies [10]. The following sections comprised the whole document: The developments in e-learning in compared to traditional approaches were examined in the introduction. The literature review segment covered the present state of e-learning during COVID-19. The methodology section included the specifics of conducting the poll, and the outcome and discussion sections came next. The results of the study and its potential future use were reviewed in the conclusion.

38.2   LITERATURE SURVEY

It is commonly known that online learning, sometimes known as e-learning, is quickly replacing the old educational system. The learning opportunities provided by the online platforms have changed and expanded. For same purpose, several platforms are available that each provide unique advantages. Among these are Google Meet, Zoom, GoTo Webinar, GoTo Meeting, Udemy, Skillshare, Coursera, and many others [11]. Such platforms are all effective means of disseminating information and skills. Each of them includes some of the fundamental functions, including video conferencing, data analysis and reporting, database administration, backups, and many more [12]. E-learning includes cognitive educational objectives, according to Bloom et al., which include "knowledge, comprehension, application, analysis, synthesis, and evaluation" [13]. Darling et al. studied the emotional educational goal of e-learning, which entails the learner's interest, attitude toward learning, and response style, among other things [14]. These e-learning objectives may enable the students to learn online and succeed in their chosen fields.

The e-learning system emerged as the most popular one during this pandemic condition, according to Adeoye et al. (COVID-19) [15]. The study found that e-learning platforms improve student and instructor knowledge and provide a number of benefits, including making it simple to access vast volumes of data and fixing the issue of inadequate infrastructure, as well as allowing students to learn at their own pace [15]. Faherty, L. J., et al. show that online education is the most practical choice in this epidemic condition. Different technologies (the internet, email, Google Classroom, etc.) aid in the efficient operation of distant learning settings [16]. Anastasiades et al. shown that online education is the most practical choice in this epidemic condition. Different technologies (the internet, email, Google Classroom, etc.) aid in the efficient operation of distant learning settings [17, 18]. Sintema investigated educational institutions with little access to technology. The study found that nations with little access to technology struggle with many issues and are not yet able to fully implement a national e-learning system [19].

38.3   METHODOLOGY

The present study included data collection from the student's survey. The research was performed on the same set of students (113 students) of different age groups (18–21). Engineering students from the first year to third year voluntarily participated in the research work and had different understanding levels, through which we accumulated our research data. The survey was designed especially with the help of professional teachers who have been working in various research programs already and could under-stand students' approach better. The COVID-19 situation was taken into account in the survey and the data collected by the questionnaire was after the students were using e-learning and online platforms for their work solely. The students participated in the survey online (through Google Forms) and had responded in a week by actually experi-encing e-learning alone. The survey was conducted to understand students' approach to e-learning pre- and post-COVID-19 teaching scenarios (traditional teaching and online teaching). The data was analysed based on gender, branch, and semester. Out of 113 students, 84 were male and 29 were female. The graphical representation of gender- and branch-wise differentiation is clearly shown in Figure 38.1 and Figure 38.2.

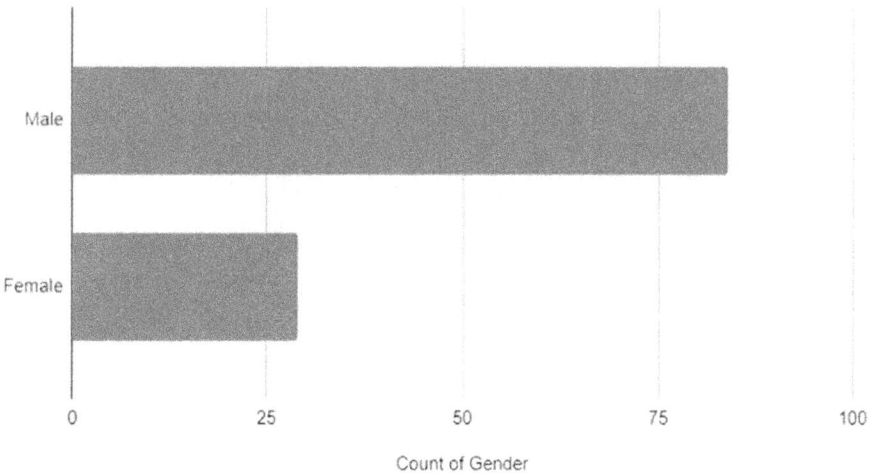

FIGURE 38.1   Gender differentiation.

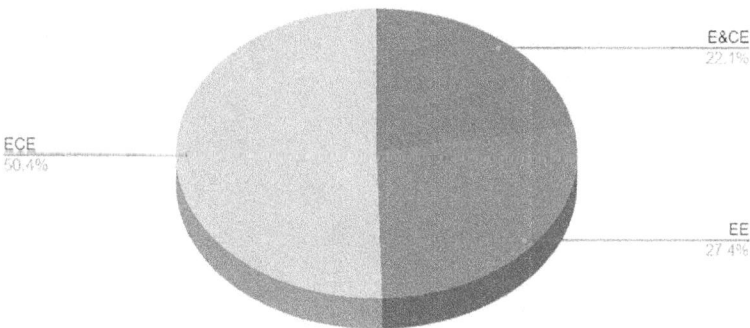

FIGURE 38.2   Branch-wise distribution.

Students from the second (51.3%), fourth (24.8%), and sixth (23.9%) semester contributed to the survey and the involvement can be read with the help of Figure 38.3. A trial on a few of the online platforms was conducted in the first week of the lockdown period. Afterward, students' preference for an online platform was asked and the response collected from the survey had shown that 60.2% of students preferred online learning using GoTo Webinar, whereas only 4.4% of students preferred using Google Meet. According to the student's preference as shown in Figure 38.4, our team decided to run all the courses on GoTo Webinar.

Count of Semester Enrolled

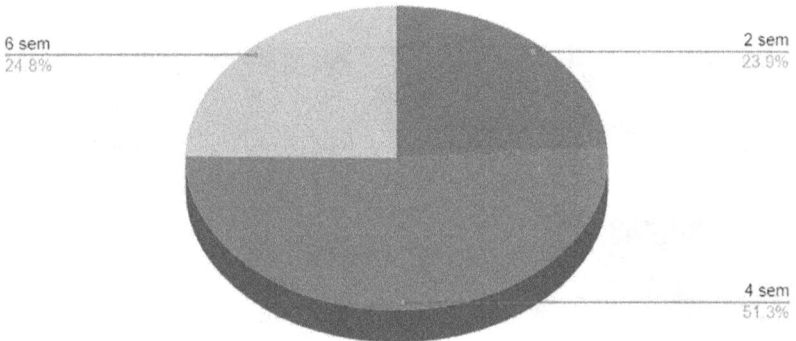

FIGURE 38.3   Semester-wise distribution.

Which online platform you like the most

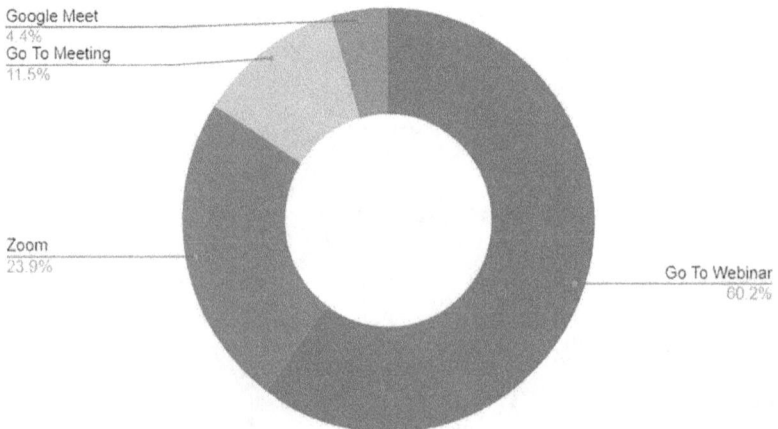

FIGURE 38.4   Choice of platform.

38.4 RESULT AND DISCUSSION

Due to COVID-19, students were learning solely through online mediums. Webinars were conducted by Chitkara University to help the students understand their course better and provide them an overall learning environment. According to the results obtained in Figure 38.5, 90.3% of students found online webinars helpful in learning their course. Figure 38.6 shows the students opinion on motivation acquired by online learning. 63.7% of the students gave positive response in this regard.

E-learning and traditional learning platforms have always been debated upon, even after various benefits of online learning. The data collected from the students had shown (Figure 38.7) that 30% students moderately disagree when asked if

I would find online webinars useful in learning my course.

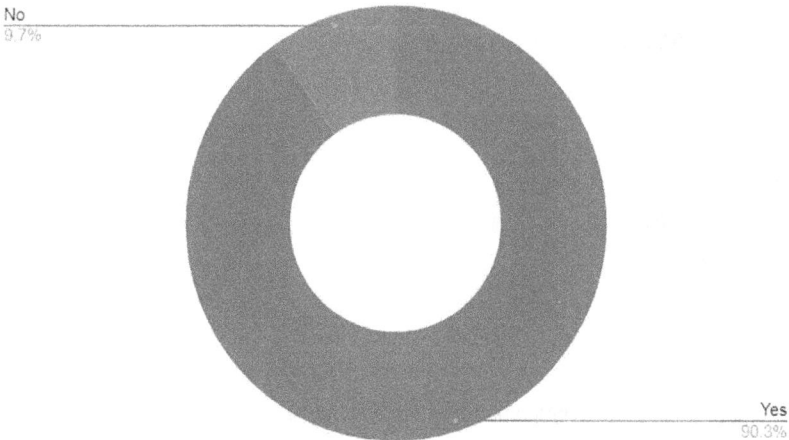

No
9.7%

Yes
90.3%

FIGURE 38.5 Course usefulness through online teaching.

Did you feel motivated after learning through online mode?

No
36.3%

Yes
63.7%

FIGURE 38.6 Students' opinion on motivation acquired by online learning.

online-learning-enabled training would help in better understanding of the course than formal methods and almost the majority of the students had a neutral opinion which presents us the fact that students were not completely satisfied by learning on online modes.

A substantial number of students, when asked if they would prefer online learning over classroom teaching, answered "No." Figure 38.8 shows that 68.1% of students did not prefer online learning platforms, and only 31.9% of them did. The survey was

Online-learning enabled training would help in better understanding of the course than formal teaching methods

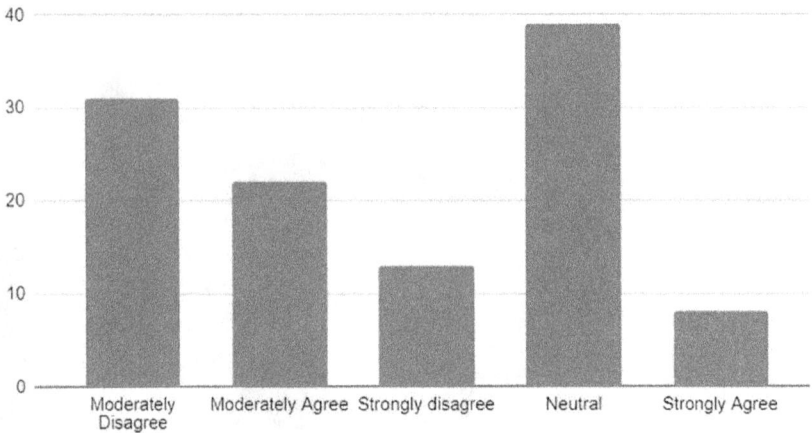

FIGURE 38.7   Students opinion on better understanding in online learning than the formal method.

Would you prefer online-learning over classroom teaching?

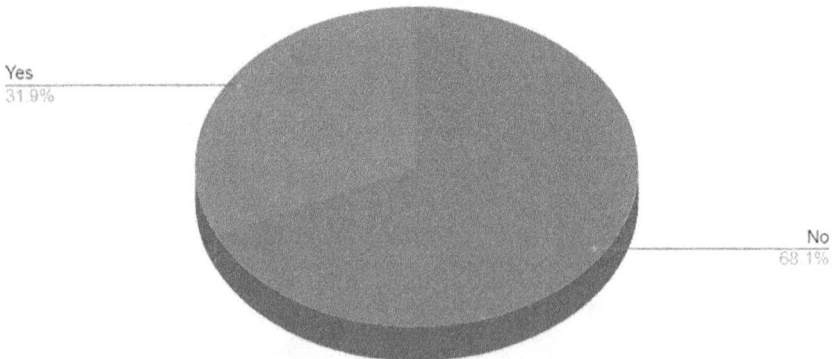

FIGURE 38.8   Students' preference for online learning over classroom teaching.

conducted with students living in different parts of the country, and each of them had a different experience of online learning mainly due to internet connectivity, the strength of signal, internet pack exhausted, and so on.

38.5 CONCLUSION

The present work focuses on the perception of students on the use of e-learning and online platforms. Motivation, stimulus, feedback and reinforcement, opportunities to apply learning, and rewards are the essential elements of effective learning. A feedback form was designed for all the points mentioned in this chapter, and students' opinions are plotted in the mentioned figures. The feedback form's results revealed that students were not currently experiencing an efficient learning environment when using online platforms but rather were encountering several difficulties. According to [20, 21], it will take 12 to 18 months for us to discover an effective vaccination or for our lives to become as stable as before the pandemic. In the meanwhile, we must develop in the habit of living with COVID-19.

In the end, it will be necessary to address the restrictions that the pupils encountered. Students must receive training in utilizing an online platform and must become used to using e-learning. By using the student input that has been obtained, the next stage of unsuccessful learning may be addressed: "Internet quality should be improved in both ends," "More clarification on some practical topics," and "Teachers should use a digital writing pad rather than writing things using a mouse." By taking into account and implementing the input from the students, we anticipate that the problems with the online learning will be fixed as soon as possible.

REFERENCES

[1] Reinke, W. M., Herman, K. C., & Newcomer, L. (2016). The brief student–teacher classroom interaction observation: Using dynamic indicators of behaviors in the classroom to predict outcomes and inform practice. *Assessment for Effective Intervention*, *42*(1), 32–42.

[2] Tularam, G. A. (2018). Traditional vs Non-traditional teaching and learning strategies-the case of e-learning!. *International Journal for Mathematics Teaching and Learning*, *19*(1), 129–158.

[3] Wilson, K., & Korn, J. H. (2007). Attention during lectures: Beyond ten minutes. *Teaching of Psychology*, *34*(2), 85–89.

[4] Horváth, K. (2004). E-learning management systems in Hungarian higher education. *Teaching Mathematics and Computer Science*, *2*, 357–384.

[5] Zamora-Polo, F., Luque Sendra, A., Aguayo-Gonzalez, F., & Sanchez-Martin, J. (2019). Conceptual framework for the use of building information modeling in engineering education. *International Journal of Engineering Education*, *35*(3), 744–755.

[6] Aparicio, M., Bacao, F., & Oliveira, T. (2016). An e-learning theoretical framework. *An e-Learning Theoretical Framework*, *1*, 292–307.

[7] Truong, H. M. (2016). Integrating learning styles and adaptive e-learning system: Current developments, problems and opportunities. *Computers in Human Behavior*, *55*, 1185–1193.

[8] Dutta, R., Malhotra, S., Kumar, A., Parashar, A., & Sharma, S. (2022, May). Effect of cognition on e-learners as compared to traditional learners during Covid-19. In *AIP Conference Proceedings* (Vol. 2357, No. 1, p. 080010). AIP Publishing LLC.

[9] Yilmaz, R. (2017). Exploring the role of e-learning readiness on student satisfaction and motivation in flipped classroom. *Computers in Human Behavior, 70,* 251–260.

[10] McCarthy, M. (2018). Innovative practices in technology and the improvement of learning and teaching in higher education: A case for private institutions (Doctoral dissertation, Botho University).

[11] Archibald, M. M., Ambagtsheer, R. C., Casey, M. G., & Lawless, M. (2019). Using zoom videoconferencing for qualitative data collection: Perceptions and experiences of researchers and participants. *International Journal of Qualitative Methods, 18,* 1609406919874596.

[12] Gautam, S. S., & Tiwari, M. K. (2016). Components and benefits of E-learning system. *International Research Journal of Computer Science (IRJCS), 3*(1), 14–17.

[13] Immetman, A., & Schneider, P. (1998). Assessing student kearning in study-abroad programs: A conceptual framework and methodology for assessing student learning in study-abroad programs. *Journal of Studies in International Education, 2*(2), 59–80.

[14] Darling-Hammond, L. (2006). Constructing 21st-century teacher education. *Journal of Teacher Education, 57*(3), 300–314.

[15] Adeoye, I. A., Adanikin, A. F., & Adanikin, A. (2020). *COVID-19 and E-learning: Nigeria Tertiary Education System Experience.* The Elizade University Institutional Repository.

[16] Faherty, L. J., Schwartz, H. L., Ahmed, F., Zheteyeva, Y., Uzicanin, A., & Uscher-Pines, L. (2019). School and preparedness officials' perspectives on social distancing practices to reduce influenza transmission during a pandemic: Considerations to guide future work. *Preventive Medicine Reports, 14,* 100871.

[17] Anastasiades, P. S., Filippousis, G., Karvunis, L., Siakas, S., Tomazinakis, A., Giza, P., & Mastoraki, H. (2010). Interactive videoconferencing for collaborative learning at a distance in the school of 21st century: A case study in elementary schools in Greece. *Computers & Education, 54*(2), 321–339.

[18] Thamarana, S. (2016). Role of e-learning and virtual learning environment in English language learning. *Teaching English Language and Literature: Innovative Methods and Practices, ELTAI Tirupati,* 61–62.

[19] Sintema, E. J. (2020). Effect of COVID-19 on the performance of grade 12 students: Implications for STEM education. *Eurasia Journal of Mathematics, Science and Technology Education, 16*(7), em1851.

[20] Sangeeta, & Tandon, U. (2021). Factors influencing adoption of online teaching by school teachers: A study during COVID-19 pandemic. *Journal of Public Affairs, 21*(4), e2503.

[21] Mahajan, P., & Kaushal, J. (2020). Epidemic trend of COVID-19 transmission in India during lockdown-1 phase. *Journal of Community Health, 45*(6), 1291–1300.

Index

For Product Safety Concerns and Information please contact our EU
representative GPSR@taylorandfrancis.com
Taylor & Francis Verlag GmbH, Kaufingerstraße 24, 80331 München, Germany